Value Addition and Utilization of Lignocellulosic Biomass

Gunjan Mukherjee • Sunny Dhiman
Editors

Value Addition and Utilization of Lignocellulosic Biomass

Through Novel Technological Interventions

Springer

Editors
Gunjan Mukherjee
University Institute of Biotechnology
Chandigarh University
Mohali, Punjab, India

Sunny Dhiman
University Institute of Biotechnology
Chandigarh University
Mohali, Punjab, India

ISBN 978-981-96-2785-1 ISBN 978-981-96-2786-8 (eBook)
https://doi.org/10.1007/978-981-96-2786-8

This Springer imprint is published by the registered company Springer Nature Singapore Pte Ltd.
The registered company address is: 152 Beach Road, #21-01/04 Gateway East, Singapore 189721, Singapore

Preface

Lignocellulosic biomass, a vast and renewable resource derived from plant-based materials, has emerged as a cornerstone in global sustainability initiatives. With its potential to replace nonrenewable resources in various industrial sectors, lignocellulosic biomass offers a pathway to achieving environmental and economic goals. However, its effective utilization poses significant challenges due to its complex structure and recalcitrant nature. This book, titled *Value Addition and Utilization of Lignocellulosic Biomass Through Novel Technological Interventions*, delves into the innovative strategies, advancements, and applications that are revolutionizing the landscape of biomass utilization.

The need for sustainable practices has never been more critical. Rapid industrialization, coupled with increasing concerns over climate change and depleting fossil fuel reserves, has necessitated a paradigm shift toward renewable resources. Lignocellulosic biomass, which comprises cellulose, hemicellulose, and lignin, is abundant in agricultural residues, forestry waste, and dedicated energy crops. Its availability, coupled with its versatility, makes it an ideal candidate for value-added applications, such as biofuels, bioplastics, biochemicals, and other high-value bioproducts. However, the inherent recalcitrance of lignocellulosic biomass to chemical and biological conversion necessitates innovative approaches to unlock its full potential. The journey of lignocellulosic biomass utilization has been marked by significant breakthroughs, driven by interdisciplinary research spanning chemistry, biology, engineering, and materials science. This book captures these advancements, providing a comprehensive exploration of the subject. It brings together contributions from leading experts and researchers, ensuring that readers gain a nuanced understanding of both the theoretical and practical aspects of biomass valorization.

The multidisciplinary approach of this book makes it a valuable resource for a wide-ranging audience, including researchers, academicians, industry professionals, and policymakers. Students and educators in fields such as biotechnology, chemical engineering, environmental science, and materials science will find it particularly beneficial, as it bridges the gap between fundamental science and practical applications. This book is a testament to the collective efforts of researchers, innovators, and industry stakeholders who are shaping the future of lignocellulosic biomass utilization. It celebrates the spirit of innovation while acknowledging the challenges that lie ahead. The editors express their gratitude to the contributors for sharing their expertise and insights, as well as to the reviewers for their constructive feedback.

As the world moves toward a sustainable and bio-based economy, the efficient utilization of lignocellulosic biomass will play a pivotal role. It is our hope that this book will inspire further research, foster collaborations, and contribute to the advancement of this vital field. By unlocking the potential of lignocellulosic biomass through novel technological interventions, we can pave the way for a greener and more sustainable future.

Mohali, Punjab, India Gunjan Mukherjee
Mohali, Punjab, India Sunny Dhiman

Contents

Advancing Green Technologies for Sustainable Bioenergy from Lignocellulosic Biomass 1
Mishika Ahuja, Gunjit Setia, Simran Kaur, Sunny Dhiman, and Gunjan Mukherjee

Bio-interventions for Valorization of Lignocellulosic Biomass: Recent Trends and Applications 25
Charu Sharma and Papiya Mukherjee

Exploring the Impact of Emerging Pretreatment Technologies on Effective Lignocellulosic Biomass Utilization 45
Vishnu Damodaran Nambissan, Saivi Singh, and Sathyanarayana N. Gummadi

Novel Approaches to Alleviate Lignocellulosic Biomass Waste for Effective and Sustainable Utilization 89
Dhruv Agrawal, Sandeep Kaur Saggu, Sapna Devi, Shiv Kumar, and Manoj Kumar

Valorization of Cotton Biomass and Processing Waste for Production of Biohydrogen and Biomethane 117
Ajinath Dukare, Anam Shaikh, Anagha Pawar, Kanika Sharma, and Nadanathangam Vigneshwaran

Microbial Fuel Cells and Sustainable Electricity Generation from Lignocellulosic Waste Biomass 139
Eduardo Dellosso Penteado, Yvan Jesus Olortiga Asencios, Anthony Andrey Ramalho Diniz, Luiza Helena da Silva Martins, Johnatt Allan Rocha de Oliveira, and Andrea Komesu

Gasification and Syngas Production from Lignocellulosic Waste 167
Yvan Jesus Olortiga Asencios, Eduardo Dellosso Penteado, Luiza Helena da Silva Martins, Johnatt Allan Rocha de Oliveira, and Andrea Komesu

Engineering Microbial Strains and Enzymes for Enhanced Productivity . . . 191
George Dzorgbenya Ametefe, Oluwabusayo Odunola Oluyide, Babamotemi Olusola Itakorode, Ismail Muritala, Omotayo Oyedara, and Divine Senanu Ametefe

Smart Recycling and Sustainable Lignocellulosic Waste Management 221
Nihar Gupta, W. P. T. Wijenayake, Deblina Roy, Rahul Kumar, Meenakshi Rangot, Priya Chugh, Gustavo Santoyo, Sourav Chattaraj, Sawant Shraddha Bhaskar, Snežana Andjelković, Beatriz Elena Guerra-Sierra, and Debasis Mitra

Application of Artificial Intelligence Techniques in Biorefinery Processes . . . 251
Protibha Nath Banerjee

Harnessing Bioproducts for a Sustainable Circular Economy . . . 263
Prasann Kumar and Joginder Singh

Recent Findings and Advances in Sustainable Conversion of Lignocellulosic Waste to Bioplastic Precursors for a Circular Economy . . . 295
Shilajit Barua, Gourab Shome, Sayak Dolai, and Jisan Sarwar

Bioethanol for "Bharat" (India): A Focused Journey Toward Sustainability . . . 335
Sibabrata Mukherjee

Editors and Contributors

About the Editors

Gunjan Mukherjee currently serves as a Professor and Director at the University Institute of Biotechnology, Chandigarh University, India. With over two decades of international recognition in the fields of microbial fermentation, bioprocess development, reactor design, and the conversion of lignocellulosic biomass into value-added chemicals and biofuels, Dr. Mukherjee embodies the epitome of expertise in industrial biotechnology. His professional trajectory took him to prestigious positions at leading organizations, including Senior Scientist roles at Praj Industries Limited, Pune, India, and TERI, New Delhi, India. His global reach extends to a visiting professorship at Deakin University, Australia. A hallmark of his career is the culmination of international patents and numerous research publications in esteemed journals. Dr. Mukherjee actively serves as an Editor, Associate Editor, and honorary member on the editorial boards of several reputable journals. His exemplary contributions have garnered him numerous honors and awards, including the PARIS Award from PRAJ Industries Limited, Pune, India, and the Rising Star Award from TERI, India.

Sunny Dhiman, is an Associate Professor at the University Institute of Biotechnology, Chandigarh University, India. He has been actively engaged in research, education, and scientific writing in his field for over 13 years. With over a decade of extensive experience in academic teaching and research, Dr. Dhiman has contributed significantly to various reputable international journals, such as *Food Production, Processing and Nutrition*, *International Microbiology*, *Journal of Food Process Engineering*, *Biologia*, *Journal of Scientific & Industrial Research*, *Mycology*, and *Research Journal of Biotechnology*, among others. His research pursuits encompass a diverse array of interests, including microbial fermentation, waste management, bioremediation, and waste valorization. His publication record includes over 40 peer-reviewed articles, encompassing 16 book chapters and two patents. Currently, Dr. Dhiman focuses on investigating microbial metabolic capabilities for waste valorization, functional food development, purification, and optimization studies of microbial enzymes and pigments. Beyond his research endeavors, Dr. Dhiman actively participates as a guest editor and reviewer for several international journals. He has guided several graduate and postgraduate dissertation projects and is currently guiding two doctoral students. He is a member of several scientific and professional societies including the Microbiologists Society of India, the American Society of Microbiology, IAENG, and IFERP.

Contributors

Dhruv Agrawal Umeå Plant Science Centre, Department of Plant Physiology, Umeå University, Umeå, Sweden

Department of Microbiology, Guru Nanak Dev University, Amritsar, Punjab, India

Mishika Ahuja School of Biosciences, Apeejay Stya University, Gurugram, Haryana, India

Divine Senanu Ametefe Wireless Communication Technology Group (WiCOT), College of Engineering, School of Electrical Engineering, Universiti Teknologi MARA (UiTM), Shah Alam, Selangor, Malaysia

George Dzorgbenya Ametefe Department of Biotechnology, Faculty of Basic and Applied Sciences, College of Science, Engineering, and Technology, Osun State University, Osogbo, Osun State, Nigeria

Snežana Andjelković Institute for Forage Crops, Kruševac, Serbia

Yvan Jesus Olortiga Asencios Universidade Federal de São Paulo, UNIFESP, Instituto do Mar, Santos, São Paulo, Brazil

Protibha Nath Banerjee Department of Chemistry, Pimpri Chinchwad University, Pune, India

Shilajit Barua Department of Microbiology, Vijaygarh Jyotish Ray College, Kolkata, India

Sawant Shraddha Bhaskar Department of Plant Pathology, Bihar Agricultural University, Bhagalpur, Bihar, India

Sourav Chattaraj Centre for Industrial Biotechnology Research, School of Pharmaceutical Sciences, Siksha 'O' Anusandhan, Deemed to be University, Bhubaneswar, Odisha, India

Priya Chugh School of Agriculture, Graphic Era Hill University, Dehradun, Uttarakhand, India

Luiza Helena da Silva Martins Universidade Federal Rural da Amazônia, UFRA, Instituto de Saúde e Produção Animal, Belém, Pará, Brazil

Johnatt Allan Rocha de Oliveira Universidade Federal do Pará, UFPA, Faculdade de Nutrição, Belém, Pará, Brazil

Sapna Devi Department of Microbiology, DAV University, Jalandhar, Punjab, India

Sunny Dhiman University Institute of Biotechnology, Chandigarh University, Mohali, Punjab, India

Anthony Andrey Ramalho Diniz Universidade Federal de São Paulo, UNIFESP, Instituto do Mar, Santos, São Paulo, Brazil

Sayak Dolai Department of Microbiology, Vijaygarh Jyotish Ray College, Kolkata, India

Ajinath Dukare ICAR-Central Institute for Research on Cotton Technology, Mumbai, Maharashtra, India

Beatriz Elena Guerra-Sierra Grupo Microbiota, Facultad de Ciencias Naturales, Universidad de Santander, Bucaramanga, Colombia

Sathyanarayana N. Gummadi Applied and Industrial Microbiology Lab, Department of Biotechnology, Bhupat and Jyoti Mehta School of Biosciences, Indian Institute of Technology Madras, Chennai, India

Nihar Gupta Department of Soil Science, Punjab Agricultural University, Ludhiana, Punjab, India

Babamotemi Olusola Itakorode Department of Biotechnology, Faculty of Basic and Applied Sciences, College of Science, Engineering, and Technology, Osun State University, Osogbo, Osun State, Nigeria

Simran Kaur School of Biosciences, Apeejay Stya University, Gurugram, Haryana, India

Andrea Komesu Universidade Federal de São Paulo, UNIFESP, Instituto do Mar, Santos, São Paulo, Brazil

Manoj Kumar Department of Microbiology, Guru Nanak Dev University, Amritsar, Punjab, India

Prasann Kumar Department of Agronomy, School of Agriculture, Lovely Professional University, Phagwara, Punjab, India

Rahul Kumar Department of Microbiology, Graphic Era (Deemed to be University), Dehradun, Uttarakhand, India

Shiv Kumar Department of Microbiology, Guru Gobind Singh Medical College & Hospital, Baba Farid University of Health Sciences, Faridkot, Punjab, India

Debasis Mitra Department of Microbiology, Graphic Era (Deemed to be University), Dehradun, Uttarakhand, India

Gunjan Mukherjee University Institute of Biotechnology, Chandigarh University, Mohali, Punjab, India

Papiya Mukherjee Department of Botany, Panjab University, Chandigarh, India

Sibabrata Mukherjee F2I Nutrition & Health Private Limited, a subsidiary of Leaf by Lesaffre, Mumbai, Maharashtra, India

Ismail Muritala Department of Biotechnology, Faculty of Basic and Applied Sciences, College of Science, Engineering, and Technology, Osun State University, Osogbo, Osun State, Nigeria

Vishnu Damodaran Nambissan Applied and Industrial Microbiology Lab, Department of Biotechnology, Bhupat and Jyoti Mehta School of Biosciences, Indian Institute of Technology Madras, Chennai, India

Oluwabusayo Odunola Oluyide Department of Biotechnology, Faculty of Basic and Applied Sciences, College of Science, Engineering, and Technology, Osun State University, Osogbo, Osun State, Nigeria

Omotayo Oyedara Department of Biotechnology, Faculty of Basic and Applied Sciences, College of Science, Engineering, and Technology, Osun State University, Osogbo, Osun State, Nigeria

Anagha Pawar ICAR-Central Institute for Research on Cotton Technology, Mumbai, Maharashtra, India

Eduardo Dellosso Penteado Universidade Federal de São Paulo, UNIFESP, Instituto do Mar, Santos, São Paulo, Brazil

Meenakshi Rangot Basic Science (Microbiology), Dr. Yashwant Singh Parmar University of Horticulture & Forestry, Nauni, Himachal Pradesh, India

Deblina Roy Department of Agronomy, Bidhan Chandra Krishi Viswavidyalaya, Nadia, West Bengal, India

Sandeep Kaur Saggu Department of Biotechnology, Kanya Maha Vidyalaya, Jalandhar, Punjab, India

Gustavo Santoyo Instituto de Investigaciones Químico Biológicas, Universidad Michoacana de San Nicolás de Hidalgo, Ciudad Universitaria, Morelia, Michoacán, Mexico

Jisan Sarwar Department of Microbiology, Vijaygarh Jyotish Ray College, Kolkata, India

Gunjit Setia School of Biosciences, Apeejay Stya University, Gurugram, Haryana, India

Anam Shaikh ICAR-Central Institute for Research on Cotton Technology, Mumbai, Maharashtra, India

Charu Sharma Department of Botany, Panjab University, Chandigarh, India

Kanika Sharma ICAR-Central Institute for Research on Cotton Technology, Mumbai, Maharashtra, India

Gourab Shome Department of Microbiology, Vijaygarh Jyotish Ray College, Kolkata, India

Joginder Singh Department of Botany, Nagaland University, Nagaland, India

Saivi Singh Applied and Industrial Microbiology Lab, Department of Biotechnology, Bhupat and Jyoti Mehta School of Biosciences, Indian Institute of Technology Madras, Chennai, India

Nadanathangam Vigneshwaran ICAR-Central Institute for Research on Cotton Technology, Mumbai, Maharashtra, India

W. P. T. Wijenayake Department of Agricultural Biology, Faculty of Agriculture, University of Peradeniya, Peradeniya, Sri Lanka

Advancing Green Technologies for Sustainable Bioenergy from Lignocellulosic Biomass

Mishika Ahuja, Gunjit Setia, Simran Kaur, Sunny Dhiman, and Gunjan Mukherjee

Abstract

For many years, fossil fuels constituted the world's primary source of energy. However, concerns about climate change and resource depletion necessitate a shift to renewable energy sources. Lignocellulosic biomass, derived from plants containing cellulose, hemicellulose, and lignin, is a promising source of sustainable bioenergy production. Its consumption reduces waste accumulation and dependency on finite fossil fuels. However, the primary challenge is to efficiently convert these materials into bioethanol while mitigating environmental impacts. To accomplish this goal, scientists are investigating different approaches including advanced pretreatment techniques, enzymatic hydrolysis, and microbial fermentation. These approaches are designed to increase bioethanol yields and reduce production costs and waste generation, advancing a more sustainable and environmentally friendly methodology. Recently, advancements in applied chemistry under extreme and unconventional conditions have led the way for potential commercial solutions in the market. High hydrostatic pressure, microwave, and ultrasound technologies are emerging as promising sustainable green pretreatment options for lignocellulosic biomass utilization in large-scale biorefineries. As the world moves toward a more sustainable energy future, advancements in bioethanol production from lignocellulosic waste materials are critical to lowering greenhouse gas emissions and mitigating environmental impacts. This book chapter provides a comprehensive overview of initiatives aimed at

M. Ahuja · G. Setia · S. Kaur
School of Biosciences, Apeejay Stya University, Gurugram, Haryana, India

S. Dhiman (✉) · G. Mukherjee
University Institute of Biotechnology, Chandigarh University, Mohali, Punjab, India
e-mail: sunny.uibt@cumail.in

G. Mukherjee, S. Dhiman (eds.), *Value Addition and Utilization of Lignocellulosic Biomass*, https://doi.org/10.1007/978-981-96-2786-8_1

improving the sustainable production of biofuel from lignocellulosic waste, emphasizing both the environmental and economic aspects of this renewable energy source.

Keywords

Sustainable Bioenergy · Lignocellulosic Biomass · Renewable Energy · Biorefinery Processes · Biofuel Production · Environmental Sustainability

1 Introduction

For more than 250 years since the onset of the Industrial Revolution, the global economy has been predominantly shaped by a linear model characterized by the processes of "take, make, and dispose," grounded in the assumption of limitless and inexpensive natural resources. However, this model has resulted in extensive environmental degradation and the impending depletion of nonrenewable resources. In contrast, the twenty-first century introduces the "Bioeconomy," a progressive economic framework that emphasizes the sustainable reuse and recovery of resources rather than their mere consumption (Jørgensen & Pedersen, 2018). A paradigm shift toward a circular economy, which emphasizes resource recovery and reuse, is imperative, with the biorefinery at its core, serving as a crucial mechanism driving the transition toward a sustainable future by transforming widely available, renewable, and nonedible lignocellulosic biomass into valuable energy, fuels, and chemicals. This biomass, which includes agricultural residues, forestry byproducts, and food industry waste, represents a significant, yet underexploited, resource that can be leveraged for sustainable energy production (Ahmed et al., 2023).

As a result, increasing the production of environmentally friendly, lignocellulosic waste materials is one way to accomplish the need for sustainable energy sources. Bioethanol, a renewable and low-carbon alternative to traditional fossil fuels, is a viable solution to mitigate the negative effects of climate change while offering a healthier and more sustainable energy source (Bušić et al., 2018). Lignocellulosic waste is a highly desirable feedstock due to its abundance and capacity to transform waste into a valuable resource that supports the fundamental concepts of the circular economy. Waste materials can be repurposed for the manufacturing of bioethanol, which will help to both economically prosper and conserve the environment by reducing the load on landfills and our dependency on finite fossil fuels (Devi et al., 2022). Yet, many challenges exist in using lignocellulosic waste to produce bioethanol. Novel and sustainable methods are needed to effectively convert complex lignocellulosic structures into bioethanol. This introduction offers the groundwork for an in-depth examination of methods, and technologies used to address these issues, with a dual emphasis on increasing bioethanol production and reducing environmental effects (Mohanty & Abdullahi, 2016). The sustainable production of bioethanol from lignocellulosic waste materials is a ray of hope, exhibiting the convergence of ecological responsibility and renewable energy innovation, as the world strives to transition toward a more sustainable and carbon-neutral future. This analysis explores the many facets of this important project, noting the exciting prospects and promising developments it offers for the production of sustainable bioenergy (Mujtaba et al., 2023).

Biomass and biomass-derived materials have been identified as some of the most promising alternatives for sustainable energy production (Zhou et al., 2008). Considering biomass, the only sustainable source of organic carbon on Earth and can be used to produce fuels and fine chemicals with net-zero carbon emissions, it is an excellent substitute for petroleum. These materials are produced by biological photosynthesis using atmospheric CO_2, water, and sunlight (Ragauskas et al., 2006). Among these, lignocellulosic biomass, the most abundant and biorenewable form of biomass on Earth, (Taarning et al., 2011) holds significant importance. Multiple studies showed the numerous potentials of lignocellulosic biomass in the sustainable production of chemicals and fuels. Additionally, it is an abundant and renewable source of feedstock (Barakat et al., 2013). Lignocellulosic biomass is projected as an abundant, carbon-neutral renewable resource that can reduce CO_2 emissions and mitigate atmospheric pollution, making it a promising alternative to crude oil for producing biofuels, biomolecules, and biomaterials (Zhou et al., 2011). Furthermore, cellulose—the main component of lignocellulosic biomass—is noticed to be an effective alternative to petroleum-based polymers due to its environmentally beneficial qualities, including biodegradability, biocompatibility, and renewability (Zhou et al., 2008).

The sustainability of using biomass to produce chemicals and fuels, however, has generated a lot of controversy. For instance, the existing method of producing bioethanol, which depends on starch and sugar crops, has serious difficulties. The scarcity of these crops could put food production in competition (Cherubini, 2010). In contrast, lignocellulosic feedstocks offer significant advantages as they constitute the nonedible portion of the plant, thus avoiding competition with food supplies (Sun & Cheng, 2002). Moreover, large quantities of lignocellulosic wastes from forestry, agriculture, and agro-industries accumulate annually. While disposing of these wastes in soil or landfills leads to serious environmental issues, they could be repurposed for producing various value-added products. From an economic standpoint, lignocellulosic biomass is substantially less expensive than crude oil and can be generated at a lower cost than other agriculturally significant biofuel feedstocks like corn starch, soybeans, and sugar cane (Mujtaba et al., 2023).

Despite these advantages, converting lignocellulosic biomass into fine chemicals and polymers remains a significant challenge (Zhou et al., 2011). Because lignocellulosic materials are resistant to enzymatic and chemical breakdown, lignocellulose has evolved to resist degradation (Barakat et al., 2013). Pretreatment of the biomass is necessary to modify the lignocellulosic matrix's chemical and physical properties; this is an expensive and energy-intensive operation. Even though lignocellulosic materials are widely available and typically inexpensive, the main obstacle is creating value-added compounds with high yield and selectivity at a reasonable price (Alonso et al., 2012). To process biomass comparable to petrochemical techniques and create renewable oil and green monomers, biorefinery and biofuel technologies are being developed. Furthermore, the quantity of biorefinery-related pilot and demonstration plants is growing (Stöcker, 2008). To produce advanced biofuels, biochemicals, and biomaterials from nonfood cellulosic biomass feedstocks, for example, firms like Lignol, Verenium, and Mascoma are leading the way in the development of biorefining technology. This study aims to explore the innovative green technologies that enable the sustainable production of bioenergy from lignocellulosic biomass, with a particular focus on bioethanol. By meticulously examining the latest advancements in this field, this review seeks to illuminate the untapped potential of lignocellulosic waste as a renewable energy resource. This exploration contributes to the broader objectives of fostering sustainability and implementing effective waste management strategies. Through a comprehensive analysis of the most recent research and developments, this chapter will provide valuable insights into the technological and operational aspects of bioenergy production from lignocellulosic biomass. Ultimately, the goal is to contribute to advancing sustainable energy solutions and promote a circular economy that minimizes waste and maximizes resource utilization.

2 Lignocellulosic Biomass: A Sustainable Feedstock

Lignocellulosic biomass (LCB), a natural carbon source derived from plant matter, represents a promising and sustainable alternative to fossil fuels. This renewable resource encompasses a wide array of substrates, including agricultural byproducts such as wheat straw and wastes from agro-food industries and paper mills (Balat, 2011; Balat & Balat, 2009; Zabed et al., 2016). LCB is a promising option for bioenergy production due to its abundance, renewability, and affordability. LCB is primarily composed of three major polymers: lignin, cellulose, and hemicellulose, along with inorganic compounds like potassium, magnesium, and calcium (Galbe & Zacchi, 2012). Cellulose and hemicellulose together constitute more than 70% of the total biomass, forming a complex, interlinked structure with lignin through covalent and hydrogen bonds. This intricate network is responsible for the high degree of recalcitrance of LCB, rendering it resistant to degradation. Within this structure, cellulose acts as the structural backbone, surrounded by a matrix of hemicellulose and lignin, which imparts additional strength and rigidity (Menon & Rao, 2012) (Fig. 1).

The four main categories of lignocellulosic biomass are agricultural wastes, grasses, hardwood, and softwood. Particularly, billions of tons of agricultural waste are produced every year all over the world. Even though these residues are plentiful, a large percentage of them are burned or thrown away, creating a large amount of trash. Furthermore, food waste can be disposed of in a variety of ways, including composting, replanting in fields, anaerobic digestion, cogeneration, incineration, sewer disposal, landfill disposal, or even throwing it into the sea. Food waste is any discarded food, including inedible parts, that is removed from the food supply chain (Bos-Brouwers et al., 2014). The biological makeup of lignocellulosic biomass is complicated and multifaceted. Mainly made of cellulose, hemicellulose, and lignin,

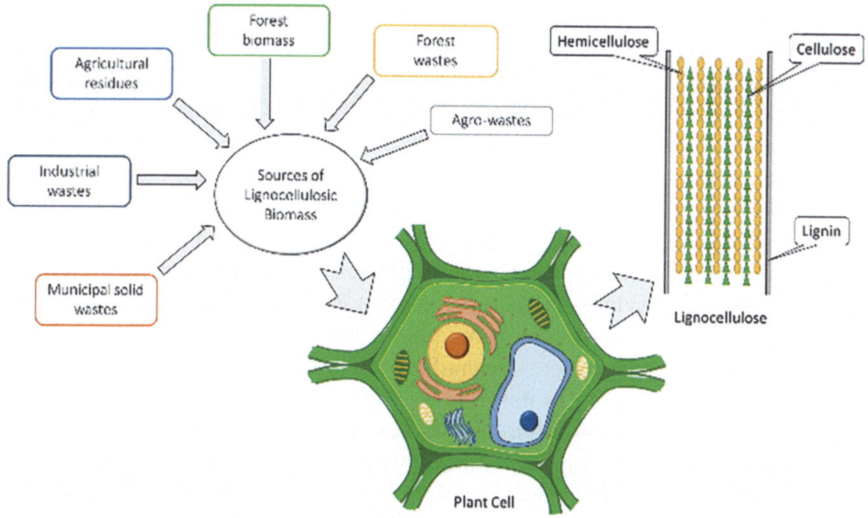

Fig. 1 Diagrammatic representation of the sources of lignocellulosic biomass and its structural makeup at the cellular level, with an emphasis on the main elements and how they are arranged inside the cell walls of plants

these polymers are also linked to smaller components including acetyl groups, minerals, and phenolic substituents. They are arranged into a three-dimensional, irregular structure inside the cell wall of plants. The relative number and order of these components vary depending on the individual type of lignocellulosic biomass, leading to its distinctive features and difficulty in processing. Cellulose, the most abundant biopolymer on Earth, is a crystalline polysaccharide composed of repeating cellobiose units, which are disaccharides formed by two glucose molecules linked by β-1,4-glycosidic bonds (Lee, 1997). The unique three-dimensional structure of cellulose is stabilized by extensive hydrogen bonding networks, contributing to its high crystallinity and mechanical strength (Sun, 2010). This inherent recalcitrance, attributed to both the crystallinity and the strong intermolecular interactions within the cellulose microfibrils, presents a considerable challenge for its efficient conversion into valuable products. The global availability of cellulose is substantial, with an estimated annual production of approximately 1.5×10^{12} tons. Despite being a linear glucose homopolymer with a molecular weight ranging from 50 kDa to 2.5×10^6 kDa—equivalent to 300 to 15,000 glucose residues—cellulose exhibits variations in composition depending on its source. While cellulose is primarily composed of glucose (over 95%), it may also contain minor amounts of other sugars such as galactose, mannose, or xylose (Lavoine et al., 2012). The microfibrillar structure of cellulose consists of highly ordered crystalline regions interspersed with less organized, amorphous areas. These amorphous regions are more susceptible to physical and chemical treatments, whereas the crystalline regions confer resistance to enzymatic hydrolysis (Bochek, 2003; Nishiyama et al., 2002). This dual nature of cellulose, with its resistant crystalline structures and more reactive amorphous areas, complicates its processing for industrial applications, highlighting the need for advanced technologies to overcome these challenges (Satgé et al., 2004).

Hemicelluloses, the second most abundant group of polysaccharides in plants, play a critical structural role within lignocellulosic biomass by facilitating the bonding between cellulose and lignin. Unlike the homopolymeric nature of cellulose, hemicelluloses comprise a diverse family of heteropolysaccharides, consisting of various monomeric units such as pentoses (primarily xylose and arabinose), hexoses (including glucose, mannose, and galactose), and carboxylic acids (mainly mannuronic and galacturonic acids) (Sun, 2010). Among these, xylan—a homopolymer made up of xylose—is the predominant component. In contrast to cellulose, hemicelluloses possess a more amorphous structure with shorter, less ordered chains, which renders them more susceptible to hydrolysis. This structural variability allows hemicelluloses to be soluble in water and extractable from plant cell walls using acidic solutions. Hemicelluloses thus contribute to the structural integrity of the plant cell wall by forming a complex network with cellulose and lignin, aiding in the overall stability and functionality of the plant's lignocellulosic matrix (Agbor et al., 2011).

Lignin, the third most abundant natural polymer following cellulose, is unique as the only biosynthesized aromatic-skeletal polymer in nature (Buranov & Mazza, 2008). The amount of lignin in plants varies greatly depending on the kind: agricultural residues have the lowest lignin content, between 5% and 15%, while hardwoods have the highest amount, between 20% and 25% of their mass, in softwoods (Abo et al., 2019). Lignin is synthesized through the oxidative coupling of three phenylpropanoid monomers—p-coumaryl alcohol, coniferyl alcohol, and sinapyl alcohol—leading to the formation of the phenolic units known as H (p-hydroxyphenyl), G (guaiacyl), and S (syringyl), respectively. For example, in wheat straw lignin, the distribution of these units is approximately H: 5%; G: 49%; and S: 46%. This complex phenolic structure endows lignin with its critical role in providing compressive strength to plants and making them resistant to degradation (Rubin, 2008).

Plant cell walls contain different amounts of cellulose, hemicellulose, and lignin in different species, tissues, and developmental stages (Barakat et al., 2013). Typically, lignocellulosic biomass is composed of 35–50% cellulose, 20–35% hemicellulose, and 10–25% lignin. Ash, proteins, and oils make up the remaining fraction (Saha, 2005). Table 1 provides a detailed breakdown of the chemical composition for specific types of lignocellulosic biomass (Cherubini, 2010).

3 Green Technologies for Sustainable Bioconversion

As the world faces the depletion of fossil resources and the urgent need for environmental sustainability, biomass emerges as a promising alternative energy source. This abundant, cost-effective, and renewable resource offers a sustainable solution for energy production and rural development (International Energy Agency, 2006). Currently, biomass plays a significant role in the global energy supply, especially in

Table 1 Comparison and contrast between the advantages and disadvantages of different pretreatment methods commonly used in the processing of lignocellulosic biomass

Pretreatment methods	Effects	Advantages	Disadvantages	Reference
Mechanical milling	Decrease the particle size and reduce the crystallinity of lignocellulosic materials	Maintain precise control over the final particle size to facilitate material handling	Significant energy usage	Maurya et al. (2015)
Acid	Separating hemicellulose and lignin	In some cases, enzymatic hydrolysis may not be necessary, as the acid used in pretreatment can hydrolyze the biomass directly, yielding fermentable sugars	High cost of reactors, corrosive and toxic chemicals, and the formation of inhibitory byproducts	Jönsson & Martín (2016)
Extrusion	Fiber shortening and separation	High biomass concentration, minimal byproduct formation, rapid processing time	Significant energy usage, ineffective without chemical additives, requires chemical assistance, primarily effective on grasses and herbaceous plants	Duque et al. (2017)
Alkaline	Lignin and hemicellulose removal	Result in less sugar breakdown compared to acid pretreatment	Generation of inhibitors	Zhang et al. (2016)
Oxidation	Removal of lignin and hemicellulose	Lower production of products	Cellulose is partly degraded, high cost	Chandel et al. (2013)
Organosolv	Lignin elimination and hemicellulose isolation	Generate lignin-reduced substrates that minimize enzyme adsorption, enabling enzyme recycling and reuse	Significant capital expenditure, exposure to hazardous chemicals, and generation of inhibitory byproducts	Nitsos et al. (2017)

developing countries. Projections suggest that its contribution will continue to grow, potentially accounting for nearly 33–50% of primary energy consumption by 2050 (Uihlein & Schebek, 2009). To fully harness biomass's potential, it is crucial to develop technologies that efficiently convert waste biomass into bioenergy and chemical products while minimizing environmental impact. Green technologies, which emphasize sustainability, resource conservation, and reduced fossil fuel dependence, provide a promising approach to achieving these goals (Emissions Scenarios : Summary for Policymakers, 2000). By integrating green technologies into various stages of biomass conversion—such as pretreatment, hydrolysis, and fermentation—efficiency can be enhanced, waste reduced, and the use of harmful chemicals minimized. In a circular economy, waste is converted into useful resources, and this strategy is in line with those ideas. Green technology is a broad term that encompasses innovative methodologies and approaches aimed at minimizing environmental impact and promoting sustainability across various sectors. These technologies are developed with the specific goals of conserving natural resources, safeguarding ecosystems, and reducing reliance on fossil fuels. By prioritizing environmental protection and human health, green technologies offer a viable alternative to traditional methods that often contribute to pollution, waste generation, and ecological degradation (Podesta et al., 2007). Green technologies are intended to improve biomass conversion into biofuels and other valuable products in the context of bioenergy production, while simultaneously lowering waste, energy consumption, and the usage of hazardous chemicals. These technologies emphasize the use of renewable resources, lower greenhouse gas emissions, and environmentally benign catalysts and solvents. By mitigating the negative consequences of human activities, green technologies contribute to a more sustainable and resilient future. Commonly referred to as environmental or clean technology, these innovations underscore a commitment to environmental stewardship throughout the production and consumption cycles (Munn, 1992; Policies — Vikaspedia, 2006). Using biomass with green technologies together offers a sustainable and promising path to meeting global energy demands while addressing environmental concerns. By leveraging the renewable nature of biomass and applying innovative green technologies, we can create a more sustainable and resilient future.

Green technologies offer substantial environmental benefits, particularly in the context of mitigating climate change. These technologies can greatly lower greenhouse gas emissions and support the worldwide effort to battle climate change by utilizing energy-efficient procedures and renewable biomass sources (Feng et al., 2024). Additionally, green technologies play a crucial role in improving air quality. By reducing reliance on fossil fuels and minimizing emissions of particulate matter and other harmful pollutants, they help create cleaner air, which in turn promotes better public health and overall well-being (Feng et al., 2024). Another critical environmental advantage of green technologies is the preservation of water resources. These technologies can minimize water pollution and significantly reduce the water footprint associated with bioenergy production (Li et al., 2023). This conservation of valuable water resources is particularly important in regions facing water scarcity. Moreover, green technologies contribute to the protection of biodiversity by reducing the need for extensive land conversion for agricultural purposes. By minimizing the negative impacts on ecosystems, they help preserve biodiversity and protect endangered species. Furthermore, green technologies promote sustainable land management practices, such as crop rotation and agroforestry, which enhance soil health, reduce erosion, and bolster ecosystem resilience (Dönmez et al., 2024). These practices contribute to the long-term sustainability of agricultural lands and natural ecosystems.

The economic benefits of green technologies are equally compelling. One of the most significant advantages is the potential for job creation and economic growth. The development and implementation of green technologies can generate new employment opportunities, particularly in rural areas, and stimulate economic activity (He et al., 2023). This can lead to revitalized local economies and improved livelihoods for many communities. Moreover, increasing reliance on renewable bioenergy sources can enhance energy independence by reducing dependence on foreign oil. This shift not only bolsters energy security but also brings economic benefits, contributing to geopolitical stability (He et al., 2023). Green technologies also play a vital role in rural development. By fostering bioenergy production, they provide economic opportunities for rural communities, leading to job creation, income generation, and overall improvements in living standards. Additionally, these technologies can lead to cost reductions in bioenergy production, making it more efficient and less reliant on fossil fuels. This efficiency can help mitigate the impact of fluctuating energy prices, providing greater financial stability for producers and consumers alike.

4 Integrating Green Technologies for Sustainable Biomass Conversion

Green technologies play a crucial role in enhancing the sustainability and efficiency of each stage of the lignocellulosic biomass conversion process. During the pretreatment stage, green technologies focus on using environmentally friendly methods such as renewable solvents and catalysts. These approaches help break down the complex structure of lignocellulosic materials without relying on harsh chemicals, reducing energy consumption, and minimizing environmental impact (Barakat et al., 2013). In the enzymatic hydrolysis stage, renewable source enzymes are employed to convert biomass into fermentable sugars. This not only improves process efficiency but also minimizes waste, aligning with green technology principles. The use of these enzymes, derived from sustainable sources, further enhances the environmental benefits of the process. Finally, during fermentation, green technologies enable the use of microorganisms that are either naturally optimized or genetically engineered for higher yields and lower byproduct formation. This optimization reduces waste and enhances the overall sustainability of the biofuel production process. The integration of green technologies throughout the biomass conversion process is essential for reducing environmental impact and promoting a circular economy. By minimizing waste and maximizing resource utilization, these technologies contribute to a more sustainable and efficient bioenergy production system.

5 Pretreatment Technologies

The fractionation of lignocellulosic biomass into its constituent components—cellulose, hemicellulose, and lignin—is a pivotal objective in biorefining. Single-step processes, such as pyrolysis, while cost-effective, often result in incomplete deconstruction of the lignocellulosic matrix due to their reliance on high temperatures. The resulting bio-oil is a complex mixture of numerous compounds, hindering the efficient separation of targeted chemicals and fuels. This necessitates additional downstream processing steps, increasing costs and complexity. Pretreatment, a crucial preparatory step, is essential to modify the supramolecular structure of the lignocellulose matrix, enhancing the accessibility and biodegradability of cellulose and hemicellulose for subsequent enzymatic or chemical treatments (Alonso et al., 2012; Barakat et al., 2013). By disrupting the natural binding characteristics of the lignocellulosic components, pretreatment facilitates the fractionation process, ultimately improving the overall efficiency and economic viability of biorefining operations.

Pretreatment methods, encompassing physical, chemical, physicochemical, and biological approaches, are crucial for disrupting the complex structure of lignocellulosic biomass and facilitating the fractionation of its constituent components (Saha, 2005). These methods aim to reduce particle size, enhance accessibility, and increase the susceptibility of cellulose and hemicellulose to enzymatic or chemical hydrolysis (Amin et al., 2017). Milling, irradiation, microwave, steam explosion, Ammonia Fiber Expansion (AFEX), supercritical CO_2 and its explosion, SO_2, alkaline hydrolysis, liquid hot-water pretreatment, organosolv processes, wet oxidation, ozonolysis, dilute- and concentrated-acid hydrolyzes, and biological pretreatments are just a few of the many pretreatment options that have been investigated to combat climate change globally (Kassim et al., 2022). Each pretreatment method exhibits its distinct advantages and disadvantages, influencing product yield, selectivity, and overall process efficiency (Table 1). While some methods have successfully transitioned from research to industrial application, significant challenges persist, such as the generation of environmentally harmful wastes and high energy consumption. The development of green technology solutions to address these challenges is essential for the sustainable and economically viable utilization of lignocellulosic biomass (Capolupo & Faraco, 2016).

Reducing the overall cost of turning lignocellulose into valuable products requires the integration of pretreatment techniques with downstream processes, such as enzymatic saccharification, detoxification, fermentation, and product recovery (Kassim et al., 2022). Reaping the commercial-scale benefits of lignocellulosic biomass conversion requires advancements in pretreatment technologies, microorganisms producing cellulolytic enzymes, holistic biomass utilization, and process integration (Da Silva et al., 2012).

6 Emerging Technologies for Lignocellulosic Biomass Pretreatment

In recent years, the concept of "Green Chemistry" has gained increasing interest as a potential solution to the challenge of developing a viable biorefinery concept. Central to this goal is the advancement of technologies that utilize raw materials more efficiently, minimize waste, and avoid the use of toxic and hazardous substances. Selected green methods currently being pursued for the pretreatment of lignocellulosic biomass exemplify this approach (Mishra et al., 2023). Although

Table 2 Key advantages and disadvantages of green pretreatment methods

Pretreatment methods	Effects	Advantages	Disadvantages	References
Steam explosion	Lignin removal and hemicellulose fractionation	Green solvent is used, which is biodegradable and biocompatible	Poor stability under higher pretreatment temperatures	Zhang et al. (2016)
Microbes	Lignin softening, particle size reduction	Low capital investment, moderate energy requirements, and low environmental impacts	It is much less effective for softwood	Pielhop et al. (2016)
Deep eutectic solvents	Cellulose crystallinity reduction and lignin removal	Green solvent is used, it does not cause degradation of sugars, method is suitable for mobile biomass processor	Total utility costs are high	Escobar et al. (2020)
Supercritical fluids	Lignin and hemicellulose degradation	Environment friendly, selective degradation of lignin and hemicelluloses	Very long pretreatment time (several weeks) due to slow yield	Sun et al. (2016)

these green methods are environmentally friendly, challenges persist, including high production costs, poor efficiency, and a lack of commercially available equipment suitable for industrial-scale processing (Kassim et al., 2022). However, the broader adoption of such technologies by the food industry, along with anticipated reductions in initial capital costs and increased operational scale, may promote their uptake for lignocellulosic biomass pretreatment (Table 2).

Simultaneously, a rapidly emerging field in minimum food processing involves chemical methods applied in harsh or unconventional environments. The pretreatment of lignocellulosic biomass is also being investigated using nonthermal food processing technologies such as high hydrostatic pressure, high-pressure homogenization, gamma rays, electron beams, pulsed electric fields, and microwaves. These novel approaches, which are in line with Green Chemistry principles, present viable substitutes for improving the sustainability and efficiency of the biorefinery activities that are outlined in Table 3 (Fig. 2).

7 Enzymatic Hydrolysis: Biomass to Biofuel

Enzymatic hydrolysis is a crucial step in the conversion of lignocellulosic biomass into biofuels and other valuable products. It involves the breakdown of the complex polysaccharides' cellulose and hemicellulose into simpler sugars, such as glucose and xylose, using specific enzymes (Vasić et al., 2021). These sugars can then be fermented by microorganisms to produce bioethanol, biogas, or other biobased chemicals. Enzymatic hydrolysis follows pretreatment, which aims to disrupt the

Table 3 Advancements in Green Chemistry and nonthermal processing technologies for sustainable pretreatment of lignocellulosic biomass

Method	Description	Advantages	Applications	Limitations	References
Microwave irradiation	Electromagnetic radiation (300 300,000 MHz) that transfers energy selectively to substances. Used for lignocellulosic pretreatment	– Fast heat transfer – Short reaction time – Uniform volumetric heating – Energy-efficient	– Microwave-assisted solvolysis for depolymerizing biomass – microwave-assisted pyrolysis for bio-oil and biogas production	– biomass is generally low lossy material – Requires additives like graphite for rapid heating	Dai et al. (2017), Huang et al. (2016)
Microwave-assisted solvolysis	Solvolysis under mild temperatures (<200 °C) for biomass depolymerization	– Enhanced enzymatic saccharification – Internal uniform heating	– Pretreatment of sweet sorghum bagasse, corn straw, rice husk, brewer's spent grain	– Limited effect at temperatures ≤100 °C	Chen et al. (2017), Ravindran et al. (2018)
Microwave-assisted pyrolysis	Pyrolysis using microwave irradiation under high temperatures (>400 °C) to convert biomass into biofuels	– Faster heating rate compared to conventional methods – High bio-oil and syn-gas yield	– Conversion of agricultural biomass into higher-value products	– Requires precise control to avoid overheating	Huang et al. (2015), Lo et al. (2017)
Ultrasound	High-frequency sound waves (20 kHz - 1 MHz) used to disrupt biomass structure and enhance chemical reactions	– Alters surface structure – Produces oxidizing radicals – Can disrupt lignin linkages	– Pretreatment of sugarcane bagasse, spent coffee waste – Enhances enzymatic saccharification	– Ultrasonic vibration energy may be too low to alter biomass surface conformation	Subhedar et al. (2018)
Combination of microwave and ultrasound	Dual-energy approach combining the effects of both microwaves and ultrasound	– Accelerates hydrolysis and biodegradability – Enhances enzymatic action on biomass	– Pretreatment of corn cobs, lignocellulosic waste – Patent applications for biomass treatment	– Limited by the efficiency of individual processes	Huaiyin Teachers College (2017), Peter Herbert North (2010)

Gamma ray	Ionizing radiation from radioisotopes (cobalt-60 or cesium-137) used to modify lignocellulosic structure	– Easily penetrates biomass – Causes cellulose breakdown – Forms free radicals	– Enhances enzymatic hydrolysis of lignocellulosic biomass	– High doses required – Potential safety concerns	Hyun Hong et al. (2014), Liu et al. (2015)
Electron beam (EB) irradiation	Accelerated beams of electrons used to disrupt cell wall polymers of lignocellulosic biomass	– Depolymerizes cellulose and hemicellulose – Increases phenolic yield	– Depolymerization of sugar maple, other lignocellulosic biomass	– Limited by penetration depth and energy requirement	Grabowski & Shields (2015), Mante et al. (2014)

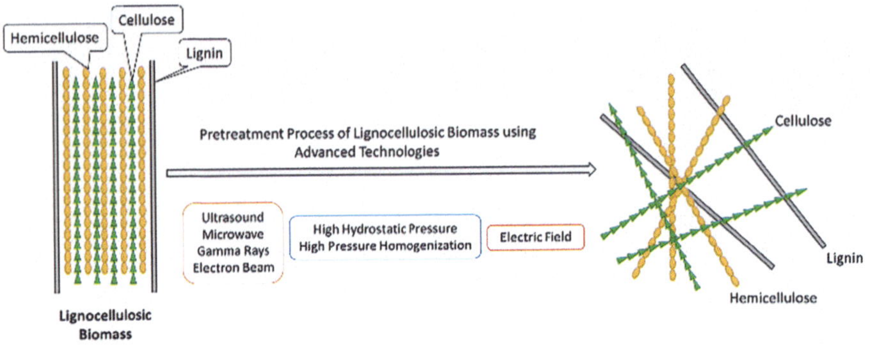

Fig. 2 Diagram illustrating the pretreatment process of lignocellulosic biomass using advanced technologies

recalcitrant structure of lignocellulosic biomass and improve the accessibility of cellulose and hemicellulose to enzymes. During enzymatic hydrolysis, cellulases and hemicellulases are added to the pretreated biomass to catalyze the breakdown of the polysaccharides. The released sugars can then be fermented to produce the desired biofuel or product (Vasić et al., 2021).

The addition of enzymes is a crucial step in the enzymatic hydrolysis of lignocellulosic biomass, where cellulases and hemicellulases play significant roles in breaking down complex polymers into simpler sugars. Cellulases, including endoglucanases, exoglucanases (cellobiohydrolases), and β-glucosidases, are responsible for the degradation of cellulose (Obeng et al., 2017). Endoglucanases cleave internal bonds within the cellulose chain, creating new chain ends and reducing the polymerization degree, while exoglucanases act on the ends of cellulose chains, releasing cellobiose, a disaccharide composed of two glucose molecules. β-Glucosidases then hydrolyze cellobiose into glucose, completing the breakdown of cellulose into fermentable sugars. Hemicellulases, on the other hand, target hemicellulose, breaking it down into its constituent sugars (Obeng et al., 2017). Xylanases degrade xylan, a major component of hemicellulose, into xylose, while mannanases target mannan, another hemicellulose component, converting it into mannose. Depending on the specific sugar composition of hemicellulose, additional enzymes may be required to fully hydrolyze the polymer (Fig. 3).

The hydrolysis reaction occurs under mild conditions, with temperatures ranging from 40 to 50°C and a slightly acidic pH of around 4.5–5, which are optimal for the activity of most cellulases and hemicellulases (Saha et al., 2017). During this process, the enzymes catalyze the breakdown of cellulose and hemicellulose by cleaving the glycosidic bonds in the polymers, a reaction that can take several hours to days depending on the biomass type and process conditions (Deng et al., 2022). The outcome of enzymatic hydrolysis is the release of monosaccharides, primarily glucose from cellulose and a mixture of sugars such as xylose, arabinose, and mannose from hemicellulose. These sugars are then available for fermentation by microorganisms to produce biofuels like ethanol or other valuable chemicals (Deng et al., 2022).

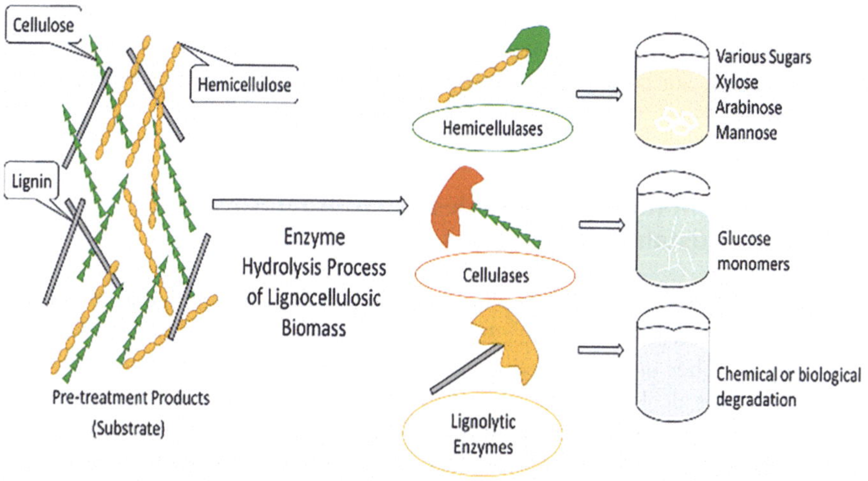

Fig. 3 Diagram illustrating the enzymatic hydrolysis process of lignocellulosic biomass, a key step in converting plant-based materials into biofuels and other valuable products

Several factors influence the efficiency of enzymatic hydrolysis, with substrate accessibility being one of the most critical. The ease with which enzymes can access cellulose and hemicellulose largely determines the effectiveness of hydrolysis. Pretreatment methods that increase the porosity of the biomass and reduce its crystallinity are particularly beneficial, as they enhance enzyme accessibility to the target substrates (Li et al., 2022). Cellulose crystallinity itself is a significant factor; highly crystalline cellulose is more resistant to enzymatic degradation. Therefore, pretreatments that reduce this crystallinity can substantially improve hydrolysis rates. Another important factor is enzyme loading. The quantity of enzyme applied during the hydrolysis process directly impacts both the rate and the overall extent of hydrolysis (Abolore et al., 2024). Although increasing enzyme loading can accelerate the process, it also raises the associated costs, necessitating a balance between efficiency and economic feasibility. Additionally, the presence of inhibitors poses a challenge to effective hydrolysis. Compounds such as phenolic derivatives, which are often released during the pretreatment of lignocellulosic biomass, can inhibit enzyme activity and diminish hydrolysis efficiency (Mathibe et al., 2020). Consequently, it is essential to remove or neutralize these inhibitors to optimize the hydrolysis process.

To optimize enzymatic hydrolysis efficiency, several strategies can be implemented. The addition of nonionic surfactants or polymers like polyethylene glycol (PEG) can significantly enhance hydrolysis by altering the surface properties of the biomass (Da Silva et al., 2020). PEG, for instance, can adsorb to lignin, preventing enzyme binding and directing enzymes toward cellulose, thereby increasing hydrolysis efficiency. Moreover, optimizing process parameters such as solid loading, enzyme concentration, shaking speed, and hydrolysis time is crucial for maximizing sugar yield (Da Silva et al., 2020). Notably, higher solid loadings can increase the concentration of sugars in the hydrolysate, improving overall process efficiency. Following the successful enzymatic hydrolysis of lignocellulosic biomass, the released sugars can be efficiently fermented to produce a range of biofuels and chemicals.

7.1 Fermentation Technologies: Microorganisms and Sustainable Processes

The pretreatment and hydrolysis processes are pivotal for optimizing the subsequent fermentation phase (Su et al., 2020). Fermentation, though a natural biochemical pathway, relies on microorganisms to convert fermentable sugars into various products, such as alcohol or lactic acid. Industrial yeasts, particularly *Saccharomyces cerevisiae*, are commonly used in industries like winemaking, brewing, and alcohol production, and are also integral to sugar-based biofuel production (Mohd Azhar et al., 2017). When cellulosic material or slurry is pretreated, it is converted into fermentable sugars, making it suitable for enzymatic or acidic hydrolysis. When this slurry is mixed with water, it forms a broth that can be fermented. In typical batch fermentation, *S. cerevisiae* ferments hexose sugars, such as glucose, into ethanol under anaerobic conditions and controlled temperatures. This fermentation process produces carbon dioxide as a byproduct, and nitrogen is often added to accelerate the reaction. *S. cerevisiae* is known for its high ethanol yield from hexose sugars, but it cannot ferment other sugars, like those found in lignocellulosic biomass, which often contains pentose sugars such as D-xylose (Ferreira et al., 2019; Naghshbandi et al., 2019) (Fig. 4).

Another microorganism, *Zymomonas mobilis*, shares a similar limitation. However, natural xylose-fermenting yeasts, such as *Pichia stipitis* and *Candida*

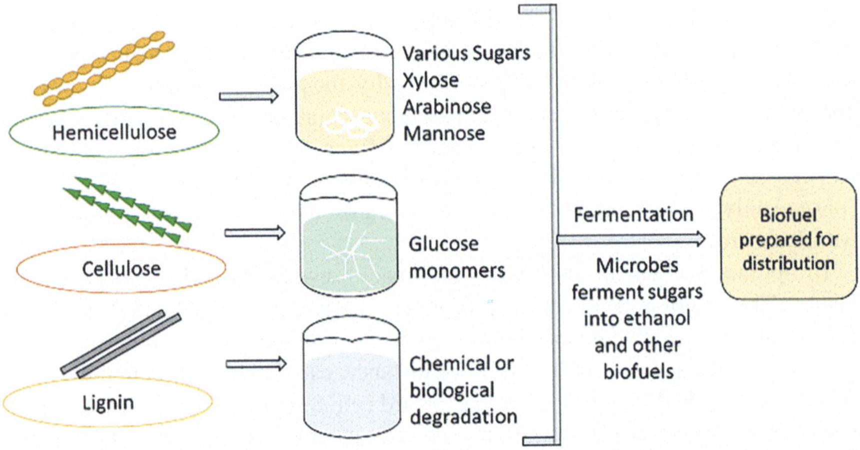

Fig. 4 Image illustrating the process of converting lignocellulosic biomass into biofuels involving fermentation

parapsilosis, can metabolize xylose through the action of xylose reductase, which converts xylose to xylitol, and xylitol dehydrogenase, which further converts xylitol to xylulose. An ideal microorganism for fermentation should tolerate high alcohol concentrations and withstand chemical inhibitors produced during pretreatment and hydrolysis. To address these challenges, genetically engineered microorganisms capable of fermenting both pentose and hexose sugars have been developed (Ko & Lee, 2018). Additionally, thermophilic anaerobic bacteria, such as *Thermoanaerobacter ethanolicus*, *Thermoanaerobacter mathranii*, and *Thermoanaerobacter brockii,* have been studied for bioethanol production. These bacteria can withstand extreme temperatures, making them suitable for pretreating low-cost biomass feedstocks. However, their low ethanol tolerance remains a significant drawback (Arora et al., 2015). Fermentation can be carried out using continuous, batch, or fed-batch processes, depending on the characteristics of the lignocellulosic hydrolysate and the kinetics of the microorganisms employed. In continuous fermentation, a substrate-containing feed and culture medium are continuously supplied to a reactor containing active microorganisms. The batch process, a simpler and closed system, begins with a limited initial nutrient supply inoculated with microorganisms that carry out the fermentation. Fed-batch reactors, commonly used in industrial settings, combine the benefits of both continuous and batch processes. They allow for increased viable cell concentrations, extended culture lifetimes, and higher product accumulation (Gutiérrez-Antonio et al., 2017) (Table 4).

7.2 Challenges and Future Directions

Despite the notable advancements in the field of bioethanol production over the past decade, including the development of sophisticated engineered microorganisms and efforts to integrate pretreatment, hydrolysis, and fermentation processes into more efficient steps, several significant challenges remain (Pascual et al., 2022). Addressing these challenges is crucial for bridging the gap between innovative research and practical industrial applications. One major challenge is the selection of a suitable pretreatment strategy. The chosen method must be cost-effective while not compromising the efficiency of enzymatic saccharification. Finding a balance between cost and efficiency is vital to optimize the overall process (Pascual et al., 2022). Additionally, improving the anaerobic digestibility of biomass is essential for enhancing the yield and efficiency of subsequent stages.

Another critical issue is the limitation of carbohydrate degradation and the generation of inhibitors during pretreatment. These factors can hinder the conversion yield, making it imperative to minimize their impact (Maurya et al., 2015). Moreover, there is a need to reduce the consumption of toxic chemicals, as well as the energy and water used in the process, to enhance environmental sustainability and reduce operational costs. Advancements in biocatalysts are crucial, as developing and applying novel biocatalysts that enhance the efficiency of saccharification can significantly impact the overall process (Pascual et al., 2022). This includes designing

Table 4 Techniques and strategies for integrating hydrolysis and fermentation in bioethanol production

Technique/strategy.	Description	Advantages	References
Separate hydrolysis and fermentation (SHF)	Enzymatic hydrolysis and fermentation occur in separate steps, each optimized for specific conditions	– each step can be optimized independently	Chow et al. (2015), Tavva et al. (2016)
Simultaneous saccharification and fermentation (SSF)	Combines diluted-acid pretreatment with high temperatures, followed by fermentation in a single process	– Faster hydrolysis rates due to continuous sugar conversion – Reduced enzyme loading. – Higher bioethanol yields – Shorter processing time – Smaller reactor volumes – Low sugar concentrations in fermenters due to immediate consumption	Chohan et al. (2020), Luiza Astolfi et al. (2020)
Direct microbial conversion (DMC)	Integrates cellulose production, hydrolysis, and glucose fermentation into a single step	– Simplifies the production process – Reduces the number of reactors required – Lowers overall costs	Prasad et al. (2019)

enzymes with enhanced specific activity, greater thermal stability, and reduced end-product inhibition to improve their performance. Reducing the overall environmental and economic footprint of the bioethanol production process is another important goal (Bhatia et al., 2012). A comprehensive understanding of the structure and composition of various biomass types, as well as the effects of different pretreatment techniques at both macro and molecular levels, is crucial. In-depth studies of the interactions between biomass, microorganisms, products, and byproducts during hydrolysis and fermentation will help establish optimal conditions for these processes.

While existing research provides a broad foundation, further interdisciplinary studies are needed to advance bioethanol production from the laboratory scale to commercial viability. Transitioning to commercial-scale production presents significant challenges, requiring extensive pilot-scale studies, process optimization, and substantial financial investment. Continued research and development are essential to make bioethanol production a profitable and widely adopted technology.

8 Conclusion

Enhancing the eco-friendly production of bioethanol from lignocellulosic waste materials is a critical endeavor that addresses sustainability, climate change mitigation, and waste management challenges. As the world transitions from finite fossil fuels to renewable energy sources, lignocellulosic biomass emerges as a valuable, abundant resource for bioethanol production. This review explores the multifaceted nature of this topic, focusing on the environmental and economic benefits of bioethanol, the challenges associated with its production, and potential solutions. Bioethanol is a renewable and eco-friendly fuel source that can reduce greenhouse gas emissions and alleviate the burden on landfills. However, producing bioethanol from lignocellulosic biomass requires overcoming significant challenges, such as biomass availability, pretreatment efficiency, enzyme costs, and microbial tolerance. Addressing these challenges necessitates innovative solutions, research advancements, and continuous technological improvements. This chapter suggests further research in areas such as enzymatic pretreatment techniques, microbial strain optimization, bioreactor design, green solvent integration, and life cycle assessment. By focusing on these areas, we can develop more efficient and environmentally friendly processes for producing bioethanol from lignocellulosic waste materials. The future of bioethanol production lies in the pursuit of green technologies, circular bioeconomics, and optimized production processes. By addressing the challenges and maximizing the potential of lignocellulosic biomass, we can contribute to a more sustainable and greener energy future.

References

Abo, B. O., Gao, M., Wang, Y., Wu, C., Ma, H., & Wang, Q. (2019). Lignocellulosic biomass for bioethanol: An overview on pretreatment, hydrolysis and fermentation processes. *Reviews on Environmental Health, 34*(1), 57–68. https://doi.org/10.1515/REVEH-2018-0054/MACHINEREADABLECITATION/RIS

Abolore, R. S., Jaiswal, S., & Jaiswal, A. K. (2024). Green and sustainable pretreatment methods for cellulose extraction from lignocellulosic biomass and its applications: A review. *Carbohydrate Polymer Technologies and Applications, 7*, 100396. https://doi.org/10.1016/J.CARPTA.2023.100396

Agbor, V. B., Cicek, N., Sparling, R., Berlin, A., & Levin, D. B. (2011). Biomass pretreatment: Fundamentals toward application. *Biotechnology Advances, 29*(6), 675–685. https://doi.org/10.1016/J.BIOTECHADV.2011.05.005

Ahmed, S. F., Kabir, M., Mehjabin, A., Oishi, F. T. Z., Ahmed, S., Mannan, S., Mofijur, M., Almomani, F., Badruddin, I. A., & Kamangar, S. (2023). Waste biorefinery to produce renewable energy: Bioconversion process and circular bioeconomy. *Energy Reports, 10*, 3073–3091. https://doi.org/10.1016/J.EGYR.2023.09.137

Alonso, D. M., Wettstein, S. G., Mellmer, M. A., Gurbuz, E. I., & Dumesic, J. A. (2012). Integrated conversion of hemicellulose and cellulose from lignocellulosic biomass. *Energy & Environmental Science, 6*(1), 76–80. https://doi.org/10.1039/C2EE23617F

Amin, F. R., Khalid, H., Zhang, H., Rahman, S., Zhang, R., Liu, G., & Chen, C. (2017). Pretreatment methods of lignocellulosic biomass for anaerobic digestion. *AMB Express, 7*(1), 1–12. https://doi.org/10.1186/S13568-017-0375-4/TABLES/2

Arora, R., Behera, S., & Kumar, S. (2015). Bioprospecting thermophilic/thermotolerant microbes for production of lignocellulosic ethanol: A future perspective. *Renewable and Sustainable Energy Reviews, 51*, 699–717. https://doi.org/10.1016/J.RSER.2015.06.050

Balat, M. (2011). Production of bioethanol from lignocellulosic materials via the biochemical pathway: A review. *Energy Conversion and Management, 52*(2), 858–875. https://doi.org/10.1016/J.ENCONMAN.2010.08.013

Balat, M., & Balat, H. (2009). Recent trends in global production and utilization of bio-ethanol fuel. *Applied Energy, 86*(11), 2273–2282. https://doi.org/10.1016/J.APENERGY.2009.03.015

Barakat, A., de Vries, H., & Rouau, X. (2013). Dry fractionation process as an important step in current and future lignocellulose biorefineries: A review. *Bioresource Technology, 134*, 362–373. https://doi.org/10.1016/J.BIORTECH.2013.01.169

Bhatia, L., Johri, S., & Ahmad, R. (2012). An economic and ecological perspective of ethanol production from renewable agro waste: A review. *AMB Express, 2*(1), 65. https://doi.org/10.1186/2191-0855-2-65/FIGURES/3

Bochek, A. M. (2003). Effect of hydrogen bonding on cellulose solubility in aqueous and nonaqueous solvents. *Russian Journal of Applied Chemistry, 76*(11), 1711–1719. https://doi.org/10.1023/B:RJAC.0000018669.88546.56/METRICS

Bos-Brouwers, H. E. J., Timmermans, A. J. M., Soethoudt, J. M., Ostergren, K., Gustavsson, J., Hansen, O. J., Möller, H., Anderson, G., O'Connor, C., & Quested, T. (2014). *FUSIONS definitional framework for food waste*. FUSIONS N FP7-KBBE-2012-6-311972. https://research.wur.nl/en/publications/fusions-definitional-framework-for-food-waste

Buranov, A. U., & Mazza, G. (2008). Lignin in straw of herbaceous crops. *Industrial Crops and Products, 28*(3), 237–259. https://doi.org/10.1016/J.INDCROP.2008.03.008

Bušić, A., Mardetko, N., Kundas, S., Morzak, G., Belskaya, H., Šantek, M. I., Komes, D., Novak, S., & Šantek, B. (2018). Proizvodnja bioetanola iz obnovljivih sirovina te njegovo odvajanje i pročišćavanje: pregled. *Food Technology and Biotechnology, 56*(3), 289–311. https://doi.org/10.17113/FTB.56.03.18.5546

Capolupo, L., & Faraco, V. (2016). Green methods of lignocellulose pretreatment for biorefinery development. *Applied Microbiology and Biotechnology, 100*(22), 9451–9467. https://doi.org/10.1007/S00253-016-7884-Y

Chandel, A. K., da Silva, S. S., Chandel, A. K., & Da Silva, S. S. (2013). Sustainable degradation of lignocellulosic biomass—Techniques. In *Applications and Commercialization. Sustainable Degradation of Lignocellulosic Biomass—Techniques, Applications and Commercialization.* IntechOpen. https://doi.org/10.5772/1490

Chen, H., Liu, J., Chang, X., Chen, D., Xue, Y., Liu, P., Lin, H., & Han, S. (2017). A review on the pretreatment of lignocellulose for high-value chemicals. *Fuel Processing Technology, 160*, 196–206. https://doi.org/10.1016/J.FUPROC.2016.12.007

Cherubini, F. (2010). The biorefinery concept: Using biomass instead of oil for producing energy and chemicals. *Energy Conversion and Management, 51*(7), 1412–1421. https://doi.org/10.1016/J.ENCONMAN.2010.01.015

Chohan, N. A., Aruwajoye, G. S., Sewsynker-Sukai, Y., & Gueguim Kana, E. B. (2020). Valorisation of potato peel wastes for bioethanol production using simultaneous saccharification and fermentation: Process optimization and kinetic assessment. *Renewable Energy, 146*, 1031–1040. https://doi.org/10.1016/J.RENENE.2019.07.042

Chow, T. J., Su, H. Y., Tsai, T. Y., Chou, H. H., Lee, T. M., & Chang, J. S. (2015). Using recombinant cyanobacterium (Synechococcus elongatus) with increased carbohydrate productivity as feedstock for bioethanol production via separate hydrolysis and fermentation process. *Bioresource Technology, 184*, 33–41. https://doi.org/10.1016/J.BIORTECH.2014.10.065

Da Silva, S. S., Chandel, A. K., Wickramasinghe, S. R., & Domínguez, J. M. G. (2012). Fermentative production of value-added products from lignocellulosic biomass. *BioMed Research International, 2012*(1), 826162. https://doi.org/10.1155/2012/826162

Da Silva, A. S. A., Espinheira, R. P., Teixeira, R. S. S., De Souza, M. F., Ferreira-Leitão, V., & Bon, E. P. S. (2020). Constraints and advances in high-solids enzymatic hydrolysis of lignocellulosic biomass: A critical review. *Biotechnology for Biofuels, 13*(1), 1–28. https://doi.org/10.1186/S13068-020-01697-W

Dai, L., He, C., Wang, Y., Liu, Y., Yu, Z., Zhou, Y., Fan, L., Duan, D., & Ruan, R. (2017). Comparative study on microwave and conventional hydrothermal pretreatment of bamboo sawdust: Hydrochar properties and its pyrolysis behaviors. *Energy Conversion and Management, 146*, 1–7. https://doi.org/10.1016/J.ENCONMAN.2017.05.007

Deng, W., Feng, Y., Fu, J., Guo, H., Guo, Y., Han, B., Jiang, Z., Kong, L., Li, C., Liu, H., Nguyen, P. T. T., Ren, P., Wang, F., Wang, S., Wang, Y., Wang, Y., Wong, S. S., Yan, K., Yan, N., et al. (2022). Catalytic conversion of lignocellulosic biomass into chemicals and fuels. *Green Energy & Environment, 8(1), 10–114.* https://doi.org/10.1016/J.GEE.2022.07.003

Devi, A., Bajar, S., Kour, H., Kothari, R., Pant, D., & Singh, A. (2022). Lignocellulosic biomass valorization for bioethanol production: A circular bioeconomy approach. *BioEnergy Research, 15*(4), 1820–1841. https://doi.org/10.1007/S12155-022-10401-9

Dönmez, D., Isak, M. A., İzgü, T., & Şimşek, Ö. (2024). Green horizons: Navigating the future of agriculture through sustainable practices. *Sustainability, 16, 3505.* https://doi.org/10.3390/SU16083505

Duque, A., Manzanares, P., & Ballesteros, M. (2017). Extrusion as a pretreatment for lignocellulosic biomass: Fundamentals and applications. *Renewable Energy, 114*, 1427–1441. https://doi.org/10.1016/J.RENENE.2017.06.050

Emissions Scenarios: Summary for Policymakers. (2000). IPCC.

Escobar, E. L. N., da Silva, T. A., Pirich, C. L., Corazza, M. L., & Pereira Ramos, L. (2020). Supercritical fluids: A promising technique for biomass pretreatment and fractionation. *Frontiers in Bioengineering and Biotechnology, 8*, 524832. https://doi.org/10.3389/FBIOE.2020.00252/BIBTEX

Feng, Q., Usman, M., Saqib, N., & Mentel, U. (2024). Modelling the contribution of green technologies, renewable energy, economic complexity, and human capital in environmental sustainability: Evidence from BRICS countries. *Gondwana Research, 132*, 168–181. https://doi.org/10.1016/J.GR.2024.04.010

Ferreira, R. M., Mota, M. J., Lopes, R. P., Sousa, S., Gomes, A. M., Delgadillo, I., & Saraiva, J. A. (2019). Adaptation of Saccharomyces cerevisiae to high pressure (15, 25 and 35 MPa) to enhance the production of bioethanol. *Food Research International (Ottawa Ont.), 115*, 352–359. https://doi.org/10.1016/J.FOODRES.2018.11.027

Galbe, M., & Zacchi, G. (2012). Pretreatment: The key to efficient utilization of lignocellulosic materials. *Biomass and Bioenergy, 46*, 70–78. https://doi.org/10.1016/J.BIOMBIOE.2012.03.026

Grabowski, C., & Shields, W. M. (2015). The impact of electron beam pretreatment on the fermentation of wood-based sugars. https://experts.esf.edu/esploro/outputs/undergraduate/The-Impact-of-Electron-Beam-Pretreatment/99870858704826

Gutiérrez-Antonio, C., Gómez-Castro, F. I., de Lira-Flores, J. A., & Hernández, S. (2017). A review on the production processes of renewable jet fuel. *Renewable and Sustainable Energy Reviews, 79*, 709–729. https://doi.org/10.1016/J.RSER.2017.05.108

He, Z., Chen, Z., & Feng, X. (2023). The role of green technology innovation on employment: Does industrial structure optimization and air quality matter? *Environmental Sciences Europe, 35*(1), 1–17. https://doi.org/10.1186/S12302-023-00773-W/TABLES/7

Huaiyin Teachers College. (2017). Method for improving lignocellulose enzymolysis efficiency by lignocellulose pretreatment through ultrasonic-microwave synchronously-assisted ionic liquid system. https://eureka.patsnap.com/patent-CN106591395A

Huang, Y. F., Te Chiueh, P., Kuan, W. H., & Lo, S. L. (2015). Effects of lignocellulosic composition and microwave power level on the gaseous product of microwave pyrolysis. *Energy, 89*, 974–981. https://doi.org/10.1016/j.energy.2015.06.035

Huang, Y. F., Te Chiueh, P., Kuan, W. H., & Lo, S. L. (2016). Microwave pyrolysis of lignocellulosic biomass: Heating performance and reaction kinetics. *Energy, 100*, 137–144. https://doi.org/10.1016/j.energy.2016.01.088

Hyun Hong, S., Taek Lee, J., Lee, S., Gon Wi, S., Ju Cho, E., Singh, S., Sik Lee, S., & Yeoup Chung, B. (2014). Improved enzymatic hydrolysis of wheat straw by combined use of gamma

ray and dilute acid for bioethanol production. *Radiation Physics and Chemistry, 94*(1), 231–235. https://doi.org/10.1016/J.RADPHYSCHEM.2013.05.056
International Energy Agency. (2006). www.iea.org/w/
Jönsson, L. J., & Martín, C. (2016). Pretreatment of lignocellulose: Formation of inhibitory by-products and strategies for minimizing their effects. *Bioresource Technology, 199*, 103–112. https://doi.org/10.1016/J.BIORTECH.2015.10.009
Jørgensen, S., & Pedersen, L. J. T. (2018). The circular rather than the linear economy. In *Palgrave studies in sustainable business in association with future earth* (Vol. F1861, pp. 103–120). Palgrave Macmillan. https://doi.org/10.1007/978-3-319-91971-3_8/FIGURES/4
Kassim, M. A., Meng, T. K., Kamaludin, R., Hussain, A. H., & Bukhari, N. A. (2022). Bioprocessing of sustainable renewable biomass for bioethanol production. In *Value-chain of biofuels: fundamentals, technology, and standardization* (pp. 195–234). Elsevier. https://doi.org/10.1016/B978-0-12-824388-6.00004-X
Ko, J. K., & Lee, S. M. (2018). Advances in cellulosic conversion to fuels: Engineering yeasts for cellulosic bioethanol and biodiesel production. *Current Opinion in Biotechnology, 50*, 72–80. https://doi.org/10.1016/J.COPBIO.2017.11.007
Lavoine, N., Desloges, I., Dufresne, A., & Bras, J. (2012). Microfibrillated cellulose—Its barrier properties and applications in cellulosic materials: A review. *Carbohydrate Polymers, 90*(2), 735–764. https://doi.org/10.1016/J.CARBPOL.2012.05.026
Lee, J. (1997). Biological conversion of lignocellulosic biomass to ethanol. *Journal of Biotechnology, 56*(1), 1–24. https://doi.org/10.1016/S0168-1656(97)00073-4
Li, X., Shi, Y., Kong, W., Wei, J., Song, W., & Wang, S. (2022). Improving enzymatic hydrolysis of lignocellulosic biomass by bio-coordinated physicochemical pretreatment—A review. *Energy Reports, 8*, 696–709. https://doi.org/10.1016/J.EGYR.2021.12.015
Li, C., Ahmad, S. F., Ahmad Ayassrah, A. Y. A. B., Irshad, M., Telba, A. A., Mahrous Awwad, E., & Imran Majid, M. (2023). Green production and green technology for sustainability: The mediating role of waste reduction and energy use. *Heliyon, 9*(12), e22496. https://doi.org/10.1016/J.HELIYON.2023.E22496
Liu, Y., Zhou, H., Wang, S., Wang, K., & Su, X. (2015). Comparison of γ-irradiation with other pretreatments followed with simultaneous saccharification and fermentation on bioconversion of microcrystalline cellulose for bioethanol production. *Bioresource Technology, 182*, 289–295. https://doi.org/10.1016/J.BIORTECH.2015.02.009
Lo, S. L., Huang, Y. F., Te Chiueh, P., & Kuan, W. H. (2017). Microwave pyrolysis of lignocellulosic biomass. *Energy Procedia, 105*, 41–46. https://doi.org/10.1016/J.EGYPRO.2017.03.277
Luiza Astolfi, A., Rempel, A., Cavanhi, V. A. F., Alves, M., Deamici, K. M., Colla, L. M., & Costa, J. A. V. (2020). Simultaneous saccharification and fermentation of spirulina sp. and corn starch for the production of bioethanol and obtaining biopeptides with high antioxidant activity. *Bioresource Technology, 301*, 122698. https://doi.org/10.1016/J.BIORTECH.2019.122698
Mante, O. D., Amidon, T. E., Stipanovic, A., & Babu, S. P. (2014). Integration of biomass pretreatment with fast pyrolysis: An evaluation of electron beam (EB) irradiation and hot-water extraction (HWE). *Journal of Analytical and Applied Pyrolysis, 110*(1), 44–54. https://doi.org/10.1016/J.JAAP.2014.08.004
Mathibe, B. N., Malgas, S., Radosavljevic, L., Kumar, V., Shukla, P., & Pletschke, B. I. (2020). Lignocellulosic pretreatment-mediated phenolic by-products generation and their effect on the inhibition of an endo-1,4-β-xylanase from Thermomyces lanuginosus VAPS-24. *3 Biotech, 10*(8), 349. https://doi.org/10.1007/S13205-020-02343-W
Maurya, D. P., Singla, A., & Negi, S. (2015). An overview of key pretreatment processes for biological conversion of lignocellulosic biomass to bioethanol. *3 Biotech, 5*(5), 597–609. https://doi.org/10.1007/S13205-015-0279-4/TABLES/1
Menon, V., & Rao, M. (2012). Trends in bioconversion of lignocellulose: Biofuels, platform chemicals & biorefinery concept. *Progress in Energy and Combustion Science, 38*(4), 522–550. https://doi.org/10.1016/J.PECS.2012.02.002

Mishra, K., Siwal, S. S., Nayaka, S. C., Guan, Z., & Thakur, V. K. (2023). Waste-to-chemicals: Green solutions for bioeconomy markets. *Science of the Total Environment, 887*, 164006. https://doi.org/10.1016/J.SCITOTENV.2023.164006

Mohanty, B., & Abdullahi, I. I. (2016). Bioethanol production from lignocellulosic waste-A review. *Biosci Biotechnol Res Asia, 13*(2), 1153–1161. https://doi.org/10.13005/BBRA/2146

Mohd Azhar, S. H., Abdulla, R., Jambo, S. A., Marbawi, H., Gansau, J. A., Mohd Faik, A. A., & Rodrigues, K. F. (2017). Yeasts in sustainable bioethanol production: A review. *Biochemistry and Biophysics Reports, 10*, 52–61. https://doi.org/10.1016/J.BBREP.2017.03.003

Mujtaba, M., Fernandes Fraceto, L., Fazeli, M., Mukherjee, S., Savassa, S. M., Araujo de Medeiros, G., do Espírito Santo Pereira, A., Mancini, S. D., Lipponen, J., & Vilaplana, F. (2023). Lignocellulosic biomass from agricultural waste to the circular economy: A review with focus on biofuels, biocomposites and bioplastics. *Journal of Cleaner Production, 402*, 136815. https://doi.org/10.1016/J.JCLEPRO.2023.136815

Munn, R. E. (1992). Towards sustainable development. *Atmos Environ, 26*(15), 2725–2731. https://doi.org/10.1016/0960-1686(92)90007-8

Naghshbandi, M. P., Tabatabaei, M., Aghbashlo, M., Gupta, V. K., Sulaiman, A., Karimi, K., Moghimi, H., & Maleki, M. (2019). Progress toward improving ethanol production through decreased glycerol generation in Saccharomyces cerevisiae by metabolic and genetic engineering approaches. *Renewable and Sustainable Energy Reviews, 115*, 109353. https://doi.org/10.1016/J.RSER.2019.109353

Nishiyama, Y., Langan, P., & Chanzy, H. (2002). Crystal structure and hydrogen-bonding system in cellulose Iβ from synchrotron X-ray and neutron fiber diffraction. *Journal of the American Chemical Society, 124*(31), 9074–9082. https://doi.org/10.1021/JA0257319/SUPPL_FILE/JA0257319_S1.CIF

Nitsos, C., Rova, U., & Christakopoulos, P. (2017). Organosolv fractionation of softwood biomass for biofuel and biorefinery applications. *Energies, 11, 50*. https://doi.org/10.3390/EN11010050

Obeng, E. M., Adam, S. N. N., Budiman, C., Ongkudon, C. M., Maas, R., & Jose, J. (2017). Lignocellulases: A review of emerging and developing enzymes, systems, and practices. *Bioresou Bioprocess, 4*(1), 1–22. https://doi.org/10.1186/S40643-017-0146-8

Pascual, A. R., Víctor, E. E., Martín, C., Broda, M., Yelle, D. J., & Serwá Nska, K. (2022). Bioethanol production from lignocellulosic biomass—Challenges and solutions. *Molecules, 27*, 8717. https://doi.org/10.3390/MOLECULES27248717

Peter Herbert North. (2010). *Method and system for fractionation of lignocellulosic biomass.*

Pielhop, T., Amgarten, J., Von Rohr, P. R., & Studer, M. H. (2016). Steam explosion pretreatment of softwood: The effect of the explosive decompression on enzymatic digestibility. *Biotechnology for Biofuels, 9*(1), 1–13. https://doi.org/10.1186/S13068-016-0567-1/FIGURES/7

Podesta, J., Stern, T., & Batten, K. (2007). *Capturing the energy opportunity creating a low-carbon economy*. Resilience.

Policies—Vikaspedia. (2006). https://vikaspedia.in/energy/policy-support/environment-1/forests/policies

Prasad, R. K., Chatterjee, S., Mazumder, P. B., Gupta, S. K., Sharma, S., Vairale, M. G., Datta, S., Dwivedi, S. K., & Gupta, D. K. (2019). Bioethanol production from waste lignocelluloses: A review on microbial degradation potential. *Chemosphere, 231*, 588–606. https://doi.org/10.1016/J.CHEMOSPHERE.2019.05.142

Ragauskas, A. J., Williams, C. K., Davison, B. H., Britovsek, G., Cairney, J., Eckert, C. A., Frederick, W. J., Hallett, J. P., Leak, D. J., Liotta, C. L., Mielenz, J. R., Murphy, R., Templer, R., & Tschaplinski, T. (2006). The path forward for biofuels and biomaterials. *Science (New York, N.Y.), 311*(5760), 484–489. https://doi.org/10.1126/SCIENCE.1114736

Ravindran, R., Jaiswal, S., Abu-Ghannam, N., & Jaiswal, A. K. (2018). A comparative analysis of pretreatment strategies on the properties and hydrolysis of brewers' spent grain. *Bioresource Technology, 248*, 272–279. https://doi.org/10.1016/J.BIORTECH.2017.06.039

Rubin, E. M. (2008). Genomics of cellulosic biofuels. *Nature, 454*(7206), 841–845. https://doi.org/10.1038/NATURE07190

Saha, B. (2005). Enzymes as biocatalysts for conversion of lignocellulosic biomass to fermentable sugars. In *Handbook of industrial biocatalysis*. CRC Press. 24–1–24–12. https://doi.org/10.1201/9781420027969.CH24

Saha, K., Uma Maheswari, R., Sikder, J., Chakraborty, S., da Silva, S. S., & dos Santos, J. C. (2017). Membranes as a tool to support biorefineries: Applications in enzymatic hydrolysis, fermentation and dehydration for bioethanol production. *Renewable and Sustainable Energy Reviews, 74*, 873–890. https://doi.org/10.1016/J.RSER.2017.03.015

Satgé, C., Granet, R., Verneuil, B., Branland, P., & Krausz, P. (2004). Synthesis and properties of biodegradable plastic films obtained by microwave-assisted cellulose acylation in homogeneous phase. *Comptes Rendus Chimie, 7*(2), 135–142. https://doi.org/10.1016/J.CRCI.2003.11.003

Stöcker, M. (2008). Biofuels and biomass-to-liquid fuels in the biorefinery: Catalytic conversion of lignocellulosic biomass using porous materials. *Angewandte Chemie International Edition, 47*(48), 9200–9211. https://doi.org/10.1002/ANIE.200801476

Su, T., Zhao, D., Khodadadi, M., & Len, C. (2020). Lignocellulosic biomass for bioethanol: Recent advances, technology trends, and barriers to industrial development. *Current Opinion in Green and Sustainable Chemistry, 24*, 56–60. https://doi.org/10.1016/J.COGSC.2020.04.005

Subhedar, P. B., Ray, P., & Gogate, P. R. (2018). Intensification of delignification and subsequent hydrolysis for the fermentable sugar production from lignocellulosic biomass using ultrasonic irradiation. *Ultrasonics Sonochemistry, 40*, 140–150. https://doi.org/10.1016/J.ULTSONCH.2017.01.030

Sun, R. C. (2010). Cereal straw as a resource for sustainable biomaterials and biofuels: Chemistry, extractives, lignins, hemicelluloses and cellulose. In *Cereal straw as a resource for sustainable biomaterials and biofuels: Chemistry, extractives, lignins, hemicelluloses and cellulose* (pp. 1–292). Elsevier. https://doi.org/10.1016/C2009-0-06759-2

Sun, Y., & Cheng, J. (2002). Hydrolysis of lignocellulosic materials for ethanol production: A review. *Bioresource Technology, 83*(1), 1–11. https://doi.org/10.1016/S0960-8524(01)00212-7

Sun, S., Sun, S., Cao, X., & Sun, R. (2016). The role of pretreatment in improving the enzymatic hydrolysis of lignocellulosic materials. *Bioresource Technology, 199*, 49–58. https://doi.org/10.1016/J.BIORTECH.2015.08.061

Taarning, E., Osmundsen, C. M., Yang, X., Voss, B., Andersen, S. I., & Christensen, C. H. (2011). Zeolite-catalyzed biomass conversion to fuels and chemicals. *Energy & Environmental Science, 4*(3), 793–804. https://doi.org/10.1039/C004518G

Tavva, S. S. M. D., Deshpande, A., Durbha, S. R., Palakollu, V. A. R., Goparaju, A. U., Yechuri, V. R., Bandaru, V. R., & Muktinutalapati, V. S. R. (2016). Bioethanol production through separate hydrolysis and fermentation of Parthenium hysterophorus biomass. *Renewable Energy, 86*, 1317–1323. https://doi.org/10.1016/J.RENENE.2015.09.074

Uihlein, A., & Schebek, L. (2009). Environmental impacts of a lignocellulose feedstock biorefinery system: An assessment. *Biomass and Bioenergy, 33*(5), 793–802. https://doi.org/10.1016/J.BIOMBIOE.2008.12.001

Vasić, K., Knez, Ž., & Leitgeb, M. (2021). Bioethanol production by enzymatic hydrolysis from different lignocellulosic sources. *Molecules, 26*(3), 753. https://doi.org/10.3390/MOLECULES26030753

Zabed, H., Sahu, J. N., Boyce, A. N., & Faruq, G. (2016). Fuel ethanol production from lignocellulosic biomass: An overview on feedstocks and technological approaches. *Renewable and Sustainable Energy Reviews, 66*, 751–774. https://doi.org/10.1016/J.RSER.2016.08.038

Zhang, C. W., Xia, S. Q., & Ma, P. S. (2016). Facile pretreatment of lignocellulosic biomass using deep eutectic solvents. *Bioresource Technology, 219*, 1–5. https://doi.org/10.1016/J.BIORTECH.2016.07.026

Zhou, C. H., Beltramini, J. N., & Lu, G. Q. (2008). Chemoselective catalytic conversion of glycerol as a biorenewable source to valuable commodity chemicals. *Chemical Society Reviews, 37*(3), 527–549. https://doi.org/10.1039/B707343G

Zhou, C. H., Xia, X., Lin, C. X., Tong, D. S., & Beltramini, J. (2011). Catalytic conversion of lignocellulosic biomass to fine chemicals and fuels. *Chemical Society Reviews, 40*(11), 5588–5617. https://doi.org/10.1039/C1CS15124J

Bio-interventions for Valorization of Lignocellulosic Biomass: Recent Trends and Applications

Charu Sharma and Papiya Mukherjee

Abstract

Lignocellulosic biomass serves as a cornerstone in the realm of renewable energy sources. Implementing a circular economy by utilizing lignocellulosic waste as a renewable resource has the potential to reduce reliance on fossil-based resources. Over the past decade, significant research has been conducted on microbial treatment by utilizing microbes, fungi, engineered microbes, and enzymes for processing lignocellulosic biomass. The biological treatment approach outperforms other methods due to its application in milder conditions, minimal corrosiveness, and minimal byproduct formation. This significantly reduces the lignin content and crystalline nature of the lignocellulosic biomass, thereby enhancing the hydrolysis and production of monomers. Additionally, the utilization of bio-based products derived from lignocellulose has emerged as a clean product to complement fossil-based resources. This chapter focuses on discussing biological agents that promote the biodegradation of wastes. The advanced utilization of lignocellulosic biomass in various industries plays a crucial role in the bio-based economy. It encompasses innovative design and advanced production strategies to promote the industrial recovery and reuse of waste materials.

Keywords

Lignocellulosic Biomass · Lignin · Feedstock · Pretreatment · Valorization · Fermentation Technologies · Biodegradation · Sustainable · Microbial processing

C. Sharma · P. Mukherjee (✉)
Department of Botany, Panjab University, Chandigarh, India

G. Mukherjee, S. Dhiman (eds.), *Value Addition and Utilization of Lignocellulosic Biomass*, https://doi.org/10.1007/978-981-96-2786-8_2

1 Introduction

Lignocellulosic biomass (LB) represents a globally abundant and sustainable resource, possessing a sugar-rich foundation that can be transformed into specialized products through an appropriate processing strategy (Singhania et al., 2008). Sustainable biomass processing presents a universal challenge, necessitating fulfillment of key criteria including economic viability, environmental friendliness, and social responsibility. The contemporary context is distinguished by a steady rise in global population and energy demands, coupled with the rapid depletion of fossil fuel reserves leading to soaring fuel costs. Emissions from burning fossil fuels release pollutants that contribute to global warming, disrupting ecological equilibrium (Satyanarayana et al., 2009). Furthermore, reliance on non-renewable energy sources threatens both natural and economic stability, underscoring the need for alternative renewable energy sources, with lignocellulosic biomass emerging as a cornerstone in this endeavor. However, significant technical challenges persist in ensuring continuous biomass supply and maximizing sugar yields through biological saccharification routes. Despite extensive research over many decades, the integrated production of chemicals and biofuels requires further optimization from technical and economic standpoints. Chemical characteristics and structural rigidity of biomass present hurdles to its viability. LB feedstocks are subjected to pretreatment to separate lignin from hemicellulose and cellulose, followed by enzymatic hydrolysis to convert carbohydrates into a renewable sugar mix. Its eco-friendliness, cost-effectiveness, widespread availability, and high carbohydrate content, conducive to yielding substantial biofuels, make LB an attractive option. Moreover, utilizing these biomass wastes helps mitigate environmental pollution associated with their disposal.

The conventional methods of pretreatment suffer from many drawbacks, i.e., high costs, process complexity, production of degraded products, less efficiency, absence of selectivity, limited sugar release, and environmental contamination. Effective pretreatment methods are planned to preserve hemicellulosic and cellulosic components, minimize the release of toxic inhibitors, be cost-effective and energy-efficient, and promote chemical reutilization. Pretreatment strategies (physical, chemical, and biological) have been designed to address lignocellulose recalcitrant nature toward anaerobic degradation, with biological methods offering advantages in milder conditions, reduced corrosiveness, and less byproduct formation. In particular, biological pretreatment stands out for its ability to selectively reduce the lignin content and crystallinity of lignocellulosic biomass, thereby enhancing hydrolysis and monomer production for subsequent anaerobic digestion. Energy output from these approaches is notably higher and less expensive in comparison to chemical and mechanical methods. This review emphasizes the application of biological agents—fungal, microbial, engineered microbial, and enzymatic—for waste biodegradation and utilization of lignocellulose for various value-added products.

2 Microbial Processing of Lignocellulosic Biomass

There exist multiple methodologies for enhancing energy production from LB, and one of the promising approaches is the implementation of microbial-based pretreatment methods. These pretreatments alter the lignocellulosic composition through partial or complete elimination of the lignin component. In the process of anaerobic digestion, biological pretreatments such as fungal, microbes, engineered microbes, and enzymatic are aimed at enhancing biomethanation (Zheng et al., 2014). In comparison to physical and chemical strategies, the above-mentioned pretreatment with certain modifications requires less energy and generates non-toxic chemicals. Additionally, biological pretreatments result in reduced carbohydrate loss and increased microbial accessibility for the cellulosic components of the feedstock, leading to enhanced hydrolysis.

2.1 Fungal Pretreatment

This focuses on assessing lignin and hemicellulose breakdown, whereas cellulose is resistant to fungal attack (Zheng et al., 2014). One of the species, white-rot fungi is considered as efficacious for producing biogas from the lignocellulosic substrate (Sun & Cheng, 2002). However, improvised fungal treatment requires the sterilization of lignocellulosic residue. The combination of chestnut leaves and hay (1:2 ratio), underwent a pretreatment with *A. auricula-judae* (wood decomposing fungus) at 37 °C for 2–3 weeks. This led to an increase in biogasification by 15% (approx.) when compared to the biomass without any treatment (Mackuľak et al., 2012). According to Zhao (2013), exposing backyard garbage to *C. subvermispora* (white-rot fungus) at a moisture level of 60% for 30 days resulted in a significant increase (154%) in methanation. This is attributed to the activity of lignin-degrading enzymes, hemicellulases, cellulases, and xylanases produced by fungus. The use of fungal treatment was also examined as an initial step in biogas production, specifically methane and hydrogen. Zhao et al. (2014) demonstrated that it is possible to produce methane by treating unsterilized yard trimmings with fungi. Yard trimmings are a significant part of the organic waste found in municipal solid waste. This method eliminates the need for sterilization, which is not economically viable in the biorefinery industry. The study utilizes *C. subvermispora* to treat yard trimmings before subjecting them to solid-state anaerobic digestion with the effluents collected from mesophilic liquid anaerobic digester. Following 30-day fungal pretreatment, lignin breakdown of 14.8–20.2% was achieved. Subsequently, during anaerobic digestion, methane output of 34.9–44.6 l/kg of volatile solids was produced, thus resulting in saving dry substrate of 501–789 kJ/kg (Salvachúa et al., 2016). The best combination was optimized for achieving high cellulase activity with the aid of *Trichoderma viride*. The delignification of municipal waste occurred with an inoculation density of 2×10^8 spores per 25 g of substrate, moisture content of 70% for 10 days. As a result, there was a rise in the breakdown of complex carbohydrates into simpler sugars and an increase in the availability of nutrients. This benefited the sequential anaerobic digestion process for methane production with diluted fermenter sludges (Mutschlechner et al., 2015). Chatzifragkou et al. (2014) developed a biorefinery system that leverages byproducts from the biodiesel industry, particularly rapeseed meal (RSM), to produce 1,3-propanediol (PDO). The process involved incubating the fungus *A. oryzae* with RSM at 30 °C for promoting (hemi)cellulolytic enzyme production. After 48 h of fermentation period, distilled water was added to the solid residues for saccharification and to obtain the RSM hydrolysate. Utilizing RSM hydrolysates as the fermentation medium, *C. butyricum* was cultivated in batch mode, yielding 65.5 g/L of PDO with a productivity of 1.15 g/L/h.

2.2 Microbial Pretreatment

It refers to the utilization of microbes obtained from the natural environment to screen and aid in the hydrolysis of biomass during anaerobic digestion. The microbial consortia that are employed for pretreatment should have the capacity to break down cellulose and hemicellulose. A group of microorganisms obtained from a landfill that decomposes straw at 55 °C demonstrated a 96% increase in methane production during the anaerobic digestion of cassava debris mixed with distillery effluent (Ziemiński et al., 2012). Furthermore, Zheng et al. (2014) suggested a combination of freeze-dried yeast and cellulolytic microbial strains as a successful approach, besides the traditional method of utilizing screened bacteria from natural habitats. The autoclaved maize straw when treated with freeze-dried microbial consortia, led to a decrease in retention time (36%) and an increase in methane production (75%). The study by Kim and Kim (2013) employed a novel three-stage fermentation process. The effluents produced during the lactate fermentation process, referred to as lactate fermentation effluents (LFE), were utilized as the substrate for cultivating *R. sphaeroides*. Later, the bacterium transformed LFE into hydrogen gas through a process called photo fermentation. In addition, the residual remnants of LFE were transformed into CH_4 by anaerobic digester that has been derived from a wastewater treatment. The procedure facilitated the conversion of energy contained in food waste into hydrogen (H_2) and methane (CH_4). Borowski and Kucner (2015) utilized sewage sludge to ferment sugar beet pulp residues (SBPR), resulting in the 512 dm^3/kg VSfed (volatile solids fed) of biogas yield and 348 dm^3 CH_4/kg VSfed of methane production. It showed that SBPR, a derivative of bioethanol production and beet pulp waste, which is usually not as profitable as animal feed due to its low protein content and expensive dehydration process, can be used to implement the biorefinery concept. A hazardous waste, deoiled Jatropha generated from biodiesel production and unsuitable for animal use was converted into hydrogen by optimizing the microbial population (Kumar et al., 2015). Among the various methodologies tested, the inoculum pretreatment method was chosen to amplify the culture of hydrogen-producing bacteria. The inoculum when subjected to 95 °C for 30 min resulted in a maximal hydrogen production rate (HPR) of 0.36 L H_2/L-d (liters of hydrogen per liter per day). Moreover, the substrate concentration, pH, and temperature were also optimized, which resulted in the greatest HPR (1.36 L H_2/L-d). In addition to biogas production, microbial digestion has been emphasized as a preliminary step to convert biomass into economically viable biogenic compounds (Liang et al., 2014; Tang et al., 2015). In comparison to the fungus, bacterial strains have a low degree of (hemi) cellulolytic activity. As a result, reduced cellulose loss is obtained which provides considerable benefits throughout the delignification process (Brown & Chang, 2014). *B*acillus. *subtilis* was used to conduct solid-state fermentation for converting the agro-industrial waste, dry shiitake mushroom residues (DSMR) into poly(c-glutamic acid) (c-PGA) (Tang et al., 2015). The monosodium glutamate production residues (MGPR) acted as an alternative of glutamate to DSMR, and industrial waste glycerol was added to enhance the carbon concentration. When DSMR to MGPR was optimized at 12:8 ratio, the yield of c-PGA reached 115.6 g/kg. Undefined microbial cultures provide the advantage of circumventing several operational limitations, such as the need for pH control, temperature regulation, sanitation, and nutrition supply. This might be possible due to their rapid and effective adaptation to varied environmental circumstances, as well as their remarkable self-evolutionary capabilities (Agler et al., 2011). A method has been devised by Liang et al. (2014) to convert potato peel waste into acetic acid, ethanol, and lactic acid. The study aimed to utilize various cultures obtained from sewage sludge. By using batch fermentations, the yield of lactic acid (0.22 g/g), acetic acid (0.06 g/g), and ethanol (0.05 g/g) was attained.

2.3 Engineered Microbes

Progress in genetic and metabolic engineering has facilitated the creation of microbial cell factories for utilizing lignocellulosic feedstock and transforming them into useful products. Nevertheless, the task of developing a highly efficient super microbial cell capable of converting lignocellulosic biomass into bioproducts poses significant challenges (Olson et al., 2012). Consolidated bioprocessing is a one-step, simultaneous process of saccharification and fermentation, efficiently carried out by microbes. Therefore, to increase production, it is imperative to concentrate on improving the effectiveness of bifunctional catalysts within a microorganism (Maleki et al., 2022). Either a genetically modified cellulolytic microbe could be designed to convert glucose into a value-added product or a bacterium that can manufacture value-added chemicals from glucose and xylose monomers can be developed for producing cellulolytic enzymes. The primary obstacle in eukaryotic designed microbes lies in achieving proper folding and glycosylation of heterologous proteins, as these processes are crucial for protein/enzyme functionality. The beta-glucosidase gene from *S. fibuligera* and the endoglucanase gene (*EGL1*) from *T. reesei* were introduced into *S. cerevisiae* by Den Haan et al. (2007). This genetic modification allowed *S. cerevisiae* to thrive on phosphoric acid swollen cellulose (PASC), demonstrating its ability to break down cellulose. Many studies have reported the significance of genetically modified microbes in consolidated bioprocessing for biofuel production and other valued products. During the experiment, two genetically modified yeast strains were employed. The genetically modified strains were Y5/XynII-xyIA and Y5/EG-CBH-BGL, which produced different xylanase and cellulase forms, respectively. The above-mentioned strains were further utilized during consolidated bioprocessing without the use of commercially available enzymes. During 144 h of incubation period, Chen et al. (2018) observed ethanol production of 1.61 g/L, which accounted for 64.7% of ethanol yield. In another study, alcohol dehydrogenase genes from *Z. mobilis* (*adhz*) and *S. cervisiae* (*adhs*) along with pyruvate decarboxylase gene (*Pdcz*) from *Z. mobilis* were coupled for creating ethanologenic operons in a lactate deficient environment. *Bacillus subtilis* generates strains through the constriction of two fusion genes. Using potato as a substrate, Maleki et al. (2021) found ethanol production of 16.3 g/L and 21.5 g/L obtained during a 96 h fermentation period. To enhance the cost-efficiency of ethanol production through centralized bioprocessing, it is crucial to focus on the development of recombinant technology. Hence, an experiment was conducted in which genes encoding α-amylase from four different species were cloned. In *S. cerevisiae*, the expression of cloned genes resulted in two strains, Y294 (Apu A) and Y294 (AteA). As a result, the α-amylase activity of 2.17 U/mL (Apu A) and 2.98 U/mL (Ate A) was obtained. Considering the previous reports, expression of Ate A when combined with *A. tubingensis* glucoamylase gene (gla A) resulted in amylolytic *S. cerevisiae* Y294 (Ate A-Gla A) strains. After 6 days of incubation, ethanol production reached 45.77 g/L (Sakwa et al., 2018). The study conducted by Li et al. (2021) focused on genetically modified strains of *S. cerevisiae*, namely LGA-1 and LGA-C. These strains have shown a remarkable ability to produce significant quantities of D-glucaric acid. The consolidated bioprocessing was conducted by utilizing a synthetic microbial consortium of *T. reesei* (Rut C 30) and *S. cerevisiae* (LGA-1) to achieve enhanced titers and yield. A similar study focusing on the generation of n-butanol by *C. cellulovorans* using consolidated bioprocessing was also conducted. The potential for n-butanol synthesis from *C. cellulovorans* is yet to be investigated. Thus, the push-pull technique of metabolic engineering was employed to enhance the ability of *C. cellulovorans* for synthesizing n-butanol from alkali-extracted corn cob. The outcome of this process was n-butanol (4.96 g/L), which is 236 times higher than the natural production (Wen et al., 2020). It is imperative to ensure that the production of biofuel and value-added products is economically efficient. Hence, it is crucial to investigate alternative bi-catalyst choices in consolidated bioprocessing by metabolic engineering to achieve practicality and cost-efficiency.

2.4 Enzyme Treatments

This has been utilized in several biotechnological procedures to turn waste materials into improved products. Enzymes provide an environmentally benign method for reducing waste production and energy consumption, which is essential for the advancement of cost-effective and sustainable biorefineries. Ligninolytic enzymes offer a viable substitute to fungal and bacterial strains for biomass delignification. Utilization of laccase, lignin peroxidase, versatile peroxidase, and manganese peroxidase enhances the accessibility of cellulose to (hemi)cellulolytic enzymes and also enables the recovery of valuable phenolic compounds (Zakzeski et al., 2009). The synergistic effect can be harnessed for developing a biorefinery system that is both economically viable and environmentally friendly. Enzymes offer the benefits of faster delignification and a more expedited procedure compared to microorganisms (Roth & Spiess, 2015). In addition, enzymatic delignification could be performed over a wide range of temperatures and pH levels. It is not dependent on nutrient supplementation and sugar requirements (Ibarra et al., 2006).

In the study, Kumar et al. (2013) found that the enzymatic treatment of de-oiled Jatropha waste using Viscozyme (an enzymatic cocktail) at a 10% loading resulted in the maximum release of reducing sugar, i.e., 12.9 g/L. In a batch process, the enzymatic hydrolysate was utilized for hydrogen fermentation, thus resulting in maximum H_2 production (2.1 l/L). Solid-state fermentation was utilized for fungal mash production from waste cake comprising high concentration of hydrolytic enzymes (Kiran et al., 2015). Later, this mash was employed to pretreat food waste. Following hydrolysis of 24 h, the anaerobic digestion of enzymatically processed food wastes resulted in biomethane yield (2.3 times) and production (3.5 times). The experiment utilizing fungal mash at the pretreatment stage was obtained from solid-state fermentation without undergoing any purification. This approach has the potential to enhance cost competitiveness of biomethane generation from food waste. This waste upgrading aligns with the biorefinery concept and circular economy, as the domestic waste containing varying amounts of lignocellulose has detrimental effects on human health and the environment. Amulya et al. (2015) highlighted the synthesis of polyhydroxyalkanoates (PHA) from food wastes, employing a comprehensive multistage procedure that aligns with the biorefinery idea. Stage I of acidogenic fermentation of food waste yielded biohydrogen and effluent rich in volatile fatty acids from H_2 reactor. Following that, the waste substances were transformed into polyhydroxyalkanoates (PHA) with a dry cell weight of 23.7% by carrying out two stages, namely culture enrichment and PHA production. Naranjo et al. (2014) investigated a biorefinery system that converts banana leftovers into glucose, polyhydroxybutyrate, and ethanol using enzymatic processes. The remaining banana was subjected to enzymatic treatment, which involved the use of glucoamylase to convert it into glucose. The glucose obtained was then utilized by *B. sacchari* to produce PHB, while *P. stipitis* was responsible for converting glucose into bioethanol. The above method successfully reduced the costs of PHB (26%), glucose (22%), and ethanol (53%) production. Additionally, it revealed that the technology could potentially fulfill upto 30.6% of the global energy demand. Furthermore, the amount of ethanol obtained from leftover bananas was similar to the amount obtained from cane molasses and corn starch. The sugar and quercetin were extracted from onion skin waste obtained from processed onions. Through experimentation with various enzyme mixtures, researchers achieved a peak conversion rate of biosugar, i.e., 98.5% by utilizing pectinase (0.16 mg), cellulase (0.72 mg), and xylanase (1.0 mg). The enzymatic treatment also enhanced the yield of quercetin by 1.61-fold, which is a value-added substance extensively utilized in the treatment of cancer (Choi et al., 2015).

Romano et al. (2009) reported hydrolysis of hemicellulose and cellulose as the rate-limiting steps during anaerobic digestion. To increase biogas production, specific enzymes, such as hemicellulases and cellulases that could break down the biomass into simpler compounds, were used. These enzymes could be added before or during the biomethanation process. In biomethanation, the utilization of enzymatic pretreatment is constrained and lacks popularity due to their substrate selectivity and cost implications (Zheng et al., 2014). A previous experiment by Yunqin et al. (2010) found the application of mushroom compost extract to paper sludge and paper pulp. The extract is enriched with laccase and carboxymethylcellulase, thus increasing methane production by 34.2% in comparison to the untreated samples. The waste hops and sugar beet pulp were partially hydrolyzed through enzymatic pretreatment, which increased their ability for biomethanation. The pretreatment facilitated organic matter reduction in the pretreated hydrolysates of waste hops and sugar beet pulp, thus increasing methanation by 13–19% (Ziemiński et al., 2012).

3 Application of Lignocellulosic Biomass

3.1 Biofuels

In the transportation and energy sectors, biofuels derived from lignocellulosic biomass are acknowledged as carbon-neutral substitutes for conventional petroleum supplies. These biofuels can be found in both liquid forms (biobutanol, biodiesel, bioethanol, bio-oil, jet fuels) and gaseous forms (biogas, biomethane, biohydrogen, syngas). Their production involves specific biochemical and thermochemical conversion processes (Nanda et al., 2014a, b, 2015a, b; Parakh et al., 2020). Temperature, rate of heating, pressure, residence time, catalysts, concentration of feedstock, reactor type, and biomass characteristics are some of the key determinants for the quantitative and qualitative production of bio-oil. Fast pyrolysis gives the highest yield of bio-oil as it heats up more quickly, produces less vapor residue time, and cools hydrocarbon vapors more quickly than slow pyrolysis, which yields more biochar. In contrast to pyrolysis, liquefaction produces oil with less oxygen, which improves the fuel's characteristics and increases its heating value.

(a) Bio-oils derived from lignocellulose contain different ratios of degradation products from cellulose, hemicellulose, and lignin. The bio-oils so obtained are oxygenated liquid products comprising diverse proportions of aldehydes, alcohols, acids, carbonyls, catechols, ketones, esters, ethers, phenols, pyranones and vanillin (Mohan et al., 2006). According to Joshi and Lawal (2012), substances such as alcohol, aldehyde, ester, furan, ketone, and sugar are produced when cellulose and hemicellulose break down. In contrast, the breakdown of lignin produces catechol, guaiacol, phenolic, and syringol. Rutkowski (2011) examined the composition of bio-oils produced by pyrolyzing cellulose, xylan, and lignin at 450 °C in a vertical Pyrex reactor with or without $ZnCl_2$ or K_2CO_3 catalysts. The yield of bio-oil without catalysts was highest from cellulose (44 wt%)

followed by xylan (32 wt%) and lignin (28.5 wt%). The addition of 10 wt% K_2CO_3 and $ZnCl_2$ increased the bio-oil yield and produced more deoxygenated oils from all three compounds. The study concluded that both, the type of bio-polymer and presence of catalysts affect bio-oil yield. Bio-oils contain significant amounts of carboxylic acids, alcohols, and phenols (Nanda et al., 2014a, b).

(b) Ethanol obtained via enzymatic hydrolysis and microbial fermentation can be mixed with gasoline for transportation fuel. Compared to traditional fuels, bio-ethanol offers several benefits. First, it contains about 35% oxygen, enhancing combustion and reducing emissions of CO, NO, NOx, particulate matter, and hydrocarbons (Kim & Dale, 2004). Second, it is made from renewable biomass materials that are plentiful. Furthermore, employing bioethanol as a fuel for transportation could decrease greenhouse gas emissions from fossil fuels. It is important to acknowledge that bioethanol, with its higher oxygen content, may have a lower energy density compared to gasoline, resulting in reduced mileage and more frequent refueling (Morone & Cottoni, 2016). Additionally, the oxygen present in bioethanol can cause engine corrosion when used in higher blending ratios (Khuong et al., 2017). The production of bioethanol from sugar or starch-based sources raises concerns regarding economics, food security, and environmental impact due to the diversion of food crops for fuel, sparking the food versus fuel debate. Conversely, utilizing lignocellulosic biomass for bio-ethanol production, which does not compete with food crops, garners significant attention for its positive environmental and economic implications (Nanda et al., 2015a, b). This method can bolster rural economies, optimize supply chain logistics, diminish reliance on fossil fuels, and improve energy security.

(c) Biobutanol is increasingly considered a viable substitute to bioethanol attributed to its superior fuel properties and higher energy content, surpassing bio-ethanol by 30% (Nanda et al., 2017a, b, 2020). Unlike bioethanol, biobutanol can be used pure or blended with gasoline in various ratios without requiring engine modifications. Additionally, biobutanol is safer to handle, non-corrosive, and has low vapor pressure (Sarangi & Nanda, 2018). Its non-hygroscopic nature facilitates blending with gasoline at refineries before storage and distribution, and its lower solubility with water prevents issues like freezing during transportation or inside pipelines and carburetor ducts. While bioalcohols like bioethanol and biobutanol show promise in reducing dependence on fossil fuels, the feasibility of their production largely hinges on the separation processes involved, which are often energy-intensive and can impact process economics (Nanda et al., 2020). While fractional distillation is a commercially used method for separating bioethanol, separating biobutanol poses greater challenges due to lower yields, necessitating more complex and costly technologies like adsorption, liquid-liquid extraction and pervaporation (Nanda et al., 2017a, b).

(d) Biomethane has emerged as a viable fuel choice for heavy duty vehicles due to its increased calorific value, reduced density, and higher thermal efficiency. It can be produced from variety of lignocellulosic biomass via anaerobic digestion or biomethanation process (Khan et al., 2017). Biomethane serves as an alternative to fossil-derived natural gas for heat generation, cooking, heating, and electricity production. Compressed biomethane, i.e., bio-compressed natural gas (bio-CNG) shares similar characteristics with CNG on the parameters of gas usage, engine operation, and efficiency when used as a vehicle fuel. Using biomethane instead of compressed natural gas (CNG) in large motor vehicles reduces greenhouse gas emissions by about 63%. However, emissions of NOx, hydrocarbons, and CO are slightly higher than those from CNG, primarily due to biomethane's higher nitrogen content. In vehicular engines powered by biomethane, low warm-up temperatures contribute to higher CO emissions because CO oxidation to CO_2 occurs at a slower rate. Similarly, elevated hydrocarbon emissions stem from incomplete combustion due to poor oxidation at low warm-up temperatures (Khan et al., 2017).

(e) Biodiesel production from lignocellulosic residues carries substantial environmental significance beyond its economic benefits. It can be blended into standard petro-diesel at the appropriate proportions or used as a sustainable, non-hazardous and environmentally sound fuel for diesel engines without modification (Nayak et al., 2019; Reddy et al., 2018). The synthesis of biodiesel from lignocellulosic wastes has been the subject of several investigations. To produce biodiesel from crude rice bran oil, Shiu et al. (2010) used an in-situ method that involved continuous esterification followed by transesterification without sporadic product separation. Similarly, Deeba et al. (2015) used paper mill sludge for producing biodiesel via transesterification.

3.2 Bioplastics

Bioplastics provide environmentally friendly substitutes for conventional plastic products and are used in daily life, the textile sector, electronics, healthcare sector, and packaging. Over recent years, bioplastic materials have emerged as potential substitutes, gradually replacing petroleum-based plastics (PET). Unlike PET, bioplastics are derived from lignocellulosic residue, which is decomposable, thereby reducing environmental harm. Petrochemical plastics, in contrast, contain dioxins and polycyclic aromatic hydrocarbons, posing significant health risks upon environmental release. Moreover, the combustion of plastic results in the release of substantial amounts of CO_2 into the atmosphere. A staggering 34 million tons of plastic waste is generated annually, with 93% ending up in landfills and oceans, highlighting the urgent need for sustainable alternatives (Mekonnen et al., 2013). Bioplastics address these environmental challenges by offering compostable and biodegradable solutions that break down naturally under appropriate conditions.

Bioplastic precursors can be sourced from various materials, including microorganisms (e.g., polyhydroxyalkanoates), cellulose (e.g., regenerated cellulose and pectin), proteins (e.g., wool, gelatin, and casein), and lipids (e.g., plant oils or animal fats) (Burgos et al., 2016). Additionally, bio-based plastics such as polybutylene succinate and polylactic acid can be chemically produced from bio-derived sources. Some biodegradable plastics, like polycaprolactone, poly (butylene succinate-co-terephthalate), and polyglycolic acid, exhibit promising biodegradability and find diverse applications. For instance, polylactic acid is used in biomedical implants, sutures, and surgical applications (Song et al., 2009). Another category is recognized as drop-in bioplastics, including bio-propylene, bio-polyethylene terephthalate, and bio-polyethylene. These materials largely utilize existing infrastructure and machinery employed in the production of petrochemical plastics, with the primary difference being their plant-based biomass precursor (Bátori et al., 2018). However, due to technological limitations and absence of economy of scale, the manufacturing costs of drop-in bioplastics remain higher than those of petrochemical plastics in the market.

3.3 Biogas

Biogas, derived from either anaerobic digestion or biomethanation processes, primarily consists of methane (CH_4) and carbon dioxide (CO_2), with minor traces of siloxanes and hydrogen sulfides (Adekunle & Okolie, 2015). Anaerobic digestion, a biochemical process, transforms organic biodegradable components into biogas by introducing microorganisms in an oxygen-free environment. The recalcitrant trait of lignocellulosic feedstock results in decreased digestion and poor biodegradation, thus emphasizing the anaerobic breakdown of lignocellulosic residues and organic wastes for biomethane production. This versatile fuel holds the potential to supersede conventional fuels, i.e., natural gas, particularly for heat and power generation. Additionally, anaerobic digestion stands globally as a well-established and sustainable technology, especially beneficial to rural areas, particularly in economically developing nations. It possesses the capacity to stimulate rural economies by creating job opportunities and generating revenue through waste management and valorization. Implementing thermophilic conditions enhances the acidogenesis and hydrolysis processes. A neutral pH level must be maintained for the production of biogas, and it usually takes 3 weeks for microorganisms to adjust to different substrates and temperatures. Therefore, a retention time of 10–15 days is recommended. The carbon-to-nitrogen (C/N) ratio plays a prominent role in anaerobic digestion, as extremes can impact ammonia nitrogen and fatty acid levels, subsequently influencing biogas production rates. Hence, maintaining a C/N ratio between 20 and 35 is advisable (Bowen et al., 2014; Zhang et al., 2013). Various studies have explored the anaerobic digestion of lignocellulosic residues and organic wastes for biomethane production, though utilizing lignocellulosic feedstock presents challenges due to its resistant nature. Strategies like co-digestion and pretreatment methods have evolved to enhance the anaerobic digestion of lignocellulose (Sambusiti et al., 2013; Zhang et al., 2011a, b). Furthermore, studies have investigated the application of additives to expedite the anaerobic digestion process. The addition of substances during anaerobic digestion causes substrates to adhere at their surface, thereby enhancing localized substrate concentration, microbial proliferation, and biogas productivity. Many additives, biological (brown-rot fungi and white-rot fungi) and chemical (sodium hydroxide and calcium oxide) have been explored for enhancing anaerobic digestion (Demirel & Scherer, 2011; Vervaeren et al., 2010; Zhang et al., 2011a, b).

3.4 Biochemicals

Lignocellulose is prioritized as a valuable source for generating chemicals with added value, leveraging the chemistry of its primary constituents, namely cellulose, hemicellulose, and lignin. Currently, 75% of the commercially utilized platform chemicals include benzene, ethylene, propylene, toluene, and xylene (Howard et al., 2015). These chemicals serve as precursors for synthesizing various organic compounds, including polymers and resins. The majority of aromatic hydrocarbons are derived from thermal lignin degradation attributed to their inherent aromatic properties, while cellulose and hemicellulose could be thermally transformed into aliphatic compounds of low molecular weight.

Cellulose can be enzymatically or chemically hydrolyzed to produce glucose, which can then be further processed into platform chemicals like levulinic acid and 5-hydroxymethylfurfural (5-HMF) (Alonso et al., 2013). Levulinic acid and 5-HMF exhibit promising potential for producing a range of high-valued organic chemicals, various flavoring agents, polymers, and fuel additives owing to their versatile functioning. Numerous studies have reported the conversion of lignocellulosic residues into 5-HMF and levulinic acid (Peng et al., 2010). For instance, Wang et al. (2010) explored the cellulose breakdown to levulinic acid by using sulfated TiO_2 as a solid acid catalyst. This process achieved a maximum yield of 27.2% at 240 °C during 15 min reaction period. Zhao et al. (2007) reported the transformation of sugars into 5-HMF by utilizing metal halides in 1-alkyl-3-methylimidazolium chloride as a catalyst. The yield resulted in nearly 70% 5-HMF with $CrCl_2$, albeit with negligible levulinic acid formation. Asghari and Yoshida (2007) investigated the dynamics of fructose dehydration to 5-HMF and then, rehydration of 5-HMF to levulinic acid and formic acid formation by using HCl as catalyst. The reaction was conducted at various temperatures from 210 °C to 270 °C, pressures from 4 MPa to 15 MPa, and reaction times from 0.5 s to 300 s. It was found that a change in pressure did not affect the decomposition rate significantly, with the primary reaction products being 5-HMF, levulinic acid, and formic acid. A C_5 sugar alcohol, i.e., Xylitol, is a significant substance derived from the degradation of hemicellulose and its derivative compounds such as arabinose, xylan, and xylose are soluble in dilute acids. Xylitol serves as a substitute for sucrose in food sweeteners, chewing gums, pharmaceuticals, and thin coatings. Additionally, xylitol holds promising applications in the odontological sector like toothpaste formulations, hardening of teeth, antimicrobial agents, mouthwashes, and chewable vitamin tablets (Rahman et al., 2007). The xylitol production from hemicellulose and its derivatives has garnered substantial curiosity among researchers. Nowadays, xylitol is majorly produced when xylose and its derivatives are chemically reduced under an alkaline environment. However, this reaction faces challenges such as the need for high pressure (up to 50 atm), employment of expensive catalysts, and techniques for separation-purification. López-Linares et al. (2018) assessed the application of rapeseed straw for xylitol production by employing two yeast strains, *D. hansenii* and *C. guilliermondii*. This process achieved maximum xylose conversion yields of 0.55 g/g for *C. guilliermondii* and 0.45 g/g for *D. hansenii*. Vanillin and gallic acid are gaining interest as starting monomers to produce various polymers such as epoxy resins, polycarbonates, and polyesters (Cetera et al., 2018; Kumar et al., 2008). Vanillin could be obtained from the depolymerization of lignin, wherein lignin solution is treated with oxidants within alkaline pH at elevated temperatures and pressures. Every year, more than 20,000 tons of vanillin is produced, out of which 15% comes from precursors of lignin. This makes vanillin a well-known pure molecular phenolic chemical that is commercially obtained from biomass (Fache et al., 2016). Vanillin finds applications in the food industry for its vanilla-like flavor and in pharmaceuticals for manufacturing anti-foaming agents, drugs, herbicides, and chemicals (Kumar & Pruthi, 2014). Gallic acid offers various benefits as an antioxidant in tea composition, paper production, dyes, and the tanning industry. Furfural is another valuable chemical that is produced annually (250,000 tons) through the hydrolysis of forestry or agricultural waste by concentrated sulfuric acid. It is commonly used for lubricating oil refining and serves as a precursor for chemicals like tetrahydrofuran, furan, and furfuryl alcohol. Innovative conversion strategies are being developed to utilize available lignocellulose and produce ecologically acceptable compounds while considering the economic viability (Gallezot, 2012; Lichtenthaler, 2002; Zanwar et al., 2013).

3.5 Biocomposites

Biocomposites derived from lignocellulosic biomass are increasingly recognized as viable substitutes to natural fiber-reinforced composites attributed to their ecologically sound, resilience, and ability to biodegrade. Essentially, biocomposite materials are a fusion of conventional plastics sourced from petrochemicals and renewable biomass, specifically lignocellulosic residues. These materials consist of natural fibers reinforcing an organic matrix or biopolymers (Trache et al., 2016). Biopolymers are used to formulate the matrix component, while other biogenic byproducts or lignocellulosic biomass can be used to make the fibers. These fibers protect the polymers from environmental deterioration and improve their mechanical qualities. Reinforcements (flax, hemp, paper, wood) and biopolymers (natural biopolymers, polylactic acid, polyvinyl alcohol) are biodegradable and recyclable, which helps to maintain the sustainability of the environment (Ramamoorthy et al., 2015). Nevertheless, there remain challenges in effectively separating components at the end of a biocomposite material's lifespan.

4 Bottlenecks and Future Perspective

The application of lignocellulosic materials to manufacture chemicals and biofuels has been the subject of numerous studies, with encouraging outcomes in terms of sustainability and the environment. However, before reaching commercial viability, several issues must be resolved. For effective and efficient trash collection and valuation, many nations lack dedicated long-term planning and regulations. Since preprocessing and postprocessing technologies are expensive, most commercial facilities that produce chemicals and biofuels are not profitable. The low cost of conventional fossil fuels and the high costs of pretreatment are two of the main obstacles. Other risks related to biomass conversion technologies include the demand for high temperatures and pressures, the use of genetically modified microorganisms, the need for costly enzymes, and challenges associated with recovering and reusing the enzymes and metal catalysts. Current biorefineries are focusing on developing reliable pretreatment strategies, effective conversion efficiency, executing techno-economic and lifecycle assessments. Future efforts should concentrate on methodologies for collecting different biomass to corroborate a continuous feedstock supply and to develop economic approaches for biomass conversion. This includes creating original, reasonable, eco-friendly, efficient catalysts, catalyst support, and hybrid frameworks. These advancements could lead to efficient conversion processes that produce multiple products rather than just one. Despite the National Renewable Energy Laboratory (NREL) making groundbreaking discoveries in LCB biorefineries, the technology remains in its nascent stages. Significant challenges in the economic development of LCB biorefineries include issues at every stage, such as the pretreatment process, fractionation, saccharification, and sustainable biomass supply chain for the transformation of sugars and lignin into chemicals and fuels. An analytical hurdle is the cost of feedstock, which in turn is influenced by supply chains, thus necessitating thorough analysis and research to improve supply chain efficiency. Additionally, diseases, transportation, land usage, and pests have made second-generation biofuels more expensive. Future advancements in LCB conversion to biofuels will likely depend on developing sustainable pretreatment technologies that enable efficient hydrolysis along with reduced enzyme doses and fewer chemicals. Despite many efforts to develop enzyme cocktails, creating sustainable enzyme mixtures remains challenging. Exploring different enzyme sources and designing customized enzymatic cocktails through metagenomics and microbial bioprospecting could be beneficial. Understanding enzymatic synergism in these cocktails is crucial for rapid biomass conversion, which can reduce enzyme loads and lower hydrolysis costs.

5 Conclusion

The pursuit of alternative energy sources has been fueled by concern about greenhouse gas emissions and the depletion of fossil fuels. In addition, the use of biobased chemicals is gradually replacing those derived from fossil fuels. Recently, several biological pretreatment protocols have been standardized to overcome the resistance of lignocellulosic biomass to anaerobic degradation, thus improving its suitability. Due to their lower byproduct generation, lower corrosiveness, and the necessity for milder operating temperatures, environmentally friendly biological pretreatment procedures outperform their competitors. Biological pretreatment is regarded as an effective and environmentally acceptable waste management technique since it uses less energy and produces no inhibiting compounds. Therefore, biological pretreatment techniques have significant potential to enhance the digestibility of lignocellulosic biomass.

References

Adekunle, K. F., & Okolie, J. A. (2015). A review of biochemical process of anaerobic digestion. *Advances in Bioscience and Biotechnology, 6*, 205–212.

Agler, M. T., Wrenn, B. A., Zinder, S. H., & Angenent, L. T. (2011). Waste to bioproduct conversion with undefined mixed cultures: the carboxylate platform. *Trends in Biotechnology, 29*(2), 70–78.

Alonso, D. M., Wettstein, S. G., & Dumesic, J. A. (2013). Gamma-valerol actone, a sustainable platform molecule derived from lignocellulosic biomass. *Green Chemistry, 15*, 584–595.

Amulya, K., Jukuri, S., & Mohan, V. (2015). Sustainable multistage process for enhanced productivity of bioplastics from waste remediation through aerobic dynamic feeding strategy: Process integration for up-scaling. *Bioresource Technology, 188*, 231–239.

Asghari, F. S., & Yoshida, H. (2007). Kinetics of the decomposition of fructose catalyzed by hydrochloric acid in subcritical water. Formation of 5-hydroxymethylfurfural, levulinic, and formic acids. *Industrial & Engineering Chemistry Research, 46*, 7703–7710.

Bátori, V., Åkesson, D., Zamani, A., Taherzadeh, M. J., & Sárvári Horváth, I. (2018). Anaerobic degradation of bioplastics: A review. *Waste Management, 80*, 406–413.

Borowski, S., & Kucner, M. (2015). Co-digestion of sewage sludge and dewatered residues from enzymatic hydrolysis of sugar beet pulp. *Journal of the Air & Waste Management Association, 65*(11), 1354–1364.

Bowen, E. J., Dolfing, J., Davenport, R. J., Read, F. L., & Curtis, T. P. (2014). Low temperature limitation of bioreactor sludge in anaerobic treatment of domestic wastewater. *Water Science Technology, 69*, 1004–1013.

Brown, M. E., & Chang, M. C. Y. (2014). Exploring bacterial lignin degradation. *Current Opinion in Chemical Biology, 19*, 1–7.

Burgos, N., Valdés, A., & Jiméne, A. (2016). Valorization of agricultural wastes for the production of protein-based biopolymers. *Journal of Renewable Materials, 4*, 165–177.

Cetera, P., D'Auria, M., Mecca, M., & Todaro, L. (2018). Gallic acid as main product in the water extractives of Quercus frainetto ten. *Natural Product Research, 14*, 1–4.

Chatzifragkou, A., Papanikolaou, S., Kopsahelis, N., Kachrimanidou, V., Dorado, M. P., & Koutinas, A. A. (2014). Biorefinery development through utilization of biodiesel industry by-products as sole fermentation feedstock for 1,3-propanediol production. *Bioresource Technology, 159*, 167–175.

Chen, C., Wu, L., Cao, Q., Shao, H., Li, X., Zhang, Y., & Tan, X. (2018). Genome comparison of different *Zymomonas mobilis* strains provides insights on conservation of the evolution. *PLoS One, 13*(4), e0195994.

Choi, I. S., Cho, E. J., Moon, J. H., & Bae, H. J. (2015). Onion skin waste as a valorization resource for the by-products quercetin and biosugar. *Food Chemistry, 188*, 537–542.

Deeba, F., Pruthi, V., & Negi, Y. S. (2015). Converting paper mill sludge into neutral lipids by oleaginous yeast *Cryptococcus vishniaccii* for biodiesel production. *Bioresource Technology, 213*, 96–102.

Demirel, B., & Scherer, P. (2011). Trace element requirements of agricultural biogas digesters during biological conversion of renewable biomass to methane. *Biomass Bioenergy., 35*, 992–998.

Den Haan, R., Rose, S. H., Lynd, L. R., & van Zyl, W. H. (2007). Hydrolysis and fermentation of amorphous cellulose by recombinant Saccharomyces cerevisiae. *Metabolic Engineering, 9*(1), 87–94.

Fache, M., Boutevin, B., & Caillol, S. (2016). Vanillin production from lignin and its use as a renewable chemical. *ACS Sustainable Chemistry & Engineering, 4*, 35–46.

Gallezot, P. (2012). Conversion of biomass to selected chemical products. *Chemical Society Reviews, 41*, 1538–1558.

Howard, R. L., Abotsi, E., van Jansen, R. E. L., & Howard, S. (2015). Lignocellulose biotechnology: issues of bioconversion and enzyme production. *African Journal of Biotechnology, 2*, 602–619.

Ibarra, D., Romero, J., Martínez, M. J., Martínez, A. T., & Camarero, S. (2006). Exploring the enzymatic parameters for optimal delignification of eucalypt pulp by laccase mediator. *Enzyme and Microbial Technology, 39*, 1319–1327.

Joshi, N., & Lawal, A. (2012). Hydrodeoxygenation of pyrolysis oil in a microreactor. *Chemical Engineering Science, 74*, 1–8.

Khan, I. U., Hafiz Dzarfan Othman, M., Hashi, H., Matsuura, T., Ismail, A. F., Rezaei-DashtArzhandi, M., & Wan Azelee, I. (2017). Biogas as a renewable energy fuel—A review of biogas upgrading utilisation and storage. *Energy Conversion and Management, 150*, 277–294.

Khuong, L. S., Masjuki, H. H., Zulkifli, N. W., Mohamad, E. M., Kalam, M. A., Alabdulkarem, A., Arslan, A., Mosarof, M., Syahir, A. Z., & Jamshaid, M. (2017). Effect of gasoline–bioethanol blends on the properties and lubrication characteristics of commercial engine oil. *RSC Advances, 7*, 15005–15019.

Kim, S., & Dale, B. E. (2004). Global potential bioethanol production from wasted crops and crop residues. *Biomass Bioenergy, 26*, 361–375.

Kim, D. H., & Kim, M. S. (2013). Development of a novel three-stage fermentation system converting food waste to hydrogen and methane. *Bioresource Technology, 127*, 267–274.

Kiran, E. U., Trzcinski, A. P., & Liu, Y. (2015). Enhancing the hydrolysis and methane production potential of mixed food waste by an effective enzymatic pretreatment. *Bioresource Technology, 183*, 47–52.

Kumar, N., & Pruthi, V. (2014). Potential applications of ferulic acid from natural sources. *Biotechnology Reports, 4*, 86–93.

Kumar, R., Singh, S., & Singh, O. V. (2008). Bioconversion of lignocellulosic biomass Biochemical and molecular perspectives. *Journal of Industrial Microbiology and Biotechnology, 35*, 377–391.

Kumar, G., Sen, B., & Lin, C. Y. (2013). Pretreatment and hydrolysis methods for recovery of fermentable sugars from de-oiled Jatropha waste. *Bioresource Technology, 145*, 275–279.

Kumar, G., Bakonyi, P., Sivagurunathan, P., Nemestóthy, N., Bélafi-Bakó, K., & Lin, C. Y. (2015). Improved microbial conversion of de-oiled Jatropha waste into biohydrogen via inoculum pretreatment: process optimization by experimental design approach. *Biofuel Research Journal, 5*, 209–214.

Li, H., Huang, J., Wang, Y., Wang, X., Ren, Y., Yue, T., & Gao, Z. (2021). Study on the nutritional characteristics and antioxidant activity of dealcoholized sequentially fermented apple juice with *Saccharomyces cerevisiae* and *Lactobacillus plantarum* fermentation. *Food Chemistry, 363*, 130351.

Liang, S., McDonald, A. G., & Coats, E. R. (2014). Lactic acid production with undefined mixed culture fermentation of potato peel waste. *Waste Management, 34*, 2022–2027.

Lichtenthaler, F. W. (2002). Unsaturated O-and N-heterocycles from carbohydrate feedstocks. *Accounts of Chemical Research, 35*, 728–737.

López-Linares, J. C., Romero, I., Cara, C., Castro, E., & Mussatto, S. I. (2018). Xylitol production by *Debaryomyces hansenii* and *Candida guilliermondii* from rapeseed straw hemicellulosic hydrolysate. *Bioresource Technology, 247*, 736–743.

Mackuľak, T., Prousek, J., Švorc, Ľ., & Drtil, M. (2012). Increase of biogas production from pretreated hay and leaves using wood-rotting fungi. *Chemical Papers, 66*(7), 649–653.

Maleki, F., Changizian, M., Zolfaghari, N., Rajaei, S., Noghabi, K. A., & Zahiri, H. S. (2021). Consolidated bioprocessing for bioethanol production by metabolically engineered Bacillus subtilis strains. *Science Reports, 11*, 1–13.

Maleki, B., Talesh, S. A., & Mansouri, M. (2022). Comparison of catalysts types performance in the generation of sustainable biodiesel via transesterification of various oil sources: a review study. *Materials Today Sustainability, 18*, 100157.

Mekonnen, T., Mussone, P., Khalil, H., & Bressler, D. (2013). Progress in bio-based plastics and plasticizing modifications. *Journal of Materials Chemistry A, 1*, 13379–13398.

Mohan, D., Pittman, C. U., & Steele, P. H. (2006). Pyrolysis of wood/bio mass for bio-oil: A critical review. *Energy Fuel., 20*, 848–889.

Morone, P., & Cottoni, L. (2016). Biofuels: Technology, economics, and policy issues. In R. Luque, C. S. K. Lin, K. Wilson, & J. Clark (Eds.), *Handbook of biofuels production: processes and technologies* (pp. 61–83). Woodhead Publishing.

Mutschlechner, M., Illmer, P., & Wagner, A. O. (2015). Biological pre-treatment: Enhancing biogas production using the highly cellulolytic fungus *Trichoderma viride*. *Waste management., 43*, 98–107.

Nanda, S., Mohammad, J., Reddy, S. N., Kozinski, J. A., & Dalai, A. K. (2014a). Pathways of lignocellulosic biomass conversion to renew able fuels. *Biomass Conversion Biorefinery, 4*, 157–191.

Nanda, S., Mohanty, P., Kozinski, J. A., & Dalai, A. K. (2014b). Physico chemical properties of bio-oils from pyrolysis of lignocellulosic biomass with high and slow heating rate. *Energy Environmental Research, 4*, 21–32.

Nanda, S., Azargohar, R., Dalai, A. K., & Kozinski, J. A. (2015a). An assessment on the sustainability of lignocellulosic biomass for biorefining. *Renewable and Sustainable Energy Reviews, 50*, 925–941.

Nanda, S., Kozinski, J. A., & Dalai, A. K. (2015b). Lignocellulosic biomass: A review of conversion technologies and fuel products. *Current Biochemical Engineering, 3*, 324–336.

Nanda, S., Dalai, A. K., & Kozinski, J. A. (2017a). Butanol from renewable biomass: Highlights on downstream processing and recovery techniques. In P. Mondal & A. K. Dalai (Eds.), *Sustainable utilization of natural resources* (pp. 187–211). CRC Press.

Nanda, S., Golemi-Kotra, D., McDermott, J. C., Dalai, A. K., Gökalp, I., & Kozinski, J. A. (2017b). Fermentative production of butanol: perspectives on synthetic biology. *New Biotechnology, 37*, 210–221.

Nanda, S., Rana, R., Vo, D. V. N., Sarangi, P. K., Nguyen, T. D., Dalai, A. K., & Kozinski, J. A. (2020). A spotlight on butanol and propanol as next-generation synthetic fuels. In S. Nanda, D.-V. N. Vo, & P. K. Sarangi (Eds.), *Biorefinery of alternative resources: Targeting green fuels and platform chemicals* (pp. 105–126). Springer.

Naranjo, J. M., Cardona, C. A., & Higuita, J. C. (2014). Use of residual banana for polyhydroxybutyrate (PHB) production: Case of study in an integrated biorefinery. *Waste Management, 34*, 2634–2640.

Nayak, S. K., Nayak, B., Mishra, P. C., Noor, M. M., & Nanda, S. (2019). Effects of biodiesel blends and producer gas flow on overall performance of a turbocharged direct injection dual-fuel engine. *Energy Sources A, 46*, 4165.

Olson, D. G., McBride, J. E., Shaw, A. J., & Lynd, L. R. (2012). Recent progress in consolidated bioprocessing. *Current Opinion in Biotechnology, 23*, 396–405.

Parakh, P. D., Nanda, S., & Kozinski, J. A. (2020). Eco-friendly transformation of waste biomass to biofuels. *Current Biochemical Engineering, 6*, 120–134.

Peng, L., Lin, L., Zhang, J., Zhuang, J., Zhang, B., & Gong, Y. (2010). Catalytic conversion of cellulose to levulinic acid by metal chlorides. *Molecules, 15*, 5258–5272.

Rahman, S. H. A., Choudhury, J. P., Ahmad, A. L., & Kamaruddin, A. H. (2007). Optimization studies on acid hydrolysis of oil palm empty fruit bunch fiber for production of xylose. *Bioresource Technology, 98*, 554–559.

Ramamoorthy, S. K., Skrifvars, M., & Persson, A. (2015). A review of natural fibers used in biocomposites: Plant, animal and regenerated cellulose fibers. *Polymer Reviews, 55*, 107–162.

Reddy, S. N., Nanda, S., & Sarangi, P. K. (2018). Applications of supercritical fluids for biodiesel production. In P. K. Sarangi, S. Nanda, & P. Mohanty (Eds.), *Recent advancements in biofuels and bioenergy utilization* (pp. 261–284). Springer.

Romano, R. T., Zhang, R., Teter, S., & McGarvey, J. A. (2009). The effect of enzyme addition on anaerobic digestion of Jose Tall Wheat Grass. *Bioresource Technology, 100*(20), 4564–4571.

Roth, S., & Spiess, A. C. (2015). Laccases for biorefinery applications: a critical review on challenges and perspectives. *Bioprocess and Biosystems Engineering, 38*, 2285–2313.

Rutkowski, P. (2011). Pyrolysis of cellulose, xylan and lignin with the K2CO3 and ZnCl2 addition for bio-oil production. *Fuel Processing Technology, 92*, 517–522.

Sakwa, L., Cripwell, R. A., Rose, S. H., & Viljoen-Bloom, M. (2018). Consolidated bioprocessing of raw starch with *Saccharomyces cerevisiae* strains expressing fungal alpha-amylase and glucoamylase combinations. *FEMS Yeast Research, 18*, foy085.

Salvachúa, D., Mohagheghi, A., Smith, H., Bradfield, M. F., Nicol, W., Black, B. A., Biddy, M. J., Dowe, N., & Beckham, G. T. (2016). Succinic acid production on xylose-enriched biorefinery streams by *Actinobacillus succinogenes* in batch fermentation. *Biotechnology Biofuels, 9*, 28.

Sambusiti, C., Monlau, F., Ficara, E., Carrère, H., & Malpei, F. (2013). A comparison of different pre-treatments to increase methane production from two agricultural substrates. *Applied Energy, 104*, 62–70.

Sarangi, P. K., & Nanda, S. (2018). Recent developments and challenges of acetone–butanol–ethanol fermentation. In P. K. Sarangi, S. Nanda, & P. Mohanty (Eds.), *Recent advancements in biofuels and bioenergy utilization* (pp. 111–123). Springer.

Satyanarayana, K. G., Arizaga, G. G., & Wypych, F. (2009). Biodegradable composites based on lignocellulosic fibers—An overview. *Progress in Polymer Science, 34*, 982–1021.

Shiu, P. J., Gunawan, S., Hsieh, W. H., Kasim, N. S., & Ju, Y. H. (2010). Bio diesel production from rice bran by a two-step in-situ process. *Bioresource Technology, 101*, 984–989.

Singhania, R. R., Binod, P., & Pandey, A. (2008). Plant-based biofuels—An introduction. In A. Pandey (Ed.), *Handbook of plant-based biofuels* (pp. 1–10). Taylor & Francis.

Song, J. H., Murphy, R. J., Narayan, R., & Davies, G. B. H. (2009). Biodegradable and compostable alternatives to conventional plastics. *Philosophical Transactions of the Royal Society B: Biological Sciences, 364*, 2127–2139.

Sun, Y., & Cheng, J. (2002). Hydrolysis of lignocellulosic materials for ethanol production: A review. *Bioresource Technology, 83*(1), 1–11.

Tang, B., Xu, H., Xu, Z., Xu, C., Xu, Z., Lei, P., Qiu, Y., Liang, J., & Feng, X. (2015). Conversion of agroindustrial residues for high poly(c-glutamic acid) production by Bacillus subtilis NX-2 via solid-state fermentation. *Bioresource Technology, 181*, 351–354.

Trache, D., Hussin, M. H., HuiChuin, C. T., Sabar, S., Fazita, M. R. N., Taiwo, O. F. A., Hassan, T. M., & Haafiz, M. K. M. (2016). Micro crystalline cellulose: Isolation, characterization and bio-composites application—A review. *International Journal of Biological Macromolecules, 93*, 789–804.

Vervaeren, H., Hostyn, K., Ghekiere, G., & Willems, B. (2010). Biological ensilage additives as pretreatment for maize to increase the biogas production renew. *Energy, 35*, 2089–2093.

Wang, P., Zhan, S. H., & Yu, H. B. (2010). Production of levulinic acid from cellulose catalyzed by environmental-friendly catalyst. *Advanced Material Research, 96*, 183–187.

Wen, Z., Ledesma-Amaro, R., Lu, M., Jin, M., & Yang, S. (2020). Metabolic engineering of Clostridium cellulovorans to improve butanol production by consolidated bioprocessing. *ACS Synthetic Biology, 9*(2), 304–315.

Yunqin, L., Dehan, W., & Lishang, W. (2010). Biological pretreatment enhances biogas production in the anaerobic digestion of pulp and paper sludge. *Waste Management & Research, 28*(9), 800–810.

Zakzeski, J., Bruijnincx, P. C., Jongerius, A. L., & Weckhuysen, B. M. (2009). The catalytic valorization of lignin for the production of renewable chemicals. *Chemical Reviews, 110*, 3552–3599.

Zanwar, A. A., Badole, S. L., Shende, P. S., Hegde, M. V., & Bod Hankar, S. L. (2013). Role of gallic acid in cardiovascular disorders. *Polyphenols in Human Health and Disease, 2*, 1045–1047.

Zhang, L., Lee, Y. W., & Jahng, D. (2011a). Anaerobic co-digestion of food waste and piggery wastewater: focusing on the role of trace elements. *Bioresource Technology, 102*, 5048–5059.

Zhang, Q., He, J., Tian, M., Mao, Z., Tang, L., Zhang, J., & Zhang, H. (2011b). Enhancement of methane production from cassava residues by biological pretreatment using a constructed microbial consortium. *Bioresource Technology, 102*(19), 8899–8906.

Zhang, T., Liu, L., Song, Z., Ren, G., Feng, Y., Han, X., & Yang, G. (2013). Biogas production by co-digestion of goat manure with three crop residues. *PloS One, 8*(6), e66845.

Zhao, J. (2013) *Enhancement of methane production from solid-state anaerobic digestion of yard trimmings by biological pretreatment*. Master thesis, The Ohio State University.

Zhao, H., Holladay, J. E., Brown, H., & Zhang, Z. C. (2007). Metal chlorides in ionic liquid solvents convert sugars to 5-hydroxymeth ylfurfural. *Science, 316*, 1597–1600.

Zhao, J., Zheng, Y., & Li, Y. (2014). Fungal pretreatment of yard trimmings for enhancement of methane yield from solid-state anaerobic digestion. *Bioresource Technology, 156*, 176–181.

Zheng, Y., Zhao, J., Xu, F., & Li, Y. (2014). Pretreatment of lignocellulosic biomass for enhanced biogas production. *Progress in Energy and Combustion Science, 42*, 35–53.

Ziemiński, K., Romanowska, I., & Kowalska, M. (2012). Enzymatic pretreatment of lignocellulosic wastes to improve biogas production. *Waste Management, 32*(6), 1131–1137.

Exploring the Impact of Emerging Pretreatment Technologies on Effective Lignocellulosic Biomass Utilization

Vishnu Damodaran Nambissan, Saivi Singh, and Sathyanarayana N. Gummadi

Abstract

With growing awareness of global warming and climate change crisis and increased efforts by global intergovernmental bodies promoting a switch to greener alternatives from petrochemicals, research into biomass as an alternate source for biofuels and chemicals has been on the rise. Pretreatment has proved to be a major bottleneck in utilizing biomass to its full potential. Many commercial plants for the production of biofuels are functional, albeit at a higher price than their conventional petrochemical counterpart. With a view to bringing down the cost of bioproduction, reducing waste generation and utilizing biomass to its full potential, a biorefinery approach similar to a petroleum refinery is being probed. Newer nonconventional pretreatment methods are developed for efficient breakdown of lignocellulose for better utilization as feedstocks for various manufacturing processes. This chapter aims to provide an overview of lignocellulose, its sources, composition and conventional pretreatment methods. It will further focus on the novel pretreatment methods being developed over the last decade regarding its scalability to commercial production, environmental impact and economics. It will further delve into the utility of the emerging pretreatment techniques in the biorefinery development of lignocellulose and the research gaps in the current technologies.

Keywords

Lignocellulose · Pretreatment · Biorefinery · AOP · Waste valorization · Circular economy · Sustainable development goals

V. D. Nambissan · S. Singh · S. N. Gummadi (✉)
Applied and Industrial Microbiology Lab, Department of Biotechnology, Bhupat and Jyoti Mehta School of Biosciences, Indian Institute of Technology Madras, Chennai, India
e-mail: gummadi@iitm.ac.in

G. Mukherjee, S. Dhiman (eds.), *Value Addition and Utilization of Lignocellulosic Biomass*, https://doi.org/10.1007/978-981-96-2786-8_3

1 Why Transition from Petrochemicals to Greener Alternatives?

The industrial revolution unlocked a whole new energy resource fossil fuels, which has been a fundamental driver of the traditional industrial sector, the petrochemical industry, characterized by resource dependence, high energy consumption and high emissions (Megía et al., 2021). The petrochemical industry accounts for a significant and growing share of oil and gas demand. These fossil fuels like coal and petroleum take millions of years to form and replenish. Apart from its utility as fuel, these petrochemicals also act as feedstock for producing organic chemicals and polymers. About 9% of all fossil fuels and about 16% of crude oil are used for these non-energy applications (de Jong et al., 2020). This “dual use” of fossil carbon (mainly in the form of petroleum) as both fuel and feedstock by the chemical product industries risks the future availability of this resource. In 2020, crude oil demand was 101 million barrels per day (Damaurai et al., 2021) and by the year 2040, fuel demands are on track to rise to a further 28% of its current value (Kumar et al., 2020).

Chemical production using fossil fuels also comes with a negative impact on the environment and ecology. When burnt, fossil fuels produce carbon dioxide (CO_2), which is the largest driver of global climate change. In 2022, the CO_2 level was estimated to be 421 ppm, as compared to 280 ppm in the pre-industrial era (Singh et al., 2023). They are also a major contributor to local air pollution, which is estimated to be linked to millions of premature deaths each year (Ajayi & Lateef, 2023).

Fuels have an uneven distribution across the globe. Since most countries source their oil from the States, Russia, Canada or the Middle East, deterioration in geopolitical tensions undermines the availability of this natural resource and underscores the frailties of the fossil fuel age. For example, Russia’s invasion of Ukraine and instability and protracted conflicts in the Middle East could lead to further disruption to energy markets and prices (Zhang et al., 2024). Extreme volatility in the petrochemical markets highlights the importance of an affordable, reliable and resilient supply of feedstock.

Shifting the dependence of chemical product industries from petrochemicals to alternative renewable fuels and feedstock is a promising solution for the current exacerbating situation. Biomass energy has attracted great interest in recent years due to its renewability and carbon-neutral nature (Bansod et al., 2024). Lignocellulose, such as forestry and agricultural residues, represents the most abundant natural biomass in the natural world, with an approximate yield of about 200 billion tons (Wang et al., 2021e). These consist of non-edible parts of plants, ruling out competition with food demand. The major chemical components of lignocellulosic biomass are cellulose, hemicelluloses and lignin, all of which have shown great promise as feedstocks and materials for energy generation.

2 Lignocellulose: Its Sources and Composition

Lignocellulose is a collective term given to plant-based biomass, including forestry waste, agro-waste, and waste from allied industries like paper, furniture, and edible oil industries. The most abundant form of biomass energy available on earth, lignocellulose contains three major components in its structure—lignin, hemicellulose and cellulose (Bilal et al., 2024; Zoghlami et al., 2019). The most abundantly found natural polymer, cellulose, is a polymer made of monomeric glucose linked by β-1,4-glycosidic linkages (Selvaraj et al., 2024). Hemicellulose is a heteropolysaccharide made up of various monomeric units like pentoses (xylose, arabinose), hexoses (glucose, galactose, mannose) and uronic acids (galacturonic acid, glucuronic acid). Lignin, the second most abundant polymer in lignocellulose, is comprised of phenyl propane derivatives linked by irregular C–C and C–O bonds (Sun et al., 2022).

Lignin, the outermost layer of lignocellulose, forms about 15–20% of the biomass dry weight, while hemicellulose contributes 20–35% and cellulose forms 35–50% in case of grass lignocellulose (Ahmad et al., 2023; Vu et al. 2020). Apart from these significant components, the lignocellulosic biomass contains inert materials like ash, volatile matter, and extractives, which account for less than 20% of dry weight (Lymperatou et al., 2023; Rego et al., 2019). Extractives can be water-soluble compounds like proteins or non-structural sugars or ethanol-soluble compounds like waxes and chlorophyll. The composition of lignin, cellulose, and hemicellulose varies with the source and type of biomass and the geographical location (Barhoum et al., 2020; Szadkowska et al., 2021). The general composition in percent dry weight of some of the lignocellulosic biomass studied worldwide is mentioned in Table 1.

Scientists worldwide have been looking at different lignocellulose sources for valorization, from agricultural wastes to grasses and forestry waste. Agro-waste, including sugarcane bagasse, corn cob, corn stover, wheat straw, paddy straw, etc., has been studied extensively for efficient lignocellulose valorization (Sooch et al., 2019). In addition, softwoods like cedar, pine, fir, and hardwoods like willow, oak, aspen, etc. have also been evaluated as sources of lignocellulose (Sun et al., 2023b). Apart from these, waste from the paper industry, food waste, newsprint, weeds, etc. have also been used as a source of lignocellulose (Pan et al., 2021).

3 Conventional Pretreatment

For the lignocellulosic biomass to be used as feedstock for biofuel and biochemical production, it must be processed to break down the complex structure to lignin, hexose and pentose sugars, and this processing step is called pretreatment. Several pretreatment methods have been widely researched for cost-effective lignocellulose utilization. These methods can be broadly classified into physical, thermochemical, chemical and biological techniques. Figure 1 shows the categories of conventional pretreatment methods used for lignocellulose depolymerization. The advantages and disadvantages of each type of pretreatment method and recent research on the same are mentioned in Table 2.

Table 1 Composition of lignocellulose from various sources

Sr. no.	Lignocellulose source	Cellulose	Hemicellulose	Lignin	Others	References
1.	Agave bagasse	43	19	15	8	Alemán-Nava et al. (2018)
2.	Bamboo	40.25	32.01	24.45	1.9	Khan et al. (2018)
3.	Banana bunch stem	79.7 (Hemicellulose + Cellulose)		7.9	37.56	Kosinkova et al. (2017)
4.	Barley hull	33.6	37.2	19:3	NA	Mood et al. (2013)
5.	Cocoa pod husk	28.86–29.5	20.3–24.56	26.97–29.38	18.3–21.02	de Sales et al. (2017)
6.	Coconut coir	44.2	22.1	32.8	NA	Subhedar et al. (2018)
7.	Corn cob	23.11–24.05	32.78–33.9	22.32–22.7	NA	Mafa et al. (2020)
8.	Corn stover	36.8–37.44	29.34–29.86	20.24–21.36	NA	Zhang et al. (2020)
9.	Hardwood	45–47	25–40	20–25	0.8	Limayem and Ricke (2012)
10.	Palm oil frond	37.32	31.89	26.05	2.25	Khan et al. (2018)
11.	Pineapple leaves	30.54–32.45	16.67–21.96	~20	~26	Zannini et al. (2021)
12.	Rice bran	45.7	14	10.9	29	Sridevi et al. (2015)
13.	Rice Husk	36.7	24.6	28.2	10.5	Taherzadeh-Ghahfarokhi et al. (2019)
14.	Rice Straw	39.4	29.8	17.9	12.9	Taherzadeh-Ghahfarokhi et al. (2019)
15.	Softwood	40–45	25–29	30–60	0.5	Limayem and Ricke (2012)
16.	Sweet sorghum bagasse	17.39–18.11	15.82–16.74	29.29–29.38	NA	Mafa et al. (2020)
17.	Sugarcane bagasse	42.1	21.5	24	9.7	Schmetz et al. (2019)
18.	Sugar beet pulp	28.5	17.5	28.1	19.7	Schmetz et al. (2019)
19.	Switch Grass	34.2–36	24.8–31.4	19–20	NA	Ge et al. (2012)
20.	Wheat straw	41.9	26.2	25.7	6.2	Taherzadeh-Ghahfarokhi et al. (2019)

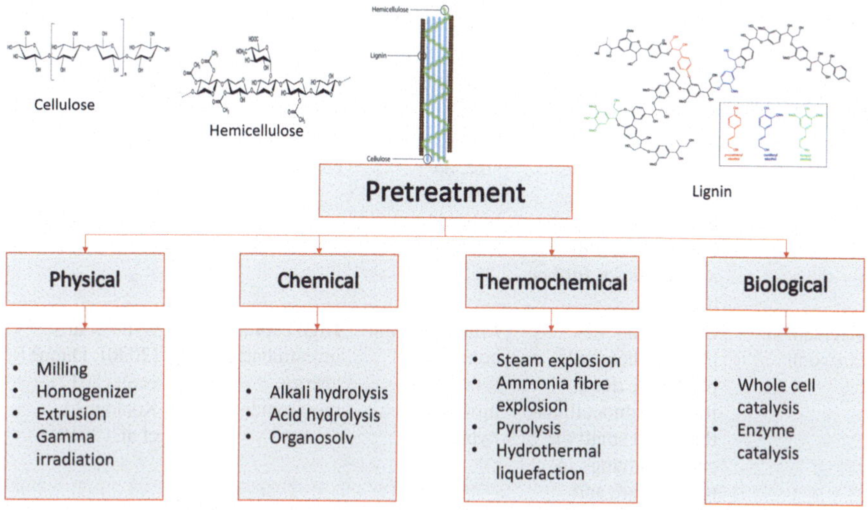

Fig. 1 Structure of lignocellulose and various conventional pretreatment methods

3.1 Physical Methods

As the name suggests, the physical pretreatment method relies on the breakdown of the recalcitrant structure of lignocellulose. It is often applied before chemical or biological pretreatment methods. It works by reducing the particle size, increasing porosity and surface area and reducing polymerization and the level of crystallinity, either of which helps in improving the digestibility of the biomass and the accessibility to enzymes, microbes, and chemicals to the sugar fractions present in lignocellulose, enhancing the utility of the feedstock (Jomnonkhaow et al., 2021; Kassim et al., 2022). Some of the common physical pretreatment methods studied include size reduction, extrusion, and high-pressure homogenizer (Rajendran et al., 2021).

Grinding, chipping, and milling are the most commonly used conventional mechanical size reduction techniques (Khan et al., 2022). These methods also do not produce any unwanted byproducts. However, the major downside of this method is the high energy consumption (Gallego-García et al., 2023). Enhanced bioethanol production was achieved by Hosseini Koupaie et al. (2019), by employing ball mill grinding of elephant grass. Similar studies were done with rice bran, where particle size reduction from 20 to 0.075 mm resulted in a 1.8-fold increase in methane production (Dai et al., 2019). Mechanical extrusion is a promising technique where lignocellulose is forced to pass through a small die of desired dimensions. On being extruded from the die, the lignocellulose experiences change in the physicochemical properties due to changes in surface area, pressure and temperature. This allows better access to enzymes and microbes for further processing of lignocellulose (Witaszek et al., 2020). This method requires optimization of temperature and compression ratio (Malik et al., 2022). Kozłowski et al. (2019) showed that mechanically extruded maize straw could help increase the fermentation yields of biogas

Table 2 Recent literature on conventional pretreatment methods, their advantages and disadvantages

Pretreatment	Principle	Advantages	Disadvantages	References
Physical pretreatment				
Size reduction	Breakdown of the structure of lignocellulose to make cellulose and hemicellulose more accessible for further treatment	Small particle size; no unwanted byproducts	High energy consumption	Akhtar et al. (2016), Dai et al. (2019), Gallegos et al. (2017)
Mechanical extrusion	Change in physicochemical properties due to forcing lignocellulose through a small die causing change in temperature and pressure	Mild operation condition; fewer unwanted byproducts	High capital investment; high energy requirement	Byun et al. (2020), Duque et al. (2017), Kozłowski et al. (2019)
Cavitation	Shockwaves released from the bubbles formed due to cavitation phenomena break the lignocellulosic structure	Mild operation conditions; high energy efficiency	Limited studies for scale up	Subhedar et al. (2018), Terán Hilares et al. (2020), Zieliński et al. (2019)
Gamma irradiation	Gamma radiation breaks Van der Waal's forces in lignocellulose; generate phenoxy radicals; breaks glycosidic bonds	Minimal unwanted byproducts; high efficiency	Scale up constraints due to environmental considerations to prevent gamma rays from to the environment	Kim et al. (2015), Loow et al. (2016), Poddar et al. (2021)
High hydrostatic pressure	Application of high pressure resulting in breakage of non-covalent bonds	Minimal process time	Scale up studies for different biomass are limited	Castañón-Rodríguez et al. (2013), Ma et al. (2018)
Thermal pretreatment				
Hydrothermal pretreatment	High heat through hot water under high pressure breaks lignocellulose structure	Minimal inhibitory compounds	Energy intensive; huge water requirement	Das et al. (2021), Lin et al. (2019)
Steam explosion	Lignocellulose subjected to high-pressure steam; sudden release of pressure; heat and temperature variation causes the breakdown of lignocellulose	Cheap; cost-effective; studied for a variety of biomass	The particle size of biomass affects efficiency; the formation of inhibitory compounds	Bhatia et al. (2020a, 2020b), Sulzenbacher et al. (2021)

(continued)

Table 2 (continued)

Pretreatment	Principle	Advantages	Disadvantages	References
Ammonium fibre explosion	Biomass soaked in ammonia and heated and depressurized; ammonia causes biomass to swell and break bonds	Less energy requirements; good for low lignin biomass	Recycling of ammonia necessary	Beig et al. (2021), Das et al. (2021), Kumar and Sharma (2017)
Chemical pretreatment				
Alkali pretreatment	Alkali breaks bonds between lignin and hemicellulose	Efficient lignin removal	High temperature for efficient hydrolysis; formation of inhibitors	Byun et al. (2020), Mafa et al. (2020), Sabeeh et al. (2020), Shi et al. (2021)
Acid pretreatment	Acid breaks hemicellulosic fraction in lignocellulose	Effective sugar fractionation	High cost; high temperature; formation of inhibitory compounds	Hickert et al. (2014), Ovejero-Pérez et al. (2020), Wang et al. (2016)
Biological pretreatment				
Whole-cell pretreatment	Microbes produce enzymes to degrade lignocellulose and utilize depolymerized fractions for metabolism	Mild operation condition	Low efficiency	Chen et al. (2020b), Gu et al. (2021)
Enzymatic pretreatment	Lignin, cellulose and hemicellulose degrading enzymes breakdown lignocellulose to release monomers	Mild operation conditions; high energy efficiency	Enzyme cost	Ferdeş et al. (2020), Vu et al. (2020)

and methane by 7.52% and 8.51%, respectively. Extrusion, even if used in combination with other methods, has relatively lower reaction conditions like pressure, temperature, pH, etc. and also produces far fewer unwanted byproducts compared to conventional chemical pretreatments (Byun et al., 2020). The main bottlenecks remain the high capital investment for the extruder and high energy consumption.

3.2 Chemical Methods

Chemical pretreatment involves the use of chemical catalysts and free radical generators that can break down bonds in the lignocellulosic structure, making them susceptible to degradation (Sai Bharadwaj et al., 2023). The quality and quantity of the products would depend not only on the pretreatment methods but also on the lignocellulose used. Chemical pretreatment includes acid hydrolysis, alkali hydrolysis, ozonolysis, organosolv pretreatment, etc.

Acid hydrolysis with inorganic and organic acids results in the breakdown of sugar fractions (hemicellulose and cellulose) of lignocellulose and separates it from the insoluble alkali lignin (Worku et al., 2024). Acids like sulfuric acid, nitric acid, phosphoric acid, hydrochloric acid, formic acid, oxalic acid and acetic acid have been studied for their utility in lignocellulose hydrolysis (Deng et al., 2023). Acid hydrolysis can be done with concentrated acids (>30% acid), which will dissolve the hemicellulose and cellulose fibres and target glycosidic bonds to release monomeric sugars. In contrast, dilute acids break the glycosidic bonds in hemicellulose, leaving the cellulose fibres susceptible to enzyme treatment (Zhou et al., 2021).

While acid hydrolysis targets the sugar fraction, alkali hydrolysis breaks down the β-O-4 ether linkages in the lignin fraction of the lignocellulosic materials. Sodium, potassium, calcium and ammonium hydroxides have been extensively studied for their effectiveness in alkali pretreatment (Deng et al., 2023). The efficiency of alkali pretreatment is reported to be better in low lignin biomass compared to high lignin ones (Worku et al., 2024). Though alkali and acid hydrolysis have high efficiency, they also have certain drawbacks like lack of full recovery of sugars, formation of inhibitory compounds, need for neutralization after the process, and a large amount of salt production (Wu et al., 2022b; Zhou & Tian, 2022).

Organosolv pretreatment is another commonly used chemical pretreatment method. In this method, solvents like ethanol, methanol, acetone, tetrahydrofuran, ethylene glycol, ethylenediamine, etc. are used for dissolving lignin and hemicellulose fractions at high temperatures of 100–250 °C to get high-purity cellulose fraction (Rabelo et al., 2023). This pretreatment method has high efficiency, low inhibitory compound formation, high selectivity of the component to be dissolved and easy solvent recovery but it suffers from high operational cost since the reactions happen at higher temperatures (Sun et al., 2023a).

Some of the other chemical pretreatment methods include advanced oxidation processes, ionic liquid pretreatment, and deep eutectic solvent pretreatment, which will be discussed in detail in Sect. 4.

3.3 Thermochemical Methods

The thermochemical pretreatment method involves applying controlled heat to biomass, which decreases the structural stability of the lignocellulose, leads to oxidation and converts them to high-energy products (Wang et al., 2021e). The two critical parameters for any thermochemical pretreatment methods are reaction time and temperature. It includes liquid hydrothermal liquefaction, steam explosion and ammonium fibre explosion, pyrolysis, torrefaction, etc.

Liquid hot water treatment, also known as hydrothermal liquefaction or aquasolv, is the process where water at high temperature and pressure is used for lignocellulose breakdown. Lin et al. (2019), in their research, showed that hydrothermal pretreatment of brown seaweed (*Saccharina latissimi*) could result in an increase in the biomethane yield at 140 °C from 281.4 mL/g volatile solid to 345.1 mL/g of volatile solid. This pretreatment technique forms minimum inhibitory compounds.

The major drawback of liquid hot water pretreatment is the requirement of energy-intensive downstream processing because of the utility of large amounts of water (Das et al., 2021).

Steam explosion is a commonly used technique where lignocellulosic biomass is subjected to high-temperature pressurized steam and suddenly depressurized. This sudden change in pressure breaks the lignocellulose structure, which can then be broken down through biological processing (Sulzenbacher et al., 2021). Bhatia et al. (2020b) performed steam explosion pretreatment of *Miscanthus* to obtain an eight- to nine-fold higher glucose yield than an untreated one. The steam explosion technique is cheap and cost-effective, and it works with a wide range of lignocellulosic biomass in combination with other pretreatment methods. The particle size of the lignocellulose affects the efficiency of sugar recovery. Also, the formation of inhibitors can affect the subsequent processing of lignocellulose.

Ammonia-based pretreatment methods include ammonia fibre explosion, ammonia recycle percolation and soaking in aqueous ammonia. For ammonia fibre explosion, lignocellulosic biomass and liquid ammonia are taken in equal amounts and heated to 60–90 °C and held for 5 min. The vessel is then suddenly depressurized, leading to a breakdown of the lignocellulose structure (Zhao et al., 2020). The ammonia causes the biomass to swell, breaking the bonds between lignin and hemicellulose and causing hemicellulose hydrolysis (Beig et al., 2021). In this case, recycling ammonia after depressurizing is critical to make the process economically viable. For ammonia recycle percolation, liquid ammonia (5–15%) is passed through a fixed bed reactor containing the lignocellulose substrate. The temperature range is usually between 140 and 210 °C. This pretreatment has helped to attain ~85% delignification (Das et al., 2021). In the case of soaking aqueous ammonia, the lignocellulose substrate is soaked in aqueous ammonia. Ammonia-based pretreatment methods are preferred for low lignin lignocellulosic biomass. All these methods are less energy intensive but require proper handling and recycling of ammonia for sustainable processing.

Pyrolysis is a thermal method where the lignocellulose feed is subjected to high temperatures of 500–800 °C without any oxidizing agent. This method is usually used for bio-oil production. Food waste can be converted to biofuel via pyrolysis. Biohydrogen is produced from lignocellulose using fast or flash pyrolysis (Bakari et al., 2024). Some of the other thermochemical methods include microwave irradiation and supercritical fluid pretreatment, which will be discussed in Sect. 4.

3.4 Biological Methods

Biological pretreatment involves microorganisms and biocatalytic enzymes like hydrolases for breaking down the lignocellulose into simpler substances. Compared to physical, thermal and chemical methods, biological methods are low cost, safe, environment-friendly and energy-efficient alternatives and have low inhibitory byproduct formation (Chen et al., 2020a; Gu et al., 2021). The biological pretreatment can be classified into two categories, viz., microbe-assisted and enzyme-assisted pretreatment. Many studies have been conducted on the utility of whole-cell monocultures or consortia of bacteria or fungi. These microbes produce extracellular ligninolytic, cellulolytic and hemicellulolytic enzymes that degrade the lignocellulose substrate to release simpler feedstock materials. Fungi studied for microbial pretreatment mainly include white rot fungi, brown rot fungi and soft rot fungi. Gu et al., in their studies, subjected corn stover, poplar and wheat straw to sequential fungal pretreatment with manganese supplement and hydrothermal liquefaction using *Ceriporiopsis subvermispora* and obtained a 163.6% increase in hemicellulose sugar yields. Other lignocelluloses used also showed a significant increase in sugar yields (Gu et al., 2021). About 33.5% and 39.6% delignification using white rot fungi *Pleurotus pulmonarius* was obtained from spruce (softwood) and alder (hardwood) sawdust, respectively (Chen et al., 2020b). Bacterial cultures are comparatively less researched compared to fungal cultures in the biological pretreatment of lignocellulose. Azman et al. provided a proof of concept for the degradation of lignin by *Stenotrophomonas* sp. and *Bacillus subtilis* by demonstrating their ability to degrade alkali lignin (Azman et al., 2019). The major bottlenecks in using whole-cell culture for biological pretreatment are its slow reaction time and the necessity to maintain sterility during the process (Galbe & Wallberg, 2019).

Enzyme-assisted pretreatment of lignocellulose employs a cocktail of various ligninolytic, cellulolytic and hemicellulolytic enzymes. Cellulolytic enzymes belong to the family of glycosyl hydrolase and include endoglucanases, exoglucanases and β-glucosidases. Hemicellulolytic enzymes include endoxylanase, exoxylanase, β-xylosidases and α-arabinofuranosidase. Laccases, lignin peroxidase, manganese peroxidase, dye decolorizing peroxidase, β-etherases, superoxide dismutases, etc. all are included in the ligninolytic enzymes (Ferdeş et al., 2020; Sharma et al., 2019; Vu et al., 2020). Matei et al. used crude enzyme extract obtained from *Trametes villosa* for pretreatment of sugarcane bagasse and rice husk. More than a ten-fold increase in reducing sugar yields compared to untreated biomass was obtained in this study (Matei et al., 2020). The high cost of the commercial enzyme cocktail is a significant shortcoming of this method. Recycling the enzyme can also increase the overall cost of the process.

4 Emerging Nonconventional Pretreatment Methods

4.1 Ionic Liquids

Ionic liquids (ILs) are molten salts made of large asymmetric organic cations and organic/inorganic anions. They have melting points below 100 °C and have high ionic conductivity (Tang et al., 2022). ILs are known as green solvents and are interesting alternatives to traditional solvents. They do not release explosive or toxic gases during their utilization and have high thermal stability with negligible vapour pressure (Khoo et al., 2021). Being stable solvents under various thermal and chemical environments, ILs can be used for an ionosolv process treatment of lignin with mild conditions and limited steps. The weaker interactions between ILs' cation and anion are responsible for the low melting point of the ILs (Sharma et al., 2024). They can achieve a wide range of acidity, basicity and polarity by changing different cations and anions in their structure. Various types of ILs, like alkyl or allyl side chain containing imidazolium-, ammonium-, phosphonium-, and pyridinium-based cations, coupled to various anions, like chloride, phosphonate and acetate, have been used in lignocellulosic biomass pretreatment. Among the different types of ILs, Imidazolium-based ILs are important and most extensively studied for lignocellulosic biomass processing. Imidazolium-based ILs, such as 1-ethyl-3-methylimidazolium chloride [Emim]Cl, 1-butyl-3-methylimidazolium chloride [Bmim]Cl and 1-ethyl-3-methylimidazolium acetate [Emim]Ac along with high pretreatment temperatures and long processing times, have been widely reported for lignocellulosic biomass pretreatment (Amini et al., 2021).

The selectivity of the IL for the dissolution of cellulose, lignin or hemicellulose can be attributed to the ions present in the ILs. Cellulose dissolution is more in the presence of more basic anions like acetate and chloride (Haykir et al., 2023). Kadokawa et al. (2021), in their study with imidazolium cation with a bulky alkyl chain and a basic carboxylate anion, were able to dissolve cellulose selectively with no dissolution of xylan. There are two conflicting views on the mechanism of dissolution. One theory suggests that the basic anion forms hydrogen bonds with hydroxyl groups of cellulose molecules, resulting in the formation of carbenes that destabilize cellulose structure and result in the dissolution of cellulose where the cation of the IL does not have any major role (Dissanayake et al., 2019; Qasim et al., 2021). Another theory suggests the active proton of the cation forms a hydrogen bond with the cellulose hydroxyl group. Accumulation of cations in this set-up results in cellulose dissolution (Lu et al., 2023) (Fig. 2).

Similarly, the selectivity of IL towards lignin dissolution can be increased by using acidic anions like hydrogen sulfate (HSO_4), methyl sulfate ($MeSO_4$) and

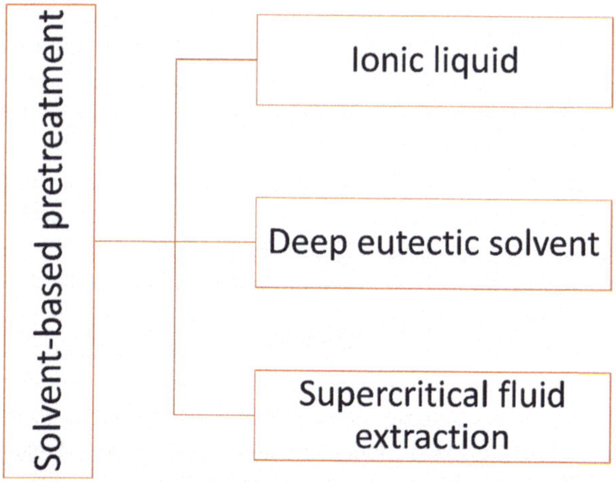

Fig. 2 Emerging solvent based lignocellulose pretreatment

trifluoromethane sulfonate (CF_3SO_3) (Amini et al., 2021). Increasing the length of the alkyl chain also results in higher interaction of the side chain with lignin, resulting in better solubility (Dias et al., 2020). In the case of lignin dissolution, the anion plays a major role, while the cation has a minor role. The anion interacts with the hydroxyl group from lignin while the alkyl side chain of the cation has shown dispersion interactions with lignin (Zubeltzu et al., 2020). Asim et al. observed an increase in lignin dissolution from wheat straw with protic acidic ionic liquids (PAIL) when a cluster of HSO_4 anions was used instead of a single anion with pyridinium cation (PyH). The lignin yield in [PyH][HSO_4], [PyH][HSO_4. (H_2SO_4)] and [PyH][HSO_4. (H_2SO_4)$_3$] was 22%, 29%, and 77%, respectively, after pretreatment for 2 h at 60 °C at 5% biomass loading of wheat straw (Asim et al., 2021). This supports the fact that increasing the basicity of the anion group enhances its hydrogen bonding forming ability with lignin, resulting in better lignin yields from biomass.

The major drawback of aprotic ionic liquid was that the preparation was expensive but protic ionic liquids are comparatively cheaper and easier to prepare. Though efficient in dissolution, the recovery and reuse of ILs are costly affairs. These ILs also tend to be toxic and non-biodegradable in some instances (Magina et al., 2021). The lack of enough scale-up studies also adds to the impediment to commercial usage.

4.2 Deep Eutectic Solvents

Like ILs, deep eutectic solvents (DESs) are also green solvents, and in recent studies, they have gained more attention than ILs as a suitable alternative solvent (Wang et al., 2021b). The discovery of a mixture of choline chloride with a range of amide compounds, which formed a liquid at room temperature and had unusual solvent properties by Abbot and coworkers. Since the melting point of choline chloride and urea in a 1:2 mole ratio resulted in a liquid with a deep drop in freezing point of the mixture to 12 °C, it was named as deep eutectic mixture, which over time became deep eutectic solvent (DES) (Abbott et al., 2003). DES are mixtures of hydrogen bond acceptors (HBA) like quaternary ammonium salts and hydrogen bond donors (HBD) like amides, amines, polyols, etc., which, when added in proper molar ratios, result in DES with low melting points (Zhou et al., 2024). The interaction of these HBDs with the HBAs reduces the anion-cation electrostatic force and results in the formation of hydrogen bonds. This causes charge delocalization leading to a significant decrease in the freezing point of DES (Ogawa & Mori, 2020). DESs share the same interesting properties as ILs with some additional advantages over traditional ILs, such as easy synthesis, preparation from comparatively cheaper substances, and easy availability (Afonso et al., 2022). An example of this can be seen in the following example. The most commonly used HBA-choline chloride is a component of vitamin B and is manufactured in bulk for cattle feed applications. Urea, commonly used as HBD, is also produced in bulk as a fertilizer (Hansen et al., 2021).

Concerning DES pretreatment of lignocellulose, the main focus of research has been to obtain cellulose-rich biomass by removing hemicellulose and lignin or to obtain lignin with minimal changes to the native structure. Acidic DES is preferred when the aim is to remove lignin and hemicellulose. It can break down lignin β-O-4 linkages with ease (Li et al., 2023). Alkaline lignin is known to break hydrogen bonds in lignin but not β-O-4 linkage limiting its utility in lignin depolymerization but better suited for cellulose dissolution (Wang et al., 2024b).

Wang et al., in their studies of the effect of DES on hydrothermal pretreated moso bamboo, found that the breakdown of the biomass was increased with increasing treatment temperature with DES. They also found that choline chloride-lactic acid gave better delignification (80.1%) compared to betaine-lactic acid (54.5%) when treated at 140 °C for 6 h with 5% solid loading (Wang et al., 2021c). This betterment in the delignification of choline chloride-lactic acid DES can be attributed to its lower viscosity compared to betaine-lactic acid. Sun et al., in their work, coupled acidic DES (choline chloride-*p*-toluene sulfonic acid) with ball milling to enhance lignin and hemicellulose removal and get high cellulose recovery. They used wheat straw, moso bamboo, poplar wood chips and pine wood shavings at 10% solid loading and milled at 380 rpm with ZrO_2 balls for 3–6 h. It was observed that lignin >95% and ~62% were removed from poplar wood chips and wheat straw, respectively, in 3 h, and an increase in treatment time did not result in any significant increase in hemicellulose or lignin removal. In all the fractions, the β-O-4 linkage of the lignin was preserved, which may lead to better value addition (Sun et al., 2023c).

Similar to ILs, DES also has limited data related to scale-up studies, which impedes its commercial utility. DES requires a relatively higher temperature than ILs for the pretreatment of biomass and recovery. Unlike ILs, DES also lacks in-depth research into the mechanism of lignocellulose pretreatment with different types of DES (Cai et al., 2023; Lobato-Rodríguez et al., 2023).

4.3 Supercritical Fluid Pretreatment

Supercritical fluids are fluids maintained above their critical temperature and pressure, due to which they acquire liquid-like densities and gas-like viscosities and diffusivities (O'Sullivan et al., 2022). This makes them excellent solvents with high mass transfers and high penetration into solid materials. Owing to their low operational cost and high energy efficiency, they are already gaining acceptance for many industrial applications (Mankar et al., 2021). Carbon dioxide, which has a low critical temperature (31.1 °C) and pressure (73.6 bar), has been widely studied for supercritical pretreatment of lignocellulosic biomass (Dharmaraja et al., 2023). Another widely studied supercritical fluid is supercritical water. Apart from CO_2 and water, the other solvents used in literature are ethanol, methanol, acetone, ammonia, butane, etc. (Badgujar et al., 2021). The selectivity and severity of supercritical fluid can be tuned by the addition of co-solvents into the system. Supercritical fluid pretreatment targets either lignin or sugar fraction as a whole. Since hemicellulose and cellulose have similar bonds, separation by this method is not studied.

Nlandu et al. observed ~100% increase in sugar yield from flax stem fibres when pretreated with supercritical CO_2 (SC-CO_2) compared to untreated flax stem fibres that were subjected to enzymatic hydrolysis. There was also a significant increase (>100%) in the total dissolved solids on treatment with SC-CO_2 (Nlandu et al., 2020). The increase can be attributed to the higher dissolving and penetration capacity of SC-CO_2. The sudden release of pressure also results in the physical breakdown of the lignocellulosic structure. Putrino et al. did not observe any significant increase in sugar yield from green coconut fibres on pretreatment with SC-CO_2 (Putrino et al., 2020). A plausible reason for this might be the amount of moisture present in the green coconut fibres. In a study comparing the effect of hot water pretreatment and a combination of hot water and SC-CO_2 on dried bamboo, Sasaki et al. found that the glucan decomposition increased from 14.9% to 25.7% on combined treatment compared to hot water treatment alone (Sasaki & Ohsawa, 2021). This increase was mainly because of the dissolution of CO_2 in water during pretreatment of resulting in carbonic acid. This leads to an acid hydrolysis effect on the biomass along with the physical effect on the biomass.

Nardella et al. tried extracting lignin monomers and dimers from oak and fir wood chips using SC-CO_2 at 60 °C at a flow rate of 2 mL/min, 350 bar pressure for 30 min with acetone, ethyl lactate and ethanol as co-solvents. They found higher yields of lignin monomers when ethyl lactate was used as cosolvent compared to ethanol and acetone. The addition of water to the mix increased the yield further lignin monomers (Nardella et al., 2023). This may be due to the chemical effects of dissolved CO_2. When pretreating corn cob with ethanol and water system, Du et al. found that the severity of the pretreatment increased with the addition of SC-CO_2. This resulted in enhanced lignin yields and some amount of furfural as well (Du et al., 2023). The increase in lignin yield is due to the improved permeability of solvents into the biomass on addition of SC-CO_2. This also leads to breakdown of lignin carbohydrate complex leading to release of sugars from the hemicellulose fraction. Yamazaki et al. tested the use of platinum-based metal oxide catalyst along with supercritical water pretreatment of Japanese cedar wood chips and found that supercritical water pretreatment with catalyst improved aromatic monomer yields with highest yield of 36% (Yamazaki et al., 2021).

Supercritical fluids aid in release of feedstock materials from lignocellulosic biomass. The requirement of high temperature and pressure for some supercritical fluid limits their economic scalability. The pretreatment conditions for each biomass has to be optimized to minimize the release of unwanted byproducts (Table 3).

4.4 High-Voltage Plasma

On supplying energy, matter changes its state—from solid to liquid, liquid to gas, and further, on higher energy supply, to plasma. Plasma is the fourth phase of matter, consisting of ions, atoms, electrons as well as metastable and excited molecules (Langmuir, 1928). It can be classified as thermal plasma or non-thermal plasma. In thermal plasma, all the components like ions, atoms, electrons and radicals are at high temperatures, being in thermal equilibrium with each other. On the contrary, components of non-thermal plasma state are not at thermal equilibrium. This plasma

Table 3 Recent studies on IL, DES, and supercritical fluid-based pretreatment

Sr. no.	Biomass	Solvent	Pretreatment condition	Results	References
1.	*Typha latifolia* stem and leaves	[Bmin][Cl]	5 g [Bmim][Cl]; 6% w/v solid loading; 1 h pretreatment at 120 °C	23% biomass dissolution, 214.14 mg sugar/g dry matter	Timoteo-Cruz et al. (2024)
2.	Rice husk/ wheat husk	[Emim][OAc]	5% solid loading; 100 °C; 3 h	Ethanol titre of 23.43 g/L and 24.28 g/L	John and Selvarajan (2024)
3.	Jerusalem artichoke stalk	EOAA	18 liquid: solid; 170 °C; 5 h	87% cellulose recovery; 75% hemicellulose recovery; 57% lignin removal; 15% increase in 2,3-butandiol production	An et al. (2022)
4.	Corn cob	$[C_2mim]$ [OAc]	15 liquid: solid; 60% IL; 6% ammonia; 2 h	Sugar yield of 0.659 g/g; ethanol yield of 0.296 g/g; 68% lignin removal; 96% IL recovery	Velmurugan et al. (2023)
5.	Sugarcane bagasse	[Emim][OAc]	6 solid: liquid; 80 °C for 24 h followed by 140 °C for 16 h after adding DMSO as cosolvent	>99% hemicellulose removal; 71% cellulose recovered; cellulose mixed esters obtained with high molecular weight	Hernandez et al. (2024)
6.	High-yield pulp from bleached eucalyptus	Choline chloride-ethylene glycol	70% solid loading; 95 °C; 6 h	Lignocellulose nanofibrils with size ranging from 42.72 to 22.91 μm	Fu et al. (2022)
7.	Poplar wood	choline chloride, glycerol, $AlCl_3 \cdot 6H_2O$	15 liquid: solid; 140 °C; 1 h	DES-based lignin, polyacrylic acid polyvinyl alcohol hydrogel network	Yan et al. (2023a)
8.	Rice straw	Choline chloride-lactic acid	20 liquid: solid ratio; 130 °C; 3 h	Lignin nanosphere and lignin-containing cellulose nanofibres	Shen et al. (2023)
9.	Moso bamboo	Choline chloride-lactic acid	10% solid loading; 120 °C; 6 h	0.271 g glucose/g biomass; lignocellulose nanofibres with high Youngs modulus	Asada et al. (2023)

(continued)

Table 3 (continued)

Sr. no.	Biomass	Solvent	Pretreatment condition	Results	References
10.	Rice straw	Choline chloride-oxalic acid-lactic acid	30 liquid: solid; 680 W microwave; 4 min	63.4% lignin removal; lignin nanoparticles (size range: 48–95 nm)	Yan et al. (2023b)
11.	Cotton stalk	Supercritical CO_2	100 bar; 180 °C; 140 min; organosolv	20% increase in methane yield; 10 days reduction in optimal digestion time	Al Afif et al. (2020)
12.	Teak wood	Supercritical ethanol	300 °C; 30 min; 5 MPa; catalyst ($MgClO_4$, $HClO_4$, $MgClO_4 + HClO_4$)	Bio-oil yield increased from 41% for untreated teakwood to 58.2% for treated teakwood	Alper et al. (2021)
13.	Japanese cedar wood flour	Supercritical methanol	150 mg biomass; 270 °C; 27 MPa; 30 min; BSA/Soy protein as catalyst	Enhanced delignification observed	Yao et al. (2022)
14.	Oil palm trunk	Supercritical CO_2	100 g biomass; 80–120 °C; 60 min; 25–45 MPa; 0–60 °C Moisture	Increase in sugar release by enzyme hydrolysis by ~2 folds	Sohni et al. (2020)
15.	Sorghum stalk	Supercritical CO_2	0.5 g biomass; 60 °C; 20 MPa; 30 h	43.57% sugar yield from enzyme hydrolysis compared to 16.8% of untreated biomass	Zhang et al. (2019)

state is achieved at much lower temperatures and atmospheric pressure under high voltage. On application of high voltage, electrons accelerate and attain high temperatures. The electrons absorb energy just like in thermal plasma but are unable to transfer that energy to other heavy species due to a lesser number of collisions due to low pressure. Only the electrons, therefore, remain at a higher temperature, while the rest of the plasma stays cold, giving it the name "cold plasma". The excited electrons on bombarding with other neutral species lead to ionization and/or dissociation, forming highly reactive species like hydronium ions or free radicals (Meoli et al., 2023). These reactive species are capable of oxidative delignification by breaking down the complex polymer structure and exposing the cellulose fibres for enzymatic hydrolysis. They also have a role in the breakdown of hydrogen bonds in cellulose crystals, causing a reduction in crystallinity. In other words, the ionized gas alters the chemical structure of biomass and the surface properties like ablation, changes in functional groups, polymerization, and cross-linking (Shaghaleh et al., 2019).

In research by Jiang et al. for pretreatment of water hyacinth by plasma electrolysis, the formation of large amounts of active molecules like HO· and H_2O_2 caused lignin decomposition and reduction in crystalline index, along with a 126.5% increase in enzymatic hydrolysis and sugar yield (Jiang et al., 2014). In another study seeking improved sugar yield from sugarcane bagasse, it was revealed that plasma-in-liquid in a pH 12 buffer could increase glucose and xylose yields to 51.8% and 38.5%, respectively, as compared to only 19.5% and 10.3% without plasma pretreatment. Lignin solubilization was also shown to increase up to 58% (Miranda et al., 2020). Around 80% of 5-HMF and 100% of furfurals were removed from sugarcane bagasse by treatment with argon plasma, along with at least a 30% reduction in formic and acetic acid (Lin et al., 2020). Lignin removal of 68% and 36%, respectively, were achieved by Ramamoorthy et al. (2020) and Ravindran et al. (2019). Souza-Corrêa et al. (2013) showed a significant 80% lignin removal with 65% glucose yield from sugarcane bagasse. In total, 84% and an over two-fold increase in sugar yield from lignocellulosic biomass were shown by Shaghaleh et al. (2019) and Ravindran et al. (2019), respectively. Heiske et al. (2013) depicted a 45% increase in methane yield from plasma-assisted pretreated wheat straw. Zhu et al. (2022) demonstrated an increase in the solubility of cold plasma-treated cellulose compared to untreated cellulose from sugarcane bagasse pulp. This increase in solubility can be attributed to the breakage of glycosidic linkages with cold plasma treatment.

Non-thermal plasma can be obtained at mild operating conditions, without requiring exceedingly high temperatures and pressure. It does not need a vacuum chamber or ionizing gas, significantly reducing the cost of operation and maintenance. Since there is no thermal dissociation, heat-related degradation of compounds, necessity of cooling, etc., are ruled out. The method needs short processing time and does not generate any waste (Meoli et al., 2023; Sidana & Yadav, 2022). This process requires high energy input to generate and sustain cold plasma, leading to higher operation expenditure. The production of free radical species and ions generated during cold plasma treatment may lead to unwanted degradation (Abolore et al., 2024) (Fig. 3).

4.5 Microwave-Assisted Pretreatment

Microwaves use non-ionizing electromagnetic radiation to heat the substrate. The frequency of radiation used in Microwave is between 300 and 300,000 MHz, with wavelength 1 mm to 1 m (Wei et al., 2019). First introduced by Ooshima et al. (1984), microwave is now extensively being studied for efficient pretreatment of lignocellulosic biomass for the production of valuable compounds. It is particularly reported to be efficient for lignin removal and reducing cellulose crystallinity, with numerous benefits over conventional heating (Gao et al., 2021). The microwave converts the energy in electromagnetic radiation to heat energy and supplies it uniformly throughout the lignocellulosic biomass. The strength is quantified in terms of power, which usually ranges from 60 to 600 W, for lignocellulose pretreatment.

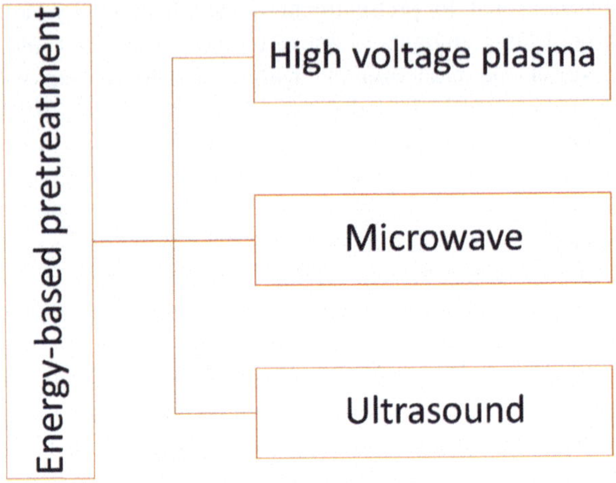

Fig. 3 Emerging energy-based pretreatment of lignocellulose

The raise in temperature is directly proportional to the power and can easily reach 800 °C in less than 30 min of treatment time (Konan et al., 2024). The pretreatment efficiency depends on water content, treatment time and temperature. Microwave pretreatment mainly works on the dielectric properties of water. Vibration of the water molecules on absorption of microwave leads to the heat transfer and subsequent degradation of lignocellulose complex (Priecel & Lopez-Sanchez, 2019). Polar solvents can also be added during the treatment to improve the effect, owing to their high dielectric property (Wang et al., 2024a). Microwave assistance in pretreatment has been reported by numerous researchers to be one of the most efficient methods for lignin extraction (Andhalkar et al., 2024; Sun et al., 2019a, 2019b). Wang et al. used microwave-pretreated corn straw for methane production and found the methane yields improved by 73.08% compared to untreated corn straw (Wang et al., 2021a). Dai and coworkers showed that the acid content of pyrolysis oil from microwave-pretreated bamboo sawdust was lower and hence of better quality (Dai et al., 2017). Microwave-assisted alkali pretreatment of wheat straw was able to remove 80% of hemicellulose and over 90% of lignin without over-degradation of saccharides or much solubilization of cellulose. The researchers also reported that lignin solubilization was highly dependent on the microwave energy input, over alkali pretreatment (Janker-Obermeier et al., 2012). Hu and Wen compared conventional heating and microwave heating for the pretreatment of switchgrass. They found high degradation of lignin structure treated with microwave energy, whereas conventionally pretreated biomass showed intact fibres. They also reported enhanced enzymatic saccharification of the microwave-pretreated switchgrass (Hu & Wen, 2008). Microwave heating paired with treatment with sulfuric acid, for the production of levulinic acid from lignocellulosic biomass took only 30 min, as compared to 8 h required by conventional heating with sulfuric acid treatment to produce the same results (Tukacs et al., 2017). When compared to conventional heating along with acidic treatment, Zhu et al. performed microwave pretreatment of wheat straw and rice straw in a domestic microwave system and reported it to be effective for further enzymatic hydrolysis or fermentation (Zhu et al., 2005, 2006).

Microwave-assisted pretreatment has many advantages like easy operation, high energy efficiency, high temperature achievability, minimum generation of inhibitory compounds, fast heat transfer and low reaction time (Gao et al., 2021; Nauman Aftab et al., 2019; Ooshima et al., 1984). It offers a controlled energy input to the biomass. However, data on the scalability of microwave pretreatment for large-scale applications is limited. Besides possible high energy requirement, the effectiveness of the microwave-assisted pretreatment method at a large scale could be questionable, as at massive biomass quality, microwave might not have deep and uniform penetration, causing nonuniformity in heating and, thus degradation (Gao et al., 2021; Konan et al., 2024).

4.6 Ultrasound-Assisted Pretreatment

The basic principle on which ultrasound pretreatment works is the cavitation phenomenon. Cavitation bubbles are formed when there is a change in the medium pressure due to the pulsating high-frequency ultrasonic waves (20 KHz–500 MHz) entering in low pressure zones (Yusaf & Al-Juboori, 2014). These bubbles containing liquid vapour and dissolved gases, penetrate within the calcitrant biomass structures, grow and break with repeated compression rarefaction cycles, or shear flows releasing high-energy shock waves in the vicinity of the bubbles. This generates localized forces that cleave the lignocellulose structure, making them more accessible and susceptible to further processing (Ravindran & Jaiswal, 2016; Zieliński et al., 2019). Apart from the mechanical force, the high temperature and pressure produced due to the breakage of the cavitation bubble lead to the dissociation of water molecules generating hydroxyl (OH^-) and superoxide (O_2^-) radicals that help in delignification and breakdown of lignocellulose (Terán Hilares et al., 2018). The pretreatment efficiency depends on parameters like amplitude, temperature, power and duration.

A study by Byun et al. on pretreatment of a variety of silver grass attained a 6.7% increase in total sugar release, decrease in chemical usage up to 26.1% and 17.0% increase in lignin removal, as compared to pretreatment without ultrasonication. The researchers revealed decrease in the requirement of chemicals for further processing as well as improved carbohydrate digestibility after ultrasound treatment (Byun et al., 2020). In a study by Wu et al. (2017), ultrasound irradiation with the frequency of 22 kHz and power of 300 W for 1 h was shown to increase the porosity and surface area of rice straw, resulting in more efficient enzymatic hydrolysis. Nakayama and Imai (2013) examined ultrasound-assisted pretreatment in kenaf for improved enzymatic hydrolysis, revealing a decreased Michaelis constant and increased enzyme adsorption on the biomass after ultra sound treatment, leading to an increased initial reaction rate. Shi et al. (2013) achieved a 27.5% increase in the hydrolysis efficiency of biomass with only 120 s ultrasonication. Jeon and coworkers evidenced improved bio-accessibility of microalgal biomass (*Scenedesmus obliquus*) by the fermentative bacteria for the production of bioenergy. Increasing the duration of sonication led to the rupture of microalgal biomass, with increased surface hydrophilicity, and static repulsion between the algal debris, leading to the enhanced accessibility to the inner spaces, as confirmed by SEM images (Jeon et al., 2013). de Farias Silva et al. (2020), also working on the same microalgal biomass, achieved a hydrolysis yield of more than 90% in a 40-min sonication time. Subhedar et al. (2018), in their study, showed at least an 80% increase in the delignification of coconut coir, pistachio shells and groundnut shells by ultrasound-assisted alkali treatment compared to only alkali pretreatment. They also showed an almost tenfold reduction in reaction time in ultrasound-assisted enzymatic hydrolysis of these lignocellulosic biomass. In total, 70% delignification paired with approximately 85% hydrolysis yield from wheat straw was obtained by coupling alkali pretreatment with ultrasonication (Saratale et al., 2020). Multi-response optimization of alkali pretreatment and ultrasonication operational conditions produced 82% and 86% of hemicellulose and lignin removal, respectively, to maximize cellulose yield for ethanol production (Brito et al., 2019). On similar account, approximately 95% cellulose was recovered from sugarcane bagasse by Ramadoss and Muthukumar (2014, 2016).

Ultrasound irradiation technique is advantageous because of its high energy efficiency at shorter processing time, mild reaction temperatures as well as reduced requirement of chemical/enzyme dosage for hydrolysis (Kassim et al., 2022). However, research into these techniques is still limited and needs to be studied further for industrial applications (Table 4).

4.7 Advanced Oxidation Processes

Traditionally, advanced oxidation processes (AOPs) were used to degrade harmful organic chemicals by total oxidation. Most of the AOPs involve reactive free radical generation, predominantly hydroxyl radicals (OH) (Rekhate & Srivastava, 2020). Recent years have seen its application in a wide variety of fields including lignocellulose breakdown, biomedical, chemical synthesis, etc. It includes methods like photocatalysis, ozonolysis, the Fenton process, hydrogen peroxide oxidation, ultraviolet irradiation, electrochemical oxidation, etc. (M'Arimi et al., 2020).

Among the AOP, the photocatalytic process (light and catalyst) has been widely studied for its ability to complete the mineralization of organic substrates at ambient conditions, "Photocatalysis" process was described employing chemical reactions accelerated by photo-induced electrons/holes, in the presence and absence of a catalyst under electromagnetic irradiation (Colmenares et al., 2009; Li et al., 2016).

Table 4 Recent studies on cold plasma, microwave irradiation and ultrasonication pretreatment methods

Sr. no.	Biomass	Pretreatment	Pretreatment condition	Results	References
1.	Maize	Cold plasma	20 g biomass in 390 mL water; 21 kHz; 16 kV; 2 h	18% increase in biogas yield	Wright et al. (2020)
2.	Spent coffee waste	Cold plasma	70 kV; 1% H_2SO_4; 2 min	Two fold increase reducing sugar yield	Ravindran et al. (2017)
3.	Brewery spent grain	Cold plasma	12.5% solid loading; 5 min	50% delignification	Pereira et al. (2021)
4.	Surgical waste cotton and waste packaging cardboard	Cold plasma	100 g biomass; 3–4.5 kV; 2–12 kHz; 15 min	68% lignin removal; 22% hemicellulose removal	Ramamoorthy et al. (2020)
5.	Rice and corn bran dietary fibres	Cold plasma	220–260 V (connected to 60 kV transformer); 15–30 min; 50 Hz	Enhances release of xylooligosaccharides compared to untreated samples	Mehta et al. (2022)
6.	Pine wood chips, beechwood chips and wheat straw	Microwave	5.3% solid loading; 600 W; 10–40% sodium cumene sulfonate; 10–60 min pretreatment time; pressure 117 psi	55%, 45% and 35% delignification for wheat straw, beech and pine respectively	Mikulski and Kłosowski (2022)
7.	Pine wood chips, beechwood chips and wheat straw	Microwave	6% solid loading; 40% sodium cumene sulfonate solution; 600 W; 117 psi; 60 min	Higher glucose release and subsequent bioethanol production	Mikulski and Kłosowski (2023)
8.	Wheat and rye stillage	Microwave	5% solid loading; 300 W; 0.2 M H_2SO_4; 15 min; 54 psi	>75% cellulose recovery for wheat stillage; >89% cellulose recovery for rye stillage	Mikulski and Kłosowski (2020)
9.	Red pine	Microwave	5% solid loading; choline chloride-lactic acid DES as solvent; 110 s; 400 W	Lignocellulose nanofibril preparation; 70% delignification; 90% hemicellulose removal	Kwon et al. (2023)

(continued)

Table 4 (continued)

Sr. no.	Biomass	Pretreatment	Pretreatment condition	Results	References
10.	Bamboo and corn cob	Microwave	10% solid loading; 14% Na_3PO_4, $12H_2O$ (for corn cob); 18% NaOH (for bamboo); 600 W; 6–8 min	0.512 g/g corn cob and 0.225 g/g bamboo sugar yields	Laltha et al. (2021)
11.	Cotton straw	Ultrasound	10% solid loading; Choline chloride-formic acid DES as solvent; 32 kHz; 30 min sonication and 3 h heating	87.7% lignin yield with 97.3% purity	Xu et al. (2023)
12.	Rice straw	Ultrasound	5% solid loading; [Amin][Cl] and [Bmim][Cl] IL-based solvent with HCl; 40 kHz; 30–240 min; 30–70 °C	Enhanced reducing sugar yield compared to untreated samples	Zhang et al. (2021)
13.	Sunflower stalks	Ultrasound	5% solid loading of ball-milled sunflower stalks in the presence of acid; 450 W; 2 h	Lignocellulose nanofibrils with high-temperature stability were produced	Ewulonu et al. (2019)
14.	Bamboo	Ultrasound	10% solid loading; acidic ethanol-water mixture; 180 °C; 55 min heating followed by 30 min ultrasonication	~66% lignin yield with >95% purity	Ji et al. (2021)
15.	Sugarcane bagasse	Ultrasound	10% solid loading; ethanol solvent; 240 W; 60 min; followed by choline chloride-glycerol-$FeCl_3$,$6H_2O$ TDES treatment for 3 h	90.31% saccharification yield; cellulose yield of 66.17% and 86.39% delignification	

Similarly, the photocatalytic lignin conversion generally includes the depolymerization or functional group transformation of lignin. Thus, lignin depolymerization by photocatalytic methods mainly aimed to cleave β-O-4 linkages (Chen et al., 2021). Primarily, there are three ways for C–O and β-O-4 linkage cleavage by photocatalysis: First by reductive cleavage begun by electrons or by reducing agents; second by oxidative cleavage initiated by holes or oxidative agents and finally by redox-neutral cleavage by electrons combined with holes/oxidative species (Xiang et al., 2020). Overall, the conversion/degradation of lignin by photocatalysis using sustainable energy indicates a novel approach. The lignin β-O-4 linkages cleavage is predominant amid the aromatic centre of lignin photocatalyzed by oxidative, reductive and redox-neutral pathways under mild conditions (Chen et al., 2021).

Ozonolysis is another AOP where ozone is used as a strong oxidizing agent for breaking lignin into smaller molecules. This pretreatment generally affects the lignin fraction and hardly affects the sugar fractions in their short treatment time (Benito-González et al., 2023). This process has minimal inhibitory byproduct formation and can be done at normal room temperature and pressure. The main bottleneck is the generation of large amounts of ozone required for the process, which requires specialized equipment and other safety considerations related to harmful ozone byproducts (Di Fraia et al., 2024).

The Fenton process is a widely studied AOP process, which has been looked at for lignocellulose breakdown as well. Fenton reagent, a mixture of ferrous ions and hydrogen peroxide, is used for the generation of free radicals which goes on to break the bonds in lignocellulose. The relative amount of hydrogen peroxide and ferrous ions in the reagent plays a major role in the LCB degradation efficiency (Wang et al., 2023a). Excess of ferrous ions leads to fast generation of free radicals, which react among themselves to form water. This leads to the wastage of oxidation power. The Fenton process is an easy-to-operate, green and sustainable pretreatment process that has potential to be used in biorefinery applications (Wang et al., 2022). This process suffers from drawbacks like long treatment time, acidic pH range and production of an iron-based sludge (Wang et al., 2021d).

Similar to the Fenton process, hydrogen peroxide in alkaline conditions splits to form hydronium and hydroxyl radicals that can aid in lignocellulose breakdown by breaking the C–C bonds in the lignin structure (Xia et al., 2022). A hydrogen peroxide-acetic acid solvent system can delignify lignocellulosic biomass >80% at lower temperatures than conventional chemical pretreatment (Wen et al., 2022). This high efficiency breakdown is due to formation of peracetic acid which is a very strong oxidant. Hydrogen peroxide recycle is a big problem in this process and leads to increased cost. Also, when some acids when used in tandem with hydrogen peroxide tend to turn into turn into explosive mixtures (Zhou et al., 2023b).

Some of the other AOP tested for lignocellulose pretreatment include wet oxidation, electrochemical oxidation and UV irradiation pretreatment (Fig. 4).

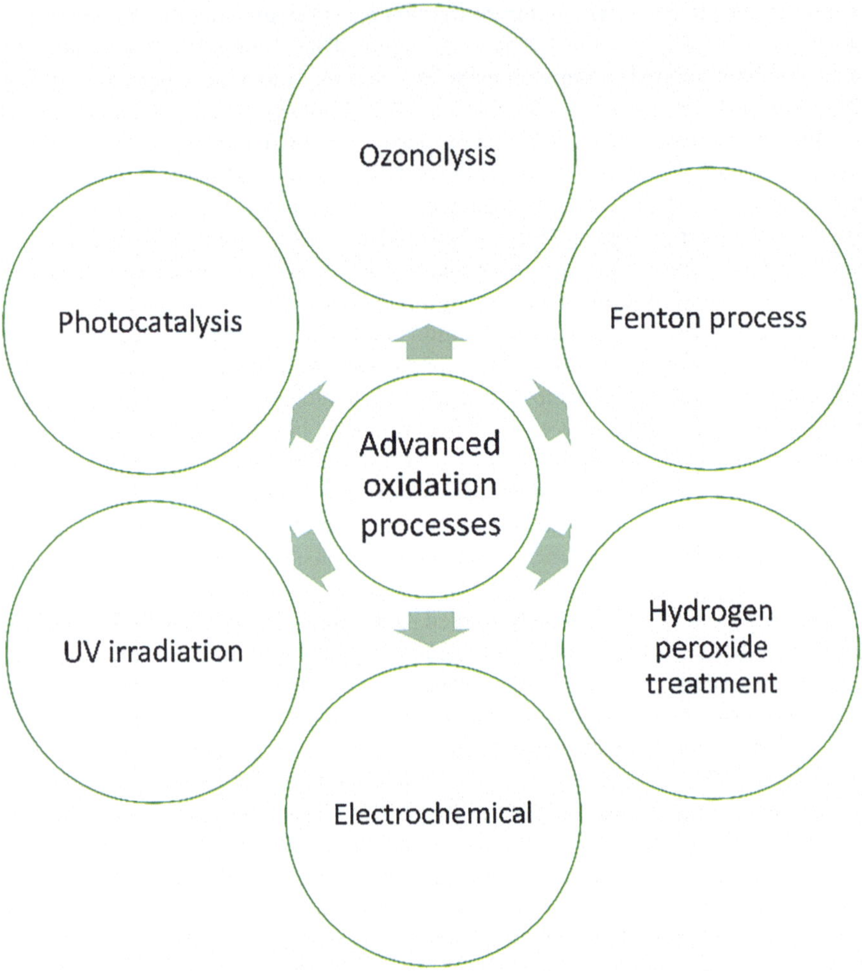

Fig. 4 Different methods in AOP for lignocellulose pretreatment

5 Lignocellulosic Biomass to Single Product-Based Process Technology

Lignocellulose, such as forestry and agricultural residues, represents the most abundant natural biomass in the natural world with an approximate yield of about 200 billion tons. These consist of non-edible parts of plants, ruling out competition with food demand. The major chemical components of lignocellulosic biomass are cellulose, hemicelluloses and lignin, all of which have shown great promise as feedstocks and materials for energy generation. Utilizing the waste biomass abundantly present in the environment offers a promising alternative to the petrochemical-based processes. These are environment-friendly, inexhaustible, cheap and readily available resources and offer a potent way to a greener and sustainable living.

5.1 Biofuel

Bio-oil is the dark brown viscous liquid fraction obtained after pyrolysis, supercritical water oxidation or hydrothermal liquefaction of lignocellulosic biomass (Baloch et al., 2018). It is known to contain over 400 compounds, including acids, alcohols, aldehydes, aromatics, ketones, phenols, esters and sugars. Bio-oil has minimal nitrogen and sulfur content, minimizing nitrogen and sulfur-based compounds in emissions when bio-oil is used as fuel. Although it has been shown to work as a substitute for petroleum-based fuels, it has certain deficiencies due to high moisture content, high oxygen content, high ash content, high viscosity and thermal instability (Baloch et al., 2018). Bio-oil composition varies with variation in the lignocellulosic feedstock used and the process parameters. Due to these reasons, refining bio-oil for extracting value-added products is widely researched.

Hydrogen is a highly combustible, colourless and odourless gas. It is the most abundant element and is a promising alternative to carbon-based fuels. It is a clean fuel that produces water on combustion. Hydrogen finds many applications in heavy oil upgrading, ammonia production, fat hydrogenation, fuel cells, etc. (Okolie et al., 2021). Through the biological route, hydrogen can be made through the gasification of lignocellulose or the reforming of bio-oil. Although much research has been carried out for biohydrogen production, none of the technologies has competitively scaled-up commercial applications. Current market demands are met by reforming natural gas and electrolysis of water (Okolie et al., 2021).

Bioethanol is another biofuel used in blending of fuel and as a fuel itself. Bioethanol has been commercially produced from lignocellulose (Zhou et al., 2021). Biomethane and biogas are studied extensively and are also deployed in the field for trials.

5.2 Other Value-Added Products

Lignin can be broken down to its units containing compounds like toluene, benzene and xylene. Large-scale production and application of these aromatic compounds provide lignin a great potential for its involvement in chemical industry. These aromatics can either be high-value biochemical like vanillin, which finds application as a flavouring agent, or a platform chemical that can be converted into high-value biochemical (Schutyser et al., 2018; Xu et al., 2019). These platform chemicals can be converted to marketable products either via chemical or biological route. The pretreatment methods used on lignocellulose usually release a mixture of compounds. The purification of compounds from this mixture increases the cost of the final product. There is a need for lowering the cost by increasing the yield of the target compound. The method to convert the mixture of aromatics into a target compound is known as funnelling (Schutyser et al., 2018), and research on the funnelling of lignin into target aromatics is still in its infancy. It needs to be studied further in terms of the range of applicability, scale-up performance, operational parameters, etc.

Vanillin is one of the most important aromatic compounds, which is used in food, pharmaceuticals, cosmetics and some other industries (Olatunde et al., 2022). Although the production of vanillin from crude oil through chemical synthesis make-up 80% of the production market but its biosynthesis from lignin as a feedstock is a potential and clean approach.

Polyhydroxyalkanoates (PHAs) are polyesters, which are synthesized by various microorganisms as carbon and energy storage materials under the nutrient imbalanced conditions. PHAs are widely used in bioplastics, biomedicine and nanotechnology for their excellent biodegradability and biocompatibility (de Souza & Gupta, 2024). Various bacteria have capability to convert PHAs from lignin by developing their metabolic pathways. Various lignin pretreatment methods have also been investigated to improve the lignin utilization capability of related bacteria for the conversion of lignin to PHA. Lignin pretreated with H_2SO_4 and NaOH showed higher PHA production with *P. putida* KT2440 (Xu et al., 2021).

Lignin nanoparticles (LNPs) with a size less than 100 nm exhibit biocompatibility, antimicrobial, anticancer, antioxidant and UV-interfering properties (Rizal et al., 2021; Wijaya et al., 2021). LNPs are being studied for potential applications in UV protective coatings, antimicrobial agents, anticorrosive agents, drug carriers, food packaging, heavy metal adsorbents, biomaterials, etc. Wu et al. (2022a) reported polyurethane film incorporated with LNPs had good antimicrobial and anti-ultraviolet properties and better hydrophobicity and hydrothermal stability. LNPs are endowed with antioxidant and photoprotective properties due to the presence of the phenolic hydroxyl group (Hussin et al., 2022).

Apart from these products, the sugar fraction has been used in fermentation for a variety of products including sugar alcohols (xylitol, arabitol, sorbitol, mannitol), organic acids (acetic acid, citric acid, succinic acid, maleic acid), furfurals (furfural and 5-hydroxymethylfurfural), etc. The fibres of lignin and cellulose have been used in the preparation of biocomposites, blending in textiles and carbon fibre production. They have also been used as adsorbents in the removal of heavy metals dyes, etc.

6 What Is Biorefinery?

Analogous to petroleum refineries, which produce multiple products from crude oil, biorefinery integrates biomass conversion processes to produce fuel or value-added products (Ubando et al., 2022). It is a sustainable step to reduce the dependency on petrochemicals, with maximum value extraction from waste biomass resources (Ahmed et al., 2023). Besides the sustainability benefits obtained by waste utilization, biorefinery offers economic viability. It often integrates renewable energy and

Table 5 Recent studies on biorefinery approach towards lignocellulose valorization

Sr. no.	Biomass	Pretreatment processes	Products targeted	References
1.	Brewers spent grain	Sequential alkali hydrolysis, bleach-alkali pretreatment, concentrated acid treatment	Lignin-nanocellulose and hemicellulose nanocellulose biocomposite	Qazanfarzadeh et al. (2024)
2.	Poplar wood	One pot ternary DES (choline chloride-bio acid-1,4-butandiol)	Furfural, lignin nanosphere and lignin-containing cellulose nanosphere	Pan et al. (2024)
3.	Corn cob	Sequential acid, alkali, enzyme hydrolysis	Ethanol, Xylose, phenol based adhesive	Pang et al. (2021)
4.	Wheat straw	Sequential subcritical water pretreatment, enzyme hydrolysis	Ethanol, biogas, organic fertilizer	Chen et al. (2022)
5.	Poplar wood	Acidic DES (choline chloride-oxalic acid)	Bioplastic	Zhou et al. (2023a)
6.	Rice straw	Sequential hydrothermal pretreatment, DES (choline chloride-lactic acid)	Activated nanocarbon, lignin nanosphere and lignin-containing cellulose nanosphere	Shen et al. (2023)
7.	Wheat straw	Sequential dilute acid pretreatment, enzyme hydrolysis	Ethanol, xylose, reactive lignin	Gu et al. (2024)
8.	Beechwood	Sequential reductive catalytic fractionation, enzyme hydrolysis	Lignin oil, sugars, carbohydrate pulp	Cooreman et al. (2023)
9.	Beechwood	Organosolv treatment, bleaching, Hydrothermal pretreatment	Base catalyzed lignin, malic acid and dissolving pulp	Nitzsche et al. (2020)
10.	Spruce sawdust	Acidic thiourea treatment, hydrogen peroxide bleaching, enzyme hydrolysis	High molecular weight lignin, carbohydrate pulp	Sirviö et al. (2023)

employs environment-friendly processes, contributing to a reduction in greenhouse gas emissions (Sidiras et al., 2022). In contrast to single-product systems, biorefineries combine various conversion technologies, allowing energy and material integration, and improving the overall process efficiency and operational expenses. It fosters the principle of circular economy, where waste is minimized, and is valorized to multiple products (Rabbi & Amin, 2024). The literature on recent biorefinery approaches towards lignocellulose pretreatment are shown in Table 5.

Pang et al in their work developed a corn cob-based biorefinery with bioethanol from the cellulose fraction, xylose from the hemicellulose fraction and a phenol formaldehyde-based adhesive from the lignin fraction (Pang et al., 2021). A conventional three-step dilute acid/dilute alkali/enzyme pretreatment was employed in this process. They used dilute acid pretreatment for recovering xylose fraction from corn cob. The biomass was then subjected to hydrolysis and got lignin in the liquid fraction while the cellulose-rich biomass was hydrolyzed and fermented for bioethanol production. The process resulted in a 70% utilization of carbon and the life cycle analysis assessment of their proposed biorefinery had lower greenhouse gas emission intensity per USD capital when compared to current manufacturing processes employed. Similar biorefinery applications with conventional pretreatments have been studied in plenty. Few of the recent literature is given in Table 5.

7 Integration of Emerging Advanced Processes for Biorefinery Applications

To make a difference in the biorefinery approach to lignocellulose valorization, the application of the emerging pretreatments has to be integrated into biorefinery applications to increase the utility of the resource and reduce waste generation. Many studies in this regard have been done in recent years.

Zhang et al. tested the efficacy of phosphoric acid/hydrogen peroxide (PHP), ammonia/hydrogen peroxide (AHP) and DES (choline chloride-ethanolamine) on licorice residue (Zhang et al., 2023). PHP 81.7% lignin was removed while AHP pretreatment resulted in 97.2% sugar recovery. DES pretreatment resulted in the highest sugar yield on enzyme treatment due to better accessibility for the enzyme to sugar fractions compared to the other two treatments. A biorefinery approach to utilization of Chinese hickory shell using AHP pretreatment was proposed by Wang et al. (2023b). The AHP pretreatment resulted in the removal of hemicellulose and lignin fractions in the liquid stream while the solid residue was cellulose-rich. The lignin from the liquid stream could be recovered by acidification of the liquid precipitating lignin for further value addition. The cellulose stream could be used for saccharification and fermentation while the liquid stream after recovery of lignin will have organic acids.

Shen et al. (2023) combined conventional hydrothermal pretreatment with DES pretreatment to produce activated nanocarbon, lignin nanosphere and lignin-containing cellulose nanofibre from rice straw. Some of the recent studies with this approach are given in Table 5.

8 Shortcomings in Current Research

The shortcomings in current research can be attributed to the requirement of an ideal pretreatment technique for lignocellulose biomass pretreatment. The properties of an ideal pretreatment process are as follows: low cost, maximum efficiency, mild pretreatment conditions, low energy requirement, environment-friendly, sustainable, low inhibitor/byproduct formation, selective fractionation, easy scale up, and reusable.

The one–shoe-fits-all approach for lignocellulose pretreatment does not work because of wide variations in the structure of lignocellulose based on the geographic location of the source plant, species of the plant, growth age and growth environment. This makes the selection of a pretreatment method very important to get the maximum recovery of resources from the biomass.

Many of the emerging pretreatment methods lack data for effectiveness on a range of biomass. Similarly, data regarding scale-up is also limited for the recently trending pretreatment processes. The process must be designed such that it can be easily scaled up at low cost. Recycle operations of the catalysts/solvent involved is another important aspect that requires the attention of the researchers. Another area that requires research efforts is the ability to pretreat high solid of biomass with minimal chemicals, additives, etc.

9 Conclusion

There is a high demand and need for sustainable, eco-friendly source to meet the energy demands and replace the existing platform chemicals produced from petroleum-based feedstocks. However, this is still a work in progress due to many hurdles in the processing of lignocellulose. The major hurdle in lignocellulose processing is the pretreatment step, which costs around 30% of the total cost. Pretreatment methods have to be carefully selected based on the lignocellulosic biomass characteristics to be treated and downstream application of the pretreated biomass. Future research is needed to design a low energy consuming, low cost, sustainable, low-toxic byproduct-producing pretreatment process. New emerging pretreatment methods require comparative efficiency data with conventional methods, which in turn requires in-depth research. A combination of pretreatment methods must be investigated to reduce the use of chemicals and increase the efficiency of the process. In this context, the emerging pretreatment methods mentioned in this chapter can be cost-effective, environment-friendly and sustainable alternatives to the conventional pretreatments for efficient biomass utilization in a biorefinery.

References

Abbott, A. P., Capper, G., Davies, D. L., Rasheed, R. K., & Tambyrajah, V. (2003). Novel solvent properties of choline chloride/urea mixtures. *Chemical Communications, 1*, 70–71. https://doi.org/10.1039/b210714g

Abolore, R. S., Jaiswal, S., & Jaiswal, A. K. (2024). Green and sustainable pretreatment methods for cellulose extraction from lignocellulosic biomass and its applications: A review. *Carbohydrate Polymer Technologies and Applications, 7*, 100396. https://doi.org/10.1016/j.carpta.2023.100396

Afonso, J., Mezzetta, A., Marrucho, I. M., & Guazzelli, L. (2022). History repeats itself again: Will the mistakes of the past for ILs be repeated for DESs? From being considered ionic liquids to becoming their alternative: The unbalanced turn of deep eutectic solvents. *Green Chemistry, 25*(1), 59–105. https://doi.org/10.1039/d2gc03198a

Ahmad, N., Aslam, S., Hussain, N., Bilal, M., & Iqbal, H. M. N. (2023). Transforming lignin biomass to value: Interplay between ligninolytic enzymes and lignocellulose depolymerization. *Bioenergy Research, 16*(3), 1246–1263. https://doi.org/10.1007/s12155-022-10541-y

Ahmed, S. F., Kabir, M., Mehjabin, A., Oishi, F. T. Z., Ahmed, S., Mannan, S., Mofijur, M., Almomani, F., Badruddin, I. A., & Kamangar, S. (2023). Waste biorefinery to produce renewable energy: Bioconversion process and circular bioeconomy. *Energy Reports, 10*, 3073–3091. https://doi.org/10.1016/j.egyr.2023.09.137

Ajayi, V. A., & Lateef, A. (2023). Biotechnological valorization of agrowastes for circular bioeconomy: Melon seed shell, groundnut shell and groundnut peel. *Cleaner and Circular Bioeconomy, 4*, 100039. https://doi.org/10.1016/j.clcb.2023.100039

Akhtar, N., Gupta, K., Goyal, D., & Goyal, A. (2016). Recent advances in pretreatment technologies for efficient hydrolysis of lignocellulosic biomass. *Environmental Progress & Sustainable Energy, 35*(2), 489–511. https://doi.org/10.1002/ep.12257

Al Afif, R., Wendland, M., Amon, T., & Pfeifer, C. (2020). Supercritical carbon dioxide enhanced pre-treatment of cotton stalks for methane production. *Energy, 194*, 116903. https://doi.org/10.1016/j.energy.2020.116903

Alemán-Nava, G. S., Gatti, I. A., Parra-Saldivar, R., Dallemand, J. F., Rittmann, B. E., & Iqbal, H. M. N. (2018). Biotechnological revalorization of Tequila waste and by-product streams for cleaner production—A review from bio-refinery perspective. *Journal of Cleaner Production, 172*, 3713–3720. https://doi.org/10.1016/j.jclepro.2017.07.134

Alper, K., Wang, Y. Y., Meng, X., Tekin, K., Karagoz, S., & Ragauskas, A. J. (2021). Use of a Lewis acid, a Brønsted acid, and their binary mixtures for the liquefaction of lignocellulose by supercritical ethanol processing. *Sustainable Energy and Fuels, 5*(21), 5445–5453. https://doi.org/10.1039/d1se01316e

Amini, E., Valls, C., & Roncero, M. B. (2021). Ionic liquid-assisted bioconversion of lignocellulosic biomass for the development of value-added products. *Journal of Cleaner Production, 326*, 129275. https://doi.org/10.1016/j.jclepro.2021.129275

An, Y. M., Zhuang, J., Li, Y., Dai, J. Y., & Xiu, Z. L. (2022). Pretreatment of Jerusalem artichoke stalk using hydroxylammonium ionic liquids and their influences on 2,3-butanediol fermentation by Bacillus subtilis. *Bioresource Technology, 354*, 127219. https://doi.org/10.1016/j.biortech.2022.127219

Andhalkar, V. V., Montané, D., Medina, F., & Constantí, M. (2024). Biorefinery design with combined deacetylation and microwave pretreatment for enhanced production of polyhydroxyalkanoates and efficient carbon utilization from lignocellulose. *Chemical Engineering Journal, 486*, 149754. https://doi.org/10.1016/j.cej.2024.149754

Asada, C., Yoshida, Y., & Nakamura, Y. (2023). Efficient conversion of moso bamboo components into glucose, lignocellulose nanofiber, and low-molecular-weight lignin through deep eutectic solvent treatment. *Biomass Conversion and Biorefinery, 13*(9), 7713–7724. https://doi.org/10.1007/s13399-021-01629-0

Asim, A. M., Uroos, M., Naz, S., & Muhammad, N. (2021). Pyridinium protic ionic liquids: Effective solvents for delignification of wheat straw. *Journal of Molecular Liquids, 325*, 115013. https://doi.org/10.1016/j.molliq.2020.115013

Azman, N. F., Megat Mohd Noor, M. J., MD Akhir, F. N., Ang, M. Y., Hashim, H., Othman, N., Zakaria, Z., & Hara, H. (2019). Depolymerization of lignocellulose of oil palm empty fruit bunch by thermophilic microorganisms from tropical climate. *Bioresource Technology, 279*, 174–180. https://doi.org/10.1016/j.biortech.2019.01.122

Badgujar, K. C., Dange, R., & Bhanage, B. M. (2021). Recent advances of use of the supercritical carbon dioxide for the biomass pre-treatment and extraction: A mini-review. *Journal of the Indian Chemical Society, 98*, 100018. https://doi.org/10.1016/j.jics.2021.100018

Bakari, R., Asha, R., Hossein, M., Huang, X., Islam, N. F., Liew, R. K., Narayan, M., Lam, S. S., & Sarma, H. (2024). Converting food waste to biofuel: A sustainable energy solution for Sub-Saharan Africa. *Sustainable Chemistry for the Environment, 7*, 100126. https://doi.org/10.1016/j.scenv.2024.100126

Baloch, H. A., Nizamuddin, S., Siddiqui, M. T. H., Riaz, S., Jatoi, A. S., Dumbre, D. K., Mubarak, N. M., Srinivasan, M. P., & Griffin, G. J. (2018). Recent advances in production and upgrading of bio-oil from biomass: A critical overview. *Journal of Environmental Chemical Engineering, 6*(4), 5101–5118. https://doi.org/10.1016/j.jece.2018.07.050

Bansod, S. P., Makwana, K., Sarangi, P. K., & Parikh, J. K. (2024). Advanced pretreatment processes for lignocellulosic biomass to biofuels production: Path towards circular bioeconomy. *Sustainable Chemistry and Pharmacy, 39*, 101514. https://doi.org/10.1016/j.scp.2024.101514

Barhoum, A., Jeevanandam, J., Rastogi, A., Samyn, P., Boluk, Y., Dufresne, A., Danquah, M. K., & Bechelany, M. (2020). Plant celluloses, hemicelluloses, lignins, and volatile oils for the synthesis of nanoparticles and nanostructured materials. *Nanoscale, 12*(45), 22845–22890. https://doi.org/10.1039/d0nr04795c

Beig, B., Riaz, M., Raza, S., Hassan, M., Zheng, Z., Karimi, K., Pugazhendhi, A., Atabani, A. E., Thuy, N., Chi, L., & Million, M. (2021). Current challenges and innovative developments in pretreatment of lignocellulosic residues for biofuel production: A review Original Equipment Manufacturer. *Fuel, 287*, 119670. https://doi.org/10.1016/j.fuel.2020.119670

Benito-González, I., Ebrahimi, M., Ortiz-Gimeno, M. M., Hoang, L., Acha, V., Aussenac, T., Martínez-Abad, A., & López-Rubio, A. (2023). Ozonation of vine shoots as a source of lignocellulosic phenolic-rich fractions. *Cellulose, 30*(17), 10795–10810. https://doi.org/10.1007/s10570-023-05558-1

Bhatia, P., Fujiwara, M., Ban, S., & Toda, T. (2020a). Effect of steam explosion pre-treatment on methane generation from Ludwigia grandiflora. *Biomass and Bioenergy, 142*, 105771. https://doi.org/10.1016/j.biombioe.2020.105771

Bhatia, R., Winters, A., Bryant, D. N., Bosch, M., Clifton-Brown, J., Leak, D., & Gallagher, J. (2020b). Pilot-scale production of xylo-oligosaccharides and fermentable sugars from Miscanthus using steam explosion pretreatment. *Bioresource Technology, 296*, 122285. https://doi.org/10.1016/j.biortech.2019.122285

Bilal, M., Qamar, S. A., Qamar, M., Yadav, V., Taherzadeh, M. J., Lam, S. S., & Iqbal, H. M. N. (2024). Bioprospecting lignin biomass into environmentally friendly polymers—Applied perspective to reconcile sustainable circular bioeconomy. *Biomass Conversion and Biorefinery, 14*(4), 4457–4483. https://doi.org/10.1007/s13399-022-02600-3

Brito, J. Q. A., Dias, F. D. S., Cunha, S., Ramos, L. P., & Teixeira, L. S. G. (2019). Multiple response optimization of alkaline pretreatment of sisal fiber (*Agave sisalana*) assisted by ultrasound. *Biotechnology Progress, 35*(3), e2802. https://doi.org/10.1002/btpr.2802

Byun, J., Cha, Y. L., Park, S. M., Kim, K. S., Lee, J. E., & Kang, Y. G. (2020). Lignocellulose pretreatment combining continuous alkaline single-screw extrusion and ultrasonication to enhance biosugar production. *Energies, 13*(21), 5636. https://doi.org/10.3390/en13215636

Cai, J., Xue, J., Li, Y., Wei, L., Lin, N., & Zha, X. (2023). Research progress on high-value-added application of lignocellulosic biomass based on deep eutectic solvent pretreatment. *Biomass Conversion and Biorefinery, 14*, 30913–30928. https://doi.org/10.1007/s13399-023-04972-6

Castañón-Rodríguez, J. F., Torrestiana-Sánchez, B., Montero-Lagunes, M., Portilla-Arias, J., de León, J. A. R., & Aguilar-Uscanga, M. G. (2013). Using high pressure processing (HPP) to pretreat sugarcane bagasse. *Carbohydrate Polymers, 98*(1), 1018–1024. https://doi.org/10.1016/j.carbpol.2013.06.068

Chen, X., He, B., Feng, M., Zhao, D., & Sun, J. (2020a). Immobilized laccase on magnetic nanoparticles for enhanced lignin model compounds degradation. *Chinese Journal of Chemical Engineering, 28*(8), 2152–2159. https://doi.org/10.1016/j.cjche.2020.02.028

Chen, F., Xiong, S., Sundelin, J., Martín, C., & Hultberg, M. (2020b). Potential for combined production of food and biofuel: Cultivation of Pleurotus pulmonarius on soft- and hardwood sawdusts. *Journal of Cleaner Production, 266*, 122011. https://doi.org/10.1016/j.jclepro.2020.122011

Chen, H., Wan, K., Zheng, F., Zhang, Z., Zhang, Y., & Long, D. (2021). Mechanism insight into photocatalytic conversion of lignin for valuable chemicals and fuels production: A state-of-the-art review. *Renewable and Sustainable Energy Reviews, 147*, 111217. https://doi.org/10.1016/j.rser.2021.111217

Chen, J., Zhang, B., Liu, B., Yi, Y., Shan, Y., Zhou, Y., Wang, X., & Lü, X. (2022). Full components conversion of lignocellulose via a closed-circuit biorefinery process on a pilot scale. *Environmental Research, 214*, 113946. https://doi.org/10.1016/j.envres.2022.113946

Colmenares, J. C., Luque, R., Campelo, J. M., Colmenares, F., Karpiński, Z., & Romero, A. A. (2009). Nanostructured photocatalysts and their applications in the photocatalytic transformation of lignocellulosic biomass: An overview. *Materials, 2*(4), 2228–2258. https://doi.org/10.3390/ma2042228

Cooreman, E., Nicolaï, T., Arts, W., Van Aelst, K., Vangeel, T., Van den Bosch, S. V., Van Aelst, J., Lagrain, B., Thiele, K., Thevelein, J., & Sels, B. F. (2023). The future biorefinery: The impact of upscaling the reductive catalytic fractionation of lignocellulose biomass on the quality of the lignin oil, carbohydrate products, and pulp. *ACS Sustainable Chemistry and Engineering, 11*(14), 5440–5450. https://doi.org/10.1021/acssuschemeng.2c06913

Dai, L., He, C., Wang, Y., Liu, Y., Yu, Z., Zhou, Y., Fan, L., Duan, D., & Ruan, R. (2017). Comparative study on microwave and conventional hydrothermal pretreatment of bamboo sawdust: Hydrochar properties and its pyrolysis behaviors. *Energy Conversion and Management, 146*, 1–7. https://doi.org/10.1016/j.enconman.2017.05.007

Dai, X., Hua, Y., Dai, L., & Cai, C. (2019). Particle size reduction of rice straw enhances methane production under anaerobic digestion. *Bioresource Technology, 293*, 122043. https://doi.org/10.1016/j.biortech.2019.122043

Damaurai, J., Preechakun, T., Raita, M., Champreda, V., & Laosiripojana, N. (2021). Investigation of alkaline hydrogen peroxide in aqueous organic solvent to enhance enzymatic hydrolysis of rice straw. *BioEnergy Research, 14*(1), 122–134. https://doi.org/10.1007/s12155-020-10152-5

Das, N., Jena, P. K., Padhi, D., Kumar Mohanty, M., & Sahoo, G. (2021). A comprehensive review of characterization, pretreatment and its applications on different lignocellulosic biomass for bioethanol production. *Biomass Conversion and Biorefinery, 13*, 1503–1527. https://doi.org/10.1007/s13399-021-01294-3

de Farias Silva, C. E., Meneghello, D., de Souza Abud, A. K., & Bertucco, A. (2020). Pretreatment of microalgal biomass to improve the enzymatic hydrolysis of carbohydrates by ultrasonication: Yield vs energy consumption. *Journal of King Saud University - Science, 32*(1), 606–613. https://doi.org/10.1016/j.jksus.2018.09.007

de Jong, E., Stichnothe, H., Bell, G., Henning Jørgensen, M., de Bari, I., Jacco van Haveren, E., Lindorfer, J., & an der Johannes Kepler, E. (2020). *Bio-Based Chemicals A 2020 update bio-based chemicals with input from: (pdf version) Published by IEA Bioenergy*. https://www.ieabioenergy.com/wp-content/uploads/2020/02/Bio-based-chemicals-a-2020-update-final-200213.pdf

de Sales, A. N., de Souza, A. C., Moutta, R. O., Ferreira-Leitão, V. S., Schwan, R. F., & Dias, D. R. (2017). Use of lignocellulose biomass for endoxylanase production by Streptomyces termitum. *Preparative Biochemistry and Biotechnology, 47*(5), 505–512. https://doi.org/10.1080/10826068.2016.1275015

de Souza, F. M., & Gupta, R. K. (2024). Bacteria for bioplastics: Progress, applications, and challenges. *ACS Omega, 9*(8), 8666–8686. https://doi.org/10.1021/acsomega.3c07372

Deng, W., Feng, Y., Fu, J., Guo, H., Guo, Y., Han, B., Jiang, Z., Kong, L., Li, C., Liu, H., Nguyen, P. T. T., Ren, P., Wang, F., Wang, S., Wang, Y., Wang, Y., Wong, S. S., Yan, K., Yan, N., … Zhou, H. (2023). Catalytic conversion of lignocellulosic biomass into chemicals and fuels. *Green Energy and Environment, 8*(1), 10–114. https://doi.org/10.1016/j.gee.2022.07.003

Dharmaraja, J., Shobana, S., Arvindnarayan, S., Francis, R. R., Jeyakumar, R. B., Saratale, R. G., Ashokkumar, V., Bhatia, S. K., Kumar, V., & Kumar, G. (2023). Lignocellulosic biomass conversion via greener pretreatment methods towards biorefinery applications. *Bioresource Technology, 369*, 128328. https://doi.org/10.1016/j.biortech.2022.128328

Di Fraia, A., Di Fraia, S., Godvin Sharmila, V., Banu J, R., & Massarotti, N. (2024). Role of advanced oxidation processes in lignocellulose pretreatment towards biorefinery applications: A review on emerging trends and economic considerations. *Green Chemistry*. https://doi.org/10.1039/d3gc05108k

Dias, R. M., Petrin, L. C. G., Sosa, F. H. B., Da Costa Lopes, A. M., Coutinho, J. A. P., & Da Costa, M. C. (2020). Investigation of kraft lignin solubility in protic ionic liquids and their aqueous solutions. *Industrial and Engineering Chemistry Research, 59*(40), 18193–18202. https://doi.org/10.1021/acs.iecr.0c02605

Dissanayake, N., Thalangamaarachchige, V. D., Thakurathi, M., Knight, M., Quitevis, E. L., & Abidi, N. (2019). Dissolution of cotton cellulose in 1:1 mixtures of 1-butyl-3-methylimidazolium methylphosphonate and 1-alkylimidazole co-solvents. *Carbohydrate Polymers, 221*, 63–72. https://doi.org/10.1016/j.carbpol.2019.05.071

Du, B., Li, W., Bai, Y., Pan, Z., Wang, Q., Wang, X., Lv, G., Ding, H., & Zhou, J. (2023). Effect of CO_2 concentration on improving yield and antioxidant activity of lignin from corn cobs. *Bioenergy Research, 16*(2), 954–966. https://doi.org/10.1007/s12155-022-10490-6

Duque, A., Manzanares, P., & Ballesteros, M. (2017). Extrusion as a pretreatment for lignocellulosic biomass: Fundamentals and applications. *Renewable Energy, 114*, 1427–1441. https://doi.org/10.1016/j.renene.2017.06.050

Ewulonu, C. M., Liu, X., Wu, M., & Huang, Y. (2019). Ultrasound-assisted mild sulphuric acid ball milling preparation of lignocellulose nanofibers (LCNFs) from sunflower stalks (SFS). *Cellulose, 26*(7), 4371–4389. https://doi.org/10.1007/s10570-019-02382-4

Ferdeş, M., Dincă, M. N., Moiceanu, G., Zabava, B. Ş., & Paraschiv, G. (2020). Microorganisms and enzymes used in the biological pretreatment of the substrate to enhance biogas production: A review. *Sustainability (Switzerland), 12*(17), 7205. https://doi.org/10.3390/su12177205

Fu, L., Fang, Z., Chen, H., Deng, W., Sun, C., Zhai, Y., Xu, G., Zhang, X., & Wen, Y. (2022). Production of lignocellulose nanofibril (LCNF) from high yield pulps by hydrated deep eutectic solvents (DES) pretreatment for fabricating biobased straw. *Industrial Crops and Products, 188*, 115738. https://doi.org/10.1016/j.indcrop.2022.115738

Galbe, M., & Wallberg, O. (2019). Pretreatment for biorefineries: A review of common methods for efficient utilisation of lignocellulosic materials. *Biotechnology for Biofuels, 12*(1), 294. https://doi.org/10.1186/s13068-019-1634-1

Gallego-García, M., Moreno, A. D., Manzanares, P., Negro, M. J., & Duque, A. (2023). Recent advances on physical technologies for the pretreatment of food waste and lignocellulosic residues. *Bioresource Technology, 369*, 128397. https://doi.org/10.1016/j.biortech.2022.128397

Gallegos, D., Wedwitschka, H., Moeller, L., Zehnsdorf, A., & Stinner, W. (2017). Effect of particle size reduction and ensiling fermentation on biogas formation and silage quality of wheat straw. *Bioresource Technology, 245*, 216–224. https://doi.org/10.1016/j.biortech.2017.08.137

Gao, Y., Remón, J., & Matharu, A. S. (2021). Microwave-assisted hydrothermal treatments for biomass valorisation: A critical review. *Green Chemistry, 23*(10), 3502–3525. https://doi.org/10.1039/D1GC00623A

Ge, X., Green, V. S., Zhang, N., Sivakumar, G., & Xu, J. (2012). Eastern gamagrass as an alternative cellulosic feedstock for bioethanol production. *Process Biochemistry, 47*(2), 335–339. https://doi.org/10.1016/j.procbio.2011.11.008

Gu, X., Fu, X., & Chen, S. (2021). Hydrothermal liquefaction conversion of lignocelluloses with enhanced fungal pretreatment. *Industrial Crops and Products, 162*, 113268. https://doi.org/10.1016/j.indcrop.2021.113268

Gu, H., Han, X., Zhang, J., & Bao, J. (2024). Upgrading dry acid pretreatment by post-hydrolysis for carbon efficient conversion of lignocellulose. *Bioresource Technology, 394*, 130261. https://doi.org/10.1016/j.biortech.2023.130261

Hansen, B. B., Spittle, S., Chen, B., Poe, D., Zhang, Y., Klein, J. M., Horton, A., Adhikari, L., Zelovich, T., Doherty, B. W., Gurkan, B., Maginn, E. J., Ragauskas, A., Dadmun, M., Zawodzinski, T. A., Baker, G. A., Tuckerman, M. E., Savinell, R. F., & Sangoro, J. R. (2021). Deep eutectic solvents: A review of fundamentals and applications. *Chemical Reviews, 121*(3), 1232–1285. https://doi.org/10.1021/acs.chemrev.0c00385

Haykir, N. I., Nizan Shikh Zahari, S. M. S., Harirchi, S., Sar, T., Awasthi, M. K., & Taherzadeh, M. J. (2023). Applications of ionic liquids for the biochemical transformation of lignocellulosic biomass into biofuels and biochemicals: A critical review. *Biochemical Engineering Journal, 193*, 108850. https://doi.org/10.1016/j.bej.2023.108850

Heiske, S., Schultz-Jensen, N., Leipold, F., & Schmidt, J. E. (2013). Improving anaerobic digestion of wheat straw by plasma-assisted pretreatment. *Journal of Atomic and Molecular Physics, 2013*, 1–7. https://doi.org/10.1155/2013/791353

Hernandez, S. C., Suzuki, S., Wada, N., & Takahashi, K. (2024). Ionic liquid pretreatment of lignocellulose for complete hemicellulose removal to produce high-purity cellulose mixed esters. *Industrial Crops and Products, 218*, 118909. https://doi.org/10.1016/j.indcrop.2024.118909

Hickert, L. R., Cruz, M. M., Dillon, A. J. P., Fontana, R. C., Rosa, C. A., & Ayub, M. A. Ô. Z. (2014). Fermentation kinetics of acid-enzymatic soybean hull hydrolysate in immobilized-cell bioreactors of Saccharomyces cerevisiae, Candida shehatae, Spathaspora arborariae, and their co-cultivations. *Biochemical Engineering Journal, 88*, 61–67. https://doi.org/10.1016/j.bej.2014.04.004

Hosseini Koupaie, E., Dahadha, S., Bazyar Lakeh, A. A., Azizi, A., & Elbeshbishy, E. (2019). Enzymatic pretreatment of lignocellulosic biomass for enhanced biomethane production—A review. *Journal of Environmental Management, 233*, 774–784. https://doi.org/10.1016/j.jenvman.2018.09.106

Hu, Z., & Wen, Z. (2008). Enhancing enzymatic digestibility of switchgrass by microwave-assisted alkali pretreatment. *Biochemical Engineering Journal, 38*(3), 369–378. https://doi.org/10.1016/j.bej.2007.08.001

Hussin, M. H., Appaturi, J. N., Poh, N. E., Latif, N. H. A., Brosse, N., Ziegler-Devin, I., Vahabi, H., Syamani, F. A., Fatriasari, W., Solihat, N. N., Karimah, A., Iswanto, A. H., Sekeri, S. H., & Ibrahim, M. N. M. (2022). A recent advancement on preparation, characterization and application of nanolignin. *International Journal of Biological Macromolecules, 200*, 303–326. https://doi.org/10.1016/j.ijbiomac.2022.01.007

Janker-Obermeier, I., Sieber, V., Faulstich, M., & Schieder, D. (2012). Solubilization of hemicellulose and lignin from wheat straw through microwave-assisted alkali treatment. *Industrial Crops and Products, 39*, 198–203. https://doi.org/10.1016/j.indcrop.2012.02.022

Jeon, B.-H., Choi, J.-A., Kim, H.-C., Hwang, J.-H., Abou-Shanab, R. A., Dempsey, B. A., Regan, J. M., & Kim, J. R. (2013). Ultrasonic disintegration of microalgal biomass and consequent improvement of bioaccessibility/bioavailability in microbial fermentation. *Biotechnology for Biofuels, 6*(1), 37. https://doi.org/10.1186/1754-6834-6-37

Ji, Q., Yu, X., Yagoub, A. E. G. A., Chen, L., Mustapha, A. T., & Zhou, C. (2021). Enhancement of lignin removal and enzymolysis of sugarcane bagasse by ultrasound-assisted ethanol synergized deep eutectic solvent pretreatment. *Renewable Energy, 172*, 304–316. https://doi.org/10.1016/j.renene.2021.03.050

Jiang, B., Zheng, J., Qiu, S., Wu, M., Zhang, Q., Yan, Z., & Xue, Q. (2014). Review on electrical discharge plasma technology for wastewater remediation. *Chemical Engineering Journal, 236*, 348–368. https://doi.org/10.1016/j.cej.2013.09.090

John, A. J., & Selvarajan, E. (2024). Ionic liquid-assisted pretreatment of lignocellulosic biomass using purified Streptomyces MS2A cellulase for bioethanol production. *International Journal of Biological Macromolecules, 270*, 132149. https://doi.org/10.1016/j.ijbiomac.2024.132149

Jomnonkhaow, U., Sittijunda, S., & Reungsang, A. (2021). Influences of size reduction, hydration, and thermal-assisted hydration pretreatment to increase the biogas production from Napier grass and Napier silage. *Bioresource Technology, 331*, 125034. https://doi.org/10.1016/j.biortech.2021.125034

Kadokawa, R., Endo, T., Yasaka, Y., Ninomiya, K., Takahashi, K., & Kuroda, K. (2021). Cellulose preferentially dissolved over xylan in ionic liquids through precise anion interaction regulated by bulky cations. *ACS Sustainable Chemistry and Engineering, 9*(26), 8686–8691. https://doi.org/10.1021/acssuschemeng.1c02492

Kassim, M. A., Meng, T. K., Kamaludin, R., Hussain, A. H., & Bukhari, N. A. (2022). Bioprocessing of sustainable renewable biomass for bioethanol production. In *Value-chain of biofuels* (pp. 195–234). Elsevier. https://doi.org/10.1016/B978-0-12-824388-6.00004-X

Khan, A. S., Man, Z., Bustam, M. A., Nasrullah, A., Ullah, Z., Sarwono, A., Shah, F. U., & Muhammad, N. (2018). Efficient conversion of lignocellulosic biomass to levulinic acid using acidic ionic liquids. *Carbohydrate Polymers, 181*, 208–214. https://doi.org/10.1016/j.carbpol.2017.10.064

Khan, M. U., Usman, M., Ashraf, M. A., Dutta, N., Luo, G., & Zhang, S. (2022). A review of recent advancements in pretreatment techniques of lignocellulosic materials for biogas production: Opportunities and limitations. *Chemical Engineering Journal Advances, 10*, 100263. https://doi.org/10.1016/j.ceja.2022.100263

Khoo, K. S., Tan, X., Ooi, C. W., Chew, K. W., Leong, W. H., Chai, Y. H., Ho, S. H., & Show, P. L. (2021). How does ionic liquid play a role in sustainability of biomass processing? *Journal of Cleaner Production, 284*, 124772. https://doi.org/10.1016/j.jclepro.2020.124772

Kim, J. Y., Na, C. S., Kim, D. S., Kim, J. B., & Seo, Y. W. (2015). The effect of chronic gamma ray irradiation on lignocellulose of Brachypodium distachyon. *Cellulose, 22*(4), 2419–2430. https://doi.org/10.1007/s10570-015-0687-y

Konan, D., Rodrigue, D., Koffi, E., Elkoun, S., Ndao, A., & Adjallé, K. (2024). Combination of technologies for biomass pretreatment: A focus on extrusion. *Waste and Biomass Valorization, 5*(8), 1–22. https://doi.org/10.1007/s12649-024-02472-w

Kosinkova, J., Ramirez, J. A., Jablonský, M., Ristovski, Z. D., Brown, R., & Rainey, T. J. (2017). Energy and chemical conversion of five Australian lignocellulosic feedstocks into bio-crude through liquefaction. *RSC Advances, 7*(44), 27707–27717. https://doi.org/10.1039/c7ra02335a

Kozłowski, K., Lewicki, A., Czekała, W., Wójtowicz, A., Kupryaniuk, K., & Dróżd, D. (2019). Extrusion pretreatment of maize straw—Case study for a Polish biogas plants. *International Agrophysics, 33*(4), 527–535. https://doi.org/10.31545/intagr/113548

Kumar, A. K., & Sharma, S. (2017). Recent updates on different methods of pretreatment of lignocellulosic feedstocks: A review. *Bioresources and Bioprocessing, 4*(1), 1–19. https://doi.org/10.1186/s40643-017-0137-9

Kumar, R., Strezov, V., Weldekidan, H., He, J., Singh, S., Kan, T., & Dastjerdi, B. (2020). Lignocellulose biomass pyrolysis for bio-oil production: A review of biomass pre-treatment methods for production of drop-in fuels. *Renewable and Sustainable Energy Reviews, 123*, 109763. https://doi.org/10.1016/j.rser.2020.109763

Kwon, G. J., Cho, S. W., Bandi, R., Yang, B. S., Dadigala, R., Han, S. Y., Ma, S. Y., Kim, J. K., Kim, N. H., & Lee, S. H. (2023). Production of lignocellulose nanofibrils by conventional and microwave-assisted deep-eutectic-solvent pretreatments: Mechanical, antioxidant, and UV-blocking properties. *Cellulose, 30*(7), 4277–4292. https://doi.org/10.1007/s10570-023-05164-1

Laltha, M., Sewsynker-Sukai, Y., & Gueguim Kana, E. B. (2021). Development of microwave-assisted alkaline pretreatment methods for enhanced sugar recovery from bamboo and corn cobs: Process optimization, chemical recyclability and kinetics of bioethanol production. *Industrial Crops and Products, 174*, 114166. https://doi.org/10.1016/j.indcrop.2021.114166

Langmuir, I. (1928). Oscillations in ionized gases. *Proceedings of the National Academy of Sciences, 14*(8), 627–637. https://doi.org/10.1073/pnas.14.8.627

Li, S. H., Liu, S., Colmenares, J. C., & Xu, Y. J. (2016). A sustainable approach for lignin valorization by heterogeneous photocatalysis. *Green Chemistry, 18*(3), 594–607. https://doi.org/10.1039/c5gc02109j

Li, P., Yang, C., Jiang, Z., Jin, Y., & Wu, W. (2023). Lignocellulose pretreatment by deep eutectic solvents and related technologies: A review. *Journal of Bioresources and Bioproducts, 8*(1), 33–44. https://doi.org/10.1016/j.jobab.2022.11.004

Limayem, A., & Ricke, S. C. (2012). Lignocellulosic biomass for bioethanol production: Current perspectives, potential issues and future prospects. *Progress in Energy and Combustion Science, 38*(4), 449–467. https://doi.org/10.1016/j.pecs.2012.03.002

Lin, R., Deng, C., Ding, L., Bose, A., & Murphy, J. D. (2019). Improving gaseous biofuel production from seaweed Saccharina latissima: The effect of hydrothermal pretreatment on energy efficiency. *Energy Conversion and Management, 196*, 1385–1394. https://doi.org/10.1016/j.enconman.2019.06.044

Lin, S.-P., Kuo, T.-C., Wang, H.-T., Ting, Y., Hsieh, C.-W., Chen, Y.-K., Hsu, H.-Y., & Cheng, K.-C. (2020). Enhanced bioethanol production using atmospheric cold plasma-assisted detoxification of sugarcane bagasse hydrolysate. *Bioresource Technology, 313*, 123704. https://doi.org/10.1016/j.biortech.2020.123704

Lobato-Rodríguez, Á., Gullón, B., Romaní, A., Ferreira-Santos, P., Garrote, G., & Del-Río, P. G. (2023). Recent advances in biorefineries based on lignin extraction using deep eutectic solvents: A review. *Bioresource Technology, 388*, 129744. https://doi.org/10.1016/j.biortech.2023.129744

Loow, Y. L., Wu, T. Y., Yang, G. H., Md. Jahim, J., Teoh, W. H., & Mohammad, A. W. (2016). Role of energy irradiation in aiding pretreatment of lignocellulosic biomass for improving reducing sugar recovery. *Cellulose, 23*(5), 2761–2789. https://doi.org/10.1007/s10570-016-1023-x

Lu, X., Xu, S., Chen, J., Ni, L., Ma, X., Cao, S., & Gao, H. (2023). Cellulose dissolution in ionic liquid from hydrogen bonding perspective: First-principles calculations. *Cellulose, 30*(7), 4181–4195. https://doi.org/10.1007/s10570-023-05140-9

Lymperatou, A., Engelsen, T. K., Skiadas, I. V., & Gavala, H. N. (2023). Prediction of methane yield and pretreatment efficiency of lignocellulosic biomass based on composition. *Waste Management, 155*, 302–310. https://doi.org/10.1016/j.wasman.2022.10.040

M'Arimi, M. M., Mecha, C. A., Kiprop, A. K., & Ramkat, R. (2020). Recent trends in applications of advanced oxidation processes (AOPs) in bioenergy production: Review. *Renewable and Sustainable Energy Reviews, 121*, 109669. https://doi.org/10.1016/j.rser.2019.109669

Ma, T., Zhao, J., Ao, L., Liao, X., Ni, Y., Hu, X., & Song, Y. (2018). Effects of different pretreatments on pumpkin (Cucurbita pepo) lignocellulose degradation. *International Journal of Biological Macromolecules, 120*, 665–672. https://doi.org/10.1016/j.ijbiomac.2018.08.124

Mafa, M. S., Malgas, S., Bhattacharya, A., & Rashamuse, K. (2020). The effects of alkaline pretreatment on agricultural biomasses (corn cob and sweet sorghum bagasse). *Agronomy, 10*(1211), 1–13.

Magina, S., Barros-Timmons, A., Ventura, S. P. M., & Evtuguin, D. V. (2021). Evaluating the hazardous impact of ionic liquids—Challenges and opportunities. *Journal of Hazardous Materials, 412*, 125215. https://doi.org/10.1016/j.jhazmat.2021.125215

Malik, K., Sharma, P., Yang, Y., Zhang, P., Zhang, L., Xing, X., Yue, J., Song, Z., Nan, L., Yujun, S., El-Dalatony, M. M., Salama, E.-S., & Li, X. (2022). Lignocellulosic biomass for bioethanol: Insight into the advanced pretreatment and fermentation approaches. *Industrial Crops and Products, 188*, 115569. https://doi.org/10.1016/j.indcrop.2022.115569

Mankar, A. R., Pandey, A., Modak, A., & Pant, K. K. (2021). Pretreatment of lignocellulosic biomass: A review on recent advances. *Bioresource Technology, 334*, 125235. https://doi.org/10.1016/j.biortech.2021.125235

Matei, J. C., Soares, M., Bonato, A. C. H., de Freitas, M. P. A., Helm, C. V., Maroldi, W. V., Magalhães, W. L. E., Haminiuk, C. W. I., & Maciel, G. M. (2020). Enzymatic delignification of sugar cane bagasse and rice husks and its effect in saccharification. *Renewable Energy, 157*, 987–997. https://doi.org/10.1016/j.renene.2020.05.028

Megía, P. J., Vizcaíno, A. J., Calles, J. A., & Carrero, A. (2021). Hydrogen production technologies: From fossil fuels toward renewable sources. A mini review. *Energy & Fuels, 35*(20), 16403–16415. https://doi.org/10.1021/acs.energyfuels.1c02501

Mehta, D., Purohit, A., Bajarh, P., Yadav, K., Shivhare, U. S., & Yadav, S. K. (2022). Cold plasma processing improved the extraction of xylooligosaccharides from dietary fibers of rice and corn bran with enhanced in-vitro digestibility and anti-inflammatory responses. *Innovative Food Science and Emerging Technologies, 78*, 103027. https://doi.org/10.1016/j.ifset.2022.103027

Meoli, C. M., Iervolino, G., & Procentese, A. (2023). Non-thermal plasma as a biomass pretreatment in biorefining processes. *Processes, 11*(2), 536. https://doi.org/10.3390/pr11020536

Mikulski, D., & Kłosowski, G. (2020). Microwave-assisted dilute acid pretreatment in bioethanol production from wheat and rye stillages. *Biomass and Bioenergy, 136*, 105528. https://doi.org/10.1016/j.biombioe.2020.105528

Mikulski, D., & Kłosowski, G. (2022). Delignification efficiency of various types of biomass using microwave-assisted hydrotropic pretreatment. *Scientific Reports, 12*(1), 4561. https://doi.org/10.1038/s41598-022-08717-9

Mikulski, D., & Kłosowski, G. (2023). Cellulose hydrolysis and bioethanol production from various types of lignocellulosic biomass after microwave-assisted hydrotropic pretreatment. *Renewable Energy, 206*, 168–179. https://doi.org/10.1016/j.renene.2023.02.061

Miranda, F. S., Rabelo, S. C., Pradella, J. G. C., Di Carli, C., Petraconi, G., Maciel, H. S., Pessoa, R. S., & Vieira, L. (2020). Plasma in-liquid using non-contact electrodes: A method of pretreatment to enhance the enzymatic hydrolysis of biomass. *Waste and Biomass Valorization, 11*(9), 4921–4931. https://doi.org/10.1007/s12649-019-00824-5

Mood, S. H., Golfeshan, A. H., Tabatabaei, M., Jouzani, G. S., Najaf, G. H., Gholami, M., & Ardjmand, M. (2013). Lignocellulosic biomass to bioethanol, a comprehensive review with a focus on pretreatment. *Renewable and Sustainable Energy Reviews, 27*, 77–93. https://doi.org/10.1016/j.rser.2013.06.033

Nakayama, R., & Imai, M. (2013). Promising ultrasonic irradiation pretreatment for enzymatic hydrolysis of Kenaf. *Journal of Environmental Chemical Engineering, 1*(4), 1131–1136. https://doi.org/10.1016/j.jece.2013.08.030

Nardella, F., Prothmann, J., Sandahl, M., Spégel, P., Ribechini, E., & Turner, C. (2023). Native lignin extraction from soft- and hardwood by green and benign sub/supercritical fluid extraction methodologies. *RSC Advances, 13*(32), 21945–21953. https://doi.org/10.1039/d3ra01873c

Nauman Aftab, M., Iqbal, I., Riaz, F., Karadag, A., & Tabatabaei, M. (2019). Different pretreatment methods of lignocellulosic biomass for use in biofuel production. In *Biomass for bioenergy—Recent trends and future challenges*. IntechOpen. https://doi.org/10.5772/intechopen.84995

Nitzsche, R., Gröngröft, A., Köchermann, J., Meisel, K., Etzold, H., Verges, M., Leschinsky, M., Bachmann, J., Saake, B., Torkler, S., Patzsch, K., Rößiger, B., Pufky-Heinrich, D., & Unkelbach, G. (2020). *Platform and fine chemicals from woody biomass: Demonstration and assessment of a novel biorefinery*. https://doi.org/10.1007/s13399-020-00769-z/Published

Nlandu, H., Belkacemi, K., Chorfa, N., Elkoun, S., Robert, M., & Hamoudi, S. (2020). Flax nanofibrils production via supercritical carbon dioxide pre-treatment and enzymatic hydrolysis. *Canadian Journal of Chemical Engineering, 98*(1), 84–95. https://doi.org/10.1002/cjce.23596

O'Sullivan, A., Ryan, K. M., & Padrela, L. (2022). Production of biopharmaceutical dried-powders using supercritical CO_2 technology. *Journal of Supercritical Fluids, 187*, 105645. https://doi.org/10.1016/j.supflu.2022.105645

Ogawa, H., & Mori, H. (2020). Lithium salt/amide-based deep eutectic electrolytes for lithium-ion batteries: Electrochemical, thermal and computational study. *Physical Chemistry Chemical Physics, 22*(16), 8853–8863. https://doi.org/10.1039/d0cp01255f

Okolie, J. A., Mukherjee, A., Nanda, S., Dalai, A. K., & Kozinski, J. A. (2021). Next-generation biofuels and platform biochemicals from lignocellulosic biomass. *International Journal of Energy Research, 45*(10), 14145–14169. https://doi.org/10.1002/er.6697

Olatunde, A., Mohammed, A., Ibrahim, M. A., Tajuddeen, N., & Shuaibu, M. N. (2022). Vanillin: A food additive with multiple biological activities. *European Journal of Medicinal Chemistry Reports, 5*, 100055. https://doi.org/10.1016/j.ejmcr.2022.100055

Ooshima, H., Aso, K., Harano, Y., & Yamamoto, T. (1984). Microwave treatment of cellulosic materials for their enzymatic hydrolysis. *Biotechnology Letters, 6*(5), 289–294. https://doi.org/10.1007/BF00129056

Ovejero-Pérez, A., Rigual, V., Domínguez, J. C., Alonso, M. V., Oliet, M., & Rodriguez, F. (2020). Acidic depolymerization vs ionic liquid solubilization in lignin extraction from eucalyptus wood using the protic ionic liquid 1-methylimidazolium chloride. *International Journal of Biological Macromolecules, 157*, 461–469. https://doi.org/10.1016/j.ijbiomac.2020.04.194

Pan, S.-Y., Tsai, C.-Y., Liu, C.-W., Wang, S.-W., Kim, H., & Fan, C. (2021). Anaerobic co-digestion of agricultural wastes toward circular bioeconomy. *iScience, 24*(7), 102704. https://doi.org/10.1016/j.isci.2021.102704

Pan, Z., Liu, X., Ashori, A., Xu, F., Barta, K., & Zhang, X. (2024). Fully upgrade lignocellulose into high-valued nanomaterials and chemicals through one-pot treatment by bio-acid tailored deep eutectic solvents. *Chemical Engineering Journal, 496*, 153952. https://doi.org/10.1016/j.cej.2024.153952

Pang, B., Sun, Z., Wang, L., Chen, W. J., Sun, Q., Cao, X. F., Shen, X. J., Xiao, L., Yan, J. L., Deuss, P. J., Yuan, T. Q., & Sun, R. C. (2021). Improved value and carbon footprint by complete utilization of corncob lignocellulose. *Chemical Engineering Journal, 419*, 129565. https://doi.org/10.1016/j.cej.2021.129565

Pereira, G. N., Cesca, K., Cubas, A. L. V., Bianchet, R. T., Junior, S. E. B., Zanella, E., Stambuk, B. U., Poletto, P., & de Oliveira, D. (2021). Non-thermal plasma as an innovative pretreatment technology in delignification of brewery by-product. *Innovative Food Science and Emerging Technologies, 74*, 102827. https://doi.org/10.1016/j.ifset.2021.102827

Poddar, B. J., Nakhate, S. P., Gupta, R. K., Chavan, A. R., Singh, A. K., Khardenavis, A. A., & Purohit, H. J. (2021). A comprehensive review on the pretreatment of lignocellulosic wastes for improved biogas production by anaerobic digestion. *International Journal of Environmental Science and Technology, 19*, 3429–3456. https://doi.org/10.1007/s13762-021-03248-8

Priecel, P., & Lopez-Sanchez, J. A. (2019). Advantages and limitations of microwave reactors: From chemical synthesis to the catalytic valorization of biobased chemicals. *ACS Sustainable Chemistry & Engineering, 7*(1), 3–21. https://doi.org/10.1021/acssuschemeng.8b03286

Putrino, F. M., Tedesco, M., Bodini, R. B., & de Oliveira, A. L. (2020). Study of supercritical carbon dioxide pretreatment processes on green coconut fiber to enhance enzymatic hydrolysis of cellulose. *Bioresource Technology, 309*, 123387. https://doi.org/10.1016/j.biortech.2020.123387

Qasim, U., Rafiq, S., Jamil, F., Ahmed, A., Ali, T., Kers, J., Khurram, M. S., Hussain, M., Inayat, A., & Park, Y. K. (2021). Processing of lignocellulose in ionic liquids: A cleaner and sustainable approach. *Journal of Cleaner Production, 323*, 129189. https://doi.org/10.1016/j.jclepro.2021.129189

Qazanfarzadeh, Z., Masek, A., Chakraborty, S., & Kumaravel, V. (2024). Development of brewer's spent grain-derived bio nanocomposites through a multiproduct biorefinery approach for food packaging. *Industrial Crops and Products, 220*, 119226. https://doi.org/10.1016/j.indcrop.2024.119226

Rabbi, M. F., & Amin, M. B. (2024). Circular economy and sustainable practices in the food industry: A comprehensive bibliometric analysis. *Cleaner and Responsible Consumption, 14*, 100206. https://doi.org/10.1016/j.clrc.2024.100206

Rabelo, S. C., Nakasu, P. Y. S., Scopel, E., Araújo, M. F., Cardoso, L. H., & da Costa, A. C. (2023). Organosolv pretreatment for biorefineries: Current status, perspectives, and challenges. *Bioresource Technology, 369*, 128331. https://doi.org/10.1016/j.biortech.2022.128331

Rajendran, N., Gurunathan, B., Han, J., Krishna, S., Ananth, A., Venugopal, K., & Sherly Priyanka, R. B. (2021). Recent advances in valorization of organic municipal waste into energy using biorefinery approach, environment and economic analysis. *Bioresource Technology, 337*, 125498. https://doi.org/10.1016/j.biortech.2021.125498

Ramadoss, G., & Muthukumar, K. (2014). Ultrasound assisted ammonia pretreatment of sugarcane bagasse for fermentable sugar production. *Biochemical Engineering Journal, 83*, 33–41. https://doi.org/10.1016/j.bej.2013.11.013

Ramadoss, G., & Muthukumar, K. (2016). Mechanistic study on ultrasound assisted pretreatment of sugarcane bagasse using metal salt with hydrogen peroxide for bioethanol production. *Ultrasonics Sonochemistry, 28*, 207–217. https://doi.org/10.1016/j.ultsonch.2015.07.006

Ramamoorthy, N. K., Nagarajan, R., Ravi, S., & Sahadevan, R. (2020). An innovative plasma pre-treatment process for lignocellulosic bio-ethanol production. *Energy Sources, Part A: Recovery, Utilization, and Environmental Effects, 46*, 16043–16057. https://doi.org/10.1080/15567036.2020.1815900

Ravindran, R., & Jaiswal, A. K. (2016). A comprehensive review on pre-treatment strategy for lignocellulosic food industry waste: Challenges and opportunities. *Bioresource Technology, 199*, 92–102. https://doi.org/10.1016/j.biortech.2015.07.106

Ravindran, R., Sarangapani, C., Jaiswal, S., Cullen, P. J., & Jaiswal, A. K. (2017). Ferric chloride assisted plasma pretreatment of lignocellulose. *Bioresource Technology, 243*, 327–334. https://doi.org/10.1016/j.biortech.2017.06.123

Ravindran, R., Sarangapani, C., Jaiswal, S., Lu, P., Cullen, P. J., Bourke, P., & Jaiswal, A. K. (2019). Improving enzymatic hydrolysis of brewer spent grain with nonthermal plasma. *Bioresource Technology, 282*, 520–524. https://doi.org/10.1016/j.biortech.2019.03.071

Rego, F., Soares Dias, A. P., Casquilho, M., Rosa, F. C., & Rodrigues, A. (2019). Fast determination of lignocellulosic composition of poplar biomass by thermogravimetry. *Biomass and Bioenergy, 122*, 375–380. https://doi.org/10.1016/j.biombioe.2019.01.037

Rekhate, C. V., & Srivastava, J. K. (2020). Recent advances in ozone-based advanced oxidation processes for treatment of wastewater—A review. *Chemical Engineering Journal Advances, 3*, 100031. https://doi.org/10.1016/j.ceja.2020.100031

Rizal, S., Alfatah, T., Abdul Khalil, H. P. S., Mistar, E. M., Abdullah, C. K., Olaiya, F. G., Sabaruddin, F. A., Ikramullah, & Muksin, U. (2021). Properties and characterization of lignin nanoparticles functionalized in macroalgae biopolymer films. *Nanomaterials, 11*(3), 1–21. https://doi.org/10.3390/nano11030637

Sabeeh, M., Zeshan, Liaquat, R., & Maryam, A. (2020). Effect of alkaline and alkaline-photocatalytic pretreatment on characteristics and biogas production of rice straw. *Bioresource Technology, 309*, 123449. https://doi.org/10.1016/j.biortech.2020.123449

Sai Bharadwaj, A. V. S. L., Dev, S., Zhuang, J., Wang, Y., Yoo, C. G., Jeon, B.-H., Aggarwal, S., Park, S. H., & Kim, T. H. (2023). Review of chemical pretreatment of lignocellulosic biomass using low-liquid and low-chemical catalysts for effective bioconversion. *Bioresource Technology, 368*, 128339. https://doi.org/10.1016/j.biortech.2022.128339

Saratale, G. D., Saratale, R. G., Varjani, S., Cho, S.-K., Ghodake, G. S., Kadam, A., Mulla, S. I., Bharagava, R. N., Kim, D.-S., & Shin, H. S. (2020). Development of ultrasound aided chemical pretreatment methods to enrich saccharification of wheat waste biomass for polyhydroxybutyrate production and its characterization. *Industrial Crops and Products, 150*, 112425. https://doi.org/10.1016/j.indcrop.2020.112425

Sasaki, M., & Ohsawa, K. (2021). Hydrolysis of lignocellulosic biomass in hot-compressed water with supercritical carbon dioxide. *ACS Omega, 6*(22), 14252–14259. https://doi.org/10.1021/acsomega.1c01026

Schmetz, Q., Teramura, H., Morita, K., Oshima, T., Richel, A., Ogino, C., & Kondo, A. (2019). Versatility of a dilute acid/butanol pretreatment investigated on various lignocellulosic biomasses to produce lignin, monosaccharides and cellulose in distinct phases. *ACS Sustainable Chemistry and Engineering, 7*, 11069–11079. https://doi.org/10.1021/acssuschemeng.8b05841

Schutyser, W., Renders, T., Van Den Bosch, S., Koelewijn, S. F., Beckham, G. T., & Sels, B. F. (2018). Chemicals from lignin: An interplay of lignocellulose fractionation, depolymerisation, and upgrading. *Chemical Society Reviews, 47*(3), 852–908. https://doi.org/10.1039/c7cs00566k

Selvaraj, S., Chauhan, A., Dutta, V., Verma, R., Rao, S. K., Radhakrishnan, A., & Ghotekar, S. (2024). A state-of-the-art review on plant-derived cellulose-based green hydrogels and their multifunctional role in advanced biomedical applications. *International Journal of Biological Macromolecules, 265*, 130991. https://doi.org/10.1016/j.ijbiomac.2024.130991

Shaghaleh, H., Xu, X., Liu, H., Wang, S., Alhaj Hamoud, Y., Dong, F., & Luo, J. (2019). The effect of atmospheric pressure plasma pretreatment with various gases on the structural characteristics and chemical composition of wheat straw and applications to enzymatic hydrolysis. *Energy, 176*, 195–210. https://doi.org/10.1016/j.energy.2019.03.182

Sharma, H. K., Xu, C., & Qin, W. (2019). Biological pretreatment of lignocellulosic biomass for biofuels and bioproducts: An overview. *Waste and Biomass Valorization, 10*(2), 235–251. https://doi.org/10.1007/s12649-017-0059-y

Sharma, P., Sharma, S., & Kumar, H. (2024). Introduction to ionic liquids, applications and micellization behaviour in presence of different additives. *Journal of Molecular Liquids, 393*, 123447. https://doi.org/10.1016/j.molliq.2023.123447

Shen, F., He, C., Wang, Y., Hu, J., Huang, M., Zhao, L., Zhang, S., Tian, D., & Shen, F. (2023). Fully upgrade lignocellulose to three nanomaterials by combinational pretreatment: Refining straw waste to pesticide nanocarrier. *Chemical Engineering Journal, 467*, 143376. https://doi.org/10.1016/j.cej.2023.143376

Shi, Z., Cai, Z., Wang, S., Zhong, Q., & Bozell, J. J. (2013). Short-time ultrasonication treatment in enzymatic hydrolysis of biomass. *Holzforschung, 67*(8), 891–897. https://doi.org/10.1515/hf-2013-0024

Shi, S., Guan, W., Blersch, D., & Li, J. (2021). Improving the enzymatic digestibility of alkaline-pretreated lignocellulosic biomass using polyDADMAC. *Industrial Crops and Products, 162*, 113244. https://doi.org/10.1016/j.indcrop.2021.113244

Sidana, A., & Yadav, S. K. (2022). Recent developments in lignocellulosic biomass pretreatment with a focus on eco-friendly, non-conventional methods. *Journal of Cleaner Production, 335*, 130286. https://doi.org/10.1016/j.jclepro.2021.130286

Sidiras, D., Politi, D., Giakoumakis, G., & Salapa, I. (2022). Simulation and optimization of organosolv based lignocellulosic biomass refinery: A review. *Bioresource Technology, 343*, 126158. https://doi.org/10.1016/j.biortech.2021.126158

Singh, R., Kumar, R., Sarangi, P. K., Kovalev, A. A., & Vivekanand, V. (2023). Effect of physical and thermal pretreatment of lignocellulosic biomass on biohydrogen production by thermochemical route: A critical review. *Bioresource Technology, 369*, 128458. https://doi.org/10.1016/j.biortech.2022.128458

Sirviö, J. A., Mikola, M., Ahola, J., Heiskanen, J. P., Filonenko, S., & Ämmälä, A. (2023). Highly effective fractionation chemistry to overcome the recalcitrance of softwood lignocellulose. *Carbohydrate Polymers, 312*, 120815. https://doi.org/10.1016/j.carbpol.2023.120815

Sohni, S., Hashim, R., Nidaullah, H., Sulaiman, O., Leh, C. P., Lamaming, J., Arai, T., & Kosugi, A. (2020). Enhancing the enzymatic digestibility of oil palm biomass using supercritical carbon dioxide-based pretreatment towards biorefinery application. *Industrial Crops and Products, 157*, 112923. https://doi.org/10.1016/j.indcrop.2020.112923

Sooch, B. S., Lugani, Y., & Singh, R. S. (2019). 2. Agro-industrial lignocellulosic residues for the production of industrial enzymes. In *Industrial biotechnology* (pp. 31–50). De Gruyter. https://doi.org/10.1515/9783110563337-002

Souza-Corrêa, J. A., Oliveira, C., Wolf, L. D., Nascimento, V. M., Rocha, G. J. M., & Amorim, J. (2013). Atmospheric pressure plasma pretreatment of sugarcane bagasse: The influence of moisture in the ozonation process. *Applied Biochemistry and Biotechnology, 171*(1), 104–116. https://doi.org/10.1007/s12010-013-0362-4

Sridevi, A., Narasimha, G., Ramanjaneyulu, G., Dileepkumar, K., Reddy, B. R., & Devi, P. S. (2015). Saccharification of pretreated sawdust by Aspergillus niger cellulase. *3 Biotech, 5*(6), 883–892. https://doi.org/10.1007/s13205-015-0284-7

Subhedar, P. B., Ray, P., & Gogate, P. R. (2018). Intensification of delignification and subsequent hydrolysis for the fermentable sugar production from lignocellulosic biomass using ultrasonic irradiation. *Ultrasonics Sonochemistry, 40*, 140–150. https://doi.org/10.1016/j.ultsonch.2017.01.030

Sulzenbacher, D., Atzmüller, D., Hawe, F., Richter, M., Cristobal-Sarramian, A., & Zwirzitz, A. (2021). Optimization of steam explosion parameters for improved biotechnological use of wheat straw. *Biomass Conversion and Biorefinery, 13*, 1035–1046. https://doi.org/10.1007/s13399-020-01266-z

Sun, D., Wang, B., Wang, H.-M., Li, M.-F., Shi, Q., Zheng, L., Wang, S.-F., Liu, S.-J., & Sun, R.-C. (2019a). Structural transformations of hybrid *Pennisetum* lignin: Effect of microwave-assisted hydrothermal pretreatment. *ACS Sustainable Chemistry & Engineering, 7*(3), 3073–3082. https://doi.org/10.1021/acssuschemeng.8b04695

Sun, Y.-C., Liu, X.-N., Wang, T.-T., Xue, B.-L., & Sun, R.-C. (2019b). Green process for extraction of lignin by the microwave-assisted ionic liquid approach: Toward biomass biorefinery and lignin characterization. *ACS Sustainable Chemistry & Engineering, 7*(15), 13062–13072. https://doi.org/10.1021/acssuschemeng.9b02166

Sun, D., Lv, Z. W., Rao, J., Tian, R., Sun, S. N., & Peng, F. (2022). Effects of hydrothermal pretreatment on the dissolution and structural evolution of hemicelluloses and lignin: A review. *Carbohydrate Polymers, 281*, 119050. https://doi.org/10.1016/j.carbpol.2021.119050

Sun, C., Song, G., Pan, Z., Tu, M., Kharaziha, M., Zhang, X., Show, P. L., & Sun, F. (2023a). Advances in organosolv modified components occurring during the organosolv pretreatment of lignocellulosic biomass. *Bioresource Technology, 368*, 128356. https://doi.org/10.1016/j.biortech.2022.128356

Sun, S., Wang, Q., Wang, X., Wu, C., Zhang, X., Bai, J., & Sun, B. (2023b). Dry torrefaction and continuous thermochemical conversion for upgrading agroforestry waste into eco-friendly energy carriers: Current progress and future prospect. *Science of the Total Environment, 905*, 167061. https://doi.org/10.1016/j.scitotenv.2023.167061

Sun, X., Zhou, Z., Tian, D., Zhao, J., Zhang, J., Deng, P., Zou, H., & Lu, C. (2023c). Acidic deep eutectic solvent assisted mechanochemical delignification of lignocellulosic biomass at room temperature. *International Journal of Biological Macromolecules, 234*, 123593. https://doi.org/10.1016/j.ijbiomac.2023.123593

Szadkowska, D., Zawadzki, J., & Kozakiewicz, P. (2021). Identification of extractives from various poplar species. *Forests, 12*(5), 647. https://doi.org/10.3390/f12050647

Taherzadeh-Ghahfarokhi, M., Panahi, R., & Mokhtarani, B. (2019). Optimizing the combination of conventional carbonaceous additives of culture media to produce lignocellulose-degrading enzymes by Trichoderma reesei in solid state fermentation of agricultural residues. *Renewable Energy, 131*, 946–955. https://doi.org/10.1016/j.renene.2018.07.130

Tang, X., Lv, S., Jiang, K., Zhou, G., & Liu, X. (2022). Recent development of ionic liquid-based electrolytes in lithium-ion batteries. *Journal of Power Sources, 542*, 231792. https://doi.org/10.1016/j.jpowsour.2022.231792

Terán Hilares, R., Kamoei, D. V., Ahmed, M. A., da Silva, S. S., Han, J.-I., & dos Santos, J. C. (2018). A new approach for bioethanol production from sugarcane bagasse using hydrodynamic cavitation assisted-pretreatment and column reactors. *Ultrasonics Sonochemistry, 43*, 219–226. https://doi.org/10.1016/j.ultsonch.2018.01.016

Terán Hilares, R., Dionízio, R. M., Sánchez Muñoz, S., Prado, C. A., de Sousa Júnior, R., da Silva, S. S., & Santos, J. C. (2020). Hydrodynamic cavitation-assisted continuous pre-treatment of sugarcane bagasse for ethanol production: Effects of geometric parameters of the cavitation device. *Ultrasonics Sonochemistry, 63*, 104931. https://doi.org/10.1016/j.ultsonch.2019.104931

Timoteo-Cruz, B., Gómez-Espinosa, R. M., Sánchez-Orozco, R., García-Sánchez, J. J., Ureña-Núñez, F., & Roa-Morales, G. (2024). Pretreatment of Typha latifolia biomass with imidazolium ionic liquid as a strategy for sugar production. *Biomass Conversion and Biorefinery, 14*(4), 4591–4604. https://doi.org/10.1007/s13399-022-02685-w

Tukacs, J. M., Holló, A. T., Rétfalvi, N., Cséfalvay, E., Dibó, G., Havasi, D., & Mika, L. T. (2017). Microwave-assisted valorization of biowastes to levulinic acid. *ChemistrySelect, 2*(4), 1375–1380. https://doi.org/10.1002/slct.201700037

Ubando, A. T., Ng, E. A. S., Chen, W. H., Culaba, A. B., & Kwon, E. E. (2022). Life cycle assessment of microalgal biorefinery: A state-of-the-art review. *Bioresource Technology, 360*, 127615. https://doi.org/10.1016/j.biortech.2022.127615

Velmurugan, R., Kuhad, R. C., Gupta, R., Chakraborty, S., Ashokkumar, V., & Incharoensakdi, A. (2023). Ionic liquid under alkaline aqueous conditions improves corncob delignification, polysaccharide recovery, enzymatic hydrolysis and ethanol fermentation. *Biomass and Bioenergy, 178*, 106980. https://doi.org/10.1016/j.biombioe.2023.106980

Vu, H. P., Nguyen, L. N., Vu, M. T., Johir, M. A. H., McLaughlan, R., & Nghiem, L. D. (2020). A comprehensive review on the framework to valorise lignocellulosic biomass as biorefinery feedstocks. *Science of the Total Environment, 743*, 140630. https://doi.org/10.1016/j.scitotenv.2020.140630

Wang, L., Liu, N., Guo, Z., Wu, D., Chen, W., Chang, Z., Yuan, Q., Hui, M., & Wang, J. (2016). Nitric acid-treated carbon fibers with enhanced hydrophilicity for Candida tropicalis immobilization in xylitol fermentation. *Materials, 9*(3), 206. https://doi.org/10.3390/ma9030206

Wang, C., Shao, Z., Qiu, L., Hao, W., Qu, Q., & Sun, G. (2021a). The solid-state physicochemical properties and biogas production of the anaerobic digestion of corn straw pretreated by microwave irradiation. *RSC Advances, 11*(6), 3575–3584. https://doi.org/10.1039/D0RA09867A

Wang, J., Zhang, S., Ma, Z., & Yan, L. (2021b). Deep eutectic solvents eutectogels: Progress and challenges. *Green Chemical Engineering, 2*(4), 359–367. https://doi.org/10.1016/j.gce.2021.06.001

Wang, R., Wang, K., Zhou, M., Xu, J., & Jiang, J. (2021c). Efficient fractionation of moso bamboo by synergistic hydrothermal-deep eutectic solvents pretreatment. *Bioresource Technology, 328*, 124873. https://doi.org/10.1016/j.biortech.2021.124873

Wang, Z., Zhang, F., Lv, D., Jiang, G., Zhang, H., Gao, G., Ning, A., Lam, S. S., & Song, A. (2021d). Iron oxychloride-based heterogeneous Fenton pretreatment of corn stover for enhanced sugars production. *Chemical Engineering Journal, 416*, 127703. https://doi.org/10.1016/j.cej.2020.127703

Wang, F., Ouyang, D., Zhou, Z., Page, S. J., Liu, D., & Zhao, X. (2021e). Lignocellulosic biomass as sustainable feedstock and materials for power generation and energy storage. *Journal of Energy Chemistry, 57*, 247–280. https://doi.org/10.1016/j.jechem.2020.08.060

Wang, Z., Huang, Y., Zhang, F., Xie, H., Jiang, G., Lv, D., Zhang, H., Lam, S. S., & Song, A. (2022). Improving enzymatic saccharification of corn stover via thioglycolic acid-mediated Fenton pretreatment. *Journal of Cleaner Production, 365*, 132804. https://doi.org/10.1016/j.jclepro.2022.132804

Wang, L., Bu, Y., Sun, L., & Chen, H. (2023a). A sequential combination of advanced oxidation and enzymatic hydrolysis reduces the enzymatic dosage for lignocellulose degradation. *Renewable Energy, 211*, 617–625. https://doi.org/10.1016/j.renene.2023.04.129

Wang, Z. K., Liu, Y., Zhong, J., Huan, W., Sheng, J., Xu, C., Chen, L., & Shen, X. (2023b). A waste-free biorefinery pathway to the valorisation of Chinese hickory shell through alkaline hydrogen peroxide pretreatment. *Chemical Engineering Journal, 476*, 146657. https://doi.org/10.1016/j.cej.2023.146657

Wang, N., Xu, A., Liu, K., Zhao, Z., Li, H., & Gao, X. (2024a). Performance of green solvents in microwave-assisted pretreatment of lignocellulose. *Chemical Engineering Journal, 482*, 148786. https://doi.org/10.1016/j.cej.2024.148786

Wang, Q., Cao, K., Yan, F., Duan, X., & Shi, J. (2024b). Alkaline deep eutectic solvent for separation and extraction of lignin from lignocellulose biomass. *Biomass Conversion and Biorefinery, 15*, 1–18. https://doi.org/10.1007/s13399-023-05108-6

Wei, W., Shao, Z., Zhang, Y., Qiao, R., & Gao, J. (2019). Fundamentals and applications of microwave energy in rock and concrete processing—A review. *Applied Thermal Engineering, 157*, 113751. https://doi.org/10.1016/j.applthermaleng.2019.113751

Wen, P., Chu, J., Zhu, J., Xu, Y., & Zhang, J. (2022). Highly selective delignification of poplar by hydrogen peroxide-ethyl acetate pretreatment at room temperature. *Renewable Energy, 188*, 1022–1028. https://doi.org/10.1016/j.renene.2022.02.094

Wijaya, C. J., Ismadji, S., & Gunawan, S. (2021). A review of lignocellulosic-derived nanoparticles for drug delivery applications: Lignin nanoparticles, xylan nanoparticles, and cellulose nanocrystals. *Molecules, 26*(3), 676. https://doi.org/10.3390/molecules26030676

Witaszek, K., Pilarski, K., Niedbała, G., Pilarska, A. A., & Herkowiak, M. (2020). Energy efficiency of comminution and extrusion of maize substrates subjected to methane fermentation. *Energies, 13*(8), 1887. https://doi.org/10.3390/en13081887

Worku, L. A., Bachheti, R. K., Bachheti, A., Milessi, T. S., & Chandel, A. K. (2024). Understanding the biochemical changes at molecular level during biomass pretreatment: A comprehensive analysis. *Cellulose, 31*, 7281–7312. https://doi.org/10.1007/s10570-024-06081-7

Wright, A., Rollinson, A., Yadav, D., Lisowski, S., Iza, F., Holdich, R., Radu, T., & Hemaka Bandulasena, H. C. (2020). Plasma-assisted pre-treatment of lignocellulosic biomass for anaerobic digestion. *Food and Bioproducts Processing, 124*, 287–295. https://doi.org/10.1016/j.fbp.2020.09.005

Wu, H., Dai, X., Zhou, S.-L., Gan, Y.-Y., Xiong, Z.-Y., Qin, Y.-H., Ma, J., Yang, L., Wu, Z.-K., Wang, T.-L., Wang, W.-G., & Wang, C.-W. (2017). Ultrasound-assisted alkaline pretreatment for enhancing the enzymatic hydrolysis of rice straw by using the heat energy dissipated from ultrasonication. *Bioresource Technology, 241*, 70–74. https://doi.org/10.1016/j.biortech.2017.05.090

Wu, L., Liu, S., Wang, Q., Wang, Y., Ji, X., Yang, G., Chen, J., Li, C., & Fatehi, P. (2022a). High strength and multifunctional polyurethane film incorporated with lignin nanoparticles. *Industrial Crops and Products, 177*, 114526. https://doi.org/10.1016/j.indcrop.2022.114526

Wu, Z., Peng, K., Zhang, Y., Wang, M., Yong, C., Chen, L., Qu, P., Huang, H., Sun, E., & Pan, M. (2022b). Lignocellulose dissociation with biological pretreatment towards the biochemical platform: A review. *Materials Today Bio, 16*, 100445. https://doi.org/10.1016/j.mtbio.2022.100445

Xia, Y., Liu, Q., Hu, X., Li, X., Huang, Y., Li, W., & Ma, L. (2022). Structural evolution during corn stalk acidic and alkaline hydrogen peroxide pretreatment. *Industrial Crops and Products, 176*, 114386. https://doi.org/10.1016/j.indcrop.2021.114386

Xiang, Z., Han, W., Deng, J., Zhu, W., Zhang, Y., & Wang, H. (2020). Photocatalytic conversion of lignin into chemicals and fuels. *ChemSusChem, 13*(17), 4199–4213. https://doi.org/10.1002/cssc.202000601

Xu, Z., Lei, P., Zhai, R., Wen, Z., & Jin, M. (2019). Recent advances in lignin valorization with bacterial cultures: Microorganisms, metabolic pathways, and bio-products. *Biotechnology for Biofuels, 12*(1), 1–19. https://doi.org/10.1186/s13068-019-1376-0

Xu, Z., Pan, C., Li, X., Hao, N., Zhang, T., Gaffrey, M. J., Pu, Y., Cort, J. R., Ragauskas, A. J., Qian, W. J., & Yang, B. (2021). Enhancement of polyhydroxyalkanoate production by co-feeding lignin derivatives with glycerol in Pseudomonas putida KT2440. *Biotechnology for Biofuels, 14*(1), 11. https://doi.org/10.1186/s13068-020-01861-2

Xu, Y., Ren, T., Wu, J., Meng, G., Yang, S., Cui, L., Liu, Z., & Guo, X. (2023). Ultrasound-assisted formic acid–choline chloride deep eutectic solvent pretreatment of cotton straw to extracted lignin. *Journal of Applied Polymer Science, 140*(30), e54082. https://doi.org/10.1002/app.54082

Yamazaki, K., Sasaki, R., Watanabe, T., Kuwano, S., Murakami, Y., Mimura, N., Sato, O., & Yamaguchi, A. (2021). Effect of catalyst support on aromatic monomer production from lignocellulosic biomass over Pt-based catalysts. *Waste and Biomass Valorization, 12*(11), 6081–6089. https://doi.org/10.1007/s12649-021-01423-z

Yan, Y., He, C., Zhang, L., Dong, H., & Zhang, X. (2023a). Freeze-resistant, rapidly polymerizable, ionic conductive hydrogel induced by Deep Eutectic Solvent (DES) after lignocellulose pretreatment for flexible sensors. *International Journal of Biological Macromolecules, 224*, 143–155. https://doi.org/10.1016/j.ijbiomac.2022.10.111

Yan, Z., Wang, Z., Chen, Y., Liu, C., Liu, Y., Li, R., Si, M., & Shi, Y. (2023b). Preparation of lignin nanoparticles via ultra-fast microwave-assisted fractionation of lignocellulose using ternary deep eutectic solvents. *Biotechnology and Bioengineering, 120*(6), 1557–1568. https://doi.org/10.1002/bit.28373

Yao, Y., Takada, M., Minami, E., & Kawamoto, H. (2022). Influence of proteins on the lignin decomposition behavior of Japanese cedar (Cryptomeria japonica) wood by supercritical methanol treatment. *ACS Omega, 7*(42), 37286–37292. https://doi.org/10.1021/acsomega.2c03716

Yusaf, T., & Al-Juboori, R. A. (2014). Alternative methods of microorganism disruption for agricultural applications. *Applied Energy, 114*, 909–923. https://doi.org/10.1016/j.apenergy.2013.08.085

Zannini, D., Dal Poggetto, G., Malinconico, M., Santagata, G., & Immirzi, B. (2021). Citrus pomace biomass as a source of pectin and lignocellulose fibers: From waste to upgraded biocomposites for mulching applications. *Polymers, 13*(8), 1280. https://doi.org/10.3390/polym13081280

Zhang, Q., Zhao, M., Xu, Q., Ren, H., & Yin, J. (2019). Enhanced enzymatic hydrolysis of sorghum stalk by supercritical carbon dioxide and ultrasonic pretreatment. *Applied Biochemistry and Biotechnology, 188*(1), 101–111. https://doi.org/10.1007/s12010-018-2909-x

Zhang, Z., Li, Y., Zhang, J., Peng, N., Liang, Y., & Zhao, S. (2020). High-titer lactic acid production by Pediococcus acidilactici PA204 from corn stover through fed-batch simultaneous saccharification and fermentation. *Microorganisms, 8*(10), 1–9. https://doi.org/10.3390/microorganisms8101491

Zhang, Y., Guo, Y., Xie, X., Chernyshev, V. M., Liu, Y., & Qi, W. (2023). Effects of phosphoric acid/hydrogen peroxide, ammonia/hydrogen peroxide and deep eutectic solvent pretreatments on component separation and enzymatic saccharification of Glycyrrhiza residue. *Industrial Crops and Products, 196*, 116525. https://doi.org/10.1016/j.indcrop.2023.116525

Zhang, Q., Hu, Y., Jiao, J., & Wang, S. (2024). The impact of Russia–Ukraine war on crude oil prices: An EMC framework. *Humanities and Social Sciences Communications, 11*(1), 8. https://doi.org/10.1057/s41599-023-02526-9

Zhang, W., Liu, J., Wang, Y., Sun, J., Huang, P., & Chang, K. (2021). *Effect of ultrasound on ionic liquid-hydrochloric acid pretreatment with rice straw.* https://doi.org/10.1007/s13399-019-00595-y/Published

Zhao, C., Shao, Q., & Chundawat, S. P. S. (2020). Recent advances on ammonia-based pretreatments of lignocellulosic biomass. *Bioresource Technology, 298*, 122446. https://doi.org/10.1016/j.biortech.2019.122446

Zhou, M., & Tian, X. (2022). Development of different pretreatments and related technologies for efficient biomass conversion of lignocellulose. *International Journal of Biological Macromolecules, 202*, 256–268. https://doi.org/10.1016/j.ijbiomac.2022.01.036

Zhou, Z., Liu, D., & Zhao, X. (2021). Conversion of lignocellulose to biofuels and chemicals via sugar platform: An updated review on chemistry and mechanisms of acid hydrolysis of lignocellulose. *Renewable and Sustainable Energy Reviews, 146*, 111169. https://doi.org/10.1016/j.rser.2021.111169

Zhou, H., Guan, Y., Yan, X., Pan, Z., Xu, J., Dai, L., Zhang, M., & Si, C. (2023a). All-lignocellulose-based hard bioplastic. *Industrial Crops and Products, 193*, 116164. https://doi.org/10.1016/j.indcrop.2022.116164

Zhou, Z., Ouyang, D., Liu, D., & Zhao, X. (2023b). Oxidative pretreatment of lignocellulosic biomass for enzymatic hydrolysis: Progress and challenges. *Bioresource Technology, 367*, 128208. https://doi.org/10.1016/j.biortech.2022.128208

Zhou, K., Dai, X., Li, P., Zhang, L., Zhang, X., Wang, C., Wen, J., Huang, G., & Xu, S. (2024). Recent advances in deep eutectic solvents for next-generation lithium batteries: Safer and greener. *Progress in Materials Science, 146*, 101338. https://doi.org/10.1016/j.pmatsci.2024.101338

Zhu, S., Wu, Y., Yu, Z., Zhang, X., Wang, C., Yu, F., Jin, S., Zhao, Y., Tu, S., & Xue, Y. (2005). Simultaneous saccharification and fermentation of microwave/alkali pre-treated rice straw to ethanol. *Biosystems Engineering, 92*(2), 229–235. https://doi.org/10.1016/j.biosystemseng.2005.06.012

Zhu, S., Wu, Y., Yu, Z., Zhang, X., Li, H., & Gao, M. (2006). The effect of microwave irradiation on enzymatic hydrolysis of rice straw. *Bioresource Technology, 97*(15), 1964–1968. https://doi.org/10.1016/j.biortech.2005.08.008

Zhu, H., Han, Z., Cheng, J. H., & Sun, D. W. (2022). Modification of cellulose from sugarcane (Saccharum officinarum) bagasse pulp by cold plasma: Dissolution, structure and surface chemistry analysis. *Food Chemistry, 374*, 131675. https://doi.org/10.1016/j.foodchem.2021.131675

Zieliński, M., Dębowski, M., Kisielewska, M., Nowicka, A., Rokicka, M., & Szwarc, K. (2019). Comparison of ultrasonic and hydrothermal cavitation pretreatments of cattle manure mixed with straw wheat on fermentative biogas production. *Waste and Biomass Valorization, 10*(4), 747–754. https://doi.org/10.1007/s12649-017-9977-y

Zoghlami, A., Paës, G., Luis Sanchez, J., Hodge, D. B., Davis, M. F., Paës, G., & Zoghlami, A. (2019). Lignocellulosic biomass: Understanding recalcitrance and predicting hydrolysis. *Frontiers in Chemistry, 7*(874), 874. https://doi.org/10.3389/fchem.2019.00874

Zubeltzu, J., Formoso, E., & Rezabal, E. (2020). Lignin solvation by ionic liquids: The role of cation. *Journal of Molecular Liquids, 303*, 112588. https://doi.org/10.1016/j.molliq.2020.112588

Novel Approaches to Alleviate Lignocellulosic Biomass Waste for Effective and Sustainable Utilization

Dhruv Agrawal, Sandeep Kaur Saggu, Sapna Devi, Shiv Kumar, and Manoj Kumar

Abstract

Lignocellulosic biomass (LCB) comprises plant dry matter holding immense potential to be used as a renewable resource material for obtaining sustainable energy. LCB can be used for the large-scale production of second-generation biofuels contributing to climate change mitigation and reducing dependence on fossil fuel. Furthermore, India, with its strong agricultural economy, generates substantial plant biomass that can be harnessed for biofuel production. However, the efficient use of LCB in various products can be challenged by processing and conversion facilities due to its complex structural architecture. In the past, various technologies have been utilized for the deconstruction of LCB structure, viz., physicochemical methods, chemical pretreatments, microbial or biological and enzymatic hydrolysis. Recently, novel treatment technologies, viz., ionic liquids, deep eutectic solvents, supercritical fluids, and ultrasound-assisted treatments,

D. Agrawal
Umeå Plant Science Centre, Department of Plant Physiology, Umeå University, Umeå, Sweden

Department of Microbiology, Guru Nanak Dev University, Amritsar, Punjab, India

S. K. Saggu
Department of Biotechnology, Kanya Maha Vidyalaya, Jalandhar, Punjab, India

S. Devi
Department of Microbiology, DAV University, Jalandhar, Punjab, India

S. Kumar
Department of Microbiology, Guru Gobind Singh Medical College & Hospital, Baba Farid University of Health Sciences, Faridkot, Punjab, India

M. Kumar (✉)
Department of Microbiology, Guru Nanak Dev University, Amritsar, Punjab, India
e-mail: manojmicro.rsh@gndu.ac.in

G. Mukherjee, S. Dhiman (eds.), *Value Addition and Utilization of Lignocellulosic Biomass*, https://doi.org/10.1007/978-981-96-2786-8_4

have been developed to achieve efficient deconstruction of LCB, thereby simplifying downstream processing. This chapter aims to provide a comprehensive outline of the structural complexity of lignocellulosic materials, chemical composition, and further discussion of various novel technologies for its efficient conversion into a vast variety of value-added products with industrial importance. The advent of novel technologies has ushered in a transformative era in lignocellulosic biomass mitigation, offering promising solutions for a circular economy and a more sustainable future.

Keywords

Lignocellulosic biomass · Biofuels · Pretreatment · Value-added products · Novel technologies for LCB management · Sustainable development goals · Renewable energy

1 Introduction

Lignocellulosic biomass (LCB) comprises plant dry matter, mainly agricultural/forest waste, making it a vast and underutilized resource for the production of renewable energy and bioproducts. This plant-based biomass is mainly composed of several molecules, viz., cellulose, hemicellulose, and lignin, which form a complex and recalcitrant structure (Bajpai, 2021; Ranjan et al., 2023). LCB is primarily found in various crop residues having 38–50% cellulose, 23–32% hemicellulose, and 15–25% lignin with small amounts of other components like acetyl groups, minerals, and phenolic substituents which are recalcitrant to degradation (Mujtaba et al., 2023). Traditional methods of processing lignocellulosic biomass often struggle with inefficiencies and high costs due to the rigid nature of this material. Therefore, novel approaches to alleviate lignocellulosic biomass waste can play a critical role in its effective and sustainable utilization (Mosier et al., 2005). The initial processing steps for the utilization of LCB waste require pretreatment which can be categorized into four methods, viz., physicochemical pretreatment, physical pretreatment, chemical pretreatment, and biological pretreatment (Dahadha et al., 2017). Recently, various novel treatment technologies, viz., the use of ionic liquids, deep eutectic solvents, supercritical fluids, and ultrasound-assisted treatments, can be used for the treatment of LCB with sustainability as a key consideration (Roy et al., 2020; Saini et al., 2022; Sandeep et al., 2024). Furthermore, the development of more efficient and versatile enzymes mainly through genetic/protein engineering techniques has significantly improved the hydrolysis of cellulose and hemicellulose into fermentable sugars (Lynd et al., 2002). In addition to this, integrating metabolic engineering with systems biology has enabled the design of microbial strains tailored for specific bioconversion pathways, thus optimizing the production of desired products. The use of LCB not only provides a renewable source of energy and materials but also helps in reducing greenhouse gas emissions and dependency on fossil fuels. Moreover, the use of LCB waste for the production of biofuels, bioplastics, and platform chemicals using integrated biorefineries ensures a more efficient and economically viable utilization of this resource (Cherubini, 2010). This chapter discusses the challenges of utilizing LCB through various pretreatment methods and novel treatment technologies, integrated with sustainable practices that hold significant promise for advancing the renewable energy and bioproduct industries. These novel approaches pave the way for a more sustainable and efficient use of biomass wastes, contributing to a circular bioeconomy.

2 Lignocellulose: Structural Complexity

Lignocellulosic biomass, primarily composed of crop residues, is a sustainable, widespread, and abundant, localized natural resource for large-scale and cost-effective production of value-added products like biofuels in biorefineries (Anwar et al., 2014; Ranjan et al., 2023). Crop residue-based LCB typically comprised of 38–50% cellulose, 23–32% hemicellulose, and 15–25% lignin with small amounts of other components like acetyl groups, minerals, and phenolic substituents (Fig. 2.1) is present as a complex assembly of recalcitrant polymers (Bajpai, 2021; Maheshwari, 2018; Zoghlami & Paës, 2019). The factors responsible for LCB recalcitrance are the crystallinity of cellulose, hydrophobicity of lignin, and encapsulation of cellulose by the lignin-hemicellulose matrix (Moraïs et al., 2012; Mujtaba et al., 2023). Some physical and/or chemical pretreatment of LCB is currently considered as an essential step to make cellulose more accessible to facilitate its conversion into fermentable platform sugars and by-products (Johannes & Xuan, 2024; Kumar & Sharma, 2017; Zhao, Cao, et al., 2012a). The varying distribution of cellulose, hemicellulose, and lignin content in different LCBs that depends on the species of the plant, tissue type, and the maturity level of the cell walls (Barakat et al., 2013; Grossmann, 2024) results in different conformations and configuration of the substrate because of pretreatment (Pal et al., 2016). Table 1 summarizes the carbohydrate composition of different types of lignocellulosic biomass (based on their origin).

2.1 Generalized Yet Complex Structure of LCB

Cellulose, the main structural component of the plant cell wall, is a linear homopolymer of D-glucose subunits (500–15,000) containing β-1,4-glycosidic linkages (Saini et al., 2015). The organized fibrous structure contains repetitive subunits of D-glucose dimers (cellobiose). The hydrogen bonds (intra and inter) formed due to β-1,4-glycosidic linkages make the native cellulose very tough, crystalline, and resistant to hydrolysis (Fig. 2.2). Depending on the source of the plant cell wall, the highly crystalline cellulose structure is interspersed with the amorphous form of unorganized branched polymeric cellulosic chains which constitutes only a small percentage of the total cellulose (Meng & Ragauskas, 2014). The crystalline cellulose can exist as different allomorphs I, II, and III, with I to be the most recalcitrant

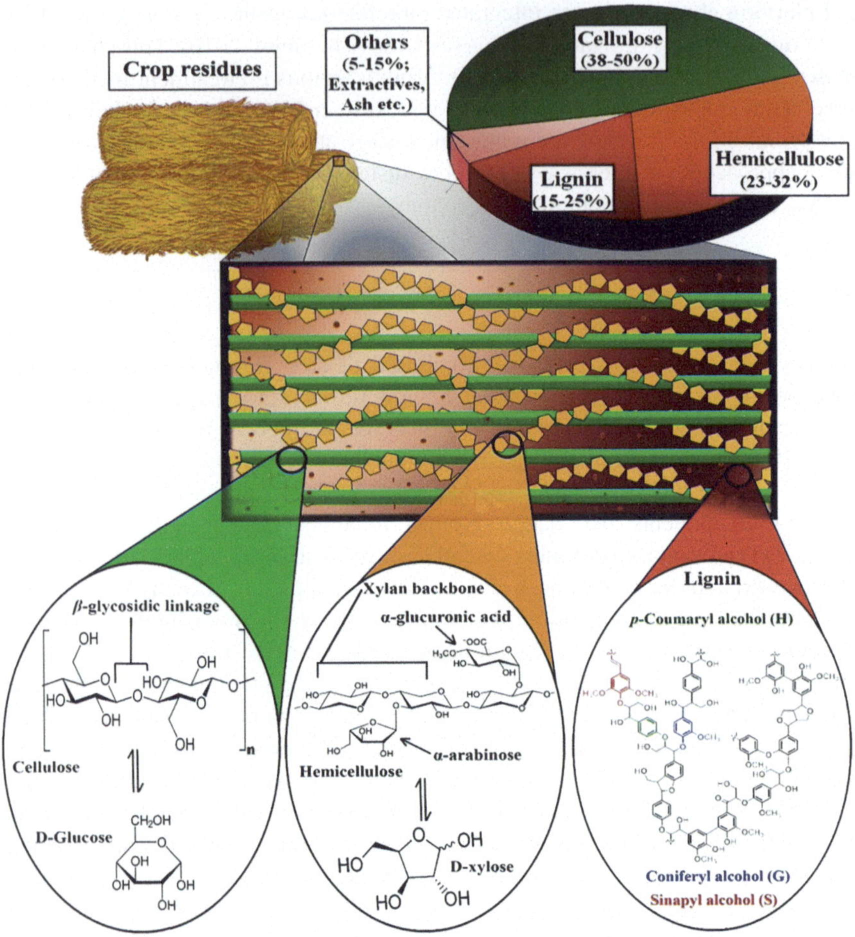

Fig. 2.1 Diagrammatic representation of the lignocellulosic biomass and its biopolymers (cellulose, hemicellulose, and lignin) with their building blocks

form and III the least (Chundawat et al., 2011). The crop residues (Table 1) primarily contain hemicellulose as an amorphous and variable hetero-polymer with xylan backbone containing β-1,4-linked D-xylose subunits (Fig. 2.3). The chains of hemicellulose comprise approximately 90% D-xylose and approximately 10% L-arabinose units that are present as substituents and are α-linked to 3 or 5 positions of xylose carbon. In addition, the heterogeneous hemicellulose may have different hexose (D-glucose, D-galactose, and D-mannose) subunits and sugar acids like uronic acids (D-glucuronic, D-galacturonic, and methyl galacturonic acids) as substituents, depending on the source of the LCB (Limayem & Ricke, 2012; Socha et al., 2014). The hemicelluloses contribute to the strengthening of the plant cell wall by forming a cross-linkage between cellulose and lignin (Alonso et al., 2013;

Table 1 Carbohydrate composition of lignocellulosic crop residue

Crop residue	Carbohydrate composition (% dry weight)			References
	Cellulose	Hemicellulose	Lignin	
Sugarcane bagasse	25.0–45.0	28.0–32.0	15.0–25.0	Alves et al. (2010)
Rice straw	29.2–34.7	23.0–25.9	17.0–19.0	Brylev et al. (2001)
Rice husks	28.7–35.6	12.0–29.3	15.4–20.0	Abbas & Ansumali (2010)
Wheat straw	35.0–39.0	23.0–30.0	12.0–16.0	Prasad et al. (2007)
Sorghum straw	32.0–35.0	24.0–27.0	15.0–21.0	Vázquez et al. (2007)
Corn Stover	35.1–39.5	20.7–24.6	11.0–19.1	Mosier et al. (2005)
Winter rye	29–30	22–26	16.1	Petersson et al. (2007)
Barley straw	36.0–43.0	24.0–33.0	6.3–9.8	García-Aparicio et al. (2006)
Barley hull	34.0	36.0	13.8–19.0	Kim et al. (2008)
Wheat bran	10.5–14.8	35.5–39.2	8.3–12.5	Miron et al. (2001)
Banana waste	13	15	14	Monsalve et al. (2006)
Nut shells	25–30	22–28	30–40	Sinner et al. (1979)
Corn cobs	33.7–41.2	31.9–36.0	6.1–15.9	Cao et al. (1997)
Jute fibers	45–53	18–21	21–26	Mosihuzzaman et al. (1989)
Oat straw	31.0–35.0	20.0–26.0	10.0–15.0	Rowell et al. (1992)

Dodd & Cann, 2009; Scheller & Ulvskov, 2010). Lignin is amorphous, irregular, three-dimensional, and a highly branched phenolic polymer composed mainly of ether-bond-linked phenylpropane units that may account for up to 20% of LCB. The chemical structure of lignin is extremely complex and hydrophobic, which largely remains unutilized during the bioconversion of cellulose and hemicellulose into fuels (Poovaiah et al., 2014; Zeng et al., 2014). Lignin is synthesized by three basic phenylpropanoid monomeric building blocks, which are p-hydroxyphenyl (H), guaiacyl (G), and syringyl (S) units, derived from p-coumaryl, coniferyl, and sinapyl alcoholic precursors, respectively (Vanholme et al., 2010). The composition of these monolignols in lignin varies according to the type of LCB and is integrated into plant cell wall structure through alkyl-aryl, alkyl-alkyl, and aryl-aryl ether bonds (Lu et al., 2017). A smaller portion of LCB composition is made up of extractives/volatiles and ash. These non-structural components (NSC) of the LCB can be extracted from the cell wall using different solvents such as water, ethanol, benzene, phenols, toluene, and their mixtures (Davison et al., 2013; Sluiter et al., 2008). These non-structural components of LCB have a major influence on the efficiency of the bioconversion processes. The extractives/volatiles include water (non-structural sugars and proteins) and ethanol (chlorophyll and waxes) soluble components (Williams et al., 2017). The inorganic content (ash) present in LCB (Reza et al., 2015) can either be intrinsic to the biomass (materials like Ca^{2+} and K^{+}) or introduced through human activity (primarily silica collected during harvest).

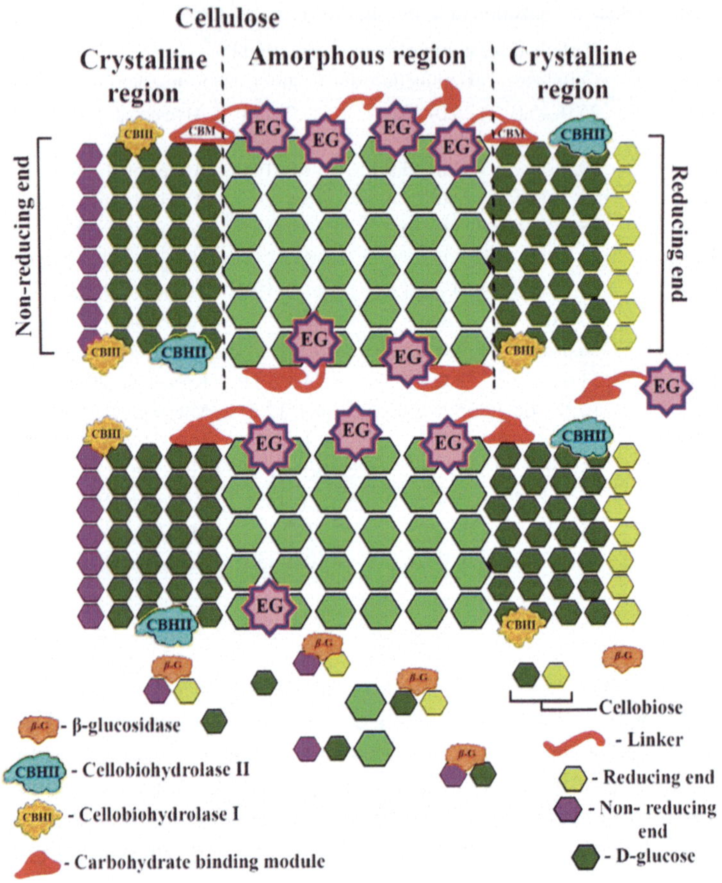

Fig. 2.2 Diagram depicting the complex structure of cellulose and various naturally occurring cellulases (endoglucanase, EG; cellobiohydrolase, CBH; and β-glucosidases, β-G), which can help in the conversion of cellulose into monomers (D-glucose) by their synergistic action

3 LCB Waste Generation: Statistics

Plant materials provide lignocellulosic biomass, which is an important source of bioenergy and biobased products. It mostly consists of forestry residues, certain energy crops, and agricultural wastes (such as straw, husks, and stalks). The production of lignocellulosic biomass is significant worldwide and is essential to the creation of biobased products and sustainable energy. An estimated 181.5 billion metric tons of lignocellulosic biomass are produced annually worldwide (Rodrigues et al., 2024). Agricultural residues constitute a significant portion of lignocellulosic biomass, including straw, husks, and stalks. Approximately 3.8 billion metric tons of

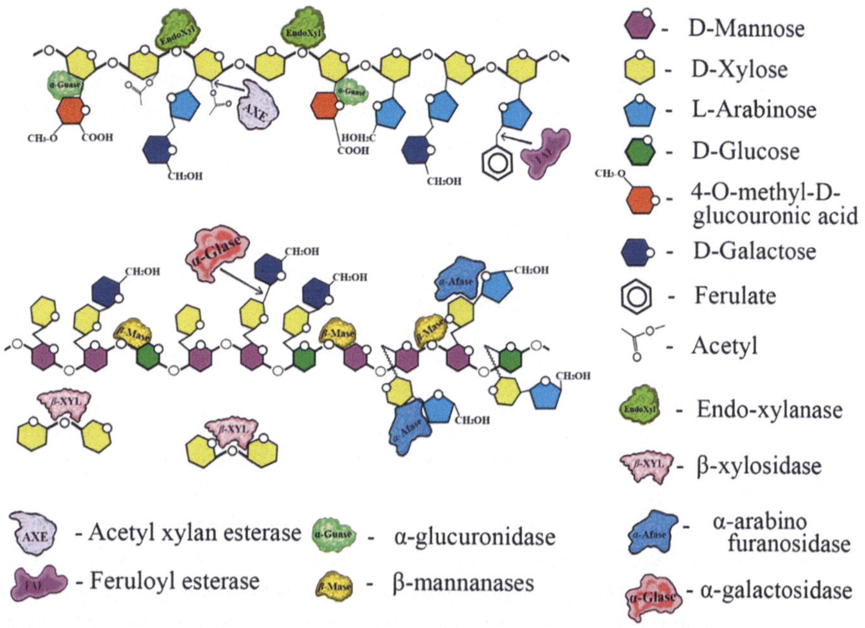

Fig. 2.3 Diagrammatic representation of the complex structure of hemicellulose and various naturally occurring hemicellulases (endo-xylanases, β-xylosidases, α-arabinofuranosidases, α-galactosidases, α-glucuronidases, β-mannanases, acetyl xylan esterases, and feruloyl esterases), which can help in the conversion of hemicellulose into monomers by their synergistic action

agricultural leftovers are produced annually worldwide. Every year, forestry operations contribute roughly 1.8 billion metric tons (Prasad et al., 2023). Mixtures of perennial energy grasses (switchgrass, miscanthus, and reed canary grass) and clover integrated into annual crop rotations at intervals of two to four years have the potential to yield 102–286 MMT (million dry metric tons) biomasses annually (Field et al., 2023). China and India lead in biomass production due to their extensive agricultural operations, while the USA and Canada contribute significantly through forestry and agricultural residues (Kumar Sarangi et al., 2023). In Europe, Sweden, Germany, and France account for a substantial portion of the yield from forestry and agricultural by-products, while Brazil notably produces significant agricultural waste, primarily from sugarcane (Sikiru et al., 2024). The total biomass on Earth is estimated to have the potential to produce around 33,000 exajoules (EJ) of energy, which is over 80 times the world's annual energy consumption (Kalak, 2023). The above statistics highlight the potential of lignocellulosic biomass as a sustainable resource for materials and energy, emphasizing its growing significance for energy security and sustainable development.

4 Conventional Pretreatment Technologies

LCB rich in pentose and hexose sugars can serve as cheap and abundant feedstock for developing a viable approach to produce value-added products (Onwosi et al., 2024; Singhvi et al., 2014). In a biorefinery, the production of second-generation biofuels from LCB involves five steps:

1. Collection and delivery of LCB to the plant
2. Pretreatment of LCB to make it vulnerable to an enzymatic attack
3. Enzymatic hydrolysis to produce fermentable sugars
4. Fermentation of the mixed sugars into ethanol
5. Product recovery

Each of these five steps must be considered critically to economize the 2G ethanol production viable in a biorefinery (Aden & Foust, 2009; Wingren et al., 2003).

The LCBs that are going to be used in the production of biofuels must be available throughout the year, and their supply from recovery field to the biorefinery must be cost-effective. The transport of low-density materials (such as wheat straw and rice straw) as a continuous supply chain and well-maintained storage facilities must be present in the biorefinery to sustain the production process throughout the year (Sanford et al., 2016). The pretreatment of LCBs is the first and most significant step as the efficacy of the subsequent process, enzymatic digestion of lignocellulose to monomeric units, is thoroughly dependent on the final product of the pretreatment process. Initially, the pretreatment process focused on the distortion of the complex LCB structure and exposing the cellulose to enzymatic attack to produce bioethanol. However, current pretreatment processes are designed to fractionate cellulose, hemicellulose, and lignin for optimal utilization of each component (Galbe & Wallberg, 2019).

4.1 Methods of Pretreatment

Cellulose and hemicellulose are tightly packed with the lignin in LCB, which makes it recalcitrance for enzymatic hydrolysis (Leonowicz et al., 1999). In biofuel production, the pretreatment process is the most significant step. The structural changes in the configuration of LCB during the pretreatment process make the subsequent LCB bioconversion more effective and easier. During pretreatment, depending upon the process strategy, operational parameters, and severity, removal of hemicellulose or/and lignin to varying degrees is attained. A number of different types of pretreatments have been developed to decrease LCB recalcitrance and enhance the cellulose accessibility for the cellulases. The main objective of these pretreatment strategies is to (1) enhance fermentable sugar yield by enzymatic hydrolysis by attaining reduction in recalcitrance of LCB, (2) decrease the degradation of sugars, and (3) decrease the formation of inhibiting compounds like furfuryl furfural and 5-hydroxymethylfurfural (Brodeur et al., 2011). Different strategies have been reported for the pretreatment of LCB, which can be categorized into four categories: (1) physical pretreatment, (2) chemical pretreatment, (3) physiochemical, and (4) biological pretreatment. The composition of cellulose, hemicellulose, lignin, and their structural complexity in a LCB defines the selection of a convenient pretreatment process for a particular type of LCB (Dahadha et al., 2017).

Usually before any pretreatment process, the physical properties of LCB are modified. The main aim of this process is to increase the surface area by reducing the size of particles and partial reduction of crystallinity and polymerization of cellulose (Rajendran et al., 2018). Milling, extrusion, microwave treatment, and ultrasonic treatment are the most widely used physical methods. These processes are environmentally safe and generate minimal harmful byproducts (Shirkavand et al., 2016). One important downside to physical pretreatment is the high usage of energy, which usually is dependent on the type of LCB used (Basak et al., 2023; Bhati et al., 2023).

4.1.1 Physicochemical Pretreatment

In these pretreatment methods, both physical changes and chemical reactions take place during the LCB pretreatment process. Various physicochemical pretreatments are steam explosion (SE), liquid hot water (LHW), and ammonia fiber explosion (AFEX) pretreatment (Jędrzejczyk et al., 2019). One of the most used physicochemical pretreatments is steam explosion (SE). It employs both physical and chemical techniques to reduce the recalcitrance of LCB (Hendriks & Zeeman, 2009). This hydrothermal process is based on the high temperature (160–240 °C) and pressure (0.7–4.8 MPa) saturated steam treatment of LCB (Agbor et al., 2011). The main purpose of SE pretreatment is the disruption of fibril structure by solubilizing hemicellulose and making cellulose more accessible for the cellulolytic enzymes for the hydrolysis process. The SE pretreatment is also an autohydrolysis process (Singh et al., 2015), such as the acetic acid produced during dechaining of acetyl moieties attached to hemicellulose further catalyzes the hemicellulose hydrolysis. Lignin is removed only to a minimal extent; however, during SE pretreatment, it undergoes melting, depolymerization, and subsequent repolymerization, leading to its redistribution on the fibril surface as droplets (Kumar et al., 2009).

A similar approach to SE is liquid hot water (LHW) pretreatment, which utilizes water at elevated temperatures (170–230 °C) and pressure (up to 5 MPa) while maintaining its liquid form. Unlike SE, LHW pretreatment does not require the sudden release of pressure, and the pressure is only used to prevent the evaporation of water (Yu et al., 2010). LHW pretreatment releases the acetyl groups, hydrolyzes hemicellulose, and eliminates lignin, thereby rendering the cellulose fibers more fragile for the enzymatic attacks (Zhuang et al., 2016). This pretreatment process does not involve any catalyst or chemical, and the formation of toxic compounds is almost absent. In addition, as the particles are broken down during the procedure, biomass size has little impact, which makes the process more advantageous (Bhutto et al., 2017). However, owing to the vast volume of water required, the procedure is very energy-consuming.

Another similar approach to SE is ammonia fiber/freeze explosion (AFEX) process in which LCB is treated with anhydrous liquid ammonia (in 1:1 ratio) under moderate temperatures (60–100 °C) and high pressures (0.7–2.7 MPa) and then subjected for quick release of pressure. Depending on the degree of saturation required for the type of LCB, the residence time can be changed from small (5–10 min) to moderate (30 min) lengths (Shirkavand et al., 2016). The rapid expansion of the ammonia gas disrupts the fibrous LCB structure, and the liquid ammonia and conditions of the process cause the swelling of the LCB, degrade hemicellulose to oligomeric sugars, change the structure of lignin, and reduce the cellulose crystallinity (Chundawat et al., 2007). A more efficient variant of the AFEX method, known as the COBRA (Compacted Biomass with Reduced Ammonia) process, has been reported by Balan et al. (2019). In this process, liquid ammonia is applied to biomass pellets at a ratio greater than 1:1, under relatively low temperatures (~70 °C) for 3–4 hours. This treatment converts native cellulose I into the more digestible cellulose III allomorph. Compared to conventional AFEX pretreatment, COBRA-treated biomass requires 60% less enzyme for enzymatic hydrolysis (Balan et al., 2019).

4.1.2 Chemical Pretreatments

4.1.2.1 Alkali Pretreatment

The alkali pretreatment method is basically a delignification process. This pretreatment disrupts the lignin-carbohydrate bond and solubilizes the lignin in the alkali solution. Commonly used alkaline reagents in alkali pretreatment are the hydroxides of ammonium, calcium, potassium, and sodium. It has been found that sodium hydroxide is the most effective alkaline reagent (Kim et al., 2016). Dissolution of lignin and hemicellulose and saponification (de-esterification) of intermolecular ester bonds are the main reactions during alkaline pretreatment (Kim, 2013). In addition, alkali pretreatment modifies the lignocellulosic structure by cellulose swelling, contributing to a decrease in crystallinity and polymerization, thereby increasing the internal surface area (Behera et al., 2014). Table 2 summarizes different alkali pretreatment reagents used for different LCBs with their reaction parameters and effects. The major drawback of this process is unfavorable mass balance, and loss of sugars along with lignin hampers its use at an industrial scale. Therefore, approaches of using diluted NaOH (0.4% w/w) are now being evaluated at the National Renewable Energy Laboratory (NREL) to overcome these drawbacks (Kuhn et al., 2016).

4.1.2.2 Acid Pretreatment

The most widely accepted technique to disrupt the LCB is the diluted-acid pretreatment process along with the pressurized steam. The process of acid pretreatment has been shown to be highly efficient in disrupting the LCB by cleaving the glycosidic bonds of hemicellulose and cellulose. This pretreatment transforms polysaccharide hemicellulose into soluble oligomeric and monomeric sugars primarily leaving intact high cellulose and lignin content (Chen et al., 2012; Jung & Kim,

Table 2 Different alkaline reagents used for the alkaline pretreatment of LCB

Alkaline reagent		Reagent used	Temperature (°C)	Time	Solid loading (%)	Major effects/remarks	References
Ammonium	NH_4OH	5–30%	30–210	5–60 min	10–50	Solubilize 10–55% hemicellulose, 0–80% delignification	Tajkarimi et al. (2008)
	NH_3	Gas	25–80	~72 h	~50	No washing required, low liquid loading rate (~50%), reaction conditions are mild	Yoo et al. (2011)
	NH_3	Liquid	70–90	~5 min	60–90	Hemicellulose and lignin are not removed, ammonia can rapidly be evaporated and then can reuse after liquefaction, AFEX	Kim & Lee (2005)
Lime		0.05–0.15 g $ca(OH)_2$/g solid	70–130	1–6 h	5–20	Solubilize 20–40% hemicellulose, 60–80% delignification, de-acetylation, require low energy	Wyman et al. (2005)
Na_2CO_3		1–30%	60–180	5–60 min	10–30	Solubilize 20–40% hemicellulose, 40–60% delignification Recovery is easier than NaOH pretreatment	Yang et al. (2012)
NaOH		0.5–10%	60–180	5–60 min	10–30	Solubilize 50% hemicellulose, 60–80% delignification, recovery of NaOH is difficult	Choi et al. (2013)

2015). This transformation of polysaccharides into oligomeric and monomeric sugars is due to the formation of hydronium ion which originates from the acid reagent being used in the pretreatment process (Lloyd & Wyman, 2005). For the acid pretreatment, both organic acids such as maleic acid (Jung et al., 2015), oxalic acid (Jeong & Lee, 2016), and formic acid (Du et al., 2016), and inorganic acids such as phosphoric acid (Nair et al., 2015), sulfuric acid (Kärcher et al., 2015), nitric acid (Kim et al., 2015), and hydrochloric acid (Zu et al., 2014) have been reported. At the industrial scale, pretreatment with dilute acid is considered appropriate as it aids in economical and environmentally sustainable conversion (Kumar & Sharma, 2017). Studies show that dilute sulfuric acid (H_2SO_4) is the most widely used for acid pretreatment of LCB. Table 3 summarizes different organic and inorganic acid pretreatment reagents used for the particular type of LCB with their reaction parameters.

4.1.2.3 Organosolv Process

In the organosolv pretreatment process, organic solvents such as ethanol, methanol, acetone, organic acid, organic peracid, and ethylene glycol or their mixture with water are used to pretreat the LCB, which leaves behind a relatively pure cellulose residue by breaking bonds between hemicellulose and lignin. This cellulose residue has increased pore volume and surface area, which enhances the convenience of enzymatic hydrolysis and saccharification (Zhang et al., 2016). To enhance the delignification rate or decrease the pretreatment temperature, some catalysts have also been employed along with the organic solvents. These catalysts usually are mineral acids (HCl, H_2SO_4, and H_3PO_4), bases (lime, NaOH, and NH_3), and some salts (Borand & Karaosmanoğlu, 2018). Table 4 summarizes different organosolv used for different LCBs with their reaction parameters.

4.1.3 Biological Pretreatments

In order to hydrolyze LCB prior to enzymatic saccharification, biological pretreatment is a low-cost and eco-friendly procedure. This approach is promising as there is no formation of inhibitors and requires less energy consumption and is environmentally sustainable (Bhatia et al., 2017). In biological pretreatment, the naturally found large taxonomic set of microbes are used to either degrade or change the LCB recalcitrant structure by secreting lignocellulolytic enzymes such as cellulase, hemicellulase, laccase, and other oxidizing enzymes (Pérez et al., 2002). Cellulose and hemicellulose are typically hydrolyzed into monomeric pentose and hexose sugars during biological pretreatment using cellulolytic and hemicellulolytic microorganisms. Lignocellulolytic bacteria and fungi are being used to pretreat LCB biologically, as whole cells or enzymes. For instance, the cellulolytic bacteria *Cellulomonas fimi* and *Thermomonospora fusca* have been extensively investigated for the development of cellulase producers. Similarly, *Paenibacillus campinasensis*, a cellulolytic bacterium which can live in harsh environments, has strong potential for LCB pretreatment (Maki et al., 2009). Some bacterial strains (such as *Azospirillum lipoferum* and *Bacillus subtilis*) have been reported to secret bacterial laccases for lignin depolymerization (Saritha et al., 2012). Fungi are ideally used for the biological pretreatment because they are capable of degrading cellulose,

Table 3 Organic and inorganic acid reagents used for the acid pretreatment of LCBs

Acid reagent		LCB	Reagent used (%)	Temperature (°C)	Time (min)	Solid loading (%)	References
Organic acids	Maleic acid ($C_4H_4O_4$)	Rice straw	1.0	190	3	10	Jung et al. (2015)
	Oxalic acid ($C_2H_2O_4$)	Yellow poplar	0.73	160	58	12.5	Jeong & Lee (2016)
	Formic acid (CH_2O_2)	Bleached softwood pulp	88.0	95	2–10	2	Du et al. (2016)
Inorganic acids	Phosphoric acid (H_3PO_4)	Wheat bran	0.5–3.0	150–210	5–20	15	Nair et al. (2015)
		Sugarcane bagasse	0.05–0.2	144–186	8–24	5	De Vasconcelos et al. (2013)
	Sulfuric acid (H_2SO_4)	Rice straw	0.5–1.0	169–190	1–25	10	Hsu et al. (2010)
	Nitric acid (HNO_3)	Corn Stover	0.2–1.0	140–160	1–10	10	Kim et al. (2015)
	Hydrochloric acid (HCl)	Corn Stover	1.0	100–130	20–40	10	Zu et al. (2014)
		Sugarcane bagasse	2–6	100–128	0–300	10	Bustos et al. (2003)

Table 4 Different organic solvents used for the organosolv pretreatment of LCB

Organosolv	Catalyst (%)	LCB	Solvent (%)	Temperature (°C)	Time	Recovery	References
Ethanol	$FeCl_3$ (0.025)	Sugarcane bagasse	60	160	72 h	93.8% glucose	Zhang et al. (2018)
Glycerol	–	Sugarcane bagasse	70	220	2 h	94% cellulose	Sun et al. (2016)
Ethanol	H_2SO_4 (1.0)	Rice straw	45	180	30 min	58.44% glucan	Asadi & Zilouei (2017)
Ethanol	H_2SO_4 (1.1)	Rice straw	65	170	1 h	70% cellulose	Sannigrahi et al. (2010)
Acetone	H_2SO_4 (0.5)	Barley straw	50	140	20 min	83.9% glucose	Salapa et al. (2018)
Butanol	H_2SO_4 (0.5)	Sorghum bagasse	25	200	1 h	84.9% cellulose	Teramura et al. (2018)
Ethanol	H_2SO_4 (0.35)	Wheat straw	50	180	40 min	89% cellulose	Salapa et al. (2017)

hemicellulose, and lignin. Biological pretreatment is used not only for the lignin removal but also for the removal of some materials, such as antimicrobials (Wan & Li, 2012). Laccases, lignin peroxidase, manganese peroxidase, and flexible peroxidase are the enzymes used for lignin degradation. For the removal of lignin and hemicellulose present in LCB, white-rot, brown-rot, and soft-rot fungi are being used. The most effective biological approach for LCB pretreatment is the use of white-rot fungal strains such as *Ceriporiopsis subvermispora*, *Cyathus stercoreus*, *Ceriporia lacerata*, *Echindodontium taxodii*, *Gloeophyllum trabeum*, *Phanerochaete chrysosporium*, *Phlebia subserialis*, *Pleurotus ostreatus*, and *Postia placenta* that can effectively metabolize lignin in different types of LCB (Kumar & Sharma, 2017). Due to its high growth rate, *Phanerochaete chrysosporium* is the most effective among the different recognized species of white-rot fungi for delignification (Yesilada et al., 2018). Table 5 summarizes the fungi used for biological pretreatment of different LCBs with their reaction parameters.

5 Novel Treatment Technologies

Lignocellulosic biomass (LCB), comprising cellulose, hemicellulose, and lignin, is an abundant and renewable resource derived from agricultural residues, forest residues, and dedicated energy crops. Its effective and sustainable utilization is crucial for producing biofuels, biochemicals, and other value-added products, contributing to a circular bioeconomy (Banu et al., 2021). Recent advancements in treatment

Table 5 Different fungi used for the biological pretreatment of LCB

Fungi	LCB	Temperature (°C)	Time (days)	Lignin degradation (%)	References
Phanerochaete chrysosporium	Wheat straw, cornstalk	29–30	12–21	28–34	Zhi & Wang 2014, Zhao et al. 2012b
Trametes hirsuta	Paddy straw, Corn Stover	30	10–42	71	Sun et al. (2011), Saritha et al. (2012)
Pycnoporus Sp. SYBC-L3	Switchgrass	30	36	30	Liu et al. (2013)
Dichomitus squalens	Rice straw	29	15	38	Bak et al. (2010)
Irpex lacteus	Wheat straw, Corn Stover	28	14–28	13–34	Salvachúa et al. (2011), Song et al. (2013)
Ceriporiopsis subvermispora	Corn Stover, Wheat straw	28	18–48	24–40	Wan & Li (2010), Cianchetta et al. (2014)
Trametes orientalis	Poplar wood	28	84	47	Wang et al. (2014)

technologies have focused on enhancing the efficiency, sustainability, and economic viability of LCB conversion (Boro et al., 2022).

Pretreatment is a critical step in LCB processing, aimed at disrupting the complex structure to enhance the accessibility of cellulose and hemicellulose for enzymatic hydrolysis. Novel pretreatment technologies include the following treatments.

5.1 Ionic Liquids

Ionic liquids (ILs) are salts in a liquid state, consisting of organic cations and either organic or inorganic anions, known for their low volatility, high thermal stability, and tunable solubility (Singh & Savoy, 2020; Trujillo-Rodríguez et al., 2018). These properties make ILs effective solvents for pretreating LCB by disrupting its complex structure, enhancing cellulose and hemicellulose accessibility for enzymatic hydrolysis. In addition to being recyclable and having a smaller environmental impact, ILs have the capacity to break down lignin and change the crystallinity of cellulose. Common cations used in ILs include imidazolium, pyridinium, ammonium, and phosphonium. The anions can range from simple halides to more complex species such as acetate, formate, or sulfate (Flieger & Flieger, 2020). Despite challenges like cost and toxicity, advances in IL technology are paving the way for their broader use in sustainable biomass conversion, contributing to efficient, eco-friendly biorefineries producing biofuels and biochemicals. Recent advances in improving the performance and sustainability of ILs in LCB pretreatment are designing task-specific ILs, the use of ILs in combination with catalysts (such as acids or enzymes), water, or other co-solvents, and integration of ILs with biorefineries (Kammoun et al., 2023).

5.2 Deep Eutectic Solvents

Deep eutectic solvents (DESs) are a class of solvents formed by mixing two or more components typically, a hydrogen bond donor, such as urea, glycerol, ethylene glycol, citric acid, lactic acid, and a hydrogen bond acceptor such as choline chloride, betaine, certain ammonium salts, which interact to form a eutectic mixture with a significantly lower melting point than that of the individual components (Hansen et al., 2020). DESs share many properties with ILs but are generally easier and cheaper to produce, often using biodegradable and non-toxic materials. DESs have the ability to specifically dissolve lignin and break down the crystalline structure of cellulose, facilitating subsequent enzymatic hydrolysis (Shen et al., 2019; Song et al., 2020). Additionally, DESs solubilize hemicellulose, making it available for enzymatic conversion. DESs, particularly choline chloride paired with acids or alcohols, can effectively solubilize lignin, which is crucial for improving the accessibility of cellulose and hemicellulose (Zhou et al., 2022). Recent research has focused on improving the performance and sustainability of DESs in LCB pretreatment by developing DESs with specific functional groups that enhance lignin dissolution or cellulose disruption while being biocompatible. Also, DESs in combination with other solvents or pretreatment methods enhance efficiency and reduce costs. The incorporation of catalysts into DESs further enhances LCB breakdown and reduces pretreatment severity. DES pretreatment enhances enzymatic hydrolysis and fermentation, improving the efficiency of biofuel and biochemical production (Guo et al., 2019; Xian et al., 2022).

5.3 Supercritical Fluids

Supercritical CO_2 and water are used to penetrate the biomass matrix effectively and offer a green and efficient alternative to traditional methods (Escobar et al., 2020; Serna et al., 2016). It complies with environmental stewardship and sustainability principles by minimizing waste, cutting energy use, and eliminating the need for hazardous solvents. Supercritical CO_2 acts as a solvent with adjustable dissolving power, enabling selective extraction of various compounds without leaving harmful residues (Ye et al., 2019). It can penetrate the biomass structure, facilitating the breakdown of lignin and hemicellulose while preserving cellulose. This results in improved enzymatic hydrolysis and subsequent fermentation.

On the other hand, supercritical water exhibits unique properties that enhance its ability to hydrolyze hemicellulose and disrupt the lignin matrix, thereby increasing the accessibility of cellulose for subsequent enzymatic hydrolysis. This pretreatment technique can significantly improve the efficiency of biomass conversion processes (Martins-Vieira et al., 2023). Additionally, supercritical water pretreatment is environmentally friendly, as it avoids the use of harsh chemicals and can be integrated into a closed-loop system for water recovery and reuse (Mankar et al., 2021).

5.4 Ultrasound-Assisted Pretreatment

Ultrasound-assisted pretreatment leverages the power of high-frequency sound waves (5–10 MHz) to enhance the efficiency and effectiveness of biomass processing (Low et al., 2022). It utilizes high-frequency sound waves to create cavitation bubbles in a liquid medium. These bubbles undergo rapid formation, growth, and collapse, generating intense localized pressure and temperature changes. This phenomenon disrupts the biomass structure, leading to several beneficial effects such as enhanced disruption of biomass structure, improved enzymatic accessibility, reduced pretreatment time, lower energy consumption, and preservation of biomass integrity (Shen et al., 2023; Sidana & Yadav, 2022). Ultrasound-assisted pretreatment is thus a promising method for enhancing the processing of lignocellulosic biomass, making it more feasible for biofuel production and other applications.

6 Role of Novel Technological Interventions

Plant-based LCB is a key biorenewable resource for the production of biofuels and biochemicals, but its complex structure poses challenges to its efficient utilization. Advanced pretreatment technologies, such as the use of ionic liquids, deep eutectic solvents, supercritical fluids, and ultrasound-assisted treatments, offer significant improvements over traditional methods by reducing energy consumption and environmental impact (Roy et al., 2020; Saini et al., 2022; Sandeep et al., 2024). The field of enzyme engineering has developed more effective enzymes for hydrolysis, optimizing enzyme cocktails, and immobilizing enzymes to enhance stability and reduce costs (Saini et al., 2022). Bioprocessing innovations, including consolidated bioprocessing and genetically engineered microbial strains, streamline conversion processes and improve yields (Chandel et al., 2020; Sandeep et al., 2024). Integrating renewable energy sources like solar and wind into biomass processing facilities further enhances sustainability by reducing the carbon footprint and operational costs (Lee et al., 2023). Waste valorization technologies transform by-products like lignin into valuable products, promoting a circular economy and maximizing resource efficiency (Adil et al., 2022). Digital and automation technologies also play a transformative role, optimizing biomass processing through advanced monitoring, control systems, data analytics, and artificial intelligence. These technologies enhance efficiency, reduce waste, lower labor costs, and improve the precision of operations, contributing to a more sustainable and effective bioeconomy (Adil et al., 2022; Ashokkumar et al., 2022).

7 Significance of LCB Mitigation

The mitigation of LCB is crucial for achieving environmental sustainability, economic viability, resource efficiency, and energy security (Nahak et al., 2022). As the world grapples with resource scarcity for energy generation and biofuel production and increasing environmental degradation, there is a need to develop and implement new approaches to reduce LCB waste crucial for a sustainable and resilient future (Ashokkumar et al., 2022). The buildup of LCB as waste can lead to environmental degradation from open burning, landfilling, and inefficient resource usage. Effective mitigation strategies reduce the environmental impact of agricultural and forestry operations, leading to lower greenhouse gas emissions, decreased deforestation pressures, and improved land-use efficiency (Kumar et al., 2023). Moreover, using LCB as a feedstock for biofuels and bioproducts can aid in transitioning to a more sustainable, circular economy by substituting non-renewable resources with renewable alternatives (Onwosi et al., 2024). The reduction of LCB is good for the environment and economically advantageous. Converting LCB into value-added products such as bioethanol, bioplastics, and biocomposites, generates new sources of income for industries and rural economies (Prasad et al., 2007). The development of scalable and affordable LCB conversion technology has the potential to boost economic growth in both developed and developing nations by increasing the competitiveness of biobased products relative to conventional fossil-based products (Adil et al., 2022).

LCB is a readily accessible and abundant resource, making it an extremely appealing option for sustainable utilization (Banu et al., 2021). However, its complex structure makes it resistant to degradation, requiring advanced pretreatment, enzymatic hydrolysis, and microbial fermentation techniques (Balan et al., 2019; Prasad et al., 2023). Mitigating lignocellulosic biomass through advanced biotechnological approaches improves resource efficiency by maximizing the conversion of biomass into fermentable sugars, which are the building blocks for various biobased products. This reduces reliance on food-based feedstocks and minimizes food and fuel competition (Sandeep et al., 2024; Sikiru et al., 2024). Energy security may be significantly improved by mitigating lignocellulosic biomass as long as global energy demands remain high. Countries may lessen their reliance on fossil fuels, diversify their energy portfolios, and increase their level of energy independence by turning LCB into biofuels like cellulosic ethanol and biogas (Saini et al., 2015). Furthermore, the decentralized structure of LCB resources permits localized energy generation, potentially improving access to energy in isolated and rural locations. Because the conversion of LCB into biofuels and bioproducts lowers carbon emissions linked to the use of fossil fuels, it helps mitigate the effects of climate change. Since the carbon dioxide absorbed by plants during growth offsets the carbon dioxide released during combustion, biofuels derived from lignocellulosic biomass (LCB) are often considered potentially carbon-neutral, depending on production methods and lifecycle emissions (Sikiru et al., 2024). Moreover, biochar, a carbon-rich by-product that can be applied as a soil supplement to promote soil health and absorb carbon, can be produced as a result of LCB mitigation (Jayakumar et al., 2023). Mitigating lignocellulosic biomass supports environmental and economic goals and advances global efforts to combat climate change.

8 Conclusion and Future Recommendations

LCB has great potential as a sustainable and renewable resource for bioenergy and bioproducts. Its convoluted structure, primarily comprised of cellulose, hemicellulose, and lignin, makes it challenging to convert into products with additional value. The intrinsic recalcitrance of LCB presents challenges for conventional techniques of processing. However, pretreatment methods, viz., comprising physical, chemical, physicochemical, and biological approaches, have significantly increased the effectiveness of LCB conversion methods. The hydrolysis of cellulose and hemicellulose has been further improved by advancements such as the use of ionic liquids, supercritical fluids, ultrasound-assisted treatment, and genetically modified enzymes, which can help in opening the door for more efficient bioconversion routes. The integration of metabolic engineering and systems biology in microbial strain engineering has optimized the production of biofuels, bioplastics, and other platform chemicals, ensuring a more efficient and economically viable utilization of LCB. The development of a circular economy is aided by the sustainable use of LCB, which also helps minimize greenhouse gas emissions and dependency on fossil fuels. By utilizing innovative and sustainable solutions for LCB processing, the bioenergy and bioproduct sectors can advance toward a more ecologically responsible and productive future in resource management. Harnessing the full potential of LCB as an essential component of renewable energy and bioproduct production requires continuous research and technological development.

References

Abbas, A., & Ansumali, S. (2010). Global potential of rice husk as a renewable feedstock for ethanol biofuel production. *Bioenergy Research, 3*, 328–334. https://doi.org/10.1007/s12155-010-9088-0

Aden, A., & Foust, T. (2009). Technoeconomic analysis of the dilute sulfuric acid and enzymatic hydrolysis process for the conversion of corn Stover to ethanol. *Cellulose, 16*, 535–545. https://doi.org/10.1007/s10570-009-9327-8

Adil, M., Nasir, A., Sikandar, A., & Khan, N. M. (2022). Recent advances in circular bioeconomy. In *Waste-to-energy* (pp. 59–84). Springer. https://doi.org/10.1007/978-3-030-91570-4_3

Agbor, V. B., Cicek, N., Sparling, R., Berlin, A., & Levin, D. B. (2011). Biomass pretreatment: Fundamentals toward application. *Biotechnology Advances, 29*(6), 675–685. https://doi.org/10.1016/j.biotechadv.2011.05.005

Alonso, D. M., Wettstein, S. G., Mellmer, M. A., Gurbuz, E. I., & Dumesic, J. A. (2013). Integrated conversion of hemicellulose and cellulose from lignocellulosic biomass. *Energy & Environmental Science, 6*(1), 76–80. https://doi.org/10.1039/C2EE23617F

Alves, E. F., Bose, S. K., Francis, R. C., Colodette, J. L., Iakovlev, M., & Van Heiningen, A. (2010). Carbohydrate composition of eucalyptus, bagasse and bamboo by a combination of methods. *Carbohydrate Polymers, 82*(4), 1097–1101. https://doi.org/10.1016/j.carbpol.2010.06.038

Anwar, Z., Gulfraz, M., & Irshad, M. (2014). Agro-industrial lignocellulosic biomass a key to unlock the future bio-energy: A brief review. *Journal of Radiation Research and Applied Science, 7*(2), 163–173. https://doi.org/10.1016/j.jrras.2014.02.003

Asadi, N., & Zilouei, H. (2017). Optimization of organosolv pretreatment of rice straw for enhanced biohydrogen production using *Enterobacter aerogenes*. *Bioresource Technology, 227*, 335–344. https://doi.org/10.1016/j.biortech.2016.12.073

Ashokkumar, V., Flora, G., Venkatkarthick, R., SenthilKannan, K., Kuppam, C., Stephy, G. M., Kamyab, H., Chen, W. H., Thomas, J., & Ngamcharussrivichai, C. (2022). Advanced technologies on the sustainable approaches for conversion of organic waste to valuable bioproducts: Emerging circular bioeconomy perspective. *Fuel, 324*, 124313. https://doi.org/10.1016/j.fuel.2022.124313

Bajpai, P. (2021). *Lignocellulosic Biomass in biotechnology*. Elsevier.

Bak, J. S., Kim, M. D., Choi, I. G., & Kim, K. H. (2010). Biological pretreatment of rice straw by fermenting with *Dichomitus squalens*. *New Biotechnology, 27*(4), 424–434. https://doi.org/10.1016/j.nbt.2010.02.021

Balan, V., Hodge, D., Zhang, J., & da Costa, S. L. (2019). Techno-economic analysis of compacted Biomass with recycled ammonia (COBRA) pretreatment process. In *41st symposium on biotechnology for fuels and chemicals*. SIMB.

Banu, J. R., Kavitha, S., Tyagi, V. K., Gunasekaran, M., Karthikeyan, O. P., & Kumar, G. (2021). Lignocellulosic biomass based biorefinery: A successful platform towards circular bioeconomy. *Fuel, 302*, 121086. https://doi.org/10.1016/j.fuel.2021.121086

Barakat, A., de Vries, H., & Rouau, X. (2013). Dry fractionation process as an important step in current and future lignocellulose biorefineries: A review. *Bioresource Technology, 134*, 362–373. https://doi.org/10.1016/j.biortech.2013.01.169

Basak, B., Kumar, R., Bharadwaj, A. S., Kim, T. H., Kim, J. R., Jang, M., Oh, S. E., Roh, H. S., & Jeon, B. H. (2023). Advances in physicochemical pretreatment strategies for lignocellulose biomass and their effectiveness in bioconversion for biofuel production. *Bioresource Technology, 369*, 128413. https://doi.org/10.1016/j.biortech.2022.128413

Behera, S., Arora, R., Nandhagopal, N., & Kumar, S. (2014). Importance of chemical pretreatment for bioconversion of lignocellulosic biomass. *Renewable and Sustainable Energy Reviews, 36*, 91–106. https://doi.org/10.1016/j.rser.2014.04.047

Bhati, N., Shreya, & Sharma, A. K. (2023). Recent trends in the pre-treatment process of lignocellulosic biomass for enhanced biofuel production. In *Production of biobutanol from biomass* (pp. 47–75). Wiley. https://doi.org/10.1002/9781394172887.ch2

Bhatia, S. K., Kim, S. H., Yoon, J. J., & Yang, Y. H. (2017). Current status and strategies for second generation biofuel production using microbial systems. *Energy Conversion and Management, 148*, 1142–1156. https://doi.org/10.1016/j.enconman.2017.06.073

Bhutto, A. W., Qureshi, K., Harijan, K., Abro, R., Abbas, T., Bazmi, A. A., Karim, S., & Yu, G. (2017). Insight into progress in pre-treatment of lignocellulosic biomass. *Energy, 122*, 724–745. https://doi.org/10.1016/j.energy.2017.01.005

Borand, M. N., & Karaosmanoğlu, F. (2018). Effects of organosolv pretreatment conditions for lignocellulosic biomass in biorefinery applications: A review. *Journal of Renewable and Sustainable Energy, 10*(3), 033104. https://doi.org/10.1063/1.5025876

Boro, M., Verma, A. K., Chettri, D., Yata, V. K., & Verma, A. K. (2022). Strategies involved in biofuel production from agro-based lignocellulose biomass. *Environmental Technology and Innovation, 28*, 102679. https://doi.org/10.1016/j.eti.2022.102679

Brodeur, G., Yau, E., Badal, K., Collier, J., Ramachandran, K. B., & Ramakrishnan, S. (2011). Chemical and physicochemical pretreatment of lignocellulosic biomass: A review. *Enzyme Research, 2011*, 787532. https://doi.org/10.4061/2011/787532

Brylev, A. N., Adylov, D. K., Tukhtaeva, G. G., Kamal'dinova, N. A., Abidova, L. D., & Rakhimov, D. A. (2001). 2001. Polysaccharides of rice straw. *Chemistry of Natural Compounds, 37*(6), 569–570. https://doi.org/10.1023/A:1014833319630

Bustos, G., Ramírez, J. A., Garrote, G., & Vázquez, M. (2003). Modeling of the hydrolysis of sugar cane bagasse with hydrochloric acid. *Applied Biochemistry and Biotechnology, 104*, 51–68. https://doi.org/10.1385/ABAB:104:1:51

Cao, N., Xia, Y., Gong, C. S., & Tsao, G. T. (1997). Production of 2, 3-butanediol from pretreated corn cob by *Klebsiella oxytoca* in the presence of fungal cellulase. In *Biotechnology for fuels and chemicals* (pp. 129–139). Springer. https://doi.org/10.1007/978-1-4612-2312-2_13

Chandel, A. K., Garlapati, V. K., Jeevan Kumar, S. P., Hans, M., Singh, A. K., & Kumar, S. (2020). The role of renewable chemicals and biofuels in building a bioeconomy. *Biofuel, Bioproducts Biorefining, 14*, 830–844. https://doi.org/10.1002/bbb.2104

Chen, W. H., Ye, S. C., & Sheen, H. K. (2012). Hydrolysis characteristics of sugarcane bagasse pretreated by dilute acid solution in a microwave irradiation environment. *Applied Energy, 93*, 237–244. https://doi.org/10.1016/j.apenergy.2011.12.014

Cherubini, F. (2010). The biorefinery concept: Using biomass instead of oil for producing energy and chemicals. *Energy Conversion and Management, 51*(7), 1412–1421. https://doi.org/10.1016/j.enconman.2010.01.015

Choi, W. I., Park, J. Y., Lee, J. P., Oh, Y. K., Park, Y. C., Kim, J. S., Park, J. M., Kim, C. H., & Lee, J. S. (2013). Optimization of NaOH-catalyzed steam pretreatment of empty fruit bunch. *Biotechnology for Biofuels, 6*, 1–8. https://doi.org/10.1186/1754-6834-6-170

Chundawat, S. P., Venkatesh, B., & Dale, B. E. (2007). Effect of particle size based separation of milled corn Stover on AFEX pretreatment and enzymatic digestibility. *Biotechnology and Bioengineering, 96*(2), 219–231. https://doi.org/10.1002/bit.21132

Chundawat, S. P., Bellesia, G., Uppugundla, N., da Costa, S. L., Gao, D., Cheh, A. M., Agarwal, U. P., Bianchetti, C. M., Phillips, G. N., Jr., Langan, P., & Balan, V. (2011). Restructuring the crystalline cellulose hydrogen bond network enhances its depolymerization rate. *Journal of the American Chemical Society, 133*(29), 11163–11174. https://doi.org/10.1021/ja2011115

Cianchetta, S., Di Maggio, B., Burzi, P. L., & Galletti, S. (2014). Evaluation of selected white-rot fungal isolates for improving the sugar yield from wheat straw. *Applied Biochemistry and Biotechnology, 173*, 609–623. https://doi.org/10.1007/s12010-014-0869-3

Dahadha, S., Amin, Z., Bazyar Lakeh, A. A., & Elbeshbishy, E. (2017). Evaluation of different pre-treatment processes of lignocellulosic biomass for enhanced biomethane production. *Energy and Fuels, 31*(10), 10335–10347. https://doi.org/10.1021/acs.energyfuels.7b02045

Davison, B. H., Parks, J., Davis, M. F., & Donohoe, B. S. (2013). Plant cell walls: Basics of structure, chemistry, accessibility and the influence on conversion. In *Aqueous pretreatment of plant biomass for biological and chemical conversion to fuels and chemicals* (pp. 23–38). Wiley. https://doi.org/10.1002/9780470975831.ch3

De Vasconcelos, S. M., Santos, A. M., Rocha, G. J., & Souto-Maior, A. M. (2013). Diluted phosphoric acid pretreatment for production of fermentable sugars in a sugarcane-based biorefinery. *Bioresource Technology, 135*, 46–52. https://doi.org/10.1016/j.biortech.2012.10.083

Dodd, D., & Cann, I. K. (2009). Enzymatic deconstruction of xylan for biofuel production. *GCB Bioenergy, 1*(1), 2–17. https://doi.org/10.1111/j.1757-1707.2009.01004.x

Du, H., Liu, C., Zhang, Y., Yu, G., Si, C., & Li, B. (2016). Preparation and characterization of functional cellulose nanofibrils via formic acid hydrolysis pretreatment and the followed high-pressure homogenization. *Industrial Crops and Products, 94*, 736–745. https://doi.org/10.1016/j.indcrop.2016.09.059

Escobar, E. L., Da Silva, T. A., Pirich, C. L., Corazza, M. L., & Pereira, R. L. (2020). Supercritical fluids: A promising technique for biomass pretreatment and fractionation. *Frontiers in Bioengineering and Biotechnology, 8*, 252. https://doi.org/10.3389/fbioe.2020.00252

Field, J., Kline, K. L., Langholtz, M., & Singh, N. (2023). Sustainably sourcing biomass feedstocks for bioenergy with carbon capture and storage in the United States. In *Energy futures initiative*. IEA Bioenergy. https://doi.org/10.23720/BT2023/2316165

Flieger, J., & Flieger, M. (2020). Ionic liquids toxicity-benefits and threats. *International Journal of Molecular Sciences, 21*(17), 6267. https://doi.org/10.3390/ijms21176267

Galbe, M., & Wallberg, O. (2019). Pretreatment for biorefineries: A review of common methods for efficient utilisation of lignocellulosic materials. *Biotechnology for Biofuels, 12*(1), 294. https://doi.org/10.1186/s13068-019-1634-1

García-Aparicio, M. P., Ballesteros, I., González, A., Oliva, J. M., Ballesteros, M., & Negro, M. J. (2006). Effect of inhibitors released during steam-explosion pretreatment of barley straw on enzymatic hydrolysis. *Applied Biochemistry and Biotechnology, 129*, 278–288. https://doi.org/10.1385/ABAB:129:1:278

Grossmann, L. (2024). Sustainable media feedstocks for cellular agriculture. *Biotechnology Advances, 73*, 108367. https://doi.org/10.1016/j.biotechadv.2024.108367

Guo, Z., Zhang, Q., You, T., Ji, Z., Zhang, X., Qin, Y., & Xu, F. (2019). Heteropoly acids enhanced neutral deep eutectic solvent pretreatment for enzymatic hydrolysis and ethanol fermentation of *Miscanthus x giganteus* under mild conditions. *Bioresource Technology, 293*, 122036. https://doi.org/10.1016/j.biortech.2019.122036

Hansen, B. B., Spittle, S., Chen, B., Poe, D., Zhang, Y., Klein, J. M., Horton, A., Adhikari, L., Zelovich, T., Doherty, B. W., & Gurkan, B. (2020). Deep eutectic solvents: A review of fundamentals and applications. *Chemical Reviews, 121*(3), 1232–1285. https://doi.org/10.1021/acs.chemrev.0c00385

Hendriks, A. T., & Zeeman, G. (2009). Pretreatments to enhance the digestibility of lignocellulosic biomass. *Bioresource Technology, 100*(1), 10–18. https://doi.org/10.1016/j.biortech.2008.05.027

Hsu, T. C., Guo, G. L., Chen, W. H., & Hwang, W. S. (2010). Effect of dilute acid pretreatment of rice straw on structural properties and enzymatic hydrolysis. *Bioresource Technology, 101*(13), 4907–4913. https://doi.org/10.1016/j.biortech.2009.10.009

Jayakumar, M., Hamda, A. S., Abo, L. D., Daba, B. J., Prabhu, S. V., Rangaraju, M., Jabesa, A., Periyasamy, S., Suresh, S., & Baskar, G. (2023). Comprehensive review on lignocellulosic biomass derived biochar production, characterization, utilization and applications. *Chemosphere, 345*, 140515. https://doi.org/10.1016/j.chemosphere.2023.140515

Jędrzejczyk, M., Soszka, E., Czapnik, M., Ruppert, A. M., & Grams, J. (2019). Physical and chemical pretreatment of lignocellulosic biomass. In *In second and third generation of feedstocks* (pp. 143–196). Elsevier. https://doi.org/10.1016/B978-0-12-815162-4.00006-9

Jeong, S. Y., & Lee, J. W. (2016). Optimization of pretreatment condition for ethanol production from oxalic acid pretreated biomass by response surface methodology. *Industrial Crops and Products, 79*, 1–6. https://doi.org/10.1016/j.indcrop.2015.10.036

Johannes, L. P., & Xuan, T. D. (2024). Comparative analysis of acidic and alkaline pretreatment techniques for bioethanol production from perennial grasses. *Energies, 17*(5), 1048. https://doi.org/10.3390/en17051048

Jung, Y. H., & Kim, K. H. (2015). Acidic pretreatment. In *Pretreatment of biomass* (pp. 27–50). Elsevier. https://doi.org/10.1016/B978-0-12-800080-9.00003-7

Jung, Y. H., Park, H. M., & Kim, K. H. (2015). Whole slurry saccharification and fermentation of maleic acid-pretreated rice straw for ethanol production. *Bioprocess and Biosystems Engineering, 38*, 1639–1644. https://doi.org/10.1007/s00449-015-1405-8

Kalak, T. (2023). Potential use of industrial biomass waste as a sustainable energy source in the future. *Energies, 16*(4), 1783. https://doi.org/10.3390/en16041783

Kammoun, M., Margellou, A., Toteva, V. B., Aladjadjiyan, A., Sousa, A. F., Luis, S. V., Garcia-Verdugo, E., Triantafyllidis, K. S., & Richel, A. (2023). The key role of pretreatment for the one-step and multi-step conversions of European lignocellulosic materials into furan compounds. *RSC Advances, 13*, 21395–21420. https://doi.org/10.1039/D3RA01533E

Kärcher, M. A., Iqbal, Y., Lewandowski, I., & Senn, T. (2015). Comparing the performance of Miscanthus x giganteus and wheat straw biomass in sulfuric acid based pretreatment. *Bioresource Technology, 180*, 360–364. https://doi.org/10.1016/j.biortech.2014.12.107

Kim, T. H. (2013). Pretreatment of lignocellulosic biomass. In *Bioprocessing technologies in biorefinery for sustainable production of fuels, chemicals, and polymers* (pp. 91–110). https://doi.org/10.1002/9781118642047.ch6

Kim, T. H., & Lee, Y. Y. (2005). Pretreatment of corn Stover by soaking in aqueous ammonia. In *Twenty-sixth symposium on biotechnology for fuels and chemicals* (pp. 1119–1131). Humana Press.

Kim, T. H., Taylor, F., & Hicks, K. B. (2008). Bioethanol production from barley hull using SAA (soaking in aqueous ammonia) pretreatment. *Bioresource Technology, 99*, 5694–5702. https://doi.org/10.1016/j.biortech.2007.10.055

Kim, I., Seo, Y. H., Kim, G. Y., & Han, J. I. (2015). Co-production of bioethanol and biodiesel from corn Stover pretreated with nitric acid. *Fuel, 143*, 285–289. https://doi.org/10.1016/j.fuel.2014.11.031

Kim, J. S., Lee, Y. Y., & Kim, T. H. (2016). A review on alkaline pretreatment technology for bioconversion of lignocellulosic biomass. *Bioresource Technology, 199*, 42–48. https://doi.org/10.1016/j.biortech.2015.08.085

Kuhn, E. M., O'Brien, M. H., Ciesielski, P. N., & Schell, D. J. (2016). Pilot-scale batch alkaline pretreatment of corn Stover. *ACS Sustainable Chemistry & Engineering, 4*(3), 944–956. https://doi.org/10.1021/acssuschemeng.5b01041

Kumar S.P., Subudhi, S., Bhatia, L., Saha, K., Mudgil, D., Prasad Shadangi, K., Srivastava, R. K., Pattnaik, B., & Arya, R. K. (2023 Jan). Utilization of agricultural waste biomass and recycling toward circular bioeconomy. *Environmental Science and Pollution Research, 30*(4), 8526–8539. https://doi.org/10.1007/s11356-022-20669-1

Kumar, A. K., & Sharma, S. (2017). Recent updates on different methods of pretreatment of lignocellulosic feedstocks: A review. *Bioresources and Bioprocessing, 4*(1), 7. https://doi.org/10.1186/s40643-017-0137-9

Kumar, P., Barrett, D. M., Delwiche, M. J., & Stroeve, P. (2009). Methods for pretreatment of lignocellulosic biomass for efficient hydrolysis and biofuel production. *Industrial and Engineering Chemistry Research, 48*(8), 3713–3729. https://doi.org/10.1021/ie801542g

Lee, J., Kim, S., You, S., & Park, Y. K. (2023). Bioenergy generation from thermochemical conversion of lignocellulosic biomass-based integrated renewable energy systems. *Renewable and Sustainable Energy Reviews, 178*, 113240. https://doi.org/10.1016/j.rser.2023.113240

Leonowicz, A., Matuszewska, A., Luterek, J., Ziegenhagen, D., Wojtaś-Wasilewska, M., Cho, N. S., Hofrichter, M., & Rogalski, J. (1999). Biodegradation of lignin by white rot fungi. *Fungal Genetics and Biology, 27*(2–3), 175–185. https://doi.org/10.1006/fgbi.1999.1150

Limayem, A., & Ricke, S. C. (2012). Lignocellulosic biomass for bioethanol production: Current perspectives, potential issues and future prospects. *Progress in Energy and Combustion Science, 38*(4), 449–467. https://doi.org/10.1016/j.pecs.2012.03.002

Liu, J., Wang, M. L., Tonnis, B., Habteselassie, M., Liao, X., & Huang, Q. (2013). Fungal pretreatment of switchgrass for improved saccharification and simultaneous enzyme production. *Bioresource Technology, 135*, 39–45. https://doi.org/10.1016/j.biortech.2012.10.095

Lloyd, T. A., & Wyman, C. E. (2005). Combined sugar yields for dilute sulfuric acid pretreatment of corn Stover followed by enzymatic hydrolysis of the remaining solids. *Bioresource Technology, 96*(18), 1967–1977. https://doi.org/10.1016/j.biortech.2005.01.011

Low, Z. L., Low, D. Y., Tang, S. Y., Manickam, S., Tan, K. W., & Ban, Z. H. (2022). Ultrasonic cavitation: An effective cleaner and greener intensification technology in the extraction and surface modification of nanocellulose. *Ultrasonics Sonochemistry, 90*, 106176. https://doi.org/10.1016/j.ultsonch.2022.106176

Lu, Y., Lu, Y. C., Hu, H. Q., Xie, F. J., Wei, X. Y., & Fan, X. (2017). Structural characterization of lignin and its degradation products with spectroscopic methods. *Journal of Spectroscopy, 2017*, 8951658. https://doi.org/10.1155/2017/8951658

Lynd, L. R., Weimer, P. J., Van Zyl, W. H., & Pretorius, I. S. (2002). Microbial cellulose utilization: Fundamentals and biotechnology. *Microbiology and Molecular Biology Reviews, 66*, 506–577. https://doi.org/10.1128/mmbr.66.3.506-577.2002

Maheshwari, N. V. (2018). Agro-industrial lignocellulosic waste: An alternative to unravel the future bioenergy. In *Biofuels: Greenhouse gas mitigation and global warming* (pp. 291–305). Springer. https://doi.org/10.1007/978-81-322-3763-1_16

Maki, M., Leung, K. T., & Qin, W. (2009). The prospects of cellulase-producing bacteria for the bioconversion of lignocellulosic biomass. *International Journal of Biological Sciences, 5*, 500. https://doi.org/10.7150/ijbs.5.500

Mankar, A. R., Pandey, A., Modak, A., & Pant, K. K. (2021). Pretreatment of lignocellulosic biomass: A review on recent advances. *Bioresource Technology, 334*, 125235. https://doi.org/10.1016/j.biortech.2021.125235

Martins-Vieira, J. C., Torres-Mayanga, P. C., & Lachos-Perez, D. (2023). Hydrothermal processing of lignocellulosic biomass: An overview of subcritical and supercritical water hydrolysis. *Bioenergy Research, 16*, 1296–1317.

Meng, X., & Ragauskas, A. J. (2014). Recent advances in understanding the role of cellulose accessibility in enzymatic hydrolysis of lignocellulosic substrates. *Current Opinion in Biotechnology, 27*, 150–158.

Miron, J., Yosef, E., & Ben-Ghedalia, D. (2001). Composition and in vitro digestibility of monosaccharide constituents of selected byproduct feeds. *Journal of Agricultural and Food Chemistry, 49*(5), 2322–2326. https://doi.org/10.1021/jf0008700

Monsalve, J. F., Medina de Perez, V. I., & Ruiz Colorado, A. A. (2006). Ethanol production of banana shell and cassava starch. *Dyna, 73*(150), 21–27.

Moraïs, S., Morag, E., Barak, Y., Goldman, D., Hadar, Y., Lamed, R., Shoham, Y., Wilson, D. B., & Bayer, E. A. (2012). Deconstruction of lignocellulose into soluble sugars by native and designer cellulosomes. *MBio, 3*(6), 10–128. https://doi.org/10.1128/mbio.00508-12

Mosier, N., Wyman, C., Dale, B., Elander, R., Lee, Y. Y., Holtzapple, M., & Ladisch, M. (2005). Features of promising technologies for pretreatment of lignocellulosic biomass. *Bioresource Technology, 96*(6), 673–686. https://doi.org/10.1016/j.biortech.2004.06.025

Mosihuzzaman, M., Quddus, A., Nahar, N., & Theander, O. (1989). Comparative study of carbohydrates in the two major species of jute (*Corchorus capsularis* and *Corchorus olitorius*). *Journal of the Science of Food and Agriculture, 48*(3), 305–310. https://doi.org/10.1002/jsfa.2740480306

Mujtaba, M., Fraceto, L. F., Fazeli, M., Mukherjee, S., Savassa, S. M., de Medeiros, G. A., Santo Pereira, A. D., Mancini, S. D., Lipponen, J., & Vilaplana, F. (2023). Lignocellulosic biomass from agricultural waste to the circular economy: A review with focus on biofuels, biocomposites and bioplastics. *Journal of Cleaner Production, 402*, 136815. https://doi.org/10.1016/j.jclepro.2023.136815

Nahak, B. K., Preetam, S., Sharma, D., Shukla, S. K., Syväjärvi, M., Toncu, D. C., & Tiwari, A. (2022). Advancements in net-zero pertinency of lignocellulosic biomass for climate neutral energy production. *Renewable and Sustainable Energy Reviews, 161*, 112393. https://doi.org/10.1016/j.rser.2022.112393

Nair, R. B., Lundin, M., Brandberg, T., Lennartsson, P. R., & Taherzadeh, M. J. (2015). Dilute phosphoric acid pretreatment of wheat bran for enzymatic hydrolysis and subsequent ethanol production by edible fungi *Neurospora intermedia*. *Industrial Crops and Products, 69*, 314–323. https://doi.org/10.1016/j.indcrop.2015.02.038

Onwosi, C. O., Ezugworie, F. N., Onyishi, C. L., & Igbokwe, V. C. (2024). Lignocellulose biomass pretreatment for efficient hydrolysis and biofuel production. In *Advances in biofuels production, optimization and applications* (pp. 1–19). Elsevier. https://doi.org/10.1016/B978-0-323-95076-3.00001-6

Pal, S., Joy, S., Trimukhe, K. D., Kumbhar, P. S., Varma, A. J., & Padmanabhan, S. (2016). Pretreatment and enzymatic process modification strategies to improve efficiency of sugar production from sugarcane bagasse. *3 Biotech, 6*, 1–4. https://doi.org/10.1007/s13205-016-0446-2

Pérez, J., Munoz-Dorado, J., De la Rubia, T. D., & Martinez, J. (2002). Biodegradation and biological treatments of cellulose, hemicellulose and lignin: An overview. *International Microbiology, 5*, 53–63. https://doi.org/10.1007/s10123-002-0062-3

Petersson, A., Thomsen, M. H., Hauggaard-Nielsen, H., & Thomsen, A. B. (2007). Potential bioethanol and biogas production using lignocellulosic biomass from winter rye, oilseed rape and faba bean. *Biomass and Bioenergy, 31*(11–12), 812–819. https://doi.org/10.1016/j.biombioe.2007.06.001

Poovaiah, C. R., Nageswara-Rao, M., Soneji, J. R., Baxter, H. L., & Stewart, C. N., Jr. (2014). Altered lignin biosynthesis using biotechnology to improve lignocellulosic biofuel feedstocks. *Plant Biotechnology Journal, 12*(9), 1163–1173. https://doi.org/10.1111/pbi.12225

Prasad, S., Singh, A., & Joshi, H. C. (2007). Ethanol as an alternative fuel from agricultural, industrial and urban residues. *Resources, Conservation and Recycling, 50*(1), 1–39. https://doi.org/10.1016/j.resconrec.2006.05.007

Prasad, B. R., Padhi, R. K., & Ghosh, G. (2023). A review on key pretreatment approaches for lignocellulosic biomass to produce biofuel and value-added products. *International journal of Environmental Science and Technology, 20*(6), 6929–6944. https://doi.org/10.1007/s13762-022-04252-2

Rajendran, K., Drielak, E., Sudarshan Varma, V., Muthusamy, S., & Kumar, G. (2018). Updates on the pretreatment of lignocellulosic feedstocks for bioenergy production–a review. *Biomass Conversion Biorefinery, 8*, 471–483. https://doi.org/10.1007/s13399-017-0269-3

Ranjan, R., Rai, R., Bhatt, S. B., & Dhar, P. (2023). Technological road map of cellulase: A comprehensive outlook to structural, computational, and industrial applications. *Biochemical Engineering Journal, 198*, 109020. https://doi.org/10.1016/j.bej.2023.109020

Reza, M. T., Emerson, R., Uddin, M. H., Gresham, G., & Coronella, C. J. (2015). Ash reduction of corn Stover by mild hydrothermal preprocessing. *Biomass Conversion Biorefinery, 5*, 21–31. https://doi.org/10.1007/s13399-014-0122-x

Rodrigues, J. S., Takeshita, V., Campos, E. V., de Freitas, A. S., de Lima, V. H., & Fraceto, L. F. (2024). Lignocellulosic biomass in nanopesticides: A path toward sustainable agriculture. *ACS Sustainable Chemistry & Engineering, 12*, 10045–10067. https://doi.org/10.1021/acssuschemeng.4c01213

Rowell, R. M., Schultz, T. P., & Narayan, R. (1992). Emerging technologies for materials and chemicals from biomass. In *ACS symposium, series no. 476*. ACS Books.

Roy, R., Rahman, M. S., & Raynie, D. E. (2020). Recent advances of greener pretreatment technologies of lignocellulose. *Current Research in Green and Sustainable Chemistry, 3*, 100035. https://doi.org/10.1016/j.crgsc.2020.100035

Saini, J. K., Saini, R., & Tewari, L. (2015). Lignocellulosic agriculture wastes as biomass feedstocks for second-generation bioethanol production: concepts and recent developments. *3 Biotech, 5*, 337–353. https://doi.org/10.1007/s13205-014-0246-5

Saini, J. K., Kaur, A., & Mathur, A. (2022). Strategies to enhance enzymatic hydrolysis of lignocellulosic biomass for biorefinery applications: A review. *Bioresource Technology, 360*, 127517. https://doi.org/10.1016/j.biortech.2022.127517

Salapa, I., Katsimpouras, C., Topakas, E., & Sidiras, D. (2017). Organosolv pretreatment of wheat straw for efficient ethanol production using various solvents. *Biomass and Bioenergy, 100*, 10–16. https://doi.org/10.1016/j.biombioe.2017.03.011

Salapa, I., Topakas, E., & Sidiras, D. (2018). Simulation and optimization of barley straw organosolv pretreatment. *Industrial Crops and Products, 113*, 80–88. https://doi.org/10.1016/j.indcrop.2018.01.018

Salvachúa, D., Prieto, A., López-Abelairas, M., Lu-Chau, T., Martínez, Á. T., & Martínez, M. J. (2011). Fungal pretreatment: An alternative in second-generation ethanol from wheat straw. *Bioresource Technology, 102*(16), 7500–7506. https://doi.org/10.1016/j.biortech.2011.05.027

Sandeep, B., Baranwal, U., & Mandpe, A. (2024). Biofuel production from microbial sources: Advances and challenges. In K. Bala, T. Ghosh, V. Kumar, & P. Sangwan (Eds.), *Harnessing microbial potential for multifarious applications. Energy, environment, and sustainability* (Vol. 10, pp. 323–357). Springer. https://doi.org/10.1007/978-981-97-1152-9_13

Sanford, K., Chotani, G., Danielson, N., & Zahn, J. A. (2016). Scaling up of renewable chemicals. *Current Opinion in Biotechnology, 38*, 112–122. https://doi.org/10.1016/j.copbio.2016.01.008

Sannigrahi, P., Miller, S. J., & Ragauskas, A. J. (2010). Effects of organosolv pretreatment and enzymatic hydrolysis on cellulose structure and crystallinity in loblolly pine. *Carbohydrate Research, 345*(7), 965–970. https://doi.org/10.1016/j.carres.2010.02.010

Saritha, M., Arora, A., & Lata. (2012). Biological pretreatment of lignocellulosic substrates for enhanced delignification and enzymatic digestibility. *Indian Journal of Microbiology, 52*, 122–130. https://doi.org/10.1007/s12088-011-0199-x

Scheller, H. V., & Ulvskov, P. (2010). Hemicelluloses. *Annual Review of Plant Biology, 61*(1), 263–289. https://doi.org/10.1146/annurev-arplant-042809-112315

Serna, L. D., Alzate, C. O., & Alzate, C. C. (2016). Supercritical fluids as a green technology for the pretreatment of lignocellulosic biomass. *Bioresource Technology, 199*, 113–120. https://doi.org/10.1016/j.biortech.2015.09.078

Shen, X. J., Wen, J. L., Mei, Q. Q., Chen, X., Sun, D., Yuan, T. Q., & Sun, R. C. (2019). Facile fractionation of lignocelluloses by biomass-derived deep eutectic solvent (DES) pretreatment for cellulose enzymatic hydrolysis and lignin valorization. *Green Chemistry, 21*(2), 275–283. https://doi.org/10.1039/C8GC03064B

Shen, L., Pang, S., Zhong, M., Sun, Y., Qayum, A., Liu, Y., Rashid, A., Xu, B., Liang, Q., Ma, H., & Ren, X. (2023). A comprehensive review of ultrasonic assisted extraction (UAE) for bioactive components: Principles, advantages, equipment, and combined technologies. *Ultrasonics Sonochemistry, 13*, 106646. https://doi.org/10.1016/j.ultsonch.2023.106646

Shirkavand, E., Baroutian, S., Gapes, D. J., & Young, B. R. (2016). Combination of fungal and physicochemical processes for lignocellulosic biomass pretreatment–a review. *Renewable and Sustainable Energy Reviews, 54*, 217–234. https://doi.org/10.1016/j.rser.2015.10.003

Sidana, A., & Yadav, S. K. (2022). Recent developments in lignocellulosic biomass pretreatment with a focus on eco-friendly, non-conventional methods. *Journal of Cleaner Production, 335*, 130286. https://doi.org/10.1016/j.jclepro.2021.130286

Sikiru, S., Abioye, K. J., Adedayo, H. B., Adebukola, S. Y., Soleimani, H., & Anar, M. (2024). Technology projection in biofuel production using agricultural waste materials as a source of energy sustainability: A comprehensive review. *Renewable and Sustainable Energy Reviews, 200*, 114535. https://doi.org/10.1016/j.rser.2024.114535

Singh, S. K., & Savoy, A. W. (2020). Ionic liquids synthesis and applications: An overview. *Journal of Molecular Liquids, 297*, 112038. https://doi.org/10.1016/j.molliq.2019.112038

Singh, J., Suhag, M., & Dhaka, A. (2015). Augmented digestion of lignocellulose by steam explosion, acid and alkaline pretreatment methods: A review. *Carbohydrate Polymers, 117*, 624–631. https://doi.org/10.1016/j.carbpol.2014.10.012

Singhvi, M. S., Chaudhari, S., & Gokhale, D. V. (2014). Lignocellulose processing: A current challenge. *RSC Advances, 4*(16), 8271–8277.

Sinner, M., Puls, J., & Dietrichs, H. (1979). Carbohydrate composition of nut shells and some other agricultural residues. *Starch-Stärke, 31*(8), 267–269. https://doi.org/10.1002/star.19790310807

Sluiter, A., Hames, B., Ruiz, R., Scarlata, C., Sluiter, J., Templeton, D., & Crocker, D. L. (2008). Determination of structural carbohydrates and lignin in biomass. *Laboratory Analytical Procedure, 1617*(1), 1–6.

Socha, A. M., Parthasarathi, R., Shi, J., Pattathil, S., Whyte, D., Bergeron, M., George, A., Tran, K., Stavila, V., Venkatachalam, S., & Hahn, M. G. (2014). Efficient biomass pretreatment using ionic liquids derived from lignin and hemicellulose. *Proceedings of the National Academy of Sciences of the United States of America, 111*(35), E3587–E3595. https://doi.org/10.1073/pnas.1405685111

Song, L., Ma, F., Zeng, Y., Zhang, X., & Yu, H. (2013). The promoting effects of manganese on biological pretreatment with *Irpex lacteus* and enzymatic hydrolysis of corn Stover. *Bioresource Technology, 135*, 89–92. https://doi.org/10.1016/j.biortech.2012.09.004

Song, Y., Chandra, R. P., Zhang, X., & Saddler, J. N. (2020). Non-productive celluase binding onto deep eutectic solvent (DES) extracted lignin from willow and corn Stover with inhibitory effects on enzymatic hydrolysis of cellulose. *Carbohydrate Polymers, 250*, 116956. https://doi.org/10.1016/j.carbpol.2020.116956

Sun, F. H., Li, J., Yuan, Y. X., Yan, Z. Y., & Liu, X. F. (2011). Effect of biological pretreatment with *Trametes hirsuta* yj9 on enzymatic hydrolysis of corn Stover. *International Biodeterioration and Biodegradation, 65*(7), 931–938. https://doi.org/10.1016/j.ibiod.2011.07.001

Sun, F. F., Zhao, X., Hong, J., Tang, Y., Wang, L., Sun, H., Li, X., & Hu, J. (2016). Industrially relevant hydrolyzability and fermentability of sugarcane bagasse improved effectively by glycerol organosolv pretreatment. *Biotechnology for Biofuels, 9*, 59. https://doi.org/10.1186/s13068-016-0472-7

Tajkarimi, M., Riemann, H. P., Hajmeer, M. N., Gomez, E. L., Razavilar, V., & Cliver, D. O. (2008). Ammonia disinfection of animal feeds—Laboratory study. *International Journal of Food Microbiology, 122*, 23–28. https://doi.org/10.1016/j.ijfoodmicro.2007.11.040

Teramura, H., Sasaki, K., Oshima, T., Kawaguchi, H., Ogino, C., Sazuka, T., & Kondo, A. (2018). Effective usage of sorghum bagasse: Optimization of organosolv pretreatment using 25%

1-butanol and subsequent nanofiltration membrane separation. *Bioresource Technology, 252*, 157–164. https://doi.org/10.1016/j.biortech.2017.12.100

Trujillo-Rodríguez, M. J., Nan, H., Varona, M., Emaus, M. N., Souza, I. D., & Anderson, J. L. (2018). Advances of ionic liquids in analytical chemistry. *Analytical Chemistry, 91*(1), 505–531. https://doi.org/10.1021/acs.analchem.8b04710

Vanholme, R., Demedts, B., Morreel, K., Ralph, J., & Boerjan, W. (2010). Lignin biosynthesis and structure. *Plant Physiology, 153*(3), 895–905. https://doi.org/10.1104/pp.110.155119

Vázquez, M., Oliva, M., Tellez-Luis, S. J., & Ramírez, J. A. (2007). Hydrolysis of sorghum straw using phosphoric acid: Evaluation of furfural production. *Bioresource Technology, 98*(16), 3053–3060. https://doi.org/10.1016/j.biortech.2006.10.017

Wan, C., & Li, Y. (2010). Microbial delignification of corn Stover by Ceriporiopsis subvermispora for improving cellulose digestibility. *Enzyme and Microbial Technology, 47*(1–2), 31–36. https://doi.org/10.1016/j.enzmictec.2010.04.001

Wan, C., & Li, Y. (2012). Fungal pretreatment of lignocellulosic biomass. *Biotechnology Advances, 30*(6), 1447–1457. https://doi.org/10.1016/j.biotechadv.2012.03.003

Wang, W., Yuan, T., & Cui, B. (2014). Biological pretreatment with white rot fungi and their co-culture to overcome lignocellulosic recalcitrance for improved enzymatic digestion. *BioResources, 9*(3), 3968–3976.

Williams, C. L., Emerson, R. M., & Tumuluru, J. S. (2017). Biomass compositional analysis for conversion to renewable fuels and chemicals. In *Biomass volume estimation and valorization for energy* (pp. 251–270). IntechOpen.

Wingren, A., Galbe, M., & Zacchi, G. (2003). Techno-economic evaluation of producing ethanol from softwood: Comparison of SSF and SHF and identification of bottlenecks. *Biotechnology Progress, 19*(4), 1109–1117. https://doi.org/10.1021/bp0340180

Wyman, C. E., Dale, B. E., Elander, R. T., Holtzapple, M., Ladisch, M. R., & Lee, Y. Y. (2005). Coordinated development of leading biomass pretreatment technologies. *Bioresource Technology, 96*(18), 1959–1966. https://doi.org/10.1016/j.biortech.2005.01.010

Xian, X., Fang, L., Zhou, Y., Li, B., Zheng, X., Liu, Y., & Lin, X. (2022). Integrated bioprocess for cellulosic ethanol production from wheat straw: New ternary deep-eutectic-solvent pretreatment, enzymatic saccharification, and fermentation. *Fermentation, 8*(8), 371. https://doi.org/10.3390/fermentation8080371

Yang, L., Cao, J., Jin, Y., Chang, H. M., Jameel, H., Phillips, R., & Li, Z. (2012). Effects of sodium carbonate pretreatment on the chemical compositions and enzymatic saccharification of rice straw. *Bioresource Technology, 124*, 283–291. https://doi.org/10.1016/j.biortech.2012.08.041

Ye, X., Xue, S. J., & Shi, J. (2019). Green separation Technology in Food Processing: Supercritical-CO2 fluid extraction. In J. Jia, D. Liu, & H. Ma (Eds.), *Advances in food processing technology* (pp. 73–99). Springer. https://doi.org/10.1007/978-981-13-6451-8_3

Yesilada, O., Birhanli, E., & Geckil, H. (2018). Bioremediation and Decolorization of textile dyes by white rot fungi and laccase enzymes. In R. Prasad (Ed.), *Mycoremediation and environmental sustainability. Fungal biology*. Springer. https://doi.org/10.1007/978-3-319-77386-5_5

Yoo, C. G., Nghiem, N. P., Hicks, K. B., & Kim, T. H. (2011). Pretreatment of corn Stover using low-moisture anhydrous ammonia (LMAA) process. *Bioresource Technology, 102*(21), 10028–10034. https://doi.org/10.1016/j.biortech.2011.08.057

Yu, G., Yano, S., Inoue, H., Inoue, S., Endo, T., & Sawayama, S. (2010). Pretreatment of rice straw by a hot-compressed water process for enzymatic hydrolysis. *Applied Biochemistry and Biotechnology, 160*, 539–551. https://doi.org/10.1007/s12010-008-8420-z

Zeng, Y., Zhao, S., Yang, S., & Ding, S. Y. (2014). Lignin plays a negative role in the biochemical process for producing lignocellulosic biofuels. *Current Opinion in Biotechnology, 27*, 38–45. https://doi.org/10.1016/j.copbio.2013.09.008

Zhang, K., Pei, Z., & Wang, D. (2016). Organic solvent pretreatment of lignocellulosic biomass for biofuels and biochemicals: A review. *Bioresource Technology, 199*, 21–33. https://doi.org/10.1016/j.biortech.2015.08.102

Zhang, H., Zhang, S., Yuan, H., Lyu, G., & Xie, J. (2018). FeCl3-catalyzed ethanol pretreatment of sugarcane bagasse boosts sugar yields with low enzyme loadings and short hydrolysis time. *Bioresource Technology, 249*, 395–401. https://doi.org/10.1016/j.biortech.2017.10.053

Zhao, L., Cao, G. L., Wang, A. J., Ren, H. Y., Dong, D., Liu, Z. N., Guan, X. Y., Xu, C. J., & Ren, N. Q. (2012a). Fungal pretreatment of cornstalk with *Phanerochaete chrysosporium* for enhancing enzymatic saccharification and hydrogen production. *Bioresource Technology, 114*, 365–369. https://doi.org/10.1016/j.biortech.2012.03.076

Zhao, X., Zhang, L., Liu, D., & Biomass recalcitrance. (2012b). Part II: Fundamentals of different pre-treatments to increase the enzymatic digestibility of lignocellulose. *Biofuels, Bioproducts and Biorefining, 6*(5), 561–579. https://doi.org/10.1002/bbb.1350

Zhi, Z., & Wang, H. (2014). White-rot fungal pretreatment of wheat straw with *Phanerochaete chrysosporium* for biohydrogen production: Simultaneous saccharification and fermentation. *Bioprocess and Biosystems Engineering, 37*, 1447–1458. https://doi.org/10.1007/s00449-013-1117-x

Zhou, M., Fakayode, O. A., Yagoub, A. E., Ji, Q., & Zhou, C. (2022). Lignin fractionation from lignocellulosic biomass using deep eutectic solvents and its valorization. *Renewable and Sustainable Energy Reviews, 156*, 111986. https://doi.org/10.1016/j.rser.2021.111986

Zhuang, X., Wang, W., Yu, Q., Qi, W., Wang, Q., Tan, X., Zhou, G., & Yuan, Z. (2016). Liquid hot water pretreatment of lignocellulosic biomass for bioethanol production accompanying with high valuable products. *Bioresource Technology, 199*, 68–75. https://doi.org/10.1016/j.biortech.2015.08.051

Zoghlami, A., & Paës, G. (2019). Lignocellulosic biomass: Understanding recalcitrance and predicting hydrolysis. *Frontiers in Chemistry, 7*, 874. https://doi.org/10.3389/fchem.2019.00874

Zu, S., Li, W. Z., Zhang, M., Li, Z., Wang, Z., Jameel, H., & Chang, H. M. (2014). Pretreatment of corn Stover for sugar production using dilute hydrochloric acid followed by lime. *Bioresource Technology, 152*, 364–370. https://doi.org/10.1016/j.biortech.2013.11.034

Valorization of Cotton Biomass and Processing Waste for Production of Biohydrogen and Biomethane

Ajinath Dukare, Anam Shaikh, Anagha Pawar, Kanika Sharma, and Nadanathangam Vigneshwaran

Abstract

Cotton-based textile industries sustain millions of people's livelihoods in India and globally. During cotton processing, a huge amount of cotton wastes and byproducts are generated and are usually landfilled or incinerated. Consequently, textile industries are facing difficulties in solid-waste management, which is the cause of environmental pollution and health hazards. Cotton wastes, mainly comprised of cellulose, hemicellulose, and lignin, represent a sustainable feedstock for the fermentative production of value-added bioproducts using microorganisms. Advances in microbial biotechnology are aiding in facilitating the effective valorization of cotton waste into valuable products. Cotton waste has been utilized for microbial production of biomethane and biohydrogen. The potential of modern metabolic engineering tools such as PCR-based DNA sequencing and assembly, genome editing, CRISPR, and cell surface engineering could develop microbial strain with enhanced hydrogen and methane yield. This book chapter highlights the aspects of cotton waste/byproducts, their structural composition, pretreatment strategies, and microbial approaches for their bioconversion into biomethane and biohydrogen. This chapter presents an innovative pathway for the cotton processing industry to achieve a waste-to-wealth solution through microbial-based biorefineries.

Keywords

Cotton biomass · Processing waste · Biomethane · Biohydrogen · Metabolic engineering

A. Dukare (✉) · A. Shaikh · A. Pawar · K. Sharma · N. Vigneshwaran
ICAR-Central Institute for Research on Cotton Technology, Mumbai, Maharashtra, India
e-mail: Ajinath.Dukare@icar.gov.in

G. Mukherjee, S. Dhiman (eds.), *Value Addition and Utilization of Lignocellulosic Biomass*, https://doi.org/10.1007/978-981-96-2786-8_5

1 Introduction

The increasing demand for cotton textiles globally has led to the generation of significant quantities of cotton biomass waste. Cotton biomass, a type of lignocellulosic biomass (LCB), is composed mainly of cellulose, hemicellulose, and lignin, which can serve as a valuable feedstock for bioconversion processes. Lignocellulosic biomass is plant dry matter, or biomass, which is the most plentiful raw material on Earth for the creation of biofuels and other bioproducts (Silva et al., 2024). In regions like India, a major hub for cotton textile production, traditional disposal methods for cotton biomass, agricultural residues, and processing waste, such as landfilling and incineration, exacerbate environmental issues by contributing to pollution and greenhouse gas emissions.

Despite these challenges, cotton waste presents a promising opportunity as a sustainable resource. Bioconversion of lignocellulosic or cotton biomass involves the transformation of this biomass into valuable products through biological processes, enhancing its valorization potential. Through bioconversion, this biomass can be transformed into various products, including bioethanol (Patel & Shah, 2021), biobutanol (Verardi et al., 2020), bioplastics, biochar (Bridgwater, 2012), bio-oil (Bridgwater, 2012), biomethane, and biohydrogen. For instance, anaerobic digestion of biomass can yield biomethane, a renewable energy source, and dark fermentation or photofermentation processes can produce biohydrogen (Alavi-Borazjani et al., 2024). Biomethane production primarily depends on several key parameters, including substrate composition, pH, temperature, and hydraulic retention time. The process generally involves anaerobic digestion, where microorganisms break down organic material in the absence of oxygen, leading to methane production. Optimizing these process parameters is critical to maximizing biomethane yields (Alavi-Borazjani et al., 2024). Similarly, biohydrogen production also requires careful control of parameters like pH, temperature, and nutrient availability. It can be produced through various methods, such as dark fermentation, photofermentation, and microbial electrolysis cells (Chandrasekhar et al., 2015; Jain et al., 2022).

Valorization of cotton waste through bioconversion not only helps in waste management but also contributes to the development of biorefineries that can produce biofuels like biomethane and biohydrogen. These biofuels are considered green fuels due to their potential to reduce greenhouse gas emissions and dependency on fossil fuels. Biomethane, a renewable natural gas, and biohydrogen, a clean energy carrier, can be produced as byproducts of the microbial fermentation process (Nebechi et al., 2024). Microorganisms play a crucial role in the biotransformation process. Specific microbial strains such as *Clostridium acetobutylicum* and *Methanosarcina barkeri* are employed for the production of hydrogen and methane, respectively (Zhang et al., 2006). With the help of these microbes, complex polysaccharides are broken down into simpler monomeric molecules, which are subsequently converted into biofuel. The increasing cotton production has led to a significant rise in biomass and lignocellulosic waste materials. This surplus presents a tremendous opportunity for the production of biofuels, which aligns with global efforts to transition to renewable energy sources (Nath, 2024). The main goal of this chapter is to demonstrate the potential of cotton biomass and other lignocellulosic feedstock for the production of biomethane and biohydrogen and to highlight its viability and contribution to sustainable energy solutions (Kumar et al., 2016).

The novelty of current research lies in its cutting-edge strategies that offer a sustainable solution to the cotton-based textile industry's waste management challenges while contributing to the growth of the biofuel sector (Elisabeth et al., 2024). By providing a thorough examination of the structural composition, pretreatment tactics, and microbiological approaches for the bioconversion of cotton biomass into biofuels, this study demonstrates the potential to convert waste into wealth through biotechnological breakthroughs (Sharma-Shivappa & Chen, 2008). The chapter focuses on the viability and value-added potential of lignocellulosic waste, specifically from cotton, in generating biofuels and other sustainable products (Adewuyi et al., 2022). This chapter intends to demonstrate the possibility of converting waste into wealth through biotechnological breakthroughs by providing a thorough examination of the structural composition, pretreatment tactics, and microbiological approaches for the bioconversion of cotton biomass into bioenergy.

2 Current Status of Cotton Processing Biomass, Processing Waste, and Byproduct Generation

Cotton is one of the most significant crops globally, contributing to nearly 40% of the world's fiber production. Throughout the various stages, from cotton harvesting to the manufacturing of textile materials, different types of cotton biomass and waste are produced. Initially, fully matured cotton bolls are harvested—either manually or mechanically—and then transported to ginneries for further processing. After harvesting, the remaining plant biomass in the field includes stalks, leaves, dried flowers, stems, and bracts (Keshav et al., 2018). On average, approximately two metric tons of dry matter per hectare of land, primarily in the form of cotton stalks, are produced each year. This biomass offers significant potential as a feedstock for biofuel production and other value-added products, underscoring the importance of finding sustainable uses for cotton waste. Despite its widespread availability, cotton field waste is often considered to have limited value, resulting in its minimal use in traditional farming methods. These post-harvest residues are typically underexploited, often ending up in landfills, being composted, or burned in fields—practices that contribute to air pollution and worsen the global climate crisis. Moreover, after the cotton is harvested, the remaining stalks can serve as breeding sites for pests like boll weevils, further complicating crop management (Silverstein et al., 2007). However, as one of the most plentiful and cost-effective lignocellulosic materials, cotton stalks offer significant potential for transformation into a variety of valuable bioproducts through advanced microbial biotechnology. By harnessing this potential, we can shift from wasteful disposal methods to more sustainable practices that turn agricultural residues into useful resources (Keshav et al., 2018).

During the yarn preparation stage, ginning is the primary and most important process. Various impurities from the field-harvested cotton fibers are separated and recovered as a processing byproduct/waste in the form of cotton gin trash (CGT). CGT is a typical byproduct of the ginning process and primarily consists of sticks, burrs, leaves, soil particles, mote, short fibers/lint, hulls, and other plant materials (Sharma-Shivappa & Chen, 2008). During the ginning process, approximately 336 kg of seed and 68–91 kg of CGT are generated from one 218 kg cotton bale (Plácido et al., 2013). CGT is a low-value waste material, and proper management is considered a financial load for most growing industries. Currently, CGT is usually disposed of as landfilling (solid waste), incineration, composting, and land application, and, to a minor extent, utilized as a feed supplement to livestock (Knox et al., 2006). Disposal options are often firmly regulated and monitored by local environmental regulations and laws, which adds further limitations. After the ginning process, cotton is stepwise processed through various processes such as spinning, warping, slashing, knitting, weaving, preparation, desizing, bleaching, dyeing/printing, and finishing to produce good-quality textile yarn. Cotton solid wastes (short and broken fibers) are mainly generated during the yarn-spinning processes (Ranjithkumar et al., 2017). In addition to CGT, the cotton processing industry also produces other significant byproducts, such as cotton exocarp and cottonseed hulls (CSH). The cotton exocarp, which is the outer layer of the cottonseed, makes up about 80% of the cotton seed and is generated during oil extraction. Annually, around 30 million tons of waste in the form of cotton exocarp and CSH are produced globally. CSH is a lignocellulosic biomass rich in cellulose, hemicellulose, and lignin (Fig. 1).

3 Cotton Biomass and Processing Waste: Structural and Composition

Among the different waste biomass (cotton stalk, gin trash, lint waste, seed hulls, seed meals, spinning mill waste, and cotton dust), the cotton stalk is the largest and most produced waste during cotton harvesting and finishing. Chemically, cotton stalks contain 32-46% cellulose, 20–28% hemicellulose (20–28%), and 26–33% lignin (Shahzad et al., 2019). According to physical characteristics, it has a moisture content of 10–20% and an ash content of 3.28–6.93% (Umesh et al., 2015). The important elements present in cotton stalks are nitrogen (1.11%), phosphorus (0.1%), and potassium chloride (3.98%), which can be as high as carbon (1.5 tons), nitrogen (20–25 kg), and potash (72 kg) per hectare. The lignin fraction in cotton stalk residues constitutes 28% w/w. The hemicellulose content is about 19.3% (Dong et al., 2022) as it is a heteropolysaccharide with 5-carbon, 6-carbon sugars, and uronic acids and strongly binds to cellulose microfibrils via hydrogen bonds and Van der Waals forces (Ranjithkumar et al., 2017). Due to its partial or wholly lignocellulosic nature, xylose content in cotton crop trash (17% w/w) and gin waste (15% w/w) is remarkably more significant than other types of cotton wastes (Agblevor et al., 2003). In general, cotton lint is primarily composed of cellulose, a glucose-based polymer with interlinked 1,4-glycosidic bonds, and is characterized

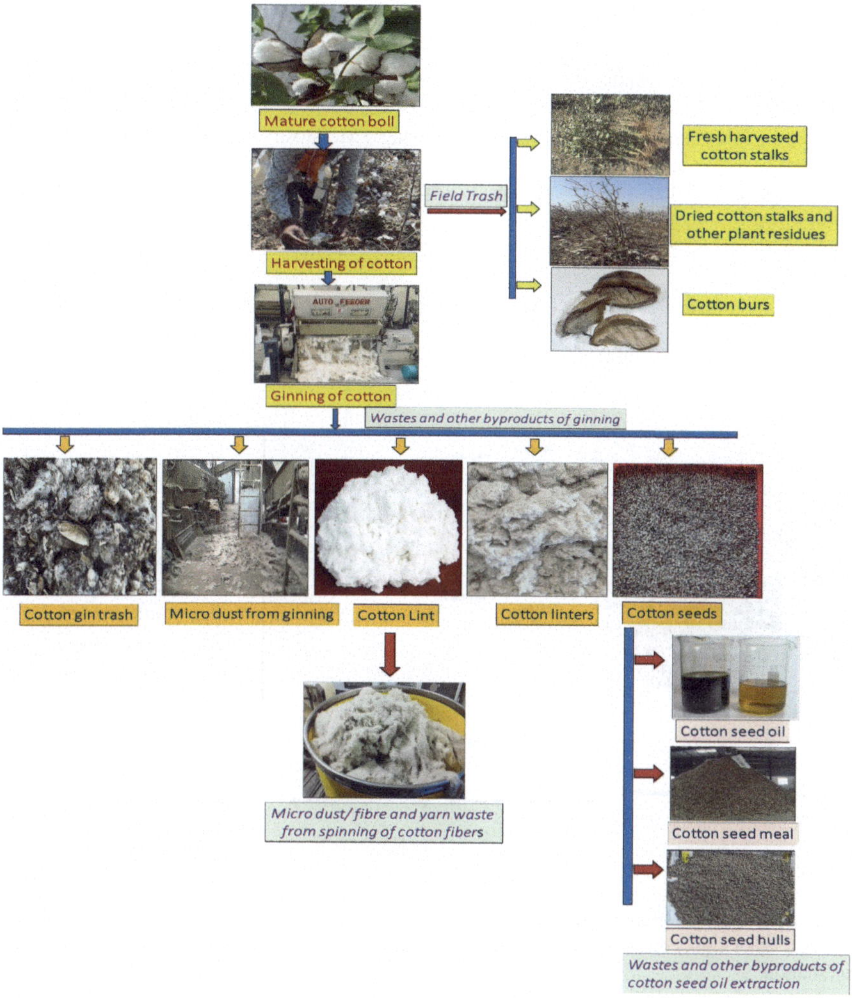

Fig. 1 Different types of cotton biomass, processing wastes, and byproducts generated as an organic material during the multiple stages of harvesting and processing of cotton lint and cottonseed

by properties such as fiber diameter (11–22 μm), length, tenacity (35 gf/tex), density (1.52 g/cm^3), and crystallinity. These properties are influenced by the maturity of the fibers and environmental conditions during the pre- and post-harvesting stages. Another important and significant cellulosic-rich waste produced during the processing is CGT, which accounts for 10% of each production bale. It consists of dry matter (90%), crude protein (7%), all digestible nutrients (46%), lignin (14%), ash (10%), high potassium content (1.2%), calcium (0.8%), phosphorus (0.2%), magnesium (0.2%), and sodium (0.1%) (Egbuta et al., 2019). Table 1 shows the typical lignocellulosic composition of cotton stalk and cotton gin trash, as reported in different studies.

Table 1 Typical lignocellulosic composition of cotton stalk and CGT, as reported in different studies

Chemical composition	Cotton stalk				Cotton gin trash			
	da Silva et al. (2019)	Shahzad et al. (2019)	Karthyani et al. (2020)	Huang et al. (2015)	McIntosh et al. (2014)	Plácido et al. (2013)	Agblevor et al. (2003)	Jordan et al. (2019)
Cellulose (%)	34.2	43.7	41.6	39.90	24–37	25	25–33	30.90
Hemicellulose (%)	26.7	12.5	21.6	17.20	14–18	10	7–16	9.74
Lignin (%)	16.5	28.6	34.0	25.60	18–23	26	20–25	19.30

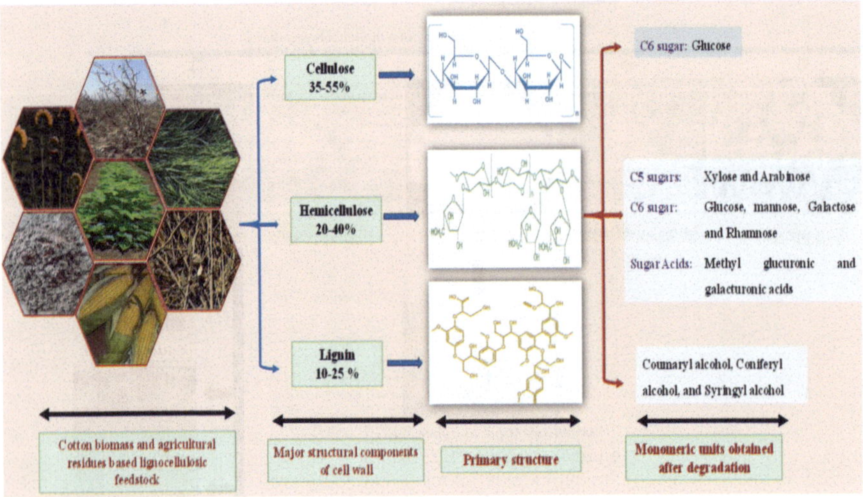

Fig. 2 Schematic diagram representing the structural and compositional aspects of cotton biomass and other lignocellulose biomass and monomeric units derived after their pretreatment and degradation

Cotton seed hull (CSH) is another LCB that contains the main components of cellulose, hemicellulose, and lignin, which mainly contains holocellulose (65.6%), alpha-cellulose (38.8%), hemicellulose (24.5%), and lignin (18.5%). It has a high amount of xylan (27–30%) in its hemicellulose polymer (Dukare, Sharma, Kautkar, et al., 2023). Likewise, denatured cotton cake, also known as cottonseed meal (CSM), is another byproduct of the cotton oil extraction industry and is mainly a protein fraction. Cotton spinning wastes and cotton micro dust are primarily cellulose-containing processing wastes (Domelsmith & Berni, 1986; Moorthy et al., 2022). Figure 2 represents the structural and compositional aspects of cotton biomass and other lignocellulose biomass and monomeric units derived after their pretreatment and degradation.

4 Pretreatment Approaches for Cotton Biomass and Processing Waste

Lignocellulosic biomass needs to be effectively pretreated due to its intricate structure so that it can be fermented into bioenergy. Several pretreatment techniques for biomass have been investigated; these may be roughly categorized as follows: (1) mechanical procedures, which pertain to the reduction of particle size; (2) chemical reactions involving diluted acids, alkalis, or organic solvents; (3) physicochemical reactions involving steam explosions and hot water; and (4) biological reactions involving microbial consortia or enzymatic techniques (Lindner et al., 2015). Pretreatment methods, such as physical, chemical, or biological treatments, are employed to increase the digestibility of biomass, making the organic material more accessible to the microbes (Guo & Wu, 2021). With peak biohydrogen and biomethane outputs observed at 33 mL H_2/g VS and 83 mL CH_4/g VS, respectively, acid pretreatment has shown to be very successful for cotton straw, yielding the best yields in both biogas creation and sugar extraction (Yoruklu et al., 2020). Cotton stalks have also been used to create hydrogen-rich gas through the use of sophisticated processes that include pyrolysis, gas-solid simultaneous gasification, and catalytic steam reforming. In addition to increasing H_2 output from 47.71 to 78.19 g/kg of cotton stalks, this integrated method also improved carbon conversion from 78.84% to 92.06% and improved the transformation of tar components and carbon deposition efficiency (Chen et al., 2015). An oxygen flow rate (OFR) of 0.8 mL/h was achieved in a different study by heat shock treatment of the inoculum at 105 °C for 0.5 h, followed by hydrolysis with 0.1M HCl and dark fermentation in a mesophilic environment, which release 43% of hydrogen (Sołowski, Konkol, & Cenian, 2020). Under varied circumstances, the anaerobic fermentation of cotton stalk wastes was optimized. The highest yields and average production rates were obtained with 15 g/L of cotton stalks soaked in 0.6 wt% HCl for 2 h at 90 °C. The mixed gas at 37 °C and initial pH of 7.0 had an H_2 content of 59.06%, and the cumulative hydrogen yield was 92.68 mL g^{-1} (Wang et al., 2017).

Furthermore, a specific hydrogen production rate of 126 ml/day of VSS and the highest hydrogen yield of 63.8 ml/gm cotton waste were attained by sulfuric acid pretreatment followed by enzymatic hydrolysis with crude cellulase from *Trichoderma reesei* (Lakshmidevi & Muthukumar, 2023). Additional techniques of pretreatment were investigated to improve cotton stalk anaerobic digestion (AD) for biomethane production. A 3% alkali hydrogen peroxide pretreatment produced the greatest cumulative methane yield (CMY) of 192.4 mL g VS^{-1}, increasing methane yield by 254.3% above untreated cotton stalks among the different agents examined (Zhang et al., 2018). With mechanical processing, cotton stalks with particle sizes ranging from 0.5 to 65 mm produced methane production that was 20.3% higher with smaller particles (3 mm and 0.5 mm), and the biogas quality ranged from 54.0% to 55.2% CH_3 (Al Afif & Pfeifer, 2021). Response surface methodology revealed a maximum biomethane potential of 114.19 ml/g VS from cotton stalk waste under thermal acid pretreatment with F/I of 0.25 and organic loading (OL) of 35 g VS/L (Oznur et al., 2023). Subsequent studies on co-digestion processes involving sludge, cow dung, and processed CGT revealed the efficacy of physical pretreatments like hot water, ultrasonication, and their combination. The combined treatment produced the highest volume of high-quality biogenic CH_4 at 52.4% and the highest biogas generation capacity of 110 L (Hanif et al., 2022). Adl et al. (2012) reported that hydrothermal pretreatment using recycled liquid from the anaerobic process and adjusted ammonia content also showed substantial methane outputs, reaching 0.200 Nm^3/kg-VS. Different C/N ratios used in the co-digestion of CGT produced high biomethane yields of 261.4 ± 4.8 and 262.6 ± 4.2 mL/g VS (Kaur & Kommalapati, 2021). Furthermore, rumen bacteria pretreatment generated 50 mL of methane in two days (Kurt Kara et al., 2021), whereas pretreatments with sodium hydroxide and ammonia solution provided the highest hydrogen production values of 17 and 15.2 mL/g VS, respectively (Ghasemian et al., 2016). These results highlight the need to choose suitable pretreatment techniques to maximize biomass conversion, increasing methane and hydrogen outputs and boosting overall process efficiency.

5 Microbial Strains and Approaches Involved in Biomethane and Hydrogen Production

Biomethane and biohydrogen production relies on specific microbial strains and advanced biotechnological methods. Methanogens like *Methanobacterium*, *Methanobrevibacter*, *Methanococcus*, *Methanosarcin*a, and *Methanosaeta* are key for biomethane production (Angelidaki & Sanders, 2004), while biohydrogen involves obligate anaerobes such as *Clostridia* species, facultative anaerobes like Enteric bacteria, archaea, cyanobacteria, green algae, and protists (Luthfi et al., 2023). The biomethane process is carried out by anaerobic digestion (AD), and biohydrogen utilizes dark fermentation (DF), which is epitomized by AD, i.e., without methanogenesis (Sołowski, Konkol, Shalaby, & Cenian, 2020). Biomethane production primarily involves four important stages: hydrolysis, acidogenesis, acetogenesis, and methanogenesis (Sharma & Melkania, 2018). Figure 3 represents simplified metabolic steps involved in the production of biomethane and biohydrogen via an anaerobic digestion method. During these stages, complex organic materials are broken down into simpler compounds, ultimately producing methane and carbon dioxide. Hydrolysis involves the breakdown of complex polymers into monomers by hydrolytic bacteria such as *Clostridium*, *Bacteroides*, and *Streptococcus*. Acidogenesis follows, where fermentative bacteria such as *Lactobacillus*, *Escherichia*, and *Enterobacter* convert these monomers into volatile fatty acids, alcohols, hydrogen, and carbon dioxide. During acetogenesis, acetogenic bacteria, including *Syntrophomonas* and *Clostridium aceticum*, convert these intermediate products into acetic acid, hydrogen, and carbon dioxide. Finally, in the methanogenesis stage, methanogens utilize acetic acid, hydrogen, and carbon dioxide to produce methane and water (Guo & Wu, 2021).

The AD process requires strictly anaerobic conditions and temperature control within either mesophilic (30–40 °C) or thermophilic (50–60 °C) ranges to ensure

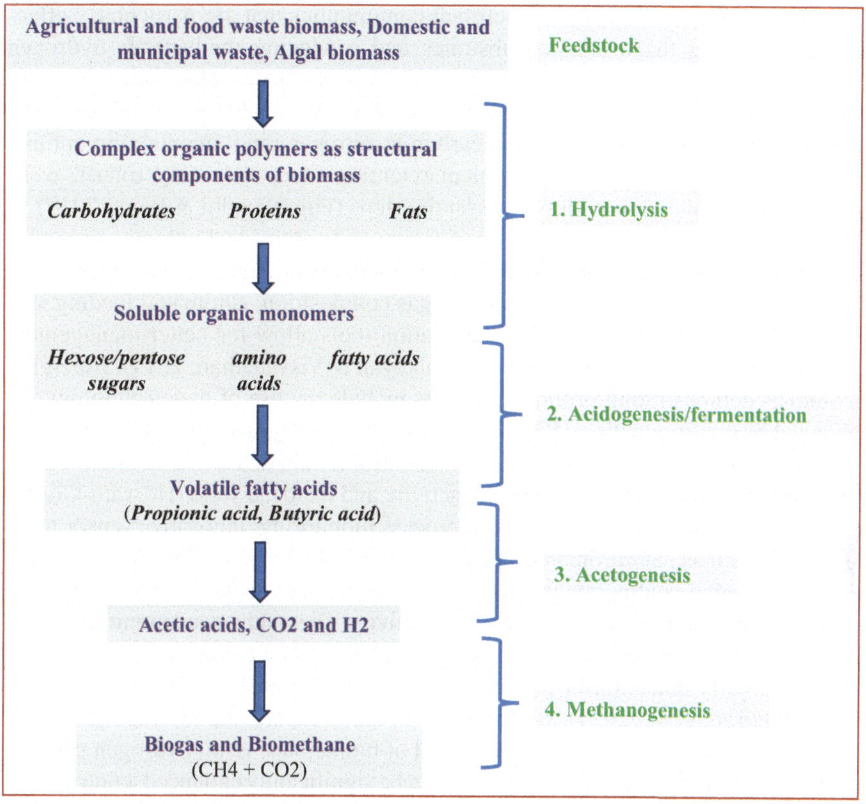

Fig. 3 Simplified metabolic steps involved in the production of biomethane and biohydrogen via anaerobic digestion method

optimal microbial activity (Mamun & Torii, 2018). Enhancing microbial activity and balancing nutrient content can be achieved through co-digestion, which involves mixing multiple types of organic waste. According to Mao et al. (2016), this method not only increases process efficiency overall but also aids in the simultaneous management of many waste streams. Soares et al. (2020) state the optimal conditions for dark fermentation typically involve mesophilic temperatures (37 °C), a neutral pH (7.0), and a hydraulic retention time of approximately 72 h. These factors significantly influence both the production and yield of biohydrogen. Common feedstocks for AD and DF include agricultural waste, animal manure, sewage sludge, algal biomass, lignocellulosic biomass, and organic municipal waste, making it a versatile and sustainable waste management solution (Guo & Wu, 2021). The intricate interplay between these diverse microbial communities is essential for efficient biomethane and biohydrogen production and highlights the importance of optimizing environmental conditions to support their activity (Angelidaki & Sanders, 2004). Selecting and adapting microbial consortia to specific feedstocks and conditions can significantly improve the efficiency of biomethane and biohydrogen production.

This involves the enrichment of microbial communities that are particularly effective at degrading the available substrates and producing methane & hydrogen (Sharma & Melkania, 2018). Optimizing various parameters such as substrate composition, pH, temperature, retention time, and inoculum quality is crucial for maximizing methane yield. Balancing the carbon-to-nitrogen ratio, maintaining optimal pH and temperature, ensuring sufficient retention time, and using robust, well-adapted microbial communities are essential steps (Pugazhendhi & Banu, 2019).

Recent advancements in process monitoring and control technologies have also contributed to the improved efficiency of biomethane production. Numerous techniques such as real-time monitoring of biogas composition, automated feeding systems, and advanced microbial characterization tools allow for better management and optimization of the AD process (Karthikeyan & Visvanathan, 2014). Emerging techniques demonstrating promising results include the use of nanotechnology for enhanced biogas production (Khan et al., 2022), the integration of artificial intelligence for predictive analytics and process control (Khan et al., 2023), and bioelectrochemical systems to boost microbial activity and methane yield (Horváth-Gönczi et al., 2023). Similarly, in biohydrogen process monitoring, innovative sensor technologies—such as gas chromatography, electrochemical sensors, and spectroscopic methods—enable continuous tracking of key parameters, including hydrogen concentration, pH, temperature, and microbial activity. These advanced monitoring systems offer high sensitivity, specificity, and real-time data acquisition capabilities, facilitating early detection of process disturbances and enabling timely corrective actions (Barrera et al., 2024). By focusing on these factors and leveraging technological innovations, the efficiency and yield of biomethane & biohydrogen production from various organic waste materials can be significantly enhanced, contributing to sustainable energy production and waste management solutions (Guo & Wu, 2021).

6 Cotton Biomass and Processing Waste as a Feedstock for Biohydrogen and Biomethane Production

With the rising demand for energy and the diminishing availability of fossil fuels, there is an urgent need to find alternative, renewable, and sustainable energy sources to meet global energy needs. Among the various renewable energy options, biohydrogen and biomethane derived from biomass offer promising alternatives as low-carbon, eco-friendly energy sources (Banu et al., 2020). The production of biological hydrogen can be achieved through several technologies, including fermentation (such as dark fermentation, photofermentation), microbial electrolysis cells, and biophotolysis. Figure 4 shows the different biological and non-biological approaches utilized for the production of biohydrogen gas. These methods represent a step toward more sustainable energy solutions, potentially reducing our reliance on fossil fuels and mitigating environmental impact (Wang, Khoo, et al., 2021).

Biogas, comprising 50–60% methane, is produced through the microbial-mediated anaerobic digestion of lignocellulosic biomass and other agricultural

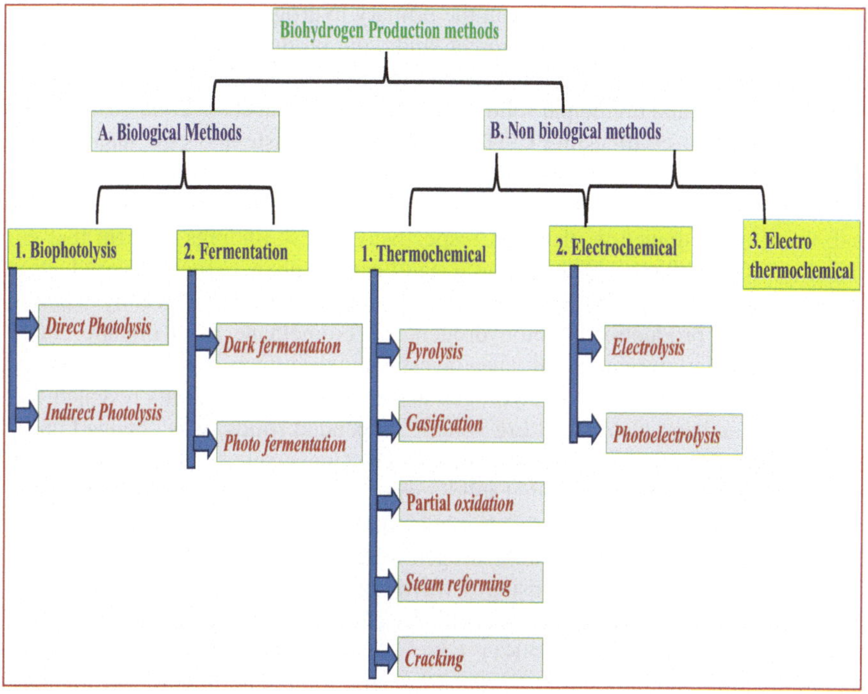

Fig. 4 Different approaches utilized for the production of hydrogen

residues (Hamawand et al., 2016). Research has indicated that various cotton wastes—including cottonseed, CGT, CSH, and cotton textile waste—are viable candidates for bioenergy production, particularly in the forms of biohydrogen and biomethane. A study by Isci and Demirer explored the biochemical methane generation potential of CS, CSH, and cotton seed meal (CSM) using anaerobic digestion. The results confirmed the effectiveness of these cotton wastes in producing methane. Specifically, 1 g of CS, CSH, and cotton oil cake yielded 65, 86, and 78 mL of methane, respectively, over 23 days in the presence of a basal medium. Moreover, the addition of basal media enhanced the biomethane yield from cotton waste, underscoring its potential for sustainable energy solutions (Isci & Demirer, 2007). Recently, a nanotechnological approach involving nanoparticles was employed to enhance biohydrogen production from cottonseed (CS) hydrolysate (Li et al., 2020). Specifically, the addition of iron (FeO) and nickel (NiO) nanoparticles (with a size of 50 nm) to the hydrogen synthesis system mediated by *Klebsiella* sp. WL1316 significantly improved bacterial substrate utilization and increased hydrogen production efficiency. The incorporation of FeO nanoparticles at 30 mg/L led to a 61% increase in hydrogen production after 96 h of fermentation compared to untreated samples. Additionally, using both nanoparticles at 30 mg/L each resulted in a 20% improvement in hydrogen synthesis. Further analysis suggested that these nanoparticles might have reduced the metabolic flux in competing pathways of hydrogen synthesis while enhancing the flux in the primary hydrogen generation pathway of the bacteria (Li et al., 2020). In another study, the production of biomethane was evaluated using a two-phase pilot-scale co-digestion process involving CGT mixed with cow manure and municipal solid waste (Macias-Corral et al., 2008). The results demonstrated that co-digesting CGT with municipal waste and cow manure produced 172 and 87 m^3 of methane/ton of dry waste, respectively. Alternatively, single anaerobic digestion of cow manure and municipal solid waste alone yielded only 62 and 37 m^3 of methane/ton of waste, respectively. These findings confirm that co-digesting CGT with other types of waste significantly enhances methane production (Macias-Corral et al., 2008). Sołowski, Konkol, and Cenian (2020) recently explored the concurrent production of biohydrogen and biomethane from organic cotton waste through dark fermentation. They investigated the effects of microaeration and pH levels on the production of hydrogen and methane. Their results revealed that the highest methane yield (0.78 L/g volatile suspended solids) was obtained under anaerobic conditions at a near-neutral pH of 7.54. Conversely, the highest hydrogen yield (1.1 mL/g volatile suspended solids) was achieved with an airflow rate of 4.6 mL/h and a pH of 6.4 (Sołowski, Konkol, Shalaby, & Cenian, 2020). Similarly, higher yields of both methane and hydrogen were observed from the anaerobic pretreatment of cottonseed (CS) using a thermophilic microbial consortium (Yuan et al., 2016). This consortium included lignocellulose-degrading microbes such as *Clostridium straminisolvens* (CSK1), *Clostridium* sp. FG4b, *Pseudoxanthomonas* sp. M1-3, *Bordetella* sp., and *Brevibacillus* sp. M1-5. The biogas and methane yields from cottonseed were significantly increased following pretreatment with this microbial consortium, underscoring its effectiveness in enhancing waste-to-energy conversion (Yuan et al., 2016). Table 2 shows the representative examples of cotton biomass and wastes, along with different approaches utilized for the biological production of biomethane and biohydrogen.

7 Strategists to Improve the Biohydrogen Production Process

Enhancing efficiency output and sustainability in the generation of biohydrogen from cotton waste requires a multidimensional strategy because of its complex structure and fiber character. Cotton waste is a potential feedstock for biohydrogen generation (Chen et al., 2015; Akyol et al., 2019), as its hydrolysis produces biofuel. This process can be enhanced by physically pretreating lignocellulosic biomass (LCB). Physical pretreatment techniques include pyrolysis, microwave irradiation, ultrasonication, ammonia fiber expansion, steam explosion, and mechanical milling (Chaitanya et al., 2016). Sharma and Melkania (2018) demonstrated that at approximately 200 °C, lignocellulosic biomass (LCB) undergoes thermochemical breakdown via pyrolysis, with phosphoric acid and glycerol enhancing lignin and xylan extraction. Meanwhile, microwave irradiation disrupts LCB structures, facilitating sugar release. Combining microwave treatment with enzymatic hydrolysis can improve yields, while ultrasonication (20–40 kHz) further enhances LCB

Table 2 Representative examples showing the potential of cotton biomass and processing waste as substrate for the production of biohydrogen and biomethane

S. No.	Substrate	Method	Biohydrogen/ biomethane	Yield	Microbial strain/inoculum	References
1.	Cotton stalk hydrolysate	Recombinant technology	Biohydrogen	0.37 ± 0.01 mol H_2/mol sugar	*Enterobacter cloacae* WL1318	Zhang et al. (2018)
2.	Cotton stalk hydrolysate	Fermentation	Biohydrogen	1.44 ± 0.08 mol H_2/mol sugar	*Klebsiella* sp. WL1316	Li et al. (2020)
3.	Cotton wastes	Dark fermentation	Biohydrogen	7.5×10^{-3} H_2/mol sugar	Sludge from an agricultural biogas plant	Sołowski, Konkol, and Cenian (2020)
4.	Cotton stalk	Enzymatically digestion and fermentation	Biohydrogen	5.63×10^{-3} H_2/mol sugar	*Citrobacter freundii*	Lakshmidevi and Muthukumar (2023)
5.	Cotton straw	Anaerobic digestion	Biohydrogen	33 mL H_2/g	*Clostridium* sp.	Yoruklu et al. (2020)
6.	Cotton wastes	Dark fermentation	Biohydrogen	4.91×10^{-5} H_2/mol sugar	Maize silage and manure	Sołowski, Konkol, Shalaby, and Cenian (2020)
7.	Cotton stalk	Anaerobic digestion	Biomethane	192.4 mL/g	Inoculum sludge of biogas	Zhang et al. (2018)
8.	Cotton gin trash	Co-digestion	Biomethane	262 ml	Goat manure	Kaur and Kommalapati (2021)
9.	Cotton stalk	Anaerobic digestion	Biomethane	52 ml after 2 days	Rumen microorganisms	Kurt Kara et al. (2021)
10.	Cotton straw	Anaerobic digestion	Biomethane	83 mL/g	Anaerobic granular sludge	Yoruklu et al. (2020)

degradation (Singh et al., 2022). The chemical pretreatment of LCB substrates is mostly done with concentrated or diluted acids, which can not only release sugars by breaking down cellulose-hemicellulose bonds but also produce toxic inhibitors and necessitate acid recovery (Amin et al., 2017). However, the overuse of alkaline pretreatment with bases like NaOH can damage anaerobic processes and have negative environmental effects (Shen et al., 2017). So, to overcome it, Amiri et al. (2014) applied the organosolv method, which is a solvent recovery technique, is environment- friendly, and effectively removes hemicellulose while preserving cellulose. While preserving cellulose, ozonolysis breaks down lignin and hemicellulose using ozone, avoiding harmful byproducts but coming at a high cost. Although they are more costly, ionic liquids that dissolve lignin and improve cellulose accessibility include 1-butyl-3-methylimidazolium chloride (García-Cubero et al., 2009; Rahim et al., 2020). A physicochemical method such as "steam explosion" uses high-pressure steam (0.6–4.9 MPa) to treat LCB at 160–210 °C (Barbanera et al., 2015). This effectively breaks down biomass but produces byproducts that are inhibitory, such as 5-HMF and furfural. LCB is also efficiently broken down by hot water pretreatment, which is frequently combined with chemicals, without producing harmful compounds (Anu et al., 2020). Ammonia fiber explosion (AFEX), a more recent technique, can achieve upto a 90% glucose yield during hydrolysis and is especially effective low-lignin substrates (Banoth et al., 2017). Because biological pretreatment methods do not produce harmful byproducts and require less energy than chemical and physical methods, they are more preferable. In this process, fungi are important players, so while using enzymes like laccase and peroxidase, white-rot fungi like *Pleurotus ostreatus* and *Pycnoporus cinnabarinus* are especially good at breaking down lignin (Suhara et al., 2012). Additionally, thermophilic cultures are used to enhance biohydrogen production by influencing temperature, pH, and CO_2 volume (mL), as well as the concentrations of Fe^{+2}, Zn^{+2}, and Ni^{+2} through the activity of the hydrogenase enzyme (Akaçin et al., 2024). For continuous production of hydrogen, Valencia-Ojeda et al. (2021) used Cellulase 50XL, which is an efficient substitute for highly specialized lignocellulosic enzymes which has demonstrated superior hydrogen production compared to other enzymes. However, pretreatment of the anaerobic inoculum is done to speed up hydrolysis, reduce rate-limiting steps, and increase process stability for production of hydrogen (Venkata et al., 2008). Further, to improve potentiality of hydrogen, metal nanoparticles (NPs) that are prone to oxidation and instability, such as nickel (Ni), iron (Fe), and cobalt (Co), are used. As NPs improve microbial metabolism by acting as cofactors for hydrogenase enzymes, they in turn increase electron transfer rates. With yields up to 4.5 times higher than controls, Fe-based NPs are particularly notable for their substantial influence on hydrogen production. Additional NPs that enhance hydrogen production include Co, Ag, and TiO_2, with green- synthesized NPs showing superior performance. Liu et al. (2017) investigated the effects of TiO_2, ZnO, and SiC nanoparticles on photo-fermentative hydrogen production by *Rhodopseudomonas* sp. nov. strain A7. The results showed that all three nanoparticles enhanced hydrogen yield, with SiC exhibiting the highest efficiency. The optimal SiC nanoparticle concentration (200 mg/L) achieved a hydrogen yield of 2.99 mol-H_2/mol-acetate, an

18.6% increase over the control. Another study done by Engliman et al. (2017) was of oxide nanoparticles used in glucose-fed anaerobic bacteria, which improved hydrogen yield by 34.38% in comparison with nickel oxide. Newly approached technology in recent years is genetic engineering, which includes methods like CRISPR/Cas9, gene editing, and cell surface engineering (Zhao et al., 2017). Kim et al. (2023) used *Clostridium acetobutylicum*, which increased hydrogen production by 66.3%, whereas Zhao et al. (2017) improved hydrogen efficiency in *Enterobacter aerogenes* by 86.2% through gene expression. Zhang et al., (2018) reported that *Enterobacter cloacae* overexpressing the formate hydrogen lyase gene (fhlA) increased hydrogen production from cotton stalks by 188%. Furthermore, significant increases in hydrogen and methane production have been observed following genetic modifications in *Caldicellulosiruptor bescii* and *Methylococcus capsulatus* (Cha et al., 2013; Tapscott et al., 2019).

8 Recent Advances in Biohydrogen/Biomethane Production from Biomass

Previous research efforts have focused on improving biomethane production efficiency through innovative approaches, including co-digestion with complementary biomass sources and advanced pretreatment strategies. Co-digestion of cotton waste with other organic substrates has shown promise in enhancing biogas yield and quality (Banu et al., 2020). Recent studies have harnessed advanced metabolic engineering tools such as PCR-based DNA sequencing and assembly, genome editing, CRISPR, and cell surface engineering to enhance methane and biohydrogen production from cotton waste (Cai & Liu, 2020). Recently, Zhao et al. (2017) demonstrated significant increases in hydrogenase activity in *E. aerogenes* through HycE and HycG gene overexpression, elevating hydrogen yields by 86.2% and 69.8%, respectively. Tapscott et al. (2019) utilized CRISPR/Cas9 gene editing in *Methylococcus capsulatus*, achieving 71% efficiency with Cas9D10A nickase, highlighting potential for industrial applications in methanotrophs and other Gram-negative microbes for high methane production. Additionally, Sang et al. (2023) engineered *Clostridium acetobutylicum* with gapC, resulting in a 66.3% increase in hydrogen production and enhanced NADH/NADPH pools, emphasizing advancements in biohydrogen technologies.

An efficient and economical method for biohydrogen production from lignocellulosic materials is the use of *Enterobacter cloacae*, which is a promising hydrogen-producing bacterium. The formate hydrogen lyase activator gene (fhlA) was positively cloned and overexpressed in *E. cloacae* WL1318, which resulted in a 188% increase in hydrogen synthesis from cotton stalk oxidation after 24 h. In addition to improved utilization of sugar and xylose, a higher hydrogen yield, and a more than 30% increase in production during fermentation, the recombinant strain also exhibited these characteristics. The overexpression of fhlA was demonstrated by metabolic analysis, which showed a positive shift from competitive pathways to hydrogen production pathways. With the first development of the fhlA gene in

E. cloacae, this work paves the way for more efficient biohydrogen production from lignocellulosic raw materials (Zhang et al., 2018). The thermophilic bacterium *Caldicellulosiruptor bescii* is suitable for consolidated bioprocessing (CBP) because it can directly utilize lignocellulosic biomass for hydrogen, acetate, and lactate production. Using genetic engineering tools, the lactate dehydrogenase (ldh) gene in *Caldicellulosiruptor bescii* can be deleted through marker replacement on a non-reversible plasmid. This deletion suppresses lactate production while increasing acetate and hydrogen yields by 21–34% (Cha et al., 2013).

9 Conclusions and Future Prospects

The present chapter has concisely summarized the colossal information about the cotton-based economy, cotton processing and the kind of waste generated, pretreatment of cotton biomass, and, most importantly, the potential of microbial biotechnology in the bioconversion of cotton wastes into bioproducts. Among the cotton wastes reviewed, CS, CGT, and CSH are typical lignocellulosic biomass, whereas cotton linters, micro dust, fabric waste, and industrial cotton waste are usually rich in cellulosic content. Various sustainable and inexpensive pretreatment methods could effectively remove lignin from lignocellulosic cotton waste to augment the accessibility of fermentable carbohydrate. Herein, we have highlighted the fruitful management of diverse cotton waste by converting it into biomethane and biohydrogen through microbial route. The use of modern genomic and synthetic biology tools such as PCR-based DNA sequencing, synthesis, and assembly, genome editing, CRISPR, cell surface engineering, and metabolic engineering for diversifying fermentative feedstock and achieving higher yield of bioproducts is briefly narrated. Recent strategies utilized for enhanced biohydrogen production are also discussed.

Biomethane and biohydrogen production is a promising technology for agriculture waste management, thereby generating sustainable sources of renewable and green energy. Similarly, the comparative production of biomethane and biohydrogen from cotton biomass and other lignocellulosic biomass may be in-depth studied to reveal the biogas production potential of cotton waste. However, to commercialize this biogas process from cotton biomass, it is a prerequisite to assess the techno-economic feasibility under different conditions and its assessment with other methods for converting biomass to methane and hydrogen. On the microbial front, extensive research on pure cultures and consortia of methanogenic and other microbial communities is needed to enhance biomethane production. Microbial Electrolysis Cell (MEC) is an alternative approach for biohydrogen production through the treatment of complex organic waste. Hence, the microbial strains that are most efficient and competitive in the MEC need to be isolated, identified, and evaluated for their enhanced hydrogen production potential in the MEC. Therefore, clear investigations of the microorganisms and their functional attributes are indispensable to utilize organic waste and cotton biomass for biohydrogen and biomethane production.

References

Adewuyi, A. (2022). Underutilized lignocellulosic waste as sources of feedstock for biofuel production in developing countries. *Frontiers in Energy Research, 10*. https://doi.org/10.3389/fenrg.2022.741570.

Adl, M., Sheng, K., & Gharibi, A. (2012). Technical assessment of bioenergy recovery from cotton stalks through anaerobic digestion process and the effects of inexpensive pre-treatments. *Applied Energy, 93(C), 251–260*. https://doi.org/10.1016/j.apenergy.2011.11.065.

Agblevor, F. A., Batz, S., & Trumbo, J. (2003). Composition and ethanol production potential of cotton gin residues. *Applied Biochemistry and Biotechnology, 105*, 219–230.

Akaçin, İ., Ersoy, Ş., Keskin, T., Abubackar, H. N., & Güngörmüşler, M. (2024). Optimizing biohydrogen production yields by employing locally isolated thermophilic bacteria from hot springs. International Journal of Hydrogen Energy, 52(Part D), 502–510. https://doi.org/10.1016/j.ijhydene.2023.07.243

Akyol, Ç., İnce, O., Bozan, M., & Özbayram, G. (2019). Biological pretreatment with Trametes versicolor to enhance methane production from lignocellulosic biomass: A metagenomic approach. *Industrial Crops and Products, 140, 111659*. https://doi.org/10.1016/j.indcrop.2019.111659

Alavi-Borazjani, S. A., da Cruz Tarelho, L. A., & Capela, M. I. (2024). Biohythane production via anaerobic digestion process: Fundamentals, scale-up challenges, and techno-economic and environmental aspects. *Environmental Science and Pollution Research, 31*, 49935–49984. https://doi.org/10.1007/s11356-024-34471-8

Al Afif, R., & Pfeifer, C. (2021). Enhancement of methane yield from cotton stalks by mechanical pre-treatment. Carbon Resources Conversion, 4, 164–168. https://doi.org/10.1016/j.crcon.2021.04.003

Amin, F. R., Khalid, H., Zhang, H., Rahman, S., Zhang, R., Liu, G., & Chen, C. (2017). Pretreatment methods of lignocellulosic biomass for anaerobic digestion. *AMB Express, 7, 72*. https://doi.org/10.1186/s13568-017-0375-4

Amiri, H., Karimi, K., & Zilouei, H. (2014). Organosolv pretreatment of rice straw for efficient acetone, butanol, and ethanol production. *Bioresource Technology, 152, 450–456*. https://doi.org/10.1016/j.biortech.2013.11.038

Anu, K. A., Rapoport, A., Kunze, G., Kumar, S., Singh, D., & Singh, B. (2020). Multifarious pretreatment strategies for lignocellulosic substrates for the generation of renewable and sustainable biofuels: A review. *Renewable Energy, 160, 1228–1252*. https://doi.org/10.1016/j.renene.2020.07.031

Angelidaki, I., & Sanders, W. (2004). Assessment of the anaerobic biodegradability of macropollutants. *Re/Views in Environmental Science and Bio/Technology, 3*(2), 117–129. https://doi.org/10.1007/s11157-004-2502-3

Banu, J. R., Kavitha, S., Kannah, R. Y., Bhosale, R. R., & Kumar, G. (2020). Industrial wastewater to biohydrogen: possibilities towards successful biorefinery route. *Bioresource Technology, 298*, 122378.

Barrera, A., Gómez-Ríos, D., & Ramírez-Malule, H. (2024). Assessment of a low-cost hydrogen sensor for detection and monitoring of biohydrogen production during sugarcane straw/vinasse co-digestion. *AgriEngineering, 6*(1), 479–490. https://doi.org/10.3390/agriengineering6010029

Barbanera, M., Buratti, C., Cotana, F., Foschini, D., & Lascaro, E. (2015). Effect of steam explosion pretreatment on sugar production by enzymatic hydrolysis of olive tree pruning. *Energy Procedia, 81, 146–154*. https://doi.org/10.1016/j.egypro.2015.12.069

Bridgwater, T. (2012). Upgrading biomass fast pyrolysis liquids. *Environmental Progress & Sustainable Energy, 31*, 11635. https://doi.org/10.1002/ep.11635

Cai, Y., & Liu, L. (2020). Anaerobic digestion of food waste and straw for biogas production: A review. *Bioresource Technology, 316*, 123994.

Chandrasekhar, K., Lee, Y. J., & Lee, D. W. (2015). Biohydrogen production: Strategies to improve process efficiency through microbial routes. *International Journal of Molecular Sciences*. https://doi.org/10.3390/ijms16048266

Cha, M., Chung, D., & Elkins, J. G. (2013). Metabolic engineering of Caldicellulosiruptor bescii yields increased hydrogen production from lignocellulosic biomass. *Biotechnology for Biofuels, 6(1), 85.* https://doi.org/10.1186/1754-6834-6-85

Chaitanya, N., Sivaramakrishna, D., Kumar, B. S., Himabindu, V., Lakshminarasu, M., & Vishwanadham, M. (2016). Selection of pretreatment method for enriching hydrogen-producing bacteria using anaerobic sewage sludge with three different substrates. *Biofuels, 7(2), 163–171.* https://doi.org/10.1080/17597269.2015.1123969

Chen, Z., Zhang, S., Chen, Z., & Ding, D. (2015). An integrated process for hydrogen-rich gas production from cotton stalks: The simultaneous gasification of pyrolysis gases and char in an entrained flow bed reactor. *Bioresource Technology, 198*, 586–592. https://doi.org/10.1016/j.biortech.2015.09.015

da Silva, J. E., de Araújo Melo, D. M., de Freitas Melo, M. A., et al. (2019). Energetic characterization and evaluation of briquettes produced from naturally colored cotton waste. *Environmental Science and Pollution Research, 26, 14259–14265.* https://doi.org/10.1007/s11356-019-04777-z

Domelsmith, L. N., & Berni, R. J. (1986). The chemical composition of standard cotton dust. *Environmental Health Perspectives, 66*, 109–112. https://doi.org/10.1289/ehp.8666109

Dong, Z., Li, N., Chu, T., Ding, J., Zhang, J., & Dong, A. (2022). High-quality natural fibers from cotton stalk bark via limited alkali penetration and simultaneous accelerated temperature rise. *Materials, 15*, 422. https://doi.org/10.3390/ma15020422

Dukare, A., Sharma, K., Kautkar, S., Dhakane-Lad, J., Yadav, R., Nadanathangam, V., & Saxena, S. (2023). Microbial xylanase aided biobleaching effect on multiple components of lignocelluloses biomass-based pulp and paper: A review. *Nordic Pulp & Paper Research Journal, 38*, 3. https://doi.org/10.1515/npprj-2023-0005

Engliman, N. S., Abdul, P. M., Wu, S.-Y., & Jahim, J. M. (2017). Influence of iron (II) oxide nanoparticle on biohydrogen production in thermophilic mixed fermentation. International Journal of Hydrogen Energy, 42(45), 27482–27493. https://doi.org/10.1016/j.ijhydene.2017.05.224

Egbuta, M. A., McIntosh, S., Waters, D. L. E., Vancov, T., & Liu, L. (2019). Chemical volatiles present in cotton gin trash: A by-product of cotton processing. *PLoS One, 14*(9), e0222146. https://doi.org/10.1371/journal.pone.0222146

Elisabeth, A. C., Pham, T.-N., Cossus, L., Abla, M., Ilc, T., & Lawrence, P. (2024). Exploring industrial lignocellulosic waste: Sources, types, and potential as high-value molecules. *Waste Management, 188*, 11–38. https://doi.org/10.1016/j.wasman.2024.07.029

García-Cubero, M. T., González-Benito, G., Indacoechea, I., Coca, M., & Bolado, S. (2009). Effect of ozonolysis pretreatment on enzymatic digestibility of wheat and rye straw. *Bioresource Technology, 100(5), 1608–1613.* https://doi.org/10.1016/j.biortech.2008.09.012

Ghasemian, M., Zilouei, H., & Asadinezhad, A. (2016). Enhanced biogas and biohydrogen production from cotton plant wastes using alkaline pretreatment. Energy & Fuels, 30(12), 10484–10493. https://doi.org/10.1021/acs.energyfuels.6b01999.

Guo, Z., & Wu, G. (2021). Enhancing methane production from lignocellulosic biomass by biological pretreatment: A review. *Renewable and Sustainable Energy Reviews, 139*, 110688.

Hamawand, I., Sandell, G., Pittaway, P., Chakrabarty, S., Yusaf, T., Chen, G., Seneweera, S., Al-Lwayzy, S., Bennett, J., & Hopf, J. (2016). Bioenergy from cotton industry wastes: A review and potential. *Renewable and Sustainable Energy Reviews, 66*, 435–448. https://doi.org/10.1016/j.rser.2016.08.033

Hanif, M. U., Zwawi, M., Algarni, M., Bahadar, A., Iqbal, H., Capareda, S. C., Hanif, M. A., Waqas, A., Hossain, N., & Siddiqui, M. T. H. (2022). The effects of using pretreated cotton gin trash on the production of biogas from anaerobic co-digestion with cow manure and sludge. *Energies, 15*(2), 490. https://doi.org/10.3390/en15020490

Horváth-Gönczi, N. N., Bagi, Z., Szuhaj, M., Rákhely, G., & Kovács, K. L. (2023). bioelectrochemical systems (bes) for biomethane production. *Fermentation, 9*, 610. https://doi.org/10.3390/fermentation9070610

Huang Y., Wei, X., Zhou, S., Liu, M., Tu, Y., Li, A., Chen, P., Wang, Y., Zhang, X., Tai, H., Peng, L., & Xia, T. (2015), Steam explosion distinctively enhances biomass enzymatic saccharification of cotton stalks by largely reducing cellulose polymerization degree in G. barbadense and G. hirsutum. *Bioresour Technol. 181:224–30.* https://doi.org/10.1016/j.biortech.2015.01.020.

Isci, A., & Demirer, G. N. (2007). Biogas production potential from cotton wastes. *Renewable Energy, 32*(5), 750–757.

Jain, R., Panwar, N. L., Jain, S., Gupta, T., Agarwal, C., & Meena, S. (2022). Bio-hydrogen production through dark fermentation: An overview. *Biomass Conversion and Biorefinery, 14*, 1–26. https://doi.org/10.1007/s13399-022-03282-7

Jordan, J. H., Easson, M. W., Dien, B., Thompson, S., & Condon, B. D. (2019). Extraction and characterization of nanocellulose crystals from cotton gin motes and cotton gin waste. *Cellulose, 26, 5959–5979.* https://doi.org/10.1007/s10570-019-02533-7

Karthikeyan, O. P., & Visvanathan, C. (2014). Bio-energy recovery from high-solid organic substrates by dry anaerobic bio-conversion processes: A review. *Reviews in Environmental Science and Bio/Technology, 12*(3), 257–284.

Kaur, H., & Kommalapati, R. R. (2021). Optimizing anaerobic co-digestion of goat manure and cotton gin trash using biochemical methane potential (BMP) test and mathematical modeling. *SN Applied Sciences, 3*(8), 724. https://doi.org/10.1007/s42452-021-04706-1

Karthyani, S., Pandey, A., & Devendra, L. P. (2020). Delignification of cotton stalks using sodium cumene sulfonate for bioethanol production. *Biofuels, 11, 431–440*

Keshav, P. K., Banoth, C., Anthappagudem, A., Rao Linga, V., & Bhukya, B. (2018). Sequential acid and enzymatic saccharification of steam exploded cotton stalk and subsequent ethanol production using Scheffersomyces stipitis NCIM 3498. *Industrial Crops and Products, 125*, 462–467. https://doi.org/10.1016/j.indcrop.2018.08.060

Khan, M., Chuenchart, W., Surendra, K. C., & Khanal, K. C. (2023). Applications of artificial intelligence in anaerobic co-digestion: Recent advances and prospects. *Bioresource Technology, 370*, 128501. https://doi.org/10.1016/j.biortech.2022.128501

Khan, S. Z., Zaidi, A. A., Naseer, M. N., & AlMohamadi, H. (2022). Nanomaterials for biogas augmentation towards renewable and sustainable energy production: A critical review. *Frontiers in Bioengineering and Biotechnology, 10*, 868454. https://doi.org/10.3389/fbioe.2022.868454

Kim, S. H., Hwang, J. H., Kim, H. J., Oh, S. J., Kim, H. J., Shin, N., Kim, S.-H., Park, J.-H., Bhatia, S. K., & Yang, Y.-H. (2023). Enhancement of biohydrogen production in Clostridium acetobutylicum ATCC 824 by overexpression of glyceraldehyde-3-phosphate dehydrogenase gene. Enzyme and Microbial Technology, 168, 110244. https://doi.org/10.1016/j.enzmictec.2023.110244.

Knox, O., Rochester, I., Vadakattu, G., & Lawrence, L. (2006). *Composting in Australian cotton production* (pp. 46–48). Aust Cotton Grow.

Kumar, A., Gautam, A., & Dutt, D. (2016). Biotechnological transformation of lignocellulosic biomass in to industrial products: An overview. *Advances in Bioscience and Biotechnology, 07*(03), 149–168. https://doi.org/10.4236/abb.2016.73014

Kurt Kara, G., Doluk, R., Civelek Yoruklu, H., Demir, A., & Ozkaya, B. (2021). Biomethane production kinetics of rumen pretreated lignocellulosic wastes. *Clean Technologies and Environmental Policy, 23*(10), 2941–2954. https://doi.org/10.1007/s10098-021-02214-9

Lakshmidevi, R., & Muthukumar, K. (2023). Biohydrogen production from enzymatically digested cotton stalks using *Citrobacter freundii*. *Journal of The Institution of Engineers (India): Series E, 104*(1), 11–18. https://doi.org/10.1007/s40034-022-00265-1

Li, Y., Zhu, Q., Ding, P., You, S., Zhang, Q., & Jiang, H. (2020). Effects of FeO and NiO nanoparticles on hydrogen production from cotton stalk hydrolysate using Klebsiella sp. WL1316: Evaluation of size and concentration of the nanoparticles. *International Journal of Hydrogen Energy, 45*(11), 6243–6253. https://doi.org/10.1016/j.ijhydene.2020.01.012

Liu, B., Jin, Y., Wang, Z., Xing, D., Ma, C., Ding, J., & Ren, N. (2017). Enhanced photo-fermentative hydrogen production of Rhodopseudomonas sp. nov. strain A7 by the addition of TiO_2, ZnO, and SiC nanoparticles. International Journal of Hydrogen Energy, 42(29), 18279–18287. https://doi.org/10.1016/j.ijhydene.2017.04.147

Lindner, J., Zielonka, S., Oechsner, H., & Lemmer, A. (2015). Effects of mechanical treatment of digestate after anaerobic digestion on the degree of degradation. *Bioresource Technology, 178,* 194–200. https://doi.org/10.1016/j.biortech.2014.09.117

Liu, B., Jin, Y., Wang, Z., Xing, D., Ma, C., Ding, J., & Ren, N. (2017). Enhanced photo-fermentative hydrogen production of Rhodopseudomonas sp. nov. strain A7 by the addition of TiO_2, ZnO, and SiC nanoparticles. International Journal of Hydrogen Energy, 42(29), 18279–18287. https://doi.org/10.1016/j.ijhydene.2017.04.147

Luthfi, A. A. I., Abdul, P. M., Jahim, J. M., Engliman, N. S., Jamali, N. S., Tan, J. P., Manaf, S. F. A., Sajab, M. S., & Bukhari, N. A. (2023). Isolation and characterization of biohydrogen-producing bacteria for biohydrogen fermentation using oil palm biomass-based carbon source. *Applied Sciences, 13*(1), 656. https://doi.org/10.3390/app13010656

Macias-Corral, M., Samani, Z., Hanson, A., Smith, G., Funk, P., Yu, H., & Longworth, J. (2008). Anaerobic digestion of municipal solid waste and agricultural waste and the effect of co-digestion with dairy cow manure. *Bioresource Technology, 99*(17), 8288–8293. https://doi.org/10.1016/j.biortech.2008.03.057

Mamun, A., & Torii, M. R. (2018). Enhancement of biogas production by co-digestion of activated sludge with food waste and fat, oil, and grease. *Energy for Sustainable Development, 45*, 82–89.

Mao, C., Feng, Y., Wang, X., & Ren, G. (2016). ChemInform abstract: Review on research achievements of biogas from anaerobic digestion. *ChemInform, 47*(28), 28297. https://doi.org/10.1002/chin.201628297

McIntosh, S., Vancov, T., Palmer, J., & Morris, S. (2014). Ethanol production from cotton gin trash using optimised dilute acid pretreatment and whole slurry fermentation processes. *Bioresource Technology, 173, 42–51.* https://doi.org/10.1016/j.biortech.2014.09.063

Moorthy, R., Ravikumar, R., Kumar, P. S., Rangasamy, G., Gurunathan, B., Ethiraj, B., & Thanabal, V. (2022). Insight into the effective utilization of cotton spinning wastes from textile mills for the production of bioethanol. *Sustainable Energy Technologies and Assessments, 53, 102770.* https://doi.org/10.1016/j.seta.2022.102770

Nath, S. (2024). Biotechnology and biofuels: Paving the way towards a sustainable and equitable energy for the future. *Discover Energy, 4*, 8. https://doi.org/10.1007/s43937-024-00032-w

Nebechi, K. O., Chika, O., Ujah, C., Asadu, O., Funsho, O., Kolawole, B., & Nnamdi, E. (2024). Production of hydrogen energy from biomass: Prospects and challenges. *Green Technologies and Sustainability, 2*(3), 100100. https://doi.org/10.1016/j.grets.2024.100100

Patel, D., & Shah, A. (2021). Integrated lignocellulosic biorefinery: Gateway for production of second-generation ethanol and value-added products. *Journal of Bioresources and Bioproducts, 6,* 108–128. https://doi.org/10.1016/j.jobab.2021.02.001

Plácido, J., Imam, T., & Capareda, S. (2013). Evaluation of ligninolytic enzymes, ultrasonication and liquid hot water as pretreatments for bioethanol production from cotton gin trash. *Bioresource Technology, 139*, 203–208. https://doi.org/10.1016/j.biortech.2013.04.012

Pugazhendhi, A., & Banu, J. R. (2019). Co-digestion of organic wastes for enhanced biogas production. *Bioresource Technology, 276*, 37–45.

Rahim, A. H. A., Man, Z., Sarwono, A., Muhammad, N., Khan, A. S., Hamzah, W. S. W., Yunus, N. M., & Elsheikh, Y. A. (2020). Probe sonication-assisted ionic liquid pretreatment for rapid dissolution of lignocellulosic biomass. *Cellulose, 27(4), 2135–2148.* https://doi.org/10.1007/s10570-019-02914-y

Ranjithkumar, M., Ravikumar, R., Sankar, M. K., Kumar, M. N., & Thanabal, V. (2017). An effective conversion of cotton waste biomass to ethanol: A critical review on pretreatment processes. *Waste and Biomass Valorization, 8*(1), 57–68. https://doi.org/10.1007/s12649-016-9563-8

Shen, J., Zhao, C., Liu, G., & Chen, C. (2017). Enhancing the performance on anaerobic digestion of vinegar residue by sodium hydroxide pretreatment. *Waste Biomass Valorization, 8(7), 1119–1126.* https://doi.org/10.1007/s12649-016-9666-2

Shahzad, K., Sohail, M., & Hamid, A. (2019). Green ethanol production from cotton stalk. *IOP Conference Series: Earth and Environmental Science, 257*, 012025. https://doi.org/10.1088/1755-1315/257/1/012025

Sharma, S., & Melkania, U. (2018). Anaerobic digestion for waste management and resource recovery. *Environmental Chemistry Letters, 16*(3), 1177–1193.

Sharma-Shivappa, R., & Chen, Y. (2008). Conversion of cotton wastes to bioenergy and value-added products. *Transactions of the ASABE, 51*, 2239–2246. https://doi.org/10.13031/2013.25377

Singh, H., Tomar, S., Qureshi, K. A., Jaremko, M., & Rai, P. K. (2022). Recent advances in biomass pretreatment technologies for biohydrogen production. *Energies, 15(3), 999.* https://doi.org/10.3390/en15030999

Silva, T. A. L., Varao, L. H. R., & Pasquini, D. (2024). Lignocellulosic biomass. In S. Thomas, M. Hosur, D. Pasquini, & C. Jose Chirayil (Eds.), *Handbook of biomass*. Springer. https://doi.org/10.1007/978-981-19-6772-6_5-1

Silverstein, R. A., Chen, Y., Sharma-Shivappa, R. R., Boyette, M. D., & Osborne, J. (2007). A comparison of chemical pretreatment methods for improving saccharification of cotton stalks. *Bioresource Technology, 98*(16), 3000–3011. https://doi.org/10.1016/j.biortech.2006.10.022

Soares, J. F., Confortin, T. C., Todero, I., Mayer, F. D., & Mazutti, M. A. (2020). Dark fermentative biohydrogen production from lignocellulosic biomass: Technological challenges and future prospects. *Renewable and Sustainable Energy Reviews, 117*, 109484. https://doi.org/10.1016/j.rser.2019.109484

Sołowski, G., Konkol, I., & Cenian, A. (2020). Methane and hydrogen production from cotton waste by dark fermentation under anaerobic and micro-aerobic conditions. *Biomass & Bioenergy, 138*, 105576. https://doi.org/10.1016/j.biombioe.2020.105576

Sołowski, G., Konkol, I., Shalaby, M., & Cenian, A. (2020). Rapid hydrogen generation from cotton wastes by mean of dark fermentation. *SN Applied Sciences, 2*(8), 1438. https://doi.org/10.1007/s42452-020-03247-3

Suhara, H., Kodama, S., Kamei, I., Maekawa, N., & Meguro, S. (2012). Screening of selective lignin-degrading basidiomycetes and biological pretreatment for enzymatic hydrolysis of bamboo culms. *International Biodeterioration & Biodegradation, 75, 176–180.* https://doi.org/10.1016/j.ibiod.2012.05.042

Tapscott, T., Guarnieri, M. T., & Henard, C. A. (2019). Development of a CRISPR/Cas9 system for Methylococcus capsulatus in vivo gene editing. *Applied and Environmental Microbiology, 85*(11), e00340. https://doi.org/10.1128/AEM.00340-19

Umesh, D., Sarsavadiya, P., Vaja, K., & Mahadeo, K. (2015). Physiochemical properties of cotton stalk biomass from agricultural residues. *Current World Environment: An International Research Journal of Environmental Sciences, 10*(1), 343–349. https://doi.org/10.12944/cwe.10.1.44

Valencia-Ojeda, C., Montoya-Rosales, J. J., Palomo-Briones, R., Montiel-Corona, V., Celis, L. B., & Razo-Flores, E. (2021). Saccharification of agave bagasse with Cellulase 50 XL is an effective alternative to highly specialized lignocellulosic enzymes for continuous hydrogen production. Journal of Environmental Chemical Engineering, 9(4), 105448. https://doi.org/10.1016/j.jece.2021.105448

Venkata, M. S., Babu, V., & Sarma, P. N. (2008). Effect of various pretreatment methods on anaerobic mixed microflora to enhance biohydrogen production utilizing dairy wastewater as substrate. *Bioresource Technology, 99*, 59–67. https://doi.org/10.1016/j.biortech.2006.12.004

Verardi, A., Lopresto, C. G., Blasi, A., Chakraborty, S., & Calabrò, V. (2020). Bioconversion of lignocellulosic biomass to bioethanol and biobutanol. In *Lignocellulosic biomass to liquid biofuels* (pp. 67–125). Elsevier. https://doi.org/10.1016/b978-0-12-815936-1.00003-4

Wang, K., Khoo, K. S., Chew, K. W., Selvarajoo, A., Chen, W.-H., Chang, J.-S., & Show, P. L. (2021). Microalgae: The future supply house of biohydrogen and biogas. *Frontiers in Energy Research, 9*, 660399. https://doi.org/10.3389/fenrg.2021.660399

Wang, J.-H., Liu, Y.-P., Wang, J.-J., & Zhang, M. (2017). Effect of acid pretreatment on hydrogen production from cotton stalk by anaerobic fermentation. Energy Sources, Part A: Recovery, Utilization, and Environmental Effects. https://doi.org/10.1080/15567036.2016.1217292.

Wang, X., Zhang, Y., Zhang, J., & Xu, H. (2021). Real-time monitoring of anaerobic digestion process using advanced sensor technologies: A review. *Bioresource Technology, 337*, 125409.

Yoruklu, H. C., Koroglu, E. O., Ozdemir, O. K., Demir, A., & Ozkaya, B. (2020). Bioenergy production from cotton straws using different pretreatment methods. International Journal of Hydrogen Energy, 45(60), 34720–34729. https://doi.org/10.1016/j.ijhydene.2020.02.104

Yildirim, O., Altinbas, M., & Ozkaya, B. (2023). Assessment of biomethane potential of cotton stalk using response surface methodology. *International Journal of Global Warming, 31*, 14–29.

Yuan, X., Ma, L., Wen, B., Zhou, D., Kuang, M., Yang, W., & Cui, Z. (2016). Enhancing anaerobic digestion of cotton stalk by pretreatment with a microbial consortium (MC1). *Bioresource Technology, 207*, 293–301. https://doi.org/10.1016/j.biortech.2016.02.037

Zhang, H., Bruns, M. A., & Logan, B. E. (2006). Biological hydrogen production by Clostridium acetobutylicum in an unsaturated flow reactor. *Water Research, 40(4), 728–734.* https://doi.org/10.1016/j.watres.2005.11.041

Zhang, H., Ning, Z., Khalid, H., Zhang, R., Liu, G., & Chen, C. (2018). Enhancement of methane production from Cotton Stalk using different pretreatment techniques. *Scientific Reports, 8*(1), 3463. https://doi.org/10.1038/s41598-018-21413-x

Zhao, J., Song, W., Cheng, J., Liu, M., Zhang, C., & Cen, K. (2017). Improvement of fermentative hydrogen production using genetically modified Enterobacter aerogenes. *International Journal of Hydrogen Energy, 42*(6), 3676–3681. https://doi.org/10.1016/j.ijhydene.2016.08.161

Microbial Fuel Cells and Sustainable Electricity Generation from Lignocellulosic Waste Biomass

Eduardo Dellosso Penteado, Yvan Jesus Olortiga Asencios, Anthony Andrey Ramalho Diniz, Luiza Helena da Silva Martins, Johnatt Allan Rocha de Oliveira, and Andrea Komesu

Abstract

The demand for energy is increasing and it is necessary to research different methods to replace the current fossil fuels approaches, which need to be more sustainable and use renewable feedstock for energy generation. In this sense, microbial fuel cell (MFC) is a cost-effective and eco-friendly power generation alternative, because it converts the chemical energy stored in organic and inorganic molecules directly into electricity, using microorganisms as biocatalysts. When lignocellulosic biomass is used as feedstock in MFC, more environmental benefits are achieved (reuse of waste and free GHG emission). However, there are a few drawbacks of using lignocellulosic biomass in microbial fuel cell, which difficult the upscaling of the system, such as the need for pretreatment and the effect on the extracellular electron transfer. Thereby, this chapter provides insight into the use of lignocellulosic biomass in MFC for power generation and future perspective.

Keywords

Microbial fuel cell · Bioelectrogenic · Lignocellulosic biomass · Electron transfers · Bioenergy · Renewable energy

---

E. D. Penteado · Y. J. O. Asencios · A. A. R. Diniz · A. Komesu (✉)
Universidade Federal de São Paulo, UNIFESP, Instituto do Mar, Santos, São Paulo, Brazil
e-mail: andrea.komesu@unifesp.br

L. H. da Silva Martins
Universidade Federal Rural da Amazônia, UFRA, Instituto de Saúde e Produção Animal, Belém, Pará, Brazil

J. A. R. de Oliveira
Universidade Federal do Pará. UFPA, Faculdade de Nutrição, Belém, Pará, Brazil

G. Mukherjee, S. Dhiman (eds.), *Value Addition and Utilization of Lignocellulosic Biomass*, https://doi.org/10.1007/978-981-96-2786-8_6

1 Introduction

According to the International Energy Agency in 2023, global energy demand increased by 1.3% in 2022. This steady growth, despite market fluctuations, underscores the fundamental trends in the energy sector, as the expanding global economy and population drive an increasing demand for energy services (IEA, 2023). A substantial share of this energy consumption is currently met by fossil fuels. However, a significant drawback of these non-renewable sources is their considerable greenhouse gas emissions, which lead to environmental pollution, climate change, and global warming. As a result, there has been a rising interest in renewable and clean energy sources to address the challenges associated with fossil fuels (Ahanchi et al., 2022).

Microbial fuel cells (MFCs) represent an emerging technology that has attracted significant interest from researchers as a renewable alternative energy source (Bazina et al., 2023). The MFC system is a bioelectrochemical device that converts organic materials directly into electricity by utilizing the metabolic activities of microorganisms.

As a typical bioelectrochemical system, MFCs consist of two main components: an anode and a cathode, which are separated by a proton exchange membrane (PEM). Electricity generation in MFCs is driven by biological oxidation at the anode and oxygen reduction at the cathode. In the anode compartment, microorganisms serve as biocatalysts, breaking down substrates to release electrons and protons through cellular respiration. The electrons are transported via an external circuit, while the protons move through the PEM, leading to a reduction reaction with oxygen in the cathode region to produce water. This energy generation method presents several benefits, such as operating under mild conditions, ease of use, and a wide variety of available biocatalysts (Wang et al., 2022).

MFCs are particularly attractive due to their ability to utilize various organic wastes, making them a sustainable solution for waste management and energy production. They can operate on a variety of substrates, such as lignocellulosic waste biomass, thus contributing to waste valorization and environmental remediation.

Lignocellulosic waste biomass is a renewable resource composed primarily of cellulose, hemicellulose, and lignin, found in plant materials such as agricultural residues, forestry by-products, and dedicated energy crops. It is regarded as one of the most plentiful renewable energy sources, with an estimated global annual production of approximately 200 billion tons (Liang et al., 2020; Shrivastava & Sharma, 2022). Because it is abundant and widely available, it makes an attractive feedstock for MFCs.

The economic feasibility and efficiency of converting waste into bioenergy are influenced by the composition and properties of the waste material. Researchers have investigated the use of lignocellulosic waste, either as hydrolysates or in powdered form, to generate bioelectricity (Verma & Mishra, 2023).

The integration of lignocellulosic biomass with MFC technology not only promotes renewable energy generation but also addresses waste management issues by valorizing agricultural and forestry residues. This dual benefit is crucial in the context of sustainable development and environmental protection.

Therefore, this work explores MFC operational principles and structures, including double, single, multi-chamber, and stacked configurations, examining factors like substrate selection, microbial communities, and operational conditions. Optimization strategies and the potential of lignocellulosic biomass in MFC applications are discussed. The physical and chemical properties of this biomass, relevant enzymatic pathways, and microbial electrogenesis mechanisms are detailed. Challenges like low power output, microbial contamination, and competition with methanogenesis are analyzed. Environmental benefits, economic feasibility, and future research directions for MFC efficiency and scalability are also addressed.

2 Microbial Fuel Cell (MFC)

Economic growth, increasing populations, and industrial and anthropogenic activities have been associated with the increase in global energy demand (Tetteh et al., 2024). All economic agents (governments, businesses, and individuals) require energy (Chen et al., 2024). According to Acheampong et al. (2021), the access to energy can improve human development, once it enhances facilities such as lightning, heating, ventilation, information communication technologies services, life-saving care in hospitals, among other general facilities.

Around the world, most countries use non-renewable energy sources, such as derivatives of petroleum and nuclear power (Abdulwahhab et al., 2023). The world dependence on non-renewable energy sources increases its vulnerability to supply instability, which also results in known adverse effects on the environment (Nugraheni et al., 2024). This high and critical demand has made all governments think about energy sources that have no environmental effect, do not pollute the environment, and are renewable (Qazani et al., 2024). Searching for sustainable sources, humans have discovered and designed new ways of generating energy, such as solar, hydro, and wind power generation methods, and also biological-based energy generation, such as biogas, bioethanol, biomass, and microbes (Khan et al., 2024). It also investigated the ability of certain types of microorganisms, known as electroactive microbes, to execute bioelectrochemical reactions to generate an electric current from chemical energy or vice versa (Abubackar et al., 2023).

Microbial fuel cell (MFC) is a novel environmental-friendly renewable energy technology, which uses microorganisms as biocatalysts, aiming at converting stocked energy, in organic and inorganic compounds, into electrical current, establishing a fascinating connection between electricity production and microbial metabolism (Ahmad et al., 2023; Fard & Wu, 2023; Penteado et al., 2015; Rebequi et al., 2023). Among other renewable sources, it uses a sustainable technology that consumes organic and inorganic waste for electricity generation, and has been continuously developed in recent decades (Khan et al., 2024). The fuel for MFCs is free, as they produce electricity using microorganisms, and they are also considered efficient devices for extracting energy from substrates in wastewater, such as nitrogen and carbon (Qazani et al., 2024). According to Safwat et al. (2023), the primary functions of MFCs are generating electricity and simultaneously reducing the organic matter in wastewater. An MFC operates based on bacterial activity, where its main general reaction is the breakdown of organic matter into carbon dioxide and water, while electric current is a byproduct of this chemical reaction, as shown in Eq. (1). Therefore, electrochemically active bacteria break down the organic substrate to produce bioelectricity, where bacteria use their metabolic pathway to transport electrons (Ahmad et al., 2023; Apollon, 2023).

$$\text{Overall reaction}: C_6H_{12}O_6 + 6H_2O + 6O_2 \rightarrow 6CO_2 + 12H_2O \tag{1}$$

According to Apollon (2023), an MFC is basically composed by an anode and a cathode compartments, where both are separated internally by a membrane. So, the anode chamber plays a prominent role in the cell, as it is where the organic matter present in substrate is oxidized through microbial activity, what produces electrons (e^-), protons (H^+), and carbon dioxide (CO_2), as shown in Eq. (2); so, the electrons are transported toward the cathode chamber (Apollon, 2023; Hamedani et al., 2024; Penteado et al., 2015; Penteado et al., 2016).

$$\text{Anode}: C_6H_{12}O_6 + 6H_2O \rightarrow 6CO_2 + 24H^+ + 24e^- \tag{2}$$

Next, thanks to the proton exchange membrane (PEM), the protons are associated with oxygen, forming the water molecule (H_2O), as shown in Eq. (3) (Apollon, 2023; Penteado et al., 2015; Penteado et al., 2016).

$$\text{Cathode}: O_2 + 4H^+ + 4e^- \rightarrow 2H_2O \tag{3}$$

There is an electrode in each chamber that serves as a bacterial support system and electron conductor. Bacteria multiply on the anode, generating a biofilm, and use their microbial metabolism to transform the organic matter from the substrate into CO_2, protons, and electrons (Hasan et al., 2023). Bacteria that can be involved in the production of electricity in MFC include *Geobacter sulfurreducens*, *Geobacter metallireducens*, *Clostridiumbutyricum*, *Rhodoferax ferrireducens*, *Aeromonas hydrophila*, and *Shewanella putrefaciens* (Mahmoudi et al., 2024).

This technology can be used to convert many types of wastewater, such as domestic, agricultural, industrial, and municipal wastewater into electrical energy, what makes MFC an appealing option for both urban and rural areas, where wastewater management and energy generation are vital problems to be solved (Hamedani et al., 2024). Its operation can be considered eco-friendly, because instead of toxic gas in oil or a diesel engine, the emissions from its operation include water, CO_2, and waste heat that can be reused (Kurniawan et al., 2022).

2.1 Types of MFC

Aiming to optimize the operational efficiency of MFCs, different types of configurations have been introduced (Roy et al., 2023). Therefore, it can be designed in a way that the anode and cathode be in the same chamber, in two different chambers, or constructed through other configurations (Safwat et al., 2023), as discussed next.

Single-chamber MFCs: This configuration consists of a single-chamber where both the anode and cathode are placed. It is characterized by a simple construction and cost-effectiveness, tending to be compact and requiring less space when compared to other configurations (Bhaduri & Behera, 2024). However, Roy et al. (2023) report that this arrangement imposes some limitations for commercial applications, such as the impossibility of increasing the volume of the chamber keeping the electrode spacing small, and what is a requirement for maintaining the power density when the arrangement is scaled up. The same authors state that this configuration is not suitable for wastewater treatment in large-scale applications.

Double-chamber MFCs: This configuration comes from the possibility of having the anode and cathode in two different chambers (Penteado et al., 2015; Safwat et al., 2023). In this configuration, both electrodes are separated by membranes (Roy et al., 2023). Electrons generated by the anaerobic digestion of the substrate by microbes in the biofilm on the anode reach the cathode through an external wire, and protons move through the membrane toward the cathodic chamber (Roy et al., 2023). The electrode spacing is crucial for the efficiency of the single-chamber MFC, what is a problem that can be eliminated in dual-chamber MFCs without membranes (Safwat et al., 2023). Therefore, according to Roy et al. (2023), the main advantage of this configuration is that it can be applied to produce energy, and for large-scale wastewater treatment.

Stacked MFCs: Multiple single MFC units can be connected in series and/or parallel, creating fuel cell stacks, what can be crucial in scaling up bioelectrochemical technology for power generation and wastewater treatment (Hassan et al., 2023). Stacked MFCs can generate greater voltage and current, in comparison to a single MFC, reducing the start-up time for bioelectrogenic activity (Shrivastava et al., 2023). According to Ida and Mandal (2023), this kind of configuration does not only increase the overall power output but it also does not affect each cell's power efficiency. The same authors state that power output produced from a parallel connection is higher.

2.2 Factors Affecting the Performance of MFCs

The bioelectricity production of MFCs can be influenced by several factors, such as the inoculated bacteria, temperature, anode material, and even substrate morphology (Xu et al., 2023). According to Sriwichai et al. (2024), factors that influence MFC operation have been extensively studied. In their opinion, the most-studied factors have been related to operating conditions, which include the types and composition of organic substrates, hydraulic retention time, and organic loading rate.

For Apollon (2023), the performance of this kind of system depends mainly on the type of substrate, electrode materials, anolyte/catholyte, bacterial communities, operating conditions, and pH of the medium. The same author points out that the performance of MFCs is also affected by factors such as chemical oxygen demand

(COD) and fuel concentration, operating temperature, and membrane thickness. Roy et al. (2023) state that the performance of MFCs is highly dependent on the construction features, and the three more critical components of an MFC are the cathode, anode, and PEM.

Additionally, Kurniawan et al. (2022) highlight a variety of bottlenecks that they understand have to be addressed before MFCs are commercialized on the market. Among the bottlenecks that they mention, they refer to high internal resistance, low electricity production, and high material cost as the major obstacles to the MFC technology implementation. Especially for the material cost, they recommend that scientists focus on research and develop suitable materials to make possible overcome the challenge of low energy production.

According to Mahmoudi et al. (2024), the full potential of MFCs for pilot-scale or real-world power generation has not yet been understood. They state that, in optimal conditions, an ideal MFC must have a maximum power density of 8.98 W/m^2, but this theoretical power density was not attained by the manufactured cells. Therefore, they conclude that there are still many technical problems that limit the commercialization of this technology. So, it is vital to identify and propose solutions for these issues, to have an MFC with extreme performance.

2.3 Optimizing the Microbial Fuel Cell for the Maximum Power Output

Literature has reported some studies highlighting improvements in power generation to overcome the low voltage output range of MFCs, focusing on increasing reactor volume and proposing different electrode materials, substrates, and diverse groups of microorganisms (Shrivastava et al., 2023).

For real-world applications, MFCs require a large and scalable design, with chambers of increased size, containing electrodes and other modules in a stack arrangement (Khan et al., 2024). For a higher power generation, Mahurede et al. (2022) state that the selection of material to be used as an electrode is crucial for the power generation, as it is important for electron transfer, microbial adhesion, and electrochemical properties. They recommend that the material cost be reduced and power densities maximized to make it possible to apply MFCs in the field.

In a general way, Ida and Mandal (2023) present some recommendations to improve the MFC performance: (1) combining MFC with other techniques commonly used for wastewater treatment; (2) investigating more electrogenic bacteria, because many electrogenic bacteria have not yet been used for wastewater treatment, and it can result in direct electron transfer mechanism; (3) promoting a better understanding of the extracellular electron transfer mechanism, by extending research; (4) improving the economic aspects and decrease bio-fouling of PEM, promoting the use of membrane-less MFC; and (5) promoting more studies about new electrode materials to improve the scalability of MFC for long-term operation.

3 Lignocellulosic Biomass

The search for renewable energy sources is crucial to meet growing global demand and mitigate the environmental impacts of fossil sources. In this context, lignocellulosic biomass (LCB) is a promising and abundant alternative, mainly cellulose, hemicellulose, and lignin. This biomass derives from agricultural and forestry residues, such as sugar cane bagasse, rice husks, corn straw, and sawdust, representing enormous unexplored energy potential (Oliveira et al., 2022; Van Meerbeek et al., 2019).

Microbial fuel cells (MFCs) emerge as an innovative technology to directly convert LCB chemical energy into electricity through the activity of electrogenic microorganisms. These cells offer several advantages over traditional biomass conversion technologies, such as high energy efficiency, lower environmental impact, and simple operation (Drendel et al., 2018; Garbini et al., 2023).

The search for renewable and sustainable alternatives becomes increasingly urgent in a world marked by growing demand for energy and the environmental impacts of fossil sources. In this scenario, lignocellulosic biomass (LCB) appears promising, offering enormous unexplored energy potential. This abundant biomass, derived from agricultural and forestry residues such as sugar cane bagasse, rice husks, corn straw, and sawdust, represents a unique opportunity to generate clean and sustainable energy (Julio et al., 2022; Souza et al., 2023).

3.1 Composition of Lignocellulosic Biomass

Lignocellulosic biomass is characterized by a complex and heterogeneous chemical composition, mainly composed of three main components: the first, cellulose is the fundamental structural component of the plant cell wall; it is a linear polymer of β-D-glucose, the most organic material abundant on the Earth, and represents around 40–60% of lignocellulosic biomass; cellulose also plays a crucial role in the biodegradability and energy potential of biomass (Zoghlami & Paës, 2019).

Next, we have hemicellulose, which is composed of pentoses (xylan, arabinan) and hexoses (mannan, galactan) linked to cellulose. Hemicellulose represents about 20–30% of LCB. Its heterogeneous structure and the presence of fermentable sugars contribute to the biodegradability of biomass and the generation of important by-products during energy conversion (Brandt et al., 2013).

Finally, there is lignin, an aromatic complex polymer that provides rigidity and resistance to the plant cell wall. Lignin, representing around 15–30% of LCB, poses a challenge in the utilization of Lignocellulosic Biomass due to its recalcitrant structure. This structure makes lignin more resistant to microbial degradation, often necessitating pre-treatments to increase its bioaccessibility, highlighting the need for further research and development in this field (Balk et al., 2023; Wu et al., 2023).

3.2 Physicochemical Properties Relevant to Microbial Fuel Cells

In addition to chemical composition, LCB has several physical properties that significantly influence its performance in microbial fuel cells (MFCs). Factors such as LCB density vary according to the origin and type of material, generally ranging between 200 and 500 kg/m^3 (Roy et al., 2023). This characteristic impacts the transport of substrates and products within the MFCs, influencing energy conversion efficiency. Furthermore, porosity is another critical factor for diffusing substrates and products from the extracellular environment to microbial cells (Sonawane et al., 2022). Adequate porosity facilitates microorganisms' access to biomass, increasing the rate of biodegradation and energy generation. Another critical factor is the surface area of the LCB, which also influences the mass transfer rate and microbial activity. A greater surface area increases the contact area between biomass and microorganisms, optimizing energy conversion (Choudhury et al., 2017).

LCB has properties that make it ideal for application in MFCs, such as biodegradability by microorganisms, which allows its direct conversion into energy through MFCs. This aspect depends on the LCB's chemical composition and physical structure, with cellulose and hemicellulose being more readily biodegradable than lignina (Silva et al., 2021). The second factor is the calorific value of LCB, which varies according to the chemical composition and moisture content, with an average value of 17–20 MJ/kg for dry LCB (Nasser et al., 2016). This characteristic contributes to the energy potentially released during biomass conversion into electricity. Finally, we find the Electronic Properties of the LCB that facilitate the transfer of electrons to the MFCs. Cellulose and hemicellulose can act as electron donors, while lignin can act as an electron transfer mediator, optimizing electricity generation (Ouyang et al., 2022).

3.3 Substrates and Microorganisms Employed in MFCs

Microbial fuel cells (MFCs) represent a promising frontier in the search for renewable and sustainable energy sources. These innovative biotechnologies take advantage of electrogenic microorganisms' ability to convert chemical energy present in various substrates into electrical energy, offering a promising alternative to conventional energy sources (Jalili et al., 2024).

The diversity of substrates that can be used in MFCs is vast, ranging from simple sugars, such as glucose, to complex polymers, such as cellulose and hemicellulose, present in lignocellulosic biomasses (Michelin et al., 2015). The choice of the ideal substrate depends on several factors, such as availability, cost, chemical composition, biodegradability, and compatibility with the microorganisms used. A wide range of microorganisms efficiently metabolize sugars such as glucose, fructose, and sucrose and are frequently used substrates in initial studies (Borrero-López et al., 2022).

However, using lignocellulosic biomass, such as agricultural and forestry residues, represents an excellent potential for producing energy from renewable and abundant sources (Blasi et al., 2023). Cellulose and hemicellulose, the majority components of lignocellulosic biomass, can be converted into fermentable sugars through pre-treatments and the action of microbial enzymes (Devi et al., 2022). Lignin, in turn, is a more recalcitrant substrate and requires more intense pretreatments and specific microorganisms for its degradation (Iram et al., 2021). In addition to sugars and polymers, various organic wastes, such as domestic, industrial, and agro-industrial effluents, can be used as substrates in MFCs, contributing to reducing environmental impacts and generating energy from waste (Oliveira et al., 2015; Ruiz-Dueñas & Martínez, 2009).

The selection of electrogenic microorganisms is another crucial factor for the success of MFCs. Various microorganisms, including bacteria and archaea, can transfer electrons to an electrode, converting chemical energy into electrical energy (Garimella et al., 2024). Proteobacteria, such as Shewanella, Geobacter, and Pseudomonas, are the most studied and used microorganisms in MFCs due to their remarkable ability to adapt to different substrates and environmental conditions (Hu et al., 2021). Firmicutes, such as Clostridium and Enterococcus, also have the potential for application in MFCs, especially in anaerobic environments (Cao et al., 2019; Garbini et al., 2023).

Methanogenic archaea, such as Methanosarcina and Methanobacterium, can be used with electrogenic microorganisms to convert substrates into electricity and methane, a combustible gas with energetic potential (Siegert et al., 2015). The choice of suitable organisms depends on the substrate used, the MFC operating conditions, and the desired characteristics, such as conversion efficiency, current production, and process stability (Enzmann et al., 2018).

Several factors influence the efficiency of MFCs, including substrate type, microbial community, fuel cell design, and operating conditions. The chemical composition and biodegradability of the substrate directly impact the energy conversion rate and current production (Ishaq et al., 2024).

The diversity and composition of the microbial community in MFCs are not to be underestimated. They are pivotal for efficient substrate degradation and electron transfer. The architecture of the MFC, including the electrode material, the distance between the electrodes, and the substrate flow configuration, also significantly influences cell performance (Molaei et al., 2023). Factors such as pH, temperature, salinity, and substrate flow rate further affect microbial activity and energy conversion efficiency (Guo et al., 2022).

Optimizing these factors is essential to maximizing MFCs' performance and achieving commercial viability. The selection of substrates with high energy density and easy biodegradability, the use of microorganisms with a high electron transfer rate, and the optimization of the fuel cell design are essential strategies to increase MFCs' efficiency (Sriwichai et al., 2024). Furthermore, the development of new materials for electrodes and membranes and the use of genetic engineering techniques to improve the characteristics of microorganisms are promising areas of research for the advancement of MFCs (Aghababaie et al., 2015).

3.4 Biomass Pretreatment for MFC Application

Microbial fuel cells (MFCs) represent a promising technology for generating clean and sustainable energy from renewable sources (Moradian et al., 2021). Lignocellulosic biomass, abundant and low cost, emerges as an attractive substrate to feed these cells. However, the complex structure of lignocellulosic biomass, mainly composed of cellulose, hemicellulose, and lignin, limits the accessibility of electrogenic microorganisms to fermentable sugars, reducing the efficiency of MFCs (Kurniawan et al., 2022). To overcome this limitation, biomass pretreatment appears to be an essential step in the bioelectrochemical conversion process (Wang et al., 2022).

Biomass pretreatment aims to disrupt the lignocellulosic matrix, increasing its specific surface, breaking the bonds between the components, and releasing fermentable sugars, making them accessible to microorganisms (Tanis et al., 2023). Several pretreatment methods can be used, each with its advantages and disadvantages, depending on the biomass type, the hydrolysate's desired characteristics, and the conversion technology (Capolupo & Faraco, 2016; Yin et al., 2021).

Pretreatment methods can be classified as physical, chemical, biological and physicochemical. Physical methods, such as grinding and extrusion, aim to increase the surface area of the biomass, facilitating the action of microorganisms and enzymes (Brodeur et al., 2011). Chemical methods, in turn, use specific chemical reagents to break the bonds between the components of biomass. These reagents can include acids, alkalis, and oxidants, each with its unique properties and applications (Lazzarini et al., 2021). Biological methods employ microorganisms or enzymes to degrade biomass more selectively. Physicochemical methods combine physical and chemical treatments to optimize process efficiency (Kucharska et al., 2018; Taylor et al., 2019).

The choice of pretreatment method depends on several factors, such as the type of biomass, the biomass's chemical composition, the process's cost, the generation of inhibitory compounds, and compatibility with subsequent processes (Kapsokalyvas et al., 2018). For example, acid pretreatment effectively removes hemicellulose and increases cellulose accessibility but can generate inhibitory compounds for microorganisms. Alkaline pretreatment, in turn, is efficient in eliminating lignin but can lead to the formation of humified substances that hinder enzymatic hydrolysis (Oliveira et al., 2014; Teixeira et al., 2014).

Biomass pretreatment also influences the composition of the hydrolysate, which in turn affects the performance of MFCs (Akhlaq & Uroos, 2023). The presence of phenolic compounds, furfurals, and organic acids in the hydrolysate can inhibit the growth of microorganisms and the transfer of electrons. Furthermore, the concentration of fermentable sugars in the hydrolysate is crucial for energy production in MFCs (Pereira et al., 2022).

Pretreatment optimization is essential to maximize the efficiency of MFCs. Several studies have been carried out to evaluate the impact of different pretreatment conditions, such as temperature, reaction time, concentration of reagents, and solid-liquid ratio, on the composition of the hydrolysate and the performance of

MFCs (Galbe & Wallberg, 2019). The combination of different pretreatment methods has also been explored to overcome the limitations of each technique (Ramaraj & Unpaprom, 2019).

In addition to pretreatment, enzymatic hydrolysis is a fundamental step to release fermentable sugars from pretreated biomass. The enzymes cellulases and hemicellulases are used to hydrolyze cellulose and hemicellulose, respectively (Yang et al., 2011). The choice of enzymes and enzymatic hydrolysis conditions influence the efficiency of biomass conversion into fermentable sugars, and your expertise is vital in making these decisions (Jahangeer et al., 2024).

In conclusion, biomass pretreatment is crucial for applying lignocellulosic biomass in microbial fuel cells. The choice of the appropriate pretreatment method depends on several factors, such as the type of biomass, the desired characteristics of the hydrolyzate, and the conversion technology to be used (Osman et al., 2023). Optimizing pretreatment and enzymatic hydrolysis is essential to maximize the efficiency of MFCs and promote the production of clean and sustainable energy from renewable sources (Sharma, Salama, et al., 2023a).

4 Conversion of Lignocellulosic Biomass into Electricity in MFCs

Microbial fuel cells (MFCs) are bioelectrochemical systems that transform chemical energy stored in substances into electricity using microorganisms (Sharma, Tsai, et al., 2023b; Shrivastava & Sharma, 2022). Because of its availability, renewability, carbon neutrality, and cost-effectiveness, lignocellulose biomass—comprising agricultural residues, forestry waste, algal biomass, and other waste biomass—has been identified as a promising feedstock for energy production in MFCs (Sani et al., 2021; Sharma, Salama, et al., 2023a; Zhao et al., 2017). The lignocellulose biomass can be used in different parts of MFC, such as in the anode and cathode chamber, for different applications (Sani et al., 2021; Zhao et al., 2017).

There are a few studies that used lignocellulose biomass directly in an anode chamber in MFC for electricity generation, as a source of nutrients and organic matter (Moradian et al., 2021; Wang et al., 2021; Zhao et al., 2017). Cattle manure fed MFC produced peak power densities of 36.6 mW m^{-2} (Lee & Nirmalakhandan, 2011). Untreated rice straw (1 g L^{-1}) was used in two-chambered microbial fuel cells generating a maximum power density of 190 mW m^{-2} (Gurung & Oh, 2015). When MFC was used to treat cassava peels, the maximum produced power was 29 mW m^{-3} (Adekunle et al., 2016). Maximum power density of 16.5 mW m^{-2} was obtained using rice bran as a fuel in MFC (Yoshimura et al., 2018). In other configurations of MFC using three chambers, the maximum power achieved was 8.78 W m^{-3} and 6.73 W m^{-3} using sugarcane bagasse and corn cobs as substrates (Krishnaraj et al., 2015). Moreover, corn bran, banana peel, and palm oil mill effluent untreated were used as fuel in membrane-less MFC and the maximum power densities were 23.75 mW m^{-2}, 12.65 mW m^{-2}, 22.03 mW m^{-2}, respectively (Mohd Zaini Makhtar & Tajarudin, 2020). These output powers are too low to be commercially viable due to the recalcitrant structure of lignocellulose (strong hydrogen bonding structure), which hinders its direct degradation by microorganisms and many microorganisms are not electrochemically active to use anode as an electron acceptor (Moradian et al., 2021; Sani et al., 2021; Wang et al., 2021; Zhao et al., 2017).

In this context, an essential step to enhance the performance of MFCs using lignocellulose biomass is the pretreatment process, as it breaks down the structure of lignocellulose and releases a variety of monosaccharides and other fermentable compounds, for example, degraded phenolic compounds, organic acids including acetic acid from hydrolysis of acetyl groups, and furan derivatives such as furfural and 5-hydroymethylfurfural (HMF) from sugar degradation (Moradian et al., 2021; Wang et al., 2021; Zhao et al., 2017). During cellulose hydrolysis, glucose can be derived, while hemicellulose hydrolysis yields hexose, pentose, uronic acid, acetic acid and other by-products (Moradian et al., 2021; Wang et al., 2021). These hydrolyzed compounds create nutritive conditions for microbial growth and also serve as electrolytes, improving the ions transference (Shrivastava & Sharma, 2022). Comparing untreated and hydrodynamic cavitation pretreated rice bran on MFC performance, the pretreated substrate enhanced in 26% the total electron transfer because of increment of biodegradability of the using lignocellulose biomass to the anode microbial community (Yoshimura et al., 2018).

Saccharides, especially glucose, are the most vital biomass-derived fuels for MFC, because a huge variety of microorganisms have developed catabolism for energy production and the complete oxidation of each glucose molecule releases 24 electrons (Wang et al., 2021). These electrons are transferred to the anode electrode surface by direct electron transfer through membrane-bound redox proteins or pili, and mediated electron transfer using an intermediate electron carrier between the microbial cells and the anode (Shrivastava & Sharma, 2022). Then, electrons flow to the cathode via an external circuit. The protons generated are migrated through an ionic exchange membrane toward the cathode, which reacts with electron-reducing compound, common oxygen (Wang et al., 2021). The difference potential between the anode and the cathode produces an electromotive force that drives current flow and generates electrical power (Sharma, Tsai, et al., 2023b). It is worthily to mention that, in glucose-fed MFC, the power density ranges from a thousandth to a fraction of mW cm^{-2} because this substrate can be used in many competing metabolisms that do not generate electricity, such as fermentation and methanogenesis (Wang et al., 2021; Zhao et al., 2017).

Pentoses, such as xylose and arabinose, are typically released from hydrolyzed hemicellulose, which is a major component of lignocellulosic biomass (Wang et al., 2021; Zhao et al., 2017). Microorganisms from the genera *Clostridium*, *Bacteroides*, and *Shewanella* can use five-carbon sugars excreting some enzymes to ferment pentoses into organic acids and other products or to directly transfer electrons to the anode (Catal et al., 2011; Huang & Angelidaki, 2008; Toczyłowska-Mamińska et al., 2018). The efficiency of pentose utilization and power density obtained by MFC depends on the microbial strains used and their metabolic pathways and period of time to adapt to the new carbon source (Catal et al., 2011; Huang & Angelidaki, 2008; Wang et al., 2021; Zhao et al., 2017).

Another byproduct of lignocellulose hydrolyzes is organic acid, such as acetate (acetic acid) and uronic acids (Wang et al., 2021; Zhao et al., 2017). Acetate is generated from the hydrolysis of acetyl groups on the hemicellulose backbone and from acidogenesis of several organic matters (Wang et al., 2021; Zhao et al., 2017). Due to its easy assimilation by electroactive bacteria, acetate is recognized as a superior substrate for electricity generation, surpassing saccharides (Chae et al., 2009; Fadzli et al., 2021; Wang et al., 2021; Zhao et al., 2017). During the dilute acid pretreatment of lignocellulose biomass, uronic acid is obtained nevertheless there has been limited research on using in MFCs (Catal et al., 2008; Wang et al., 2021; Zhao et al., 2017).

Phenol compounds, obtained from lignin degradation, such as vanillin, syringaldehyde, catechol and 4-hydroxybenzoic acid, and furans, such as furfural and 5-hydroymethylfurfural (HMF), are usually considered inhibitors in MFC, because they can suppress the microorganism's growth, particularly in pure culture (Sharma, Salama, et al., 2023a; Wang et al., 2021; Zhao et al., 2017). On the other hand, some microorganisms can use these compounds to generate electricity, excreting enzymes such as monooxygenases that can degrade aromatic compounds (Sharma, Tsai, et al., 2023b; Wang et al., 2021; Zhao et al., 2017). Furthermore, under complete oxidation, phenolic compounds and furfural can theoretically yield more electrons and greater Gibbs free energy changes than carbohydrates (Zhao et al., 2017). Moreover, phenolic compounds in the substrate of MFC also serve as electron transfer media, enhancing the rate of electron transport from microorganisms to the anode (Wang et al., 2021; Zhao et al., 2017). Up to 300 mg L^{-1}, it is possible to see the benefits of presence of phenolic compounds in MFC, over this concentration the inhibitory effects are observed in MFC, not improving the maximum power density (Hejazi et al., 2019). The inhibitory effects of hydrolysate by-products can differ based on the bacterial consortium, operational conditional (temperature, hydraulic retention time, temperature, design of MFC) (Wang et al., 2021; Zhao et al., 2017). Additional research is required to explore this further.

Ash of biomass and salts formed hydrolysate neutralization can be others inhibitors on MFC performance. The ions presented in these compounds, such as Na^+, K^+, Ca^{2+}, Mg^{2+}, Mn^{2+}, Al^{3+}, Fe^{3+}, and Ti^4, show negative effects on microorganisms' growths, due to high osmotic pressure, and on proton exchange membrane/ion exchange membrane. An inorganic salt could precipitate in the membrane, fouling it and increasing the internal resistance of MFC and decreasing the electrical power generated (Zhao et al., 2017).

In addition to being fed the anode chamber of MFC, no pretreated lignocellulose biomass has been used cathode chamber once the microbial community present in this compartment can produce biomolecules, such as lytic and oxidoreductase enzymes, that can produce simpler compounds from complex substrate (Shrivastava & Sharma, 2022). These sugar, proteins, and enzymes released during hydrolysis of lignocellulose on the cathode act as mediators, such as laccase produced by lignocellulolytic white rot fungi, catalyzing the reduction reaction and enhancing the performance of the catholyte and improve the power density produced by MFC (Ottoni et al., 2023;Pal & Sharma, 2020; Shrivastava & Sharma, 2022). Comparing the presence and absence of fungus in the cathode chamber, the current density measured in the former MFC was twice higher (42.81 ± 4.91 mA m^{-2}) than obtained in MFC control without fungus in the cathode chamber (20.31 ± 4.30 mA m^{-2}) because laccases enzymes produced mediated the reduction of oxygen to water (Ottoni et al., 2023). In this sense, the biomolecules used in the cathode of MFC can also decrease the cost of MFC technology once metal catalysts are expensive, possibly dissolver under moist conditions and danger to microbial community being impracticable for continuous-flow systems (Moradian et al., 2021; Ottoni et al., 2023; Pal & Sharma, 2020; Shrivastava & Sharma, 2022). Moreover, these substances can also reduce the toxicity of hydrolysate for microorganism once the oxidate injurious compounds (Pal & Sharma, 2020; Shrivastava & Sharma, 2022). For example, laccase can degrade phenolic compounds that are harmful to cellulolytic enzymes during the fermentation process (Pal & Sharma, 2020).

A novelty approach of lignocellulose biomass in MFC is the co-producing of electricity with other chemical products (Moradian et al., 2021; Shrivastava & Sharma, 2022). For example, 9 g of ethanol and 65.09 mW m^{-2} of power density were produced from 100 g of wheat straw in MFC once the biomass waste was firstly used in the cathode to release the simpler compounds and then was used in anode (Shrivastava & Sharma, 2023). In another study, 100 g dry seaweed biomass obtained 5.1 g of bioethanol and 0.5 W m^{-3} of power density (Offei et al., 2019). Besides bioethanol, biohydrogen can be produced together with the bioelectricity generated from MFC (Moradian et al., 2021; Shrivastava & Sharma, 2022). In a recent work, xylose, the second most abundant sugar derived from biomass, was used as substrate in a MFC and a power density of 67 mW m^{-2} and hydrogen gas production of 23 L m^{-3} were obtained (Moradian et al., 2020).

Lignocellulose biomass is also applied as the electrode for MFC because of its relatively high carbon content and its porous nature making a potential electrode material with good electrical conductivity after simple carbonization (Katam et al., 2023; Wang et al., 2021). In this sense, the new materials will have a large surface area which are essential to enhance microorganisms' adhesion, electrode reaction and electron transfer (Katam et al., 2023; Wang et al., 2021). Water hyacinth (*Eichhornia crassipes*) biochar obtained from pyrolysis was used as the cathode electrode of a single-chamber MFC obtaining 12.3 mW m^{-2} of maximum power density (Allam et al., 2020). Similarly, porous activated carbon from the silver grass (*Miscanthus sacchariflorus*) was used as an electrode in an anode chamber of MFC and 963 mW m^{-2} was obtained as maximum power density, as a consequence of the synergistic effect created by a highly porous structure with a very high surface area that enhances extracellular electron transfer (Rethinasabapathy et al., 2020). In a recent study that compared a MFC using macroporous-mesoporous carbon derived from palm fiber and carbon cloth as anode electrode, the maximum power density of 718.15 mW m^{-2} was obtained in the former material, which was 3.89 times higher than that obtained in the control experiment (146.78 mW m^{-2}) (Song et al., 2024).

5 Challenges of Using Lignocellulosic Biomass as Feedstock in MFC

Despite the challenges, the potential of agricultural waste in bioenergy production is promising. Several microbial consortia have shown the ability to transform agricultural waste into biofuels. Approximately 5% of the biomass used for energy and biofuels is agricultural waste, including bagasse, straws, ears, and stalks. This opens up several avenues for producing bioenergy from waste from the field and the processing of agricultural products (Sani et al., 2021).

It is a challenge when planning agricultural waste collection. This is similar to the fact that most approaches used during the collection process require equipment transfers, typically resulting in the loss of almost half of the biomass. Wheat straw is another significant agricultural residue that serves as a promising substrate. It is essential for the industrial production of bioethanol and other biofuels. At the same time, rice straw is widely used to produce bioethanol. Furthermore, bagasse, a byproduct of sugarcane processing, can create almost 3.6% of electrical energy (Sani et al., 2021).

The concept of using bacteria to generate electricity is old. Initially, the systems used exogenous mediators to transfer electrons from cells to the outside, enabling energy production. In newer systems, however, there is no need for exogenous mediators, which has contributed to a significant increase in the energy efficiency of microbial fuel cells (MFCs) in recent years. This advancement is due, in part, to designs that reduce the reactor's internal resistance (Logan & Regan, 2006).

Microbial fuel cells (MFCs) are bioelectrochemical systems (BES) that play a crucial role in the conversion of chemical energy from organic compounds present in biomass into electrical energy. In these systems, electrogenic microorganisms present in the anode oxidize lignocellulosic components, such as cellulose and lignin, releasing electrons that are captured by electrodes to produce an electric current (Virdis et al., 2011).

Using lignocellulosic biomass as a feedstock for microbial fuel cells (MFCs) presents various scientific and technical challenges, with one of the main challenges being economic viability. The efficiency of MFCs in converting lignocellulosic biomass into electricity is generally limited by the complexity of the lignocellulose structure, which is mainly composed of cellulose, hemicellulose, and lignin. The use of special enzymes and often extensive pretreatment to break down these components can result in a more expensive and less economically viable process (ElMekawy et al., 2015; Pandit et al., 2021).

Sani et al. (2021) identified three common obstacles in the bioelectrochemical systems (BES): the presence of methanogens, growth rate, and competition for the substrate. The focus is increasing electrochemical reactions and reducing the number of substrates used by methanogens. By purifying and isolating cultures from a mixed culture, we can address these challenges and improve electrochemical reactions and BES productivity.

Research over the past hundred years has demonstrated that bioelectrochemical systems often achieve energy efficiencies of less than 40%. This indicates that more than half of the oxidized substrate charge is consumed by non-electrogenic reactions, predominantly methanogenesis.

Taking the pure culture from a mixed culture is essential to avoid methanogenic activity before inoculation. Methanogenic activity can increase redox reactions in BES, reducing hydrogen production and, consequently, decreasing the electrochemical performance of the BES (Jadhav et al., 2019).

The processes of electrogenesis and methanogenesis compete, unlike what happens with electrogenic microorganisms, which transfer electrons to the anode and produce electrical current, methanogenic microorganisms make methane using the same substrates but do not directly contribute to electricity generation. Therefore, the presence of methanogenesis may decrease the energy efficiency of MFCs. Strategies such as pH and temperature regulation to suppress methanogenic activity are significant, as these measures are necessary to maintain the energy efficiency of MFCs despite the complications they may introduce to the system (Jung & Regan, 2011).

Several microbiological, electrochemical, and technological challenges need to be addressed if BES is to be considered a sustainable and effective technology. One of the most important microbiological problems that still needs solutions and considerations is the competition of methanogenesis for organic substrate (Jadhav et al., 2017).

Extracellular electron transfer is another aspect of lignocellulosic biomass that affects the efficiency of MFCs. Microorganisms cannot reach fermentable carbohydrates because lignin is resistant to biodegradation and can form a physical barrier. Furthermore, lignin degradation products can prevent electrogenic organisms from reproducing, and developing effective methods to remove or alter lignin before fermentation is an ongoing subject (Yang et al., 2024).

Microbial contamination also is a big obstacle to the process. MFCs rely on certain microbial consortia to decompose biomass and send electrons to the anode. However, undesirable microorganisms can compete with electrogenic microorganisms and harm the system. The crucial necessity of methods to suppress competing microorganisms or maintain a sterile environment is evident, as these factors are essential for the successful operation and scalability of MFCs (Sund et al., 2009).

It is challenging to scale up MFCs using lignocellulosic biomass. Controlling environmental and operational conditions more precisely on a laboratory scale is possible. However, with growth to pilot or industrial scales come problems such as maintaining anaerobic conditions, regulating temperature and pH, and evenly distributing substrates (Shrivastava et al., 2023). Furthermore, the complexity and costs associated with energy recovery, waste management, and biomass pretreatment have increased.

6 Environmental and Economic Considerations Microbial fuel cell (MFC)

The largest amount of lignocellulosic biomass (LCB) comes from woody biomass, agricultural residues (ex. rice straws, corn stover, sugarcane bagasse), grasses, energy crops (ex. switchgrass, miscanthus), cellulosic wastes (municipal solid waste such lumber mill as pulp mill wastes), etc. LCB has the potential for the production of biofuels, and bioelectricity, among bioproducts such as chemicals, fibers, biopulps, etc. (Mujtaba et al., 2023).

Lignocellulosic biomass (LCB) is one of the most abundant renewable energy sources, at no cost or very low cost (in many cases). Hence, it is very attractive to utilize lignocellulose-rich biomass for energy production (Shrivastava & Sharma, 2022).

Among various pathways to produce bioelectricity, microbial fuel cell (MFC) stands out because of its high energy efficiency and mild operation conditions; additionally, bioelectricity production can work along with bioremediation to form a co-generation system (Gajda et al., 2018). For example, the MFCs are capable of treating wastewater and converting different organic wastes (including LCB) simultaneously, into electricity (Sani et al., 2021).

MFCs that utilize LCB as a source play a vital role in green "carbon–neutral" bioenergy production. This carbon–neutral renewable technology can replace fossil fuels and can help for the sustainable development of society (Thulasinathan et al., 2022).

Most of the papers found during the preparation of the present manuscript were from a basic science lab-based perspective, under controlled conditions, and conducive to technology stage development. As usual for any technology when applied to the real world, one of the great challenges is its suitability for manufacturing, involving economic studies and scale-up research (Gajda et al., 2018).

In general, cost-effective green energy and fuels based on cellulosic and lignocellulosic biomass as a source require an easily available 2nd generation feedstock (the source must not be suitable for human consumption). Additionally, it is important to consider that the initial investment for the implementation of MFCs (that includes electrodes, a DC/AC converter, membranes, motors, and fans) is much higher than the initial investment for a conventional waste-treatment process, which only contains pumps, fans, and a biological tank (Sani et al., 2021).

On the other hand, according to Gajda et al. (2018), most of the principal parts and components can be bespoke, resulting in expensive devices (even at the prototype level). Another aspect is identifying alternatives that (1) perform equally well and for prolonged working periods, and principally (2) be widely available and inexpensive. This last aspect involves the system components such as membranes, electrodes, wiring, etc., to assemble and implement MFC on a larger scale.

Some researchers have studied alternative low-cost materials including ceramic, cardboard, and plant-derived electrodes (Chen et al., 2012). Ge and He (2016) reported a pilot study of a wastewater treatment plant in long-term operation, according to them, the cation exchange membrane represented over 60% of the material cost of a MFC. Thus inexpensive separators (such as including ceramic, cardboard, etc.) are very important for the economic feasibility of this technology (Ge & He, 2016).

MFCs generate electricity by using microbes that degrade cellulose with the incorporation of exogenous cellulases and electrogens enzymes. For achieving high performance in MFCs, microbes must hydrolyze LCBs into simple sugars (furthermore, acidic, and/or mechanical methods can also be used to extract fermentable sugars from LCB feedstock). The wide diversity of microorganisms capable of performing this role in MFCs allows the possibility to use almost all kinds of biomass materials, including lignin. LCBs have been demonstrated to be very complex in terms of biodegradability within MFC devices, thus several pretreatment and extraction techniques need to be employed for efficient utilization by microorganisms. This also influences directly the cost of the final device. Furthermore, in many cases, reducing the pretreatment process of LCBs reduces the process cost and ensures sustainability (Sani et al., 2021).

Regarding the cost and environment, MFC reactors are very beneficial and eco-friendly since they can, simultaneously, degrade pollutants and generate electricity. Thus decreasing energy consumption and treatment cycle, all lead to decreasing operation costs. The lifecycle assessment and techno-economics analysis of MFC are crucial for the commercialization of this technology.

7 Future Perspective

The set of knowledge collected from the lab-scale experiments needs to be explored at a pilot scale to facilitate MFC technology, aiming for its viability and its implementation in the real world. More studies under various environmental conditions are needed; parameters such as humidity/temperature/organic loading, pH, and various operation conditions such as flow rate, batch, or continuous flow, among others, need attention (Gajda et al., 2018).

More studies for the understanding of ion transport selectivity and low-cost membranes (e.g., through economic preparation methods) are vital to enable the employment of ion exchange membranes in MFC, aiming a sustainable development (Radeef & Najim, 2024).

The implementation of MFC is a promising alternative to traditional wastewater treatment, and it has potential environmental and economic benefits.

Finally, the principal issues to be addressed (before the commercialization of MFCs) include selecting a suitable substrate and its ionic strength to achieve higher power density. Moreover, factors such as flow mode, operational conditions, internal resistance, organic loading rate, pH, temperature, etc., need to be studied in more detail for future advancement. Furthermore, membranes of low cost and that are fouling resistant need to be studied on future research (Radeef & Najim, 2024; Thulasinathan et al., 2022).

8 Conclusion

Lignocellulosic biomass offers a promising renewable resource for sustainable energy generation, particularly in microbial fuel cells (MFCs). MFCs convert organic substrates into electricity through bioelectrochemical processes, with various configurations like double, single, multi-chamber, and stacked systems each affecting efficiency and scalability. Key factors influencing MFC performance include microbial consortia, substrates, and operational conditions such as pH, temperature, and electrode materials.

Lignocellulosic biomass, comprising cellulose, hemicellulose, and lignin, requires effective pretreatment for microbial degradation and electricity generation. Microorganisms such as exoelectrogens are crucial for breaking down biomass and facilitating electron transfer. Selecting and optimizing microbial consortia and understanding enzymatic pathways are vital for improving MFC efficiency.

Challenges remain, including low power outputs, microbial contamination, and competition between electrogenesis and methanogenesis. Innovative approaches in microbial selection, reactor design, and system optimization are essential to address these issues. Scaling up MFC technology for commercial viability presents additional hurdles.

Environmental and economic considerations highlight the significance of lignocellulosic biomass in MFCs, with carbon–neutral benefits and waste valorization potential. Economic feasibility is crucial for adoption. Future research should focus on microbial interactions, advanced pretreatment methods, and improved MFC design. Innovations in electrodes, reactor configurations, and operational strategies will enhance the potential of lignocellulosic biomass in MFCs, offering a viable solution for renewable energy and sustainable waste management.

References

Abdulwahhab, M. A., Najim, S. T., & Abdulwahhab, M. A. (2023). Microbial fuel cell review: thermodynamic and influencing factors, exoelectrogenic bacteria, anode and cathode configuration. *Journal of Chemical Technology & Biotechnology, 98*, 1559–1583. https://doi.org/10.1002/jctb.7371

Abubackar, H. N., Biryol, I., & Ayol, A. (2023). Yeast industry wastewater treatment with microbial fuel cells: effect of electrode materials and reactor configurations. *International Journal of Hydrogen Energy, 48*, 12424–12432. https://doi.org/10.1016/j.ijhydene.2022.05.277

Acheampong, A. O., Erdiaw-Kwasie, M. O., & Abunyewah, M. (2021). Does energy accessibility improve human development? Evidence from energy-poor regions. *Energy Economics, 96*, 105165. https://doi.org/10.1016/j.eneco.2021.105165

Adekunle, A., Gariepy, Y., Lyew, D., et al. (2016). Energy recovery from cassava peels in a single-chamber microbial fuel cell. *Energy Sources, Part A: Recovery, Utilization, and Environmental Effects, 38*(17), 2495–2502. https://doi.org/10.1080/15567036.2015.1086909

Aghababaie, M., Farhadian, M., Jeihanipour, A., & Biria, D. (2015). Effective factors on the performance of microbial fuel cells in wastewater treatment – A review. *Environmental Technology Reviews, 4*(1), 71–89. https://doi.org/10.1080/09593330.2015.1077896

Ahanchi, M., Jafary, T., Yeneneh, A. M., Rupani, P. F., Shafizadeh, A., Shahbeik, H., Pan, J., Tabatabaei, M., & Aghbashlo, M. (2022). Review on waste biomass valorization and power management systems for microbial fuel cell application. *Journal of Cleaner Production, 380*(1), 134994. https://doi.org/10.1016/j.jclepro.2022.134994

Ahmad, D., Haroon, M., Naeemullah, et al. (2023). Microbial fuel cell, their type, working principle and different factors affecting their performance: A review. *ACS Sustainable Chemistry & Engineering, 4*(2), 121–145. https://doi.org/10.37256/sce.4220233125

Akhlaq, M., & Uroos, M. (2023). Evaluating the impact of cellulose extraction via traditional and ionosolv pretreatments from domestic matchstick waste on the properties of carboxymethyl cellulose. *ACS Omega, 8*(9), 8722–8731. https://doi.org/10.1021/acsomega.2c08118

Allam, F., Elnouby, M., El-Khatib, K. M., et al. (2020). Water hyacinth (Eichhornia crassipes) biochar as an alternative cathode electrocatalyst in an air-cathode single chamber microbial fuel cell. *International Journal of Hydrogen Energy, 45*(10), 5911–5927. https://doi.org/10.1016/j.ijhydene.2019.09.164

Apollon, W. (2023). An overview of microbial fuel cell technology for sustainable electricity production. *Membranes, 13*(11), 884. https://doi.org/10.3390/membranes13110884

Balk, M., Sofia, P., Neffe, A. T., & Tirelli, N. (2023). Lignin, the lignification process, and advanced, lignin-based materials. *International Journal of Molecular Sciences, 24*(14), 11668. https://doi.org/10.3390/ijms241411668

Bazina, N., Ahmed, T. G., Almdaaf, M., Jibia, S., & Sarker, M. (2023). Power generation from wastewater using microbial fuel cells: A review. *Journal of Biotechnology*. https://doi.org/10.1016/j.jbiotec.2023.07.006

Bhaduri, S., & Behera, M. (2024). From single-chamber to multi-anodic microbial fuel cells: A review. *Journal of Environmental Management, 355*, 120465. https://doi.org/10.1016/j.jenvman.2024.120465

Blasi, A., Verardi, A., Lopresto, C. G., Siciliano, S., & Sangiorgio, P. (2023). Lignocellulosic agricultural waste valorization to obtain valuable products: An overview. *Recycling, 8*(4), 61. https://doi.org/10.3390/recycling8040061

Borrero-López, A. M., Valencia, C., & Franco, J. M. (2022). Lignocellulosic materials for the production of biofuels, biochemicals and biomaterials and applications of lignocellulose-based polyurethanes: A review. *Polymers, 14*(5), 881. https://doi.org/10.3390/polym14050881

Brandt, A., Gräsvik, J., Hallett, J. P., & Welton, T. (2013). Deconstruction of lignocellulosic biomass with ionic liquids. *Green Chemistry, 15*(3), 550–583. https://doi.org/10.1039/c2gc36364j

Brodeur, G., Yau, E., Badal, K., Collier, J., Ramachandran, K. B., & Ramakrishnan, S. (2011). Chemical and physicochemical pretreatment of lignocellulosic biomass: A review. *Enzyme Research, 2011*, 787532. https://doi.org/10.4061/2011/787532

Cao, Y., Mu, H., Liu, W., Zhang, R., Guo, J., Xian, M., & Liu, H. (2019). Electricigens in the anode of microbial fuel cells: Pure cultures versus mixed communities. *Microbial Cell Factories, 18*(1). https://doi.org/10.1186/s12934-019-1087-z

Capolupo, L., & Faraco, V. (2016). Green methods of lignocellulose pretreatment for biorefinery development. *Applied Microbiology and Biotechnology, 100*, 9451–9467. https://doi.org/10.1007/s00253-016-7884-y

Catal, T., Fan, Y., Li, K., et al. (2011). Utilization of mixed monosaccharides for power generation in microbial fuel cells. *Journal of Chemical Technology and Biotechnology, 86*, 570–574. https://doi.org/10.1002/jctb.2554

Catal, T., Li, K., Bermek, H., et al. (2008). Electricity production from twelve monosaccharides using microbial fuel cells. *Journal of Power Sources, 175*(1), 196–200. https://doi.org/10.1016/j.jpowsour.2007.09.083

Chae, K.-J., Choi, M.-J., Lee, J.-W., et al. (2009). Effect of different substrates on the performance, bacterial diversity, and bacterial viability in microbial fuel cells. *Bioresource Technology, 100*(14), 3518–3525. https://doi.org/10.1016/j.biortech.2009.02.065

Chen, S., He, G., Hu, X., Xie, M., Wang, S., Zeng, D., Hou, H., & Schröder, U. (2012). A three-dimensionally ordered macroporous carbon derived from a natural resource as anode for microbial bioelectrochemical systems. *ChemSusChem, 5*(6), 783. https://doi.org/10.1002/cssc.201100783

Chen, W., Alharthi, M., Zhang, J., et al. (2024). The need for energy efficiency and economic prosperity in a sustainable environment. *Gondwana Research, 127*, 22–35. https://doi.org/10.1016/j.gr.2023.03.025

Choudhury, P., Prasad Uday, U. S., Bandyopadhyay, T. K., Ray, R. N., & Bhunia, B. (2017). Performance improvement of microbial fuel cell (MFC) using suitable electrode and bioengineered organisms: A review. *Bioengineered, 8*(5), 471–487. https://doi.org/10.1080/21655979.2016.1267883

Devi, A., Bajar, S., Kour, H., Kothari, R., Pant, D., & Singh, A. (2022). Lignocellulosic biomass valorization for bioethanol production: A circular bioeconomy approach. *Bioenergy Research, 15*(4), 1820–1841. https://doi.org/10.1007/s12155-022-10401-9

Drendel, G., Mathews, E. R., Semenec, L., & Franks, A. E. (2018). Microbial fuel cells, related technologies, and their applications. *Applied Sciences, 8*(12), 2384. https://doi.org/10.3390/app8122384

ElMekawy, A., Srikanth, S., Bajracharya, S., Hegab, H. M., Nigam, P. S., Singh, A., & Pant, D. (2015). Food and agricultural wastes as substrates for bioelectrochemical system (BES): The synchronized recovery of sustainable energy and waste treatment. *Food Research International, 73*, 213–225. https://doi.org/10.1016/j.foodres.2014.11.045

Enzmann, F., Mayer, F., Rother, M., & Holtmann, D. (2018). Methanogens: Biochemical background and biotechnological applications. *AMB Express, 8*(1), 1. https://doi.org/10.1186/s13568-017-0531-x

Fadzli, F. S., Bhawani, S. A., & Adam Mohammad, R. E. (2021). Microbial fuel cell: Recent developments in organic substrate use and bacterial electrode interaction. *Journal of Chemistry, 1*, 4570388. https://doi.org/10.1155/2021/4570388

Fard, M. B., & Wu, D. (2023). Potential interactive effect on biomass and bio-polymeric substances of microalgal-bacterial aerobic granular sludge as a valuable resource for sustainable development. *Bioresource Technology, 376*, 128929. https://doi.org/10.1016/j.biortech.2023.128929

Gajda, I., Greenman, J., & Ieropoulos, I. A. (2018). Recent advancements in real-world microbial fuel cell applications. In *Current opinion in electrochemistry* (Vol. 11). Elsevier. https://doi.org/10.1016/j.coelec.2018.09.006

Galbe, M., & Wallberg, O. (2019). Pretreatment for biorefineries: A review of common methods for efficient utilisation of lignocellulosic materials. *Biotechnology for Biofuels, 12*, 294. https://doi.org/10.1186/s13068-019-1634-1

Garbini, G. L., Barra Caracciolo, A., & Grenni, P. (2023). Electroactive bacteria in natural ecosystems and their applications in microbial fuel cells for bioremediation: A review. *Microorganisms, 11*(5), 1255. https://doi.org/10.3390/microorganisms11051255

Garimella, S. S. S., Rachakonda, S. V., Pratapa, S. S., et al. (2024). From cells to power cells: Harnessing bacterial electron transport for microbial fuel cells (MFCs). *Annales de Microbiologie, 74*, 19. https://doi.org/10.1186/s13213-024-01761-y

Ge, Z., & He, Z. (2016). Long-term performance of a 200 liter modularized microbial fuel cell system treating municipal wastewater: Treatment, energy, and cost. *Environmental Science: Water Research & Technology, 2*(2), 274–281. https://doi.org/10.1039/c6ew00020g

Guo, Z., Zhang, L., Cui, M.-H., & Wang, A. (2022). Electrode microbial communities associated with electron donor source types in a bioelectrochemical system treating azo-dye wastewater. *Water, 14*, 1505. https://doi.org/10.3390/w14091505

Gurung, A., & Oh, S.-E. (2015). Rice straw as a potential biomass for generation of bioelectrical energy using microbial fuel cells (MFCs). *Energy Sources, Part A: Recovery, Utilization, and Environmental Effects, 37*(24), 2625–2631. https://doi.org/10.1080/15567036.2012.728678

Hamedani, E. A., Abasalt, A., & Talebi, S. (2024). Application of microbial fuel cells in wastewater treatment and green energy production: A comprehensive review of technology fundamentals and challenges. *Fuel, 370*, 131855. https://doi.org/10.1016/j.fuel.2024.131855

Hasan, M. R., Anzar, N., Sharma, P., et al. (2023). Converting biowaste into sustainable bioenergy through various processes. *Bioresource Technology Reports, 23*, 101542. https://doi.org/10.1016/j.biteb.2023.101542

Hassan, M., Kanwal, S., Singh, R. S., et al. (2023). Current challenges and future perspectives associated with configuration of microbial fuel cell for simultaneous energy generation and wastewater treatment. *International Journal of Hydrogen Energy, 50*, 323–350. https://doi.org/10.1016/j.ijhydene.2023.08.134

Hejazi, F., Ghoreyshi, A. A., & Rahimneja, M. (2019). Simultaneous phenol removal and electricity generation using a hybrid granular activated carbon adsorption-biodegradation process in a batch recycled tubular microbial fuel cell. *Biomass and Bioenergy, 129*, 105336. https://doi.org/10.1016/j.biombioe.2019.105336

Hu, Y., Wang, Y., Han, X., Shan, Y., Li, F., & Shi, L. (2021). Biofilm biology and engineering of Geobacter and Shewanella spp. for energy applications. *Frontiers in Bioengineering and Biotechnology, 9*, 786416. https://doi.org/10.3389/fbioe.2021.786416

Huang, L., & Angelidaki, I. (2008). Effect of humic acids on electricity generation integrated with xylose degradation in microbial fuel cells. *Biotechnology and Bioengineering, 3*, 413–422. https://doi.org/10.1002/bit.21786

Ida, T. K., & Mandal, B. (2023). Microbial fuel cell design, application and performance: A review. *Materials Today Proceedings, 76*, 88–94. https://doi.org/10.1016/j.matpr.2022.10.131

IEA. (2023). *World energy outlook 2023*. IEA. https://www.iea.org/reports/world-energy-outlook-2023, Licence: CC BY 4.0 (report); CC BY NC SA 4.0 (Annex A)

Iram, A., Berenjian, A., & Demirci, A. (2021). A review on the utilization of lignin as a fermentation substrate to produce lignin-modifying enzymes and other value-added products. *Molecules, 26*(10), 2960. https://doi.org/10.3390/molecules26102960

Ishaq, A., Said, M. I. M., Azman, S. B., et al. (2024). Utilization of microbial fuel cells as a dual approach for landfill leachate treatment and power production: A review. *Environmental Science and Pollution Research, 31*, 41683–41733. https://doi.org/10.1007/s11356-023-30841-w

Jadhav, D. A., Chendake, A. D., Schievano, A., & Pant, D. (2019). Suppressing methanogens and enriching electrogens in bioelectrochemical systems. *Bioresource Technology, 277*, 148–156. https://doi.org/10.1016/j.biortech.2018.12.098

Jadhav, D. A., Ray, S. G., & Ghangrekar, M. M. (2017). Third generation in bio-electrochemical system research–A systematic review on mechanisms for recovery of valuable by-products from wastewater. *Renewable and Sustainable Energy Reviews, 76*, 1022–1031. https://doi.org/10.1016/j.rser.2017.03.096

Jahangeer, M., Rehman, M. U., Nelofer, R., et al. (2024). Biotransformation of lignocellulosic biomass to value-added bioproducts: Insights into bio-saccharification strategies and potential concerns. In *Topics in catalysis*. Springer. https://doi.org/10.1007/s11244-024-01941-9

Jalili, P., Ala, A., Nazari, P., Jalili, B., & Ganji, D. D. (2024). A comprehensive review of microbial fuel cells considering materials, methods, structures, and microorganisms. *Heliyon*. https://doi.org/10.1016/j.heliyon.2024.e25439

Julio, A. A. V., Milessi, T. S., Batlle, E. A. O., Lora, E. E. S., Maya, D. M. Y., & Palácio, J. C. E. (2022). Techno-economic and environmental potential of renewable diesel as complementation for diesel and biodiesel in Brazil: A comprehensive review and perspectives. *Journal of Cleaner Production*. https://doi.org/10.1016/j.jclepro.2022.133431

Jung, S., & Regan, J. M. (2011). Influence of external resistance on electrogenesis, methanogenesis, and anode prokaryotic communities in microbial fuel cells. *Applied and Environmental Microbiology, 77*(2), 564–571. https://doi.org/10.1128/AEM.01392-10

Kapsokalyvas, D., Wilbers, A., Boogers, I. A., Appeldoorn, M. M., Kabel, M. A., Loos, J., & Van Zandvoort, M. A. (2018). Biomass pretreatment and enzymatic hydrolysis dynamics analysis based on particle size imaging. *Microscopy and Microanalysis, 24*(5), 517–525. https://doi.org/10.1017/S1431927618015143

Katam, K., Saisriyoot, M., Roddecha, S., et al. (2023). Conversion of lignocellulose residue obtained from biorefinery stream to electricity by microbial fuel cell. *E3S Web of Conferences, 428*, 01011. https://doi.org/10.1051/e3sconf/202342801011

Khan, S. S., Amjad, M., Shareef, H., et al. (2024). Review of microbial fuel cell from a techno-economic perspective. *Energy Exploration & Exploitation, 42*(1), 373–398. https://doi.org/10.1177/01445987231208510

Krishnaraj, R. N., Berchmans, S., & Pal, P. (2015). The three-compartment microbial fuel cell: A new sustainable approach to bioelectricity generation from lignocellulosic biomass. *Cellulose, 22*, 655–662. https://doi.org/10.1007/s10570-014-0463-4

Kucharska, K., Rybarczyk, P., Hołowacz, I., Łukajtis, R., Glinka, M., & Kamiński, M. (2018). Pretreatment of lignocellulosic materials as substrates for fermentation processes. *Molecules, 23*(11), 2937. https://doi.org/10.3390/molecules23112937

Kurniawan, T. A., Othman, M. H. D., Liang, X., et al. (2022). Microbial fuel cells (MFC): A potential game-changer in renewable energy development. *Sustainability, 14*(24), 1684714. https://doi.org/10.3390/su142416847

Lazzarini, A., Colaiezzi, R., Gabriele, F., & Crucianelli, M. (2021). Support-activity relationship in heterogeneous catalysis for biomass valorization and fine-chemicals production. *Materials, 14*(22), 6796. https://doi.org/10.3390/ma14226796

Lee, Y., & Nirmalakhandan, N. (2011). Electricity production in membrane-less microbial fuel cell fed with livestock organic solid waste. *Bioresource Technology, 102*(10), 5831–5835. https://doi.org/10.1016/j.biortech.2011.02.090

Liang, J., Nabi, M., Zhang, P., Zhang, G., Cai, Y., Wang, Q., Zhou, Z., & Ding, Y. (2020). Promising biological conversion of lignocellulosic biomass to renewable energy with rumen microorganisms: A comprehensive review. *Renewable and Sustainable Energy Reviews, 134*, 110335. https://doi.org/10.1016/j.rser.2020.110335

Logan, B. E., & Regan, J. M. (2006). Microbial fuel cells—challenges and applications. *Environmental Science & Technology, 40*(17), 5172–5180.

Mahmoudi, A., Mousavi, S. A., & Darvishi, P. (2024). Performance and recent development in sewage sludge-to-bioenergy using microbial fuel cells: A comprehensive review. *International Journal of Hydrogen Energy, 50*, 1432–1455. https://doi.org/10.1016/j.ijhydene.2023.10.338

Mahurede, T. P., Chihobo, C. H., Utete, B., et al. (2022). A review of microbial fuel cell prototypes, their efficacy in wastewater treatment and the contextual situation for Zimbabwe. *Fuel Communications, 17*, 100094. https://doi.org/10.1016/j.jfueco.2023.100094

Michelin, M., Ruiz, H. A., Silva, D. P., Ruzene, D. S., Teixeira, J. A., & Polizeli, M. L. T. M. (2015). Cellulose from lignocellulosic waste. In K. Ramawat & J. M. Mérillon (Eds.), *Polysaccharides*. Springer. https://doi.org/10.1007/978-3-319-16298-0_52

Mohd Zaini Makhtar, M., & Tajarudin, H. A. (2020). Electricity generation using membrane-less microbial fuel cell powered by sludge supplemented with lignocellulosic waste. *International Journal of Energy Research, 44*, 3260–3265. https://doi.org/10.1002/er.5151

Molaei, A., Ahmed, F., Ail, U., et al. (2023). New low-cost, flow-through carbon electrodes characterized in brackish water. *Chemical Papers, 77*, 1941–1950. https://doi.org/10.1007/s11696-022-02596-0

Moradian, J. M., Fang, Z., & Yong, Y. C. (2021). Recent advances on biomass-fueled microbial fuel cell. *Bioresources and Bioprocessing, 8*, 14. https://doi.org/10.1186/s40643-021-00365-7

Moradian, J. M., Xu, Z. A., Shi, Y. T., et al. (2020). Efficient biohydrogen and bioelectricity production from xylose by microbial fuel cell with newly isolated yeast of Cystobasidium slooffiae. *International Journal of Energy Research, 44*, 325–333. https://doi.org/10.1002/er.4922

Mujtaba, M., Fernandes Fraceto, L., Fazeli, M., Mukherjee, S., Savassa, S. M., Araujo de Medeiros, G., do Espírito Santo Pereira, A., Mancini, S. D., Lipponen, J., & Vilaplana, F. (2023). Lignocellulosic biomass from agricultural waste to the circular economy: A review with focus on biofuels, biocomposites and bioplastics. *Journal of Cleaner Production, 402*, 136815. https://doi.org/10.1016/j.jclepro.2023.136815

Nasser, R., Salem, M., Hiziroglu, S., Al-Mefarrej, H., Mohareb, A., Alam, M., & Aref, I. (2016). Chemical analysis of different parts of date palm (Phoenix dactylifera L.) using ultimate, proximate and thermo-gravimetric techniques for energy production. *Energies, 9*(5), 374. https://doi.org/10.3390/en9050374

Nugraheni, K., Soemarsono, A. R., Millah, N., et al. (2024). Optimal control of mathematical models in bioenergy systems as empowerment of sustainable energy sources. *Fundamental Journal of Mathematics and Applications, 7*(1), 59–67. https://doi.org/10.14710/jfma.v7i1.22482

Offei, F., Mensah, M., Kemausuor, F., & Thygesen, A. (2019). A biorefinery approach to bioethanol and bioelectricity co-production from tropical seaweeds. *Journal of Applied Phycology, 31*, 3899–3913. https://doi.org/10.1007/s10811-019-01887-6

Oliveira, J. A., Martins, L. H., Komesu, A., & Maciel Filho, R. (2015). Evaluation of alkaline delignification (NaOH) of açaí seeds (Eutherpe oleracea) treated with H_2SO_4 dilute and effect on enzymatic hydrolysis. *Chemical Engineering, 43*, 499. https://doi.org/10.3303/CET1543084

Oliveira, J. A. R., de Paula Menezes Barbosa, P., & Macêdo, G. A. (2022). High concentrate flavonoids extract from citrus pomace using enzymatic and deep eutectic solvents extraction. *Food, 11*, 3205. https://doi.org/10.3390/foods11203205

Oliveira, J. A. R., Komesu, A., & Filho, R. M. (2014). Hydrothermal pretreatment for enhancing enzymatic hydrolysis of seeds of açaí (Euterpe oleracea) and sugar recovery. *Chemical Engineering Transactions, 37*, 787–792. https://doi.org/10.3303/CET1437132

Osman, A., Lai, Z. Y., Farghali, M., et al. (2023). Optimizing biomass pathways to bioenergy and biochar application in electricity generation, biodiesel production, and biohydrogen production. *Environmental Chemistry Letters, 21*, 2639–2705. https://doi.org/10.1007/s10311-023-01613-2

Ottoni, C., do Valle Trotta, C., Martins, G., et al. (2023). In situ trametes versicolor laccase biocathode performance assessment in dual-chamber microbial fuel cells. *Bioenergy Research, 16*, 2616–2624. https://doi.org/10.1007/s12155-023-10594-7

Ouyang, D., Wang, F., Gao, D., Han, W., Hu, X., Qiao, D., & Zhao, X. (2022). Light-driven lignocellulosic biomass conversion for production of energy and chemicals. *Iscience, 25*(10), 105221. https://doi.org/10.1016/j.isci.2022.105221

Pal, M., & Sharma, R. K. (2020). Development of wheat straw based catholyte for power generation in microbial fuel cell. *Biomass and Bioenergy, 138*, 105591. https://doi.org/10.1016/j.biombioe.2020.105591

Pandit, S., Savla, N., Sonawane, J. M., Sani, A. M. D., Gupta, P. K., Mathuriya, A. S., & Prasad, R. (2021). Agricultural waste and wastewater as feedstock for bioelectricity generation using microbial fuel cells: Recent advances. *Fermentation, 7*(3), 169. https://doi.org/10.3390/fermentation7030169

Penteado, E. D., Fernandez-Marchante, C. M., Zaiat, M., et al. (2015). Energy recovery from winery wastewater using a dual chamber microbial fuel cell. *Journal of Chemical Technology and Biotechnology, 91*, 1802–1808. https://doi.org/10.1002/jctb.4771

Penteado, E. D., Fernandez-Marchante, C. M., Zaiat, M., et al. (2016). Influence of carbon electrode material on energy recovery from winery wastewater using a dual-chamber microbial fuel cell. *Environmental Technology, 38*(11), 1333–1341. https://doi.org/10.1080/09593330.2016.1226961

Pereira, J., de Melo, M. M., Silva, C. M., Lemos, P. C., & Serafim, L. S. (2022). Impact of a pretreatment step on the acidogenic fermentation of spent coffee grounds. *Bioengineering, 9*(8), 362. https://doi.org/10.3390/bioengineering9080362

Qazani, M. R., Ghasemi, M., & Asadi, H. (2024). Optimizing microbial fuel cells with multiple-objectives PSO and type-2 fuzzy neural networks. *Fuel, 372*, 132090. https://doi.org/10.1016/j.fuel.2024.132090

Radeef, A. Y., & Najim, A. A. (2024). Microbial fuel cell: The renewable and sustainable magical system for wastewater treatment and bioenergy recovery. *Energy, 360*(1), 100001. https://doi.org/10.1016/j.energ.2024.100001

Ramaraj, R., & Unpaprom, Y. (2019). Optimization of pretreatment condition for ethanol production from Cyperus difformis by response surface methodology. *3 Biotech, 9*(6), 218. https://doi.org/10.1007/s13205-019-1754-0

Rebequi, T. T., Pio, Y. C., Penteado, C. F., et al. (2023). Waste treatment and sustainable bioelectricity generation using microbial fuel cell. *Journal of Sustainable Engineering and Sciences, 1*, 9–17. https://doi.org/10.34024/jsse.2023.v1.15460

Rethinasabapathy, M., Lee, J. H., Roh, K. C., et al. (2020). Silver grass-derived activated carbon with coexisting micro-, meso- and macropores as excellent bioanodes for microbial colonization and power generation in sustainable microbial fuel cells. *Bioresource Technology, 300*, 122646. https://doi.org/10.1016/j.biortech.2019.122646

Roy, H., Rahman, T. U., Tasnim, N., et al. (2023). Microbial fuel cell construction features and application for sustainable wastewater treatment. *Membranes, 13*(5), 490. https://doi.org/10.3390/membranes13050490

Ruiz-Dueñas, F. J., & Martínez, Á. T. (2009). Microbial degradation of lignin: How a bulky recalcitrant polymer is efficiently recycled in nature and how we can take advantage of this. *Microbial Biotechnology, 2*(2), 164–177. https://doi.org/10.1111/j.1751-7915.2008.00078.x

Safwat, S. M., Khaled, A., Elawwad, A., et al. (2023). Dual-chamber microbial fuel cells as biosensors for the toxicity detection of benzene, phenol, chromium, and copper in wastewater: applicability investigation, effect of various catholyte solutions, and life cycle assessment. *Process Safety and Environment Protection, 170*, 1121–1136. https://doi.org/10.1016/j.psep.2022.12.088

Sani, A., Salva, N., Pandit, S., et al. (2021). Recent advances in bioelectricity generation through the simultaneous valorization of lignocellulosic biomass and wastewater treatment in microbial fuel cell. *Sustainable Energy Technologies and Assessments, 48*, 101572. https://doi.org/10.1016/j.seta.2021.101572

Sharma, M., Salama, E., Thakur, N., et al. (2023a). Advances in the biomass valorization in bioelectrochemical systems: A sustainable approach for microbial-aided electricity and hydrogen production. *Journal of Chemical Engineering, 465*, 142546. https://doi.org/10.1016/j.cej.2023.142546

Sharma, S., Tsai, M.-L., Sharma, V., Sun, P.-P., Nargotra, P., Bajaj, B. K., Chen, C.-W., & Dong, C.-D. (2023b). Environment friendly pretreatment approaches for the bioconversion of lignocellulosic biomass into biofuels and value-added products. *Environments, 10*, 6. https://doi.org/10.3390/environments10010006

Shrivastava, A., & Sharma, R. K. (2022). Lignocellulosic biomass based microbial fuel cells: Performance and applications. *Journal of Cleaner Production, 361*, 132269. https://doi.org/10.1016/j.jclepro.2022.132269

Shrivastava, A., & Sharma, R. K. (2023). Conversion of lignocellulosic biomass: Production of bioethanol and bioelectricity using wheat straw hydrolysate in electrochemical bioreactor. *Heliyon, 9*(1), e12951. https://doi.org/10.1016/j.heliyon.2023.e12951

Shrivastava, A., Shrivastava, D. R., & Sharma, R. K. (2023). Microbial fuel cell-based co-generation system for bioethanol production and power management solution. *Fuel, 350*, 128843. https://doi.org/10.1016/j.fuel.2023.128843

Siegert, M., Yates, M. D., Spormann, A. M., & Logan, B. E. (2015). Methanobacterium dominates biocathodic archaeal communities in methanogenic microbial electrolysis cells. *ACS Sustainable Chemistry & Engineering, 3*(7), 1668–1676. https://doi.org/10.1021/acssuschemeng.5b00367

Silva, J. P., Ticona, A. R. P., Hamann, P. R. V., Quirino, B. F., & Noronha, E. F. (2021). Deconstruction of lignin: From enzymes to microorganisms. *Molecules, 26*, 2299. https://doi.org/10.3390/molecules26082299

Sonawane, J. M., Mahadevan, R., Pandey, A., & Greener, J. (2022). Recent progress in microbial fuel cells using substrates from diverse sources. *Heliyon, 8*(12), e12353. https://doi.org/10.1016/j.heliyon.2022.e12353

Song, J., Tan, Z., Nong, Z., et al. (2024). Easily-prepared macroporous-mesoporous carbon from palm fiber as a sustainable anode for high-performance microbial fuel cells. *Fuel, 360*, 130514. https://doi.org/10.1016/j.fuel.2023.130514

Souza, A. B., Ochoa, A. A. V., Costa, J. Â. P., Leite, G. N. P., Silva, H. C. N., Tómas, A. C. C., Barbosa, D. C., & Michima, P. S. A. (2023). A review of tropical organic materials for biodiesel as a substitute energy source in internal combustion engines: A viable solution? *Energies, 16*, 3736. https://doi.org/10.3390/en16093736

Sriwichai, N., Sangcharoen, R., Saithong, T., et al. (2024). Optimization of microbial fuel cell performance application to high sulfide industrial wastewater treatment by modulating microbial function. *PLoS ONE, 19*(6), 0305673. https://doi.org/10.1371/journal.pone.0305673

Sund, C. J., Wong, M. S., & Sumner, J. J. (2009). Mitigation of the effect of catholyte contamination in microbial fuel cells using a wicking air cathode. *Biosensors and Bioelectronics, 24*(10), 3144–3147. https://doi.org/10.1016/j.bios.2009.03.019

Tanis, M. H., Wallberg, O., Galbe, M., & Al-Rudainy, B. (2023). Lignin extraction by using two-step fractionation: A review. *Molecules, 29*(1), 98. https://doi.org/10.3390/molecules29010098

Taylor, M. J., Alabdrabalameer, H. A., & Skoulou, V. (2019). Choosing physical, physicochemical and chemical methods of pre-treating lignocellulosic wastes to repurpose into solid fuels. *Sustainability, 11*, 3604. https://doi.org/10.3390/su11133604

Teixeira, R. S., Silva, A. S., Moutta, R. O., Ferreira-Leitão, V. S., Barros, R. R., Ferrara, M. A., & Bon, E. P. (2014). Biomass pretreatment: A critical choice for biomass utilization via biotechnological routes. *BMC Proceedings, 8*(Suppl 4), 34. https://doi.org/10.1186/1753-6561-8-S4-O34

Tetteh, E. K., Sijadu, N. G., & Rathilal, S. (2024). An overview of non-carbonaceous and renewable-powered technologies for green hydrogen production in South Africa: Keywords occurrence analysis. *Energy Strategy Reviews, 54*, 101486. https://doi.org/10.1016/j.esr.2024.101486

Thulasinathan, B., Jayabalan, T., Arumugam, N., Rasu Kulanthaisamy, M., Kim, W., Kumar, P., Govarthanan, M., & Alagarsamy, A. (2022). Wastewater substrates in microbial fuel cell systems for carbon-neutral bioelectricity generation: An overview. *Fuel, 317*, 123369. https://doi.org/10.1016/j.fuel.2022.123369

Toczyłowska-Mamińska, R., Szymona, K., Król, P., et al. (2018). Evolving microbial communities in cellulose-fed microbial fuel cell. *Energies, 11*(1), 124. https://doi.org/10.3390/en11010124

Van Meerbeek, K., Muys, B., & Hermy, M. (2019). Lignocellulosic biomass for bioenergy beyond intensive cropland and forests. *Renewable and Sustainable Energy Reviews, 102*, 139–149. https://doi.org/10.1016/j.rser.2018.12.009

Verma, M., & Mishra, V. (2023). Bioelectricity generation by microbial degradation of banana peel waste biomass in a dual-chamber S. cerevisiae-based microbial fuel cell. *Biomass and Bioenergy, 168*, 106677. https://doi.org/10.1016/j.biombioe.2022.106677

Virdis, B., Freguia, S., Rozendal, R., Rabaey, K., Yuan, Z., & Keller, J. (2011). Microbial fuel cells. In *Treatise on water science, vol. 4: Water-quality engineering* (pp. 641–665). Academic Press.

Wang, F., Ouyang, D., Zhou, Z., et al. (2021). Lignocellulosic biomass as sustainable feedstock and materials for power generation and energy storage. *Journal of Energy Chemistry, 57*, 247–280. https://doi.org/10.1016/j.jechem.2020.08.060

Wang, J., Ren, K., Zhu, Y., Huang, J., & Liu, S. (2022). A review of recent advances in microbial fuel cells: Preparation, operation, and application. *Biotech, 11*(4), 44. https://doi.org/10.3390/biotech11040044

Wu, W., Li, P., Huang, L., Wei, Y., Li, J., Zhang, L., & Jin, Y. (2023). The role of lignin structure on cellulase adsorption and enzymatic hydrolysis. *Biomass, 3*, 96–107. https://doi.org/10.3390/biomass3010007

Xu, G., Zhao, J., Shi, K., et al. (2023). Trends in valorization of citrus by-products from the net-zero perspective: Green processing innovation combined with applications in emission reduction. *Trends in Food Science and Technology, 137*, 124–141. https://doi.org/10.1016/j.tifs.2023.05.012

Yang, B., Dai, Z., Ding, S.-Y., & Wyman, C. E. (2011). Enzymatic hydrolysis of cellulosic biomass. *Biofuels, 2*(4), 421–449. https://doi.org/10.4155/bfs.11.116

Yang, P., Han, Y., Xue, L., Gao, Y., Liu, J., He, W., & Feng, Y. (2024). Effect of lignocellulosic biomass components on the extracellular electron transfer of biochar-based microbe-electrode in microbial electrochemical systems. *Journal of Water Process Engineering, 59*, 105013. https://doi.org/10.1016/j.jwpe.2024.105013

Yin, X., Wei, L., Pan, X., Liu, C., Jiang, J., & Wang, K. (2021). The pretreatment of lignocelluloses with green solvent as biorefinery preprocess: A minor review. *Frontiers in Plant Science, 12*, 670061. https://doi.org/10.3389/fpls.2021.670061

Yoshimura, Y., Nakashima, K., Kato, M., et al. (2018). Electricity generation from rice bran by a microbial fuel cell and the influence of hydrodynamic cavitation pretreatment. *ACS Omega, 3*(11), 15267–15271. https://doi.org/10.1021/acsomega.8b02077

Zhao, X., Liu, W., Deng, Y., et al. (2017). Low-temperature microbial and direct conversion of lignocellulosic biomass to electricity: Advances and challenges. *Renewable and Sustainable Energy Reviews, 71*, 268–282. https://doi.org/10.1016/j.rser.2016.12.055

Zoghlami, A., & Paës, G. (2019). Lignocellulosic biomass: Understanding recalcitrance and predicting hydrolysis. *Frontiers in Chemistry, 7*, 874. https://doi.org/10.3389/fchem.2019.00874

Gasification and Syngas Production from Lignocellulosic Waste

Yvan Jesus Olortiga Asencios, Eduardo Dellosso Penteado, Luiza Helena da Silva Martins, Johnatt Allan Rocha de Oliveira, and Andrea Komesu

Abstract

This chapter focuses on the thermochemical gasification of lignocellulosic waste. Lignocellulosic materials are considered renewable and eco-friendly; lignocellulosic waste is produced from various human activities such as agriculture, forestry, and municipal operations. Moreover, most food and food processing industries produce lignocellulosic waste. Lignocellulosic waste can be transformed into high-value products such as syngas through gasification processes. The introduction of lignocellulosic waste in the current economy leads to the establishment of a circular economy and obeys the green chemistry principles; all these aspects are very important in the present world scenario. Syngas obtained by this pathway is considered a source of renewable hydrogen; moreover, syngas can also be converted into higher hydrocarbons (HCs) by the Fischer–Tropsch (FT) processes (e.g., gasoline and diesel), as well as into other chemicals such as dimethyl ether (DME), higher alcohols, and ammonia. The upstream and downstream processes involved in the production of syngas from lignocellulosic waste are described in this chapter, highlighting their principal challenges for viable commercial applications.

Y. J. O. Asencios (✉) · E. D. Penteado · A. Komesu
Universidade Federal de São Paulo, UNIFESP, Instituto do Mar, Santos, São Paulo, Brazil
e-mail: yvan.jesus@unifesp.br

L. H. da Silva Martins
Universidade Federal Rural da Amazônia, UFRA, Instituto de Saúde e Produção Animal, Belém, Pará, Brazil

J. A. R. de Oliveira
Universidade Federal do Pará, UFPA, Faculdade de Nutrição, Belém, Pará, Brazil

G. Mukherjee, S. Dhiman (eds.), *Value Addition and Utilization of Lignocellulosic Biomass*, https://doi.org/10.1007/978-981-96-2786-8_7

Keywords

Syngas · Gasification · Lignocellulosic waste · Thermochemical process · Catalysis

1 Introduction

The global energy demand is rising because of the increasing population and improved living standards. Most of this energy is provided by burning fossil fuels, such as coal, natural gas, and petroleum. Using these fuels to meet the growing energy needs leads to the rapid exhaustion of fossil fuel reserves and causes environmental issues like acid rain, smog, global warming, and ozone depletion and various health problems (Islam & Dincer, 2018). Therefore, exploring alternative energy sources that are clean and sustainable is regarded as a vital step toward achieving an eco-friendly economy in the near future.

Lignocellulosic waste biomass, which consists of by-products from agricultural, forestry, and industrial processes, is an increasingly promising and rapidly growing renewable energy source for bioenergy production. It typically contains 40 to 50% cellulose, 25 to 35% hemicellulose, and 16 to 33% lignin by weight (Monir et al., 2018). Recent statistics show that the global annual production of lignocellulosic biomass is about 181.5 billion tons, but only 8.2 billion tons of this biomass is utilized across various applications (Mujtaba et al., 2023; Singh et al., 2022). This highlights the untapped potential of plant biomass for conversion into a wide range of bioproducts applicable across various sectors (Mujtaba et al., 2023), such as syngas production.

Syngas, or synthesis gas, is a versatile fuel and chemical feedstock produced from the gasification of carbon-containing materials. It is primarily composed of hydrogen (H_2), carbon monoxide (CO), carbon dioxide (CO_2), methane (CH_4), and light hydrocarbons (HCs, C_2–C_8) (Quiroga et al., 2022; Shayan et al., 2018). Syngas serves as a versatile precursor for various chemical processes, including the production of methanol (MeOH), dimethyl ether (DME), and higher hydrocarbons through Fischer–Tropsch (FT) synthesis (Caballero et al., 2022).

Biomass gasification is a thermochemical technology that involves the partial oxidation of organic materials at high temperatures (and sometimes high pressures). This process uses various reaction atmospheres, such as air, water (H_2O), oxygen (O_2), carbon dioxide (CO_2), or combinations of these gases, and can be carried out in different types of reactors. When air is the gasifying agent, the process is known as air gasification. When water is used, it is called steam reforming, and when carbon dioxide is used, it is referred to as dry reforming (Moreira et al., 2021).

Many factors are known to affect gas yield and composition, including the original biomass's composition, raw material processing, its moisture content, temperature, pressure, gasifying agent, and catalyst used (Pacioni et al., 2016). The process generally demands a significant amount of energy and, if it is catalytic, requires an active and stable catalyst. To make the process commercially viable, the primary challenges that need to be addressed are improving energy efficiency and ensuring process stability. In recent years, research has concentrated on how to tackle these issues (Caballero et al., 2022).

The integration of lignocellulosic waste biomass and syngas production has several advantages. Firstly, it provides a way to utilize abundant and renewable biomass resources, reducing dependence on fossil fuels. Secondly, it offers a potential method for waste management, as agricultural and forestry residues can be converted into valuable products. Finally, the process can be designed to be carbon-neutral or even carbon-negative, contributing to greenhouse gas (GHG) reduction goals.

Therefore, this work explores the gasification process of lignocellulosic waste, including its classification, preprocessing, and the various operational parameters affecting efficiency. In this work, we examine the stages of gasification, from heating and chemical reactions to syngas cleanup and reforming. Additionally, this work addresses the key challenges in making lignocellulosic waste gasification commercially viable and discusses its potential future impact on sustainable energy solutions.

2 Definition, Classification, and Main Sources of Lignocellulosic Waste

Lignocellulosic biomass mainly comprises three main components: cellulose, hemicellulose, and lignin. Cellulose and hemicellulose are polysaccharides, polymers of simple sugars, while lignin is a complex aromatic polymer present in the plant's cell wall, giving it its rigidity and resistance (Oliveira et al., 2015; Broda et al., 2022). The proportion of these blocks varies with the plant species, plant growth time, and growing conditions, and it affects the properties of the raw material and, ultimately, the suitability for gasification (Oliveira et al., 2014; Gao et al., 2023).

The heterogeneity of lignocellulosic biomass is another critical challenge in using this fuel for thermochemical processes. Furthermore, the complex chemical composition, presence of minerals, and variability in physical properties of lignocellulosic waste impose additional requirements for pretreatment and gasification processes (Rangabhashiyam & Balasubramanian, 2019). Additional procedures are required to optimize gasification processes to maximize synthesis gas production. Several ways of classifying lignocellulosic waste depend on parameters such as origin, physical properties, and chemical composition (Ciliberti et al., 2020). Understanding these classifications helps identify the most suitable feedstock for gasification, which in turn improves process efficiency and syngas yield.

The primary sources of lignocellulosic biomass for gasification are diverse and include:

- *Agricultural Waste*: This category includes sugarcane bagasse, corn straw, rice husk, corn stalks, and fruit and vegetable waste. These wastes are abundant and generally low cost, making them an attractive feedstock for energy production (Kalak, 2023).

- *Forestry Residues*: This category includes sawdust, wood chips, pruning residues, and tree barks. The wood industry generates large quantities of lignocellulosic waste that can be used as raw material for gasification, contributing to the sustainable management of forests (Blasi et al., 2023).
- *Industrial Waste*: This category includes waste from the paper and cellulose industry, waste from the food industry, and waste from the biofuel industry. These residues, which may have high humidity and contamination by other compounds, require specific pretreatments to ensure their suitability for gasification (Or-Chen et al., 2024).
- *Municipal Waste*: Municipal organic waste, such as food waste and paper, can be co-processed with other lignocellulosic waste in the production of synthesis gas, contributing to the integrated management of solid urban waste (Gabbar & Ahmad, 2024).

Several factors influence the effectiveness of synthesis gas made from lignocellulosic waste, which are:

- *Type of Biomass*: The quality of the synthetic gas formed depends on the biomass's chemical configuration and physical characteristics (Jideani et al., 2023). Variations in lignin, cellulose, and hemicellulose content can significantly impact the gasification process and the composition of the resulting syngas.
- *Operating Conditions*: Parameters such as temperature, pressure, biomass feed rate, and air–fuel ratio are the main factors that affect the gasification process (Hussain et al., 2023). Optimal operating conditions ensure maximum syngas yield and minimize the production of unwanted by-products.
- *Type of Gasifier*: Different gasifiers, each with unique peculiarities and applications (Rauch et al., 2013), influence the gasification process. The choice of gasifier depends on the specific type of biomass and desired end products, making it a crucial factor in optimizing synthesis gas production.

The interest in the gasification process lies in the potential for improving efficiency through biomass pretreatment. This step, which involves reducing tar formation and increasing synthetic gas formation, is a significant factor that calls our curiosity and drives us to explore further.

3 Biomass Gasification Process: Overview, Upstream and Downstream Stages

Gasification is a thermochemical process that converts biomass into a combustible gas mixture known as synthesis gas or syngas. The gasification process occurs at high temperatures, typically between 700 °C and 1500 °C, in the presence of a gasifying agent such as air, oxygen, water vapor, or carbon dioxide. During gasification, biomass undergoes a series of chemical reactions, including pyrolysis, partial oxidation, and steam reforming, resulting in the production of syngas, composed mainly of carbon monoxide (CO), hydrogen (H_2), methane (CH_4), and carbon dioxide (CO_2). The efficiency and composition of the syngas produced may vary depending on the type of biomass, reactor conditions, and gasifying agent used (Basu, 2018).

The gasification process can be divided into three main stages: upstream processing, gasification, and downstream processing (Kumar et al., 2009).

Upstream aspects relate to pre-gasification biomass preparation and handling. This includes biomass harvesting, transportation, storage, and pretreatment. Pretreatment is a critical stage that may involve drying, grinding, and torrefaction to improve biomass properties such as moisture content, energy density, and reactivity (Kumar et al., 2009). In particular, torrefaction is a process of moderate-temperature heating (200–300 °C) in the absence of oxygen that enhances the stability and efficiency of the biomass during gasification (Prins et al., 2006).

Logistics and biomass management are also critical aspects of the upstream, since the availability and transport efficiency of the biomass will determine the cost-effectiveness and sustainability of the gasification process. Among the strategies to optimize supply chains of biomass are siting gasification plants near sources of biomass and using densification techniques like briquetting and pelletizing (González et al., 2015).

Gasification is the core of the process. Key operating parameters of the gasifier include the type and design of the gasifier, gasification temperature, flow rates of biomass and oxidizing agents (such as air or steam), the type and quantity of catalysts, and the type and properties of the biomass used (Kumar et al., 2009).

Downstream aspects involve the processing and use of syngas after gasification. Unrefined syngas contains contaminants like solid particles, tars, and sulfur compounds. These impurities need to be removed to avoid operational issues and enable high-quality applications (Higman & van der Burgt, 2008). Syngas cleaning methods include cyclones, fabric filters, wet scrubbers, and chemical absorption processes to eliminate specific pollutants.

After purification, syngas has many uses. It can be burned to produce heat and electricity in internal combustion engines, gas turbines, or boilers. Another option is to use syngas as a raw material to make liquid fuels and chemicals, which happens through Fischer–Tropsch synthesis, methanol production, or ammonia synthesis (Tijmensen et al., 2002).

Turning plant-based materials into gas shows promise for making sustainable energy and chemicals. This process taps into renewable resources and helps cut down on greenhouse gases. However, some hurdles still exist. These include making the process more effective in dealing with waste and working alongside other green technologies. To make this gas-making method a real contender in the world energy market, we need to keep improving. This means getting better at cleaning and preparing the gas mix and finding smarter ways to handle the plant materials.

4 Operational Parameters: Reactor Types, Raw Material Impact, Key Operating Conditions, and Product Quality

Operating parameters are crucial for the efficiency and quality of the final product in terms of the conversion of the residues and the composition of the syngas. The type of reactor used directly influences the process. The most common reactors include fixed-bed, fluidized-bed, and plasma reactors, each with specific characteristics regarding temperature control, residence time, and heat distribution (Caballero et al., 2022).

Fixed-bed, fluidized-bed, and plasma reactors are distinct technologies for converting biomass into gas. Fixed-bed reactors consist of a stationary bed of solid material where biomass is fed, heated, and converted into gas with the passage of a gasifying agent (Alauddin et al., 2010). In fluidized-bed reactors, solid particles are suspended in an upward gas flow, providing a homogeneous mixture and improved heat transfer, increasing biomass conversion efficiency (Alauddin et al., 2010). Plasma reactors use a high-temperature plasma discharge to decompose biomass into its gaseous components, resulting in a fast and efficient conversion capable of treating various materials. Each type of reactor has its own specific advantages and challenges, influencing the choice according to the application and the nature of the biomass used (Dimitrakellis et al., 2022).

Schmid et al. (2021) investigated double fluidized-bed steam gasification to convert waste from biorefineries and industrial processes into valuable product gas. Using a 100 kW pilot plant at TU Wien, the experiments showed that the nitrogen-free gas generated is useful for heat and power production and is especially suitable for synthesizing renewable chemicals such as methanol and Fischer–Tropsch liquids. The experimental results, validated by mass and energy balances, highlight the influence of operating parameters on the gas composition, providing a basis for the design of gas cleaning systems for future biorefineries.

Shie et al. (2014) evaluated the efficiency of plasma torch gasification of municipal solid waste mixed with raw wood, previously treated by steam mechanical heat treatment (SMHT). Using a 10-kW plasma reactor, it was possible to generate syngas (CO and H_2) with almost 90% of the gaseous products appearing in just 2 min of reaction, and syngas yields ranged from 88.59% to 91.84% by volume. The gasification residue is minimized and transformed into non-hazardous vitrified lava. Steam methane reforming, hydrogasification, and Boudouard reactions increase the syngas yield, demonstrating that this method can efficiently convert municipal solid waste into bioenergy.

An innovative and interesting study by Abanades et al. (2021) explored the solar thermochemical conversion of biomass and waste to produce high-value fuels via pyro-gasification. Using concentrated solar energy to provide the necessary heat increases syngas production and quality while reducing dependence on conventional resources and CO_2 emissions. The research reviewed solar concentrating technologies, solar gasifiers from laboratory to industrial scale, and strategies for managing solar energy variations. Spouted bed technology is highlighted as a promising technology for gasification, in addition to addressing pyro-gasification modeling and kinetics for gasifier simulations.

The feedstock used in gasification is a critical factor in determining the efficiency of the process and the composition of the synthesis gas produced. The proper selection and preparation of the feedstock, such as lignocellulosic residues like sugarcane bagasse, wheat straw, and wood residues, are not only critical tasks but also play crucial roles in optimizing the process. These materials exhibit significant chemical composition, moisture content, and particle size variations, directly affecting the reaction rate, production, and the need for pretreatment. Therefore, optimizing these factors is essential to minimize the formation of undesirable by-products (Gani et al., 2023).

The main operating parameters in gasification include temperature, pressure, air or oxygen feed rate, and residence time. Temperature is one of the most critical parameters, as it influences the conversion of waste and the composition of the synthesis gas. Higher temperatures favor the production of carbon monoxide and hydrogen, the main components of synthesis gas, while lower temperatures can result in greater tar formation. Although less influential than temperature, pressure can affect gas density and reactor efficiency (Kamble et al., 2019).

The quality of the final product, synthesis gas, is determined by its composition, purity, and energy content. High-quality synthesis gas has a high proportion of carbon monoxide and hydrogen, with low contaminants such as tar, solid particles, and nitrogen compounds. Efficient removal of tar is a significant challenge, as these compounds can cause fouling and blockages in downstream equipment. Using catalysts and gas cleaning systems is a common method for improving the quality of synthesis gas (Harb et al., 2020).

5 Biomass Processing and Pretreatment of Lignocellulosic Waste: Size Reduction, Drying, and Gasifying Agent Preparation

The gasification of lignocellulosic residues to produce syngas is a promising alternative for using residual biomass. However, the efficiency of this process depends significantly on the pretreatment of the biomass, which includes size reduction, drying, and the preparation of gasifying agents. These steps are essential to ensure an efficient conversion of biomass into syngas, a mixture of carbon monoxide (CO) and hydrogen (H_2) with small amounts of other gases (Cortazar et al., 2023).

Size reduction is a critical step in biomass processing. Lignocellulosic residues, such as straw, sugarcane bagasse, and sawdust, are generally found in irregular shapes and varying sizes. Uniformizing the particle size improves the homogeneity of the gasification process, facilitating heat transfer and the reaction between the biomass and the gasifying agents. Hammer mills, crushers, and choppers are commonly used equipment to reduce the size of biomass, transforming it into fine particles or chips (Moreno-Chocontá et al., 2024).

Drying the biomass is another fundamental step in pretreatment. Biomass with high moisture content can reduce the efficiency of the gasification process, as a significant amount of energy is consumed to evaporate the water present. This can lower the temperature in the reactor, negatively affecting the conversion of biomass to syngas. Drying techniques such as air drying, fluidized-bed drying, and rotary drying are employed to reduce the moisture content of the biomass to optimum levels, typically below 20% (Cortazar et al., 2023).

Preparing gasifying agents is essential for efficient biomass gasification. Common gasifying agents include air, pure oxygen, water vapor, and carbon dioxide. The choice of gasifying agent directly influences the composition of the syngas produced. For example, the use of pure oxygen can increase the proportion of carbon monoxide and hydrogen in syngas, while the use of water vapor can promote the formation of hydrogen through the water displacement reaction (Segurado et al., 2019).

In addition to the choice of gasifying agent, its preparation and injection into the reactor need to be carefully controlled. The feed rate of the gasifying agent, its uniform distribution, and its adequate mixing with the biomass are critical factors for the efficiency of the process. Well-designed air and steam injection systems can improve the penetration of gasifying agents into the biomass, increasing the conversion efficiency and the quality of the syngas (Shi et al., 2016).

Pretreatment of lignocellulosic residues not only improves the efficiency of the gasification process but also reduces the formation of unwanted by-products such as tar and ash. Removing impurities and contaminants prior to gasification can prevent clogging and equipment wear and improve the purity of the syngas. Implementation of biomass washing and classification steps may be necessary to remove inert materials and contaminants prior to processing (Taherzadeh & Karimi, 2008).

Pretreatment of lignocellulosic biomass is crucial for efficient syngas production. It helps break down the complex structure of plant fibers and increases reactivity during thermochemical conversion. Physical methods include grinding and crushing, which reduce particle size and increase the surface area available for subsequent reactions (Kucharska et al., 2018).

Thermal pretreatment, such as torrefaction, heats the biomass to moderate temperatures (200–300 °C) without oxygen, removing moisture and volatile components, resulting in a more homogeneous and energy-dense material. These physical processes facilitate the handling and feeding of biomass into reactors (Hidalgo et al., 2023).

Chemical pretreatment methods are also widely used to modify the lignocellulosic structure and improve syngas production efficiency. Acid or alkaline hydrolysis breaks the lignin–cellulose–hemicellulose bonds, releasing fermentable sugars and increasing the material's porosity.

Steam pretreatment or steam explosion uses high pressures and temperatures to disrupt biomass, promoting the separation of structural components (Taherzadeh & Karimi, 2008).

Another effective method is using organic or inorganic solvents that solubilize lignin and hemicellulose, making cellulose more accessible for gasification (Espinoza-Acosta et al., 2014).

Biological pretreatment of lignocellulosic biomass involves the use of microorganisms, such as fungi and bacteria, to degrade lignin and hemicellulose, making the cellulose more accessible for syngas production. White-rot fungi, for example, are effective in decomposing lignin due to their ligninolytic enzymes. This method is environmentally friendly as it uses natural processes and does not require harsh chemicals. However, biological pretreatment is generally slower than physical and chemical methods and may require controlled temperature, pH, and humidity conditions to optimize microbial activity. Despite this, it offers a sustainable and low-environmental-impact approach to preparing lignocellulosic biomass for conversion to syngas (Ummalyma et al., 2019).

Each of these chemical pretreatments aims to maximize biomass conversion into syngas, facilitating the subsequent steps of the thermochemical process.

6 Gasification of Lignocellulosic Waste: Heating, Chemical Reactions, and Catalysis

Lignocellulosic waste can be transformed into high-value products such as biofuels, bioelectricity, and bioproducts such as chemicals, fibers, biopulps, and syngas (synthesis gas = H_2/CO); the last one is obtained through gasification processes. Syngas obtained by this pathway is considered a source of renewable hydrogen; moreover, syngas can also be converted into higher hydrocarbons by Fischer–Tropsch processes (e.g., gasoline and higher hydrocarbons), as well as into other chemicals such as dimethyl ether, higher alcohols, and ammonia (Mujtaba et al., 2023).

The characteristics and properties of lignocellulosic waste will influence the efficiency of its processing, the type of gasifier, and, mainly, the quality of syngas (Kumar et al., 2009; Maitlo et al., 2022).

Biomass gasification is a fundamental step, as it involves the production of syngas, which will be used as raw material for hydrogen and/or Fischer–Tropsch synthesis, among other products. Fluidized-bed gasifiers are generally used for this process. Gasifiers are classified into three main types: fixed-bed, fluidized-bed, and entrained-flow gasifiers. Gasification is carried out in the presence of oxidizing agents such as H_2O (steam), O_2 (usually coming from air), carbon dioxide, or a combination of these. The reaction temperature ranges from 600 to 1000 °C. The no-stoichiometric reaction is represented by Eq. (1) (adapted from Kumar et al., 2009):

$$\begin{aligned} CH_x\mathrm{O}_y(\mathrm{s}) + \mathrm{O}_2(\mathrm{g}) + \mathrm{H}_2\mathrm{O}(\mathrm{v}) = CH_4(\mathrm{g}) + CO(\mathrm{g}) + CO_2(\mathrm{g}) + \mathrm{H}_2(\mathrm{g}) \\ + \mathrm{H}_2\mathrm{O}(\mathrm{v}) + \mathrm{Char}(\mathrm{s}) + \mathrm{tar}(\mathrm{l}) + \mathrm{soot}(\mathrm{s}) \end{aligned} \quad (1)$$

In parentheses, the physical state of each component is detailed (solid, liquid, gas, or vapor in the case of water). CH_xO_y represents the main elemental components of biomass, O_2 is usually supplied by air (21% O_2), H_2O is supplied as steam, and H_2O in the products represents the unreacted water. Char in the products is mainly composed of carbon. Tar is a liquid or highly viscous to solid deposits mainly formed by monocyclic and polycyclic aromatic hydrocarbon (PAH) substances. By-products may also contain ash and metal oxides.

$$\text{Partial Oxidation reaction}: \mathrm{C} + \mathrm{O}_2 \rightarrow 2CO \tag{2}$$

$$\text{Total Oxidation reaction}: \mathrm{C} + \mathrm{O}_2 \rightarrow CO_2 \tag{3}$$

$$\text{Water gas shift reaction}: CO + \mathrm{H_2O} = CO_2 + \mathrm{H}_2 \tag{4}$$

$$\text{Steam reforming reaction}: CH_4 + \mathrm{H_2O} \rightarrow CO + 3\mathrm{H}_2 \tag{5}$$

$$\text{Water gas reaction } \mathrm{C} + \mathrm{H_2O} \rightarrow CO + \mathrm{H}_2 \tag{6}$$

$$\text{Hydrogasification reaction}: \mathrm{C} + 2\mathrm{H}_2 \rightarrow CH_4 \tag{7}$$

$$\text{Boudouard reaction}: \quad \mathrm{C} + CO_2 \rightarrow 2CO \tag{8}$$

The common reactions involved during gasification are represented in Eqs. 2–8. Equations (4)–(7) occur when steam is used as an oxidizing agent for gasification. The energy needed for these reactions can be supplied directly or indirectly. Furthermore, aiming for the sustainability of this process, the heat necessary for this gasification should come from renewable energy.

The degradation kinetics of various lignocellulosic biomass can be studied using thermogravimetric analyses (TGAs), as detailed by Chen et al. (2015) and Xiao et al. (2020). In the thermal degradation profile of biomass, the weight loss (in a nitrogen atmosphere) presents three stages: The first stage below 125 °C is related to dehydration (loss of water), the second stage ranging from 125 to 500 °C is related to active pyrolysis (the loss of the hemicellulose, cellulose, and part of lignin), and the third stage above 500 °C is known as passive pyrolysis (loss of remaining lignin). The composition of the biomass, the rate of heat transfer of the gasifier, and the type of oxidizing primarily influence the kinetics of the degradations and therefore the temperature ranges of these three stages.

7 Gas Cleanup and Reforming: Cleaning of Tar from Syngas, Reforming of Syngas, and Catalysis

The gaseous products obtained from lignocellulosic biomass gasification require further processing for their final destiny. The processes involved in this step are also known as downstream processing.

The principal contaminants found in syngas produced by the gasification of lignocellulosic biomass are particulate matter, tar, sulfur compounds, nitrogenous compounds, alkali metals, chlorine, and carbon dioxide. However, the level of contaminants largely depends on the raw material and gasification process. Once syngas is obtained through the gasification process, it is then conditioned to remove CO_2 and other impurities inherent in the lignocellulosic biomass. The H_2/CO ratio of syngas is adjusted at this stage. This stage can also include reactors for reforming in the presence of a catalyst (e.g., Ni-, Co-, or Fe-based catalyst), aiming to increase the conversion of hydrocarbons present (e.g., methane).

Reforming reactions, as that represented in reaction 5, change the composition of the gaseous products (principally the syngas composition, CO/H_2 ratio), aiming the specific composition for syngas utilization (e.g., for Fischer–Tropsch synthesis, methanol synthesis, and hydrogen for fuel cells).

Cleaning and purification processes may include synthesis gas cooling; filtering and debugging; water–gas shift reaction (WGSR); and removal of acid gases, CO_2, H_2S, and traces of impurities.

7.1 Tar Formation

Tar is defined as a mixture of condensable hydrocarbons with one- to five-ring aromatic compounds, along with other oxygenated hydrocarbons and polycyclic aromatic hydrocarbons (PAHs). Tar formation takes place during the gasification process (e.g., reaction 1), as the result of thermochemical reactions such as chemolysis, de-polymerization, oxidation, polymerization, and cycloaddition (Narnaware & Panwar, 2021).

The major drawback of the gasification of biomass is tar generation. Tar concentration depends on biomass feedstock, type of gasifier, and operation conditions such as temperature and moisture content. The formation of tar affects the purification of gaseous products and other processes involved in syngas conditioning. Tar in syngas causes serious problems such as catalyst deactivation and the release of carcinogenic elements. Furthermore, tar formation also affects the operation on a large scale, becoming a big barrier to overcome before the commercialization of this technology.

7.2 Catalysts

Catalysts are used in reactions involved in the gasification of biomass such as tar cracking, steam reforming, thermal cracking, dry reforming, and steam dealkylation reactions (Shahbaz et al., 2017).

Catalysts are also used for altering the gas composition to achieve the requirement for a specific use, for example, in the Fischer–Tropsch process, fuel cells, methanol synthesis, and ammonia. Aiming for the commercialization of the gasification of biomass technology, the development of low-cost and effective catalysts is needed (Narnaware & Panwar, 2021).

The decomposition of tar produces gaseous products such as H_2, CO, CH_4, and CO_2. According to some authors (Islam, 2020; Miccio et al., 2009; Shahbaz et al., 2017), the gasification of biomass with steam is the most convenient way to get an improved gaseous product gas. When the purpose of biomass gasification is to produce syngas with high H_2 content, it can be achieved by steam gasification (where the presence of steam triggers the WGSR) or catalytic steam gasification.

Catalytic reduction of tar formation has been demonstrated to be efficient in achieving clean and tar-free gaseous products. In this case, catalysts can be used in in situ mode or as a secondary catalyst. In the first mode, the catalyst is mixed with biomass feedstock to achieve catalytic gasification; through this method, tar is converted inside the reactor. In the second mode, the catalyst is used in a separate reactor located downstream of the gasifier, thus converting tar outside the gasifier. Although the second mode is highly effective in removing tar, it is expensive and brings complex implementation for small- and medium-scale systems. The in situ mode is cost-effective and promising (Miccio et al., 2009; Shahbaz et al., 2017).

Catalysts for biomass gasification can be classified into three main groups: (1) natural catalysts (dolomite, olivine, and CaO), (2) alkali-based catalysts (Li, Na, K, Rb, Cs, and Fr), and (3) transition metal-based catalysts (e.g., Ni- and Fe-based catalysts).

Natural catalysts are of lower cost and accessible, whereas metal-based catalysts are synthesized by chemical treatment, thus involving high cost. Ni- and Fe-based catalysts are highly active in biomass gasification (Narnaware & Panwar, 2021).

Among many studies reported in the literature, it is very relevant to cite the studies by Hanaoka et al. (2015), Li et al. (2016), Yang et al. (2019), and Zhang et al. (2023); they explored the gasification of woody biomass (rich in lignin) to produce biogas, which was later used in the Fischer–Tropsch synthesis to produce bio-jet, over different types of Fe-/Co-based catalysts. Some disadvantages mentioned on a large scale are difficulties in the logistics of raw materials and the cleaning of syngas gas and high capital investment.

7.3 General Precautions for Biomass Gasification

The sustainable supply of biomass allows for competition for raw biomass from other sectors. In general, it is highly recommended that the source of lignocellulosic biomass is not suitable for human consumption (the source must be a second-generation feedstock). For example, some forestry products have an established market and destination, so the raw material for syngas production should come from other low-cost and abundant alternative lignocellulosic biomass waste (Sani et al., 2021).

The heat produced by the gasification of biomass and the corresponding by-products should be used to increase the economic performance and thermal efficiency of the process.

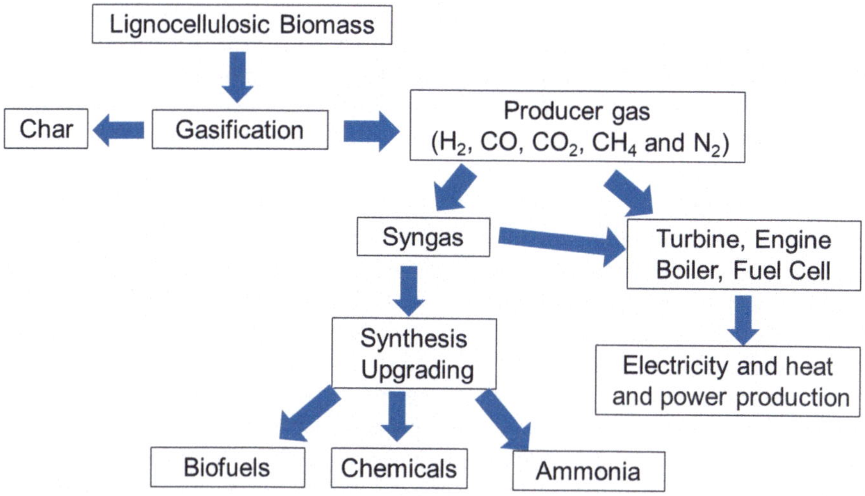

Fig. 1 Gasification of lignocellulosic biomass and products and uses. (Adapted from Molino et al., 2016)

8 Syngas Utilization: Definition and Principal Uses

A mixture of different gases (mainly H_2, CO, CO_2, CH_4, and N_2), called producer gas, is obtained from the gasification of lignocellulosic biomass, which also contains solid particulate matter, alkali metals, ammonia, sulfur, and hydrochloric acid, which are unwanted substances and chemicals (Sansaniwal et al., 2017; Ciliberti et al., 2020). Another by-product from the gasification of lignocellulosic biomass is a char, which represents 6–10% of the total weight of the biomass, and it contains significant amounts of inorganic salts in molten form (Azadi et al., 2013; Akbarian et al., 2022; Jamil et al., 2024). The producer gas can be used to produce olefin, CH_4, H_2, energy, diesel, alcohol, power, heat, gasoline, and other fine and commodity chemicals, as shown in Fig. 1 (Molino et al., 2016; Trninić et al., 2016; Ciliberti et al., 2020). The final product will depend on the quality of the producer gas, which varies with the gasifier design, the biomass feedstock, the gasification medium, and the process parameters like temperature, pressure, and feed rate (Sansaniwal et al., 2017; Ciliberti et al., 2020; Jamil et al., 2024).

The producer gas can be blended with conventional fuels to be burned in a co-combustion process in power plants, being the simplest application of gasification and its products (Molino et al., 2016; Krishna et al., 2019; Jamil et al., 2024). Co-firing is limited to 5–15% (wt) without implying any technical problem, keeping the overall electricity energy efficiency around 35% (Molino et al., 2016; Trninić et al., 2016; Sansaniwal et al., 2017). However, a critical concern in co-firing is the influence of biomass ash on the quality of boiler fly ash and bottom ash (Trninić et al., 2016; Krishna et al., 2019). There are several examples of biomass co-firing plants, such as a circulating fluidized-bed (CFB) gasifier in the Essent power plant in Geertruidenberg (the Netherlands) and the Foster Wheeler CFB gasifiers in Lahti (Finland) and Ruien (Electrabel power plant, Belgium) (Trninić et al., 2016).

Another way to use the producer gas is to burn it in internal combustion engines or gas turbines, reaching 35–45% of electrical conversion efficiency (Molino et al., 2016; Krishna et al., 2019; Situmorang et al., 2020). Because of this, these are the most effective methods for combined heat and power (CHP) production from the gasification of lignocellulosic biomass (Trninić et al., 2016; Situmorang et al., 2020). However, the producer gas needs to have a calorific value greater than 4 MJ/Nm^3 and a low concentration of pollutant contents (metals = 0.025–0.1 ppmw; H_2S = 20 ppmv) to decrease the maintenance and to avoid the rapid deterioration due to tar aerosols and deposits (Molino et al., 2016; Krishna et al., 2019; Situmorang et al., 2020; Jamil et al., 2024). Comparing gas turbines and internal combustion engines, the first one needs a higher purity producer gas in terms of tar and particulate contents (Molino et al., 2016; Situmorang et al., 2020). The maximum concentrations of tars and particulates in gas turbines are 10 mg/Nm^3 and 2.4 mg/Nm^3, while those for internal combustion engines are 100 mg/Nm^3 and 50 mg/Nm^3 (Molino et al., 2016). Other benefits of internal combustion engines include affordability (low costs), reliability, and versatility for both mobile and stationary uses (Sansaniwal et al., 2017; Situmorang et al., 2020). The utilization of biomass for CHP has been experiencing significant growth in countries such as Austria, Finland, and Germany (Trninić et al., 2016; Situmorang et al., 2020).

Producer gas can be used in fuel cells, which are electrochemical systems that produce electricity through chemical energy conversion (Molino et al., 2016; Krishna et al., 2019; Jamil et al., 2024). The main advantage of this small equipment is the high electrical efficiency, greater than 40%, which is higher than that obtained by previous technology (combustion and internal combustion engines or gas turbines) (Molino et al., 2016; Trninić et al., 2016). Two kinds of fuel cells [solid oxide fuel cells (SOFCs) and molten carbonate fuel cells (MCFCs)] can be fed with syngas, after a cleaning phase, because H_2S, tars, and particulates have to be removed to avoid catalyst poisoning and rapid deterioration (Molino et al., 2016; Krishna et al., 2019; Jamil et al., 2024). Due to the low technology readiness level (TRL), the use of gas fuel cells for electricity generation from the gasification of lignocellulosic biomass has not been applied on a large scale (Molino et al., 2016; Trninić et al., 2016).

The gas producer can be cleaned to obtain syngas, a versatile mixture primarily composed of carbon monoxide (CO) and hydrogen (H_2), which holds immense potential to be the feedstock of a variety of valuable products through established processes such as Fischer–Tropsch synthesis, methanol synthesis, and hydrogen production via water-gas shift reactions (Molino et al., 2016; Krishna et al., 2019; Jamil et al., 2024). These products include transportation fuels, like synthetic diesel and gasoline substitutes, and chemicals used in various industries.

Fischer–Tropsch (FT) synthesis uses syngas (H_2 and CO) to produce high-quality liquid hydrocarbons with no aromatics and with a high cetane number (>70) in the presence of a catalyst (Molino et al., 2016; Trninić et al., 2016). Different kinds of catalysts are being used in the FT process; however, the most common are the transition metals (cobalt, iron, and ruthenium) (Rauch et al., 2013; Molino et al., 2016). The FT process can be carried out in different temperatures: Low-temperature Fischer-Tropsch reaction (LTFT) process, 200–260 °C and High-temperature Fischer-Tropsch reaction (HTFT) process, 300–350 °C (Rauch et al., 2013). The main difference is that during the LTFT process hydrocarbons of long chains/waxes and paraffins (higher-boiling hydrocarbons) are produced, while the HTFT process results in hydrocarbons of short chains, olefins, and gasoline, among other chemicals (Rauch et al., 2013). Another difference is that cobalt- or iron-based catalysts are mostly used in the LTFT process, whereas in the HTFT process, only iron-based catalysts are used (Rauch et al., 2013). The main products of FT synthesis are synthetic diesel, gasoline substitutes, and jet fuels (Molino et al., 2016; Krishna et al., 2019; Jamil et al., 2024).

Methanol can be produced from syngas by catalytic reaction of CO and some CO_2 with H_2 (Rauch et al., 2013; Molino et al., 2016; Krishna et al., 2019; Jamil et al., 2024). Both reactions release heat and involve volume reduction; therefore, they are favored by low temperatures and high pressures (Rauch et al., 2013; Trninić et al., 2016). Because of being strongly exothermic, these reactions can lead to the formation of by-products such as methane, higher alcohols, or dimethyl ether (DME) if heat is not removed to keep optimum catalyst life and reaction rate (Rauch et al., 2013; Trninić et al., 2016). The most suitable catalysts are copper oxide, zinc oxide, or chromium oxide (Rauch et al., 2013). Zinc oxide–/chromium oxide-based catalysts have poor selectivity and require extremely vigorous conditions—pressures ranging from 300 to 1000 bar and temperatures of about 400 °C (Rauch et al., 2013). Copper- and aluminum-based catalysts are more energy-efficient and cost-effective plants, because the pressure and temperature are lower than in the former, between 50 and 100 bar and 220 and 350 °C (Rauch et al., 2013; Krishna et al., 2019). Methanol can be used to produce gasoline, diesel, and olefin (Trninić et al., 2016; Pandey et al., 2022).

Dimethylether (DME) can be obtained directly from syngas (direct process) using bifunctional catalysts such as copper–ZSM-5 zeolite and hybrid copper–alumina-based catalysts from dehydration of methanol (indirect process) using catalysts based on silica–alumina (Krishna et al., 2019; Pandey et al., 2022; Jamil et al., 2024). The direct synthesis process is more cost-effective than the indirect process because the first one eliminates the need for a separate system to produce pure methanol and requires only one reactor operating at low temperature and pressure (Molino et al., 2016; Krishna et al., 2019; Pandey et al., 2022). DME can be used as a substitute for liquefied petroleum gas, gasoline, diesel, and aviation (jet) fuels once it is nontoxic, noncarcinogenic, nonteratogenic, and nonmutagenic (Molino et al., 2016; Krishna et al., 2019).

Syngas fermentation and syngas thermochemical catalyst conversion (Fisher–Tropsch catalysis) can be used to produce ethanol from syngas obtained from the gasification of lignocellulosic biomass (Molino et al., 2016; Krishna et al., 2019). The first process is a biochemical process led by acetogenic bacteria that anaerobically transform CO and/or CO_2 and H_2 into acetyl-CoA through the reductive acetyl-CoA pathway, and then, this is converted into fuels or chemicals (acetic acid, ethanol, 2,3-butanediol, butyric acid, and butanol) via acetaldehyde or via acetate and subsequent reduction to acetaldehyde (Molino et al., 2016; Krishna et al., 2019; Ciliberti et al., 2020). The syngas thermochemical catalyst conversion into ethanol can be obtained using several catalysts, such as copper-based catalysts and rhodium-based catalysts, which are ideal for ethanol production due to their high selectivity, but the costs are prohibitive (Molino et al., 2016; Krishna et al., 2019). The syngas fermentation has several advantages over the thermochemical catalyst conversion: The temperature and the pressure are low, microorganisms are tolerant to several syngas impurities (H_2S and tar), and they can use a variety of conditions of syngas compositions, eliminating the need for an extensive gas cleanup (Trninić et al., 2016; Ciliberti et al., 2020). However, there are some drawbacks to commercializing the syngas fermentation process such as the low rates of growth and production by anaerobic microorganisms, product inhibition, problems in keeping anaerobic conditions, and problems in mass transfer between gas and liquid phases (Trninić et al., 2016).

Another by-product obtained from the gasification of lignocellulosic biomass is the char, which is a solid material formed from the mineral material present in the feedstock (ash) and the unreacted carbon (Molino et al., 2016; Trninić et al., 2016). Gasifier design, the gasification medium, and the process parameters like temperature, pressure, and feed rate affect the amount of unconverted organic fraction and the lowest heat value, which can oscillate from 25 to 30 MJ/kg (Molino et al., 2016). The biomass feedstock influences the amount of ash (Molino et al., 2016; Ahmed et al., 2024). Moreover, the char produced from lignocellulosic biomass gasification has a high specific surface area (over 215 m^2/g), high adsorption capacity, high capacity of ion exchange, and high microporosity (Ahmed et al., 2024). Because of these reasons, the char can be used in a variety of situations such as wastewater treatment, soil remediation, biofuel production, energy storage, carbon sequestration, and pollutant remediation (Ahmed et al., 2024). Although it has many uses and advantages, one of the drawbacks of this by-product of lignocellulosic biomass gasification is the presence of some toxic elements, including chemical substances and poly-aromatic hydrocarbons, which make environmental remediation more difficult (Ahmed et al., 2024).

In conclusion, the utilization of producer gas and syngas obtained from the gasification of lignocellulosic biomass represents a crucial step toward achieving sustainable energy solutions, because it diversifies the energy matrix and reduces environmental impacts. Moreover, the use of char of lignocellulosic biomass gasification can be applied in a variety of situations, improving the environmental condition due to high adsorption capacity.

9 Principal Challenges in the Gasification of Lignocellulosic Biomass for Commercial Applications

The gasification of lignocellulosic biomass has been recognized as a well-established technology for a renewable energy source, but there are several challenges that make commercial applications impossible (Sansaniwal et al., 2017; Pandey et al.,

2022). These challenges can be explored in technological barriers, economic considerations, and social–environmental impacts.

One of the main technical problems is the considerable variability in the composition of lignocellulosic biomass, which affects the gasification efficiency and the quality of the products (Sansaniwal et al., 2017; Narnaware & Panwar, 2022; Pandey et al., 2022). Various operational issues, such as gasifier shutdowns, stem from dehydration and aggregation induced by alkali and alkaline earth metals found in lignocellulosic biomass (Akbarian et al., 2022). Moreover, the concentrations of mineral compounds (Si, Na, K, P, Cl, and Ca), moisture content, and O_2 vary geographically, requiring a preprocessing step, such as size reduction, drying, and potentially pretreatment (e.g., to remove ash or alkali metals) to optimize feedstock suitability and gasification performance (Sansaniwal et al., 2017; Narnaware & Panwar, 2022). However, these preprocessing steps add complexity and cost to the overall process (Sansaniwal et al., 2017).

Gasification reactors must operate under controlled conditions of temperature, pressure, and residence time to achieve high by-product yield and quality in the gasification of lignocellulosic biomass (Sansaniwal et al., 2017; Narnaware & Panwar, 2022; Akbarian et al., 2022). Different types of gasifiers (e.g., fixed-bed, fluidized-bed, and entrained-flow) offer varying advantages in terms of feedstock flexibility, reaction kinetics, and tar reduction, as discussed previously in this chapter (Sansaniwal et al., 2017; Narnaware & Panwar, 2022). Designing gasifiers that can handle the diverse composition of lignocellulosic biomass while minimizing tar formation and maximizing syngas conversion efficiency remains a significant technological challenge (Chambon et al., 2020; Akbarian et al., 2022).

Other technological barriers are tar management and syngas cleaning (Sansaniwal et al., 2017; Narnaware & Panwar, 2022; Akbarian et al., 2022). Tar compounds, formed during lignocellulosic biomass gasification, can cause fouling and corrosion in downstream equipment and decrease the efficiency of syngas utilization (Narnaware & Panwar, 2022; Akbarian et al., 2022). Effective tar removal and cleanup systems are essential to ensure the reliability and longevity of lignocellulosic biomass gasification plants, as mentioned in the previous section (Narnaware & Panwar, 2022; Akbarian et al., 2022). Techniques such as catalytic tar cracking, thermal cracking, and tar filtration are under development to address these challenges, but they add complexity and cost to the overall gasification process (Akbarian et al., 2022).

The capital costs associated with building and commissioning a lignocellulosic biomass gasification plant are significant, primarily due to the complexity of the technology and the need for specialized equipment, being one of the financial barriers (Sansaniwal et al., 2017; Narnaware & Panwar, 2022; Akbarian et al., 2022). The expenses related to equipment and installation in lignocellulosic biomass gasification are more substantial, exceeding 30% of total plant costs, compared to other biofuel production methods, where capital costs typically account for less than 15% (Guimarães et al., 2021). In some cases, the investment required to generate clean syngas from lignocellulosic biomass gasification may represent nearly half of the total plant costs (Bressanin et al., 2020; Akbarian et al., 2022). Moreover, many of the equipment used in lignocellulosic biomass gasification must be imported in many cases, increasing the capital costs (Guimarães et al., 2021). Because of this, achieving economies of scale is crucial to reducing these costs and making the lignocellulosic biomass gasification process economically competitive with conventional fossil fuel-based energy generation technologies. However, scaling up from pilot projects to commercial-scale operations often requires substantial investment and poses risks associated with technology performance and market uncertainties (Narnaware & Panwar, 2022; Akbarian et al., 2022). Many lignocellulosic biomass gasification projects rely on self-financing and grants from diverse funding sources, as banks and other financial institutions often exhibit cautiousness due to uncertainties surrounding sustainable biomass supply, biomass pricing, and policy gaps (Narnaware & Panwar, 2022).

Another economic challenge is the feedstock supply chain of lignocellulosic biomass gasification projects (Sansaniwal et al., 2017; Narnaware & Panwar, 2022; Akbarian et al., 2022). The logistics of procuring and transporting lignocellulosic biomass feedstocks from diverse and often dispersed sources to gasification facilities can impact the overall economics of biomass gasification (Rosenfeld et al., 2020; Narnaware & Panwar, 2022; Akbarian et al., 2022). Lignocellulosic biomass suffers from low bulk density, resulting in a greater number of truck movements to transport the feedstock of gasification, making the logistics operations a challenge (Guimarães et al., 2021; Akbarian et al., 2022). Moreover, larger lignocellulosic biomass gasification plants need huge biomass supplies, which may not be sourced from a single location, requiring collection efforts that could raise transportation costs by 5 to 15% as capacity scales from 500 kW to 10 MW (Narnaware & Panwar, 2022). Increased petroleum fuel prices adversely affect transportation and other operational costs within the biomass supply chain of lignocellulosic biomass gasification (Akbarian et al., 2022). Despite biomass being a viable source of power production, financial institutions remain cautious about biomass projects due to uncertainties in biomass supply, pricing, and logistics (Narnaware & Panwar, 2022; Akbarian et al., 2022). Therefore, efficient feedstock supply chains are essential to ensure reliable and cost-effective operation of lignocellulosic biomass gasification plants.

The economic viability of lignocellulosic biomass gasification is influenced by energy market dynamics, including the price of competing energy sources (e.g., natural gas, coal, and renewable electricity) and regulatory policies supporting renewable energy development and biomass utilization (Sansaniwal et al., 2017; Narnaware & Panwar, 2022; Akbarian et al., 2022). Numerous bioenergy programs have been launched by various countries, yet their progress has been hindered by numerous challenges such as delays in obtaining planning permissions and technical approvals, sluggish processing and accountability, and excessive bureaucratic paperwork (Sansaniwal et al., 2017). Uncertainties in energy prices and regulatory frameworks can affect investment decisions and the long-term profitability of lignocellulosic biomass gasification projects (Akbarian et al., 2022; Jamil et al., 2024).

Biomass gasification is often promoted as a low-carbon or carbon-neutral technology because it utilizes renewable biomass feedstocks that can potentially offset carbon emissions from fossil fuel combustion (Akbarian et al., 2022; Jamil et al., 2024). However, the sustainability of biomass feedstock sourcing, land-use impacts, and lifecycle greenhouse gas emissions (including emissions from feedstock production, transportation, and gasification) must be carefully considered. Greenhouse gas (GHG) emission of a biomass gasification CHP plant was 7–15 g CO_2 eq/kWhe, 90% lower than when natural gas and diesel generators were used, which emit 240–270 g CO_2 eq/kWhe and 310–360 g/CO_2 eq/kWhe, respectively (Safarian et al., 2020; Akbarian et al., 2022; Jamil et al., 2024). Ensuring that biomass gasification contributes positively to overall greenhouse gas mitigation goals requires comprehensive lifecycle assessments and sustainable biomass sourcing practices.

Another environmental barrier is the air pollutant and emission control (Sansaniwal et al., 2017; Narnaware & Panwar, 2022; Akbarian et al., 2022). The gasification of lignocellulosic biomass can produce emissions of particulate matter, nitrogen oxides (NOx), sulfur compounds, and other air pollutants, depending on the feedstock composition and gasification conditions (Akbarian et al., 2022). Effective emission control technologies, such as particulate filters, selective catalytic reduction (SCR), and scrubbers, are essential to comply with environmental regulations and minimize the environmental footprint of gasification plants (Narnaware & Panwar, 2022; Akbarian et al., 2022). Comparing biomass gasification and incineration, the former has 50% of the acidification potential, which is a result of gases like NO, NO_2, SO_2, and HCL emissions to the atmosphere (Safarian et al., 2020). Balancing environmental performance with economic feasibility remains a challenge for biomass gasification technology developers and operators.

The gasification of lignocellulosic biomass holds promise as a sustainable and renewable energy solution, offering opportunities to diversify energy sources and reduce greenhouse gas emissions. However, realizing the full commercial potential of biomass gasification requires overcoming significant technological, economic, and environmental challenges. Continued research, innovation, and policy support are essential to address these challenges and enable the widespread deployment of biomass gasification technologies in the global energy landscape. By addressing these challenges effectively, biomass gasification can play a crucial role in achieving a more sustainable and resilient energy future.

10 Conclusions

The exploration of lignocellulosic waste for bioenergy production highlights its potential as a renewable and abundant resource. Lignocellulosic waste, from sources like agricultural residues and forestry by-products, offers a sustainable feedstock for bioenergy. The gasification process, which includes upstream preparation (size reduction, drying, and gasifying agent preparation) and downstream stages (heating, chemical reactions, and catalysis), is crucial for converting this waste into valuable syngas.

Operational parameters, including reactor types and raw material characteristics, significantly impact the process efficiency and product quality. Pretreatment steps like size reduction and drying are essential to optimize the gasification process. Post-gasification, cleaning, and reforming of syngas through catalytic processes ensure high-quality output suitable for various applications, including clean energy and chemical production.

Despite its promise, commercializing lignocellulosic waste gasification faces challenges such as maintaining consistent product quality, economic feasibility, and technological advancements. Addressing these issues through ongoing research and innovation is vital. The future of lignocellulosic waste gasification is promising, with its potential to contribute to renewable energy and waste management solutions, provided these challenges are met.

References

Abanades, S., Rodat, S., & Boujjat, H. (2021). Solar thermochemical green fuels production: A review of biomass pyro-gasification, solar reactor concepts and modelling methods. *Energies, 14*(5), 1494. https://doi.org/10.3390/en14051494

Ahmed, S. F., Mehejabin, F., Chowdhury, A. A., Almomani, F., Khan, N. A., Badruddin, I. A., & Kamangar, S. (2024). Biochar produced from waste-based feedstocks: Mechanisms, affecting factors, economy, utilization, challenges, and prospects. *GCB Bioenergy, 16*, e13175. https://doi.org/10.1111/gcbb.13175

Akbarian, A., Andooz, A., Kowsari, E., Ramakrishna, S., Asgari, S., & Cheshmeh, Z. A. (2022). Challenges and opportunities of lignocellulosic biomass gasification in the path of circular bioeconomy. *Bioresource Technology, 362*, 127774. https://doi.org/10.1016/j.biortech.2022.127774

Alauddin, Z. A. B. Z., Lahijani, P., Mohammadi, M., & Mohamed, A. R. (2010). Gasification of lignocellulosic biomass in fluidized beds for renewable energy development: A review. *Renewable and Sustainable Energy Reviews, 14*(9), 2852–2862. https://doi.org/10.1016/j.rser.2010.07.026

Azadi, P., Inderwildi, O. L., Farnood, R., & King, D. A. (2013). Liquid fuels, hydrogen and chemicals from lignin: A critical review. *Renewable and Sustainable Energy Reviews, 21*, 506–523. https://doi.org/10.1016/j.rser.2012.12.022

Basu, P. (2018). *Biomass gasification, pyrolysis and torrefaction: Practical design and theory*. Academic Press.

Blasi, A., Verardi, A., Lopresto, C. G., Siciliano, S., & Sangiorgio, P. (2023). Lignocellulosic agricultural waste valorization to obtain valuable products: An overview. *Recycling, 8*, 61. https://doi.org/10.3390/recycling8040061

Bressanin, J. M., Klein, B. C., Chagas, M. F., Watanabe, M. D. B., Sampaio, I. L. M., Bonomi, A., Morais, E. R., & Cavalett, O. (2020). Techno-economic and environmental assessment of biomass gasification and Fischer–Tropsch synthesis integrated to sugarcane biorefineries. *Energies, 13*, 4576. https://doi.org/10.3390/en13174576

Broda, M., Yelle, D. J., & Serwańska, K. (2022). Bioethanol production from lignocellulosic biomass-challenges and solutions. *Molecules, 27*(24), 8717. https://doi.org/10.3390/molecules27248717

Caballero, J. J. B., Zaini, I. N., & Yang, W. (2022). Reforming processes for syngas production: A mini-review on the current status, challenges, and prospects for biomass conversion to fuels. *Applications in Energy and Combustion Science, 10*, 100064. https://doi.org/10.1016/j.jaecs.2022.100064

Chambon, C. L., Karia, T., Sandwell, P., & Hallett, J. P. (2020). Techno-economic assessment of biomass gasification-based mini-grids for productive energy applications: The case of rural India. *Renewable Energy, 154*, 432–444.

Chen, Z., Hu, M., Zhu, X., Guo, D., Liu, S., Hu, Z., Xiao, B., Wang, J., & Laghari, M. (2015). Characteristics and kinetic study on pyrolysis of five lignocellulosic biomass via thermogravimetric analysis. *Bioresource Technology, 192*, 441. https://doi.org/10.1016/j.biortech.2015.05.062

Ciliberti, C., Biundo, A., Albergo, R., Agrimi, G., Braccio, G., de Bari, I., & Pisano, I. (2020). Syngas derived from lignocellulosic biomass gasification as an alternative resource for innovative bioprocesses. *PRO, 8*, 1567. https://doi.org/10.3390/pr8121567

Cortazar, M., Santamaria, L., Lopez, G., Alvarez, J. D. R. A., Zhang, L., Wang, R., & Olazar, M. (2023). A comprehensive review of primary strategies for tar removal in biomass gasification. *Energy Conversion and Management, 276*, 116496. https://doi.org/10.1016/j.enconman.2022.116496

Dimitrakellis, P., Delikonstantis, E., Stefanidis, G. D., & Vlachos, D. G. (2022). Plasma technology for lignocellulosic biomass conversion toward an electrified biorefinery. *Green Chemistry, 24*(7), 2680–2721.

Espinoza-Acosta, J. L., Torres-Chávez, P. I., Carvajal-Millán, E., Ramírez-Wong, B., Bello-Pérez, L. A., & Montaño-Leyva, B. (2014). Ionic liquids and organic solvents for recovering lignin from lignocellulosic biomass. *BioResources, 9*(2), 3660.

Gabbar, H., & Ahmad, M. S. (2024). Integrated waste-to-energy process optimization for municipal solid waste. *Energies, 17*, 497. https://doi.org/10.3390/en17020497

Gani, A., Zaki, M., Mamat, R., Nizar, M., Rosdi, S. M., Yana, S., & Sarjono, R. E. (2023). Analysis of technological developments and potential of biomass gasification as a viable industrial process: A review. *Case Studies in Chemical and Environmental Engineering, 8*, 100439. https://doi.org/10.1016/j.cscee.2023.100439

Gao, Y., Wang, M., Raheem, A., Wang, F., Wei, J., Xu, D., Song, X., Bao, W., Huang, A., Zhang, S., & Zhang, H. (2023). Syngas production from biomass gasification: Influences of feedstock properties, reactor type, and reaction parameters. *ACS Omega, 8*(35), 31620–31631. https://doi.org/10.1021/acsomega.3c03050

González, R., Phillips, R. B., Saloni, D., Jameel, H., Abt, R., & Pirraglia, A. (2015). Biomass to energy in the southern United States: Supply chain and delivered cost. *BioResources, 6*(3), 2954–2976.

Guimarães, H. R., Bressanin, J. M., Motta, I. L., Chagas, M. F., Klein, B. C., Bonomi, A., Maciel Filho, R., & Watanabe, M. D. B. (2021). Bottlenecks and potentials for the gasification of lignocellulosic biomasses and Fischer-Tropsch synthesis: A case study on the production of advanced liquid biofuels in Brazil. *Energy Conversion and Management, 245*, 114629. https://doi.org/10.1016/j.enconman.2021.114629

Hanaoka, T., Miyazawa, T., Shimura, K., & Hirata, S. (2015). Jet fuel synthesis from Fischer-Tropsch product under mild hydrocracking conditions using Pt-loaded catalysts. *Chemical Engineering Journal, 263*, 178. https://doi.org/10.1016/j.cej.2014.11.042

Harb, R., Rivera-Tinoco, R., Nemer, M., Zeghondy, B., & Bouallou, C. (2020). Towards synthetic fuels production from biomass gasification: Tar content at low temperatures. *Biomass and Bioenergy, 137*, 105540. https://doi.org/10.1016/j.biombioe.2020.105540

Hidalgo, D., Castro, J., Díez, D., Martín-Marroquín, J. M., Gómez, M., & Pérez, E. (2023). Torrefaction at low temperature as a promising pretreatment of lignocellulosic biomass in anaerobic digestion. *Energy, 263*, 125822. https://doi.org/10.1016/j.energy.2022.125822

Higman, C., & van der Burgt, M. (2008). *Gasification*. Gulf Professional Publishing.

Hussain, M., Ali, O., Raza, N., Zabiri, H., Ahmed, A., & Ali, I. (2023). Recent advances in dynamic modeling and control studies of biomass gasification for production of hydrogen rich syngas. *RSC Advances, 13*(34), 23796–23811. https://doi.org/10.1039/d3ra01219k

Islam, M. W. (2020). A review of dolomite catalyst for biomass gasification tar removal. *Fuel, 267*, 117095. https://doi.org/10.1016/j.fuel.2020.117095

Islam, S., & Dincer, I. (2018). A comparative study of syngas production from two types of biomass feedstocks with waste heat recovery. *Journal of Energy Resources Technology, 140*(9), 092002. https://doi.org/10.1115/1.4039873

Jamil, F., Inayat, A., Hussain, M., Akhter, P., Abideen, Z., Ghenai, C., Shanableh, A., & Abdellatief, T. M. M. (2024). Valorization of waste biomass to biofuels for power production and transportation in optimized way: A comprehensive review. *Advanced Energy and Sustainability Research, 5*, 2400104. https://doi.org/10.1002/aesr.202400104

Jideani, T., Chukwuchendo, E., & Khotseng, L. (2023). An approach towards the conversion of biomass feedstocks into biofuel using a zeolite Socony Mobil-5-based catalysts via the hydrothermal liquefaction process: A review. *Catalysts, 13*, 1425. https://doi.org/10.3390/catal13111425

Kalak, T. (2023). Potential use of industrial biomass waste as a sustainable energy source in the future. *Energies, 16*, 1783. https://doi.org/10.3390/en16041783

Kamble, A. D., Saxena, V. K., Chavan, P. D., & Mendhe, V. A. (2019). Co-gasification of coal and biomass an emerging clean energy technology: Status and prospects of development in Indian context. *International Journal of Mining Science and Technology, 29*(2), 171–186. https://doi.org/10.1016/j.ijmst.2018.03.011

Krishna, B. B., Biswas, B., & Bhaskar, T. (2019). Chapter 12. Gasification of lignocellulosic biomass. In A. Pandey, C. Larroche, C.-G. Dussap, E. Gnansounou, S. K. Khanal, & S. Ricke (Eds.), *Biomass, biofuels, biochemicals, biofuels: Alternative feedstocks and conversion processes for the production of liquid and gaseous biofuels* (2nd ed., pp. 285–300). Academic Press. https://doi.org/10.1016/B978-0-12-816856-1.00012-9

Kucharska, K., Rybarczyk, P., Hołowacz, I., Łukajtis, R., Glinka, M., & Kamiński, M. (2018). Pretreatment of lignocellulosic materials as substrates for fermentation processes. *Molecules, 23*(11), 2937. https://doi.org/10.3390/molecules23112937

Kumar, A., Jones, D. D., & Hanna, M. A. (2009). Thermochemical biomass gasification: A review of the current status of the technology. *Energies, 2*, 3. https://doi.org/10.3390/en20300556

Li, J., Yang, G., Yoneyama, Y., Vitidsant, T., & Tsubaki, N. (2016). Jet fuel synthesis via Fischer-Tropsch synthesis with varied 1-olefins as additives using co/ZrO2-SiO2 bimodal catalyst. *Fuel, 171*, 159. https://doi.org/10.1016/j.fuel.2015.12.062

Maitlo, G., Ali, I., Mangi, K. H., Ali, S., Maitlo, H. A., Unar, I. N., & Pirzada, A. M. (2022). Thermochemical conversion of biomass for syngas production: Current status and future trends. *Sustainability, 14*, 5. https://doi.org/10.3390/su14052596

Miccio, F., Piriou, B., Ruoppolo, G., & Chirone, R. (2009). Biomass gasification in a catalytic fluidized reactor with beds of different materials. *Chemical Engineering Journal, 154*(1–3), 369. https://doi.org/10.1016/j.cej.2009.04.002

Molino, A., Chianese, S., & Musmarra, D. (2016). Biomass gasification technology: The state of the art overview. *Journal of Energy Chemistry, 25*(1), 10–25. https://doi.org/10.1016/j.jechem.2015.11.005

Monir, M. U., Abd Aziz, A., Kristanti, R. A., & Yousuf, A. (2018). Gasification of lignocellulosic biomass to produce syngas in a 50 kW downdraft reactor. *Biomass and Bioenergy, 119*, 335–345. https://doi.org/10.1016/j.biombioe.2018.10.006

Moreira, R., Bimbela, F., Gil-Lalaguna, N., Sánchez, J. L., & Portugal, A. (2021). Clean syngas production by gasification of lignocellulosic char: State of the art and future prospects. *Journal of Industrial and Engineering Chemistry, 101*, 1–20. https://doi.org/10.1016/j.jiec.2021.05.040

Moreno-Chocontá, L. N., Lozano-Pérez, A. S., & Guerrero-Fajardo, C. A. (2024). Evaluation of the effect of particle size and biomass-to-water ratio on the hydrothermal carbonization of sugarcane Bagasse. *ChemEngineering, 8*(2), 43. https://doi.org/10.3390/chemengineering8020043

Mujtaba, M., Fraceto, L. F., Fazeli, M., Mukherjee, S., Savassa, S. M., de Medeiros, G. A., & Vilaplana, F. (2023). Lignocellulosic biomass from agricultural waste to the circular economy: A review with focus on biofuels, biocomposites and bioplastics. *Journal of Cleaner Production, 402*, 136815. https://doi.org/10.1016/j.jclepro.2023.136815

Narnaware, S. L., & Panwar, N. L. (2021). Catalysts and their role in biomass gasification and tar abetment: A review. *Biomass Conversion and Biorefinery*. https://doi.org/10.1007/s13399-021-01981-1

Narnaware, S. L., & Panwar, N. L. (2022). Biomass gasification for climate change mitigation and policy framework in India: A review. *Bioresource Technology Reports, 17*, 100892. https://doi.org/10.1016/j.biteb.2021.100892

Oliveira, J. A. R., Komesu, A., & Filho, R. M. (2014). Hydrothermal pretreatment for enhancing enzymatic hydrolysis of seeds of açaí (Euterpe oleracea) and sugar recovery. *Chemical Engineering Transactions, 37*, 787–792. https://doi.org/10.3303/CET1437132

Oliveira, J. A., Martins, L. H., Komesu, A., & Maciel Filho, R. (2015). Evaluation of alkaline delignification (NaOH) of açaí seeds (Eutherpe oleracea) treated with H2SO4 dilute and effect on enzymatic hydrolysis. *Chemical Engineering Transactions, 43*, 499–504. https://doi.org/10.3303/CET1543084

Or-Chen, D., Gerchman, Y., Mamane, H., & Peretz, R. (2024). Paper-mill wastes for bioethanol production in relation to circular economy concepts: A review. *Applied Sciences, 14*, 1081. https://doi.org/10.3390/app14031081

Pacioni, T. R., Soares, D., Di Domenico, M., Rosa, M. F., Moreira, R. D. F. P. M., & José, H. J. (2016). Bio-syngas production from agro-industrial biomass residues by steam gasification. *Waste Management, 58*, 221–229. https://doi.org/10.1016/j.wasman.2016.08.021

Pandey, P. H.; Ukarde, T.M.; Mahale, J. S.; Vasishta, A.; Pawar, H. S. (2022). Chapter 9. Syngas production via biomass gasification. Biomass, biofuels, biochemicals biochemicals and materials production from sustainable biomass resources. Hu Li, S. Saravanamurugan, Ashok Pandey, Sasikumar Elumalai. Elsevier, 211-261. https://doi.org/10.1016/B978-0-12-824419-7.00019-4.

Prins, M. J., Ptasinski, K. J., & Janssen, F. J. J. G. (2006). Torrefaction of wood: Part 1. Weight loss kinetics. *Journal of Analytical and Applied Pyrolysis, 77*(1), 28–34. https://doi.org/10.1016/j.jaap.2006.01.002

Quiroga, E., Cifuentes, B., Moltó, J., Ortuño, N., Conesa, J., Davó-Quiñonero, A., & Cobo, M. (2022). Integration of steam gasification and catalytic reforming of lignocellulosic biomass as a strategy to improve syngas quality and pollutants removal. *Waste Management, 147*, 48–59. https://doi.org/10.1016/j.wasman.2022.05.012

Rangabhashiyam, S., & Balasubramanian, P. (2019). The potential of lignocellulosic biomass precursors for biochar production: Performance, mechanism and wastewater application—A review. *Industrial Crops and Products, 128*, 405–423. https://doi.org/10.1016/j.indcrop.2018.11.041

Rauch, R., Hrbek, J., & Hofbauer, H. (2013). Biomass gasification for synthesis gas production and applications of the syngas. *Wiley Interdisciplinary Reviews: Energy and Environment, 3*(4), 343–362. https://doi.org/10.1002/wene.97

Rosenfeld, D. C., Böhm, H., Lindorfer, J., & Lehner, M. (2020). Scenario analysis of implementing a power-to-gas and biomass gasification system in an integrated steel plant: A techno-economic and environmental study. *Renewable Energy, 147*, 1511–1524. https://doi.org/10.1016/j.renene.2019.09.053

Safarian, S., Unnthorsson, R., & Richter, C. (2020). Performance analysis and environmental assessment of small-scale waste biomass gasification integrated CHP in Iceland. *Energy, 197*, 117268. https://doi.org/10.1016/j.energy.2020.117268

Sani, A. M., Savla, N., Pandit, S., Singh Mathuriya, A., Gupta, P. K., Khanna, N., Pramod Babu, R., & Kumar, S. (2021). Recent advances in bioelectricity generation through the simultaneous valorization of lignocellulosic biomass and wastewater treatment in microbial fuel cell. *Sustainable Energy Technologies and Assessments, 48*, 101572. https://doi.org/10.1016/j.seta.2021.101572

Sansaniwal, S. K., Rosen, M. A., & Tyagi, S. K. (2017). Global challenges in the sustainable development of biomass gasification: An overview. *Renewable and Sustainable Energy Reviews, 80*, 23–43. https://doi.org/10.1016/j.rser.2017.05.215

Schmid, J. C., Benedikt, F., Fuchs, J., Mauerhofer, A. M., Müller, S., & Hofbauer, H. (2021). Syngas for biorefineries from thermochemical gasification of lignocellulosic fuels and residues—5 years' experience with an advanced dual fluidized bed gasifier design. *Biomass Conversion and Biorefinery, 11*, 2405–2442.

Segurado, R., Pereira, S., Correia, D., & Costa, M. (2019). Techno-economic analysis of a trigeneration system based on biomass gasification. *Renewable and Sustainable Energy Reviews, 103*, 501–514. https://doi.org/10.1016/j.rser.2019.01.008

Shahbaz, M., Yusup, S., Inayat, A., Patrick, D. O., & Ammar, M. (2017). The influence of catalysts in biomass steam gasification and catalytic potential of coal bottom ash in biomass steam gasification: A review. *Renewable and Sustainable Energy Reviews, 73*, 468. https://doi.org/10.1016/j.rser.2017.01.153

Shayan, E., Zare, V., & Mirzaee, I. (2018). Hydrogen production from biomass gasification; a theoretical comparison of using different gasification agents. *Energy Conversion and Management, 159*, 30–41. https://doi.org/10.1016/j.enconman.2017.12.096

Shi, H., Si, W., & Li, X. (2016). The concept, design and performance of a novel rotary kiln type air-staged biomass gasifier. *Energies, 9*(2), 67. https://doi.org/10.3390/en9020067

Shie, J. L., Chen, L. X., Lin, K. L., & Chang, C. Y. (2014). Plasmatron gasification of biomass lignocellulosic waste materials derived from municipal solid waste. *Energy, 66*, 82–89. https://doi.org/10.1016/j.energy.2013.12.042

Singh, N., Singhania, R. R., Nigam, P. S., Dong, C. D., Patel, A. K., & Puri, M. (2022). Global status of lignocellulosic biorefinery: Challenges and perspectives. *Bioresource Technology, 344*, 126415. https://doi.org/10.1016/j.biortech.2021.126415

Situmorang, Y. A., Zhao, Z., Yoshida, A., Abudula, A., & Guan, G. (2020). Small-scale biomass gasification systems for power generation (<200 kW class): A review. *Renewable and Sustainable Energy Reviews, 117*, 109486. https://doi.org/10.1016/j.rser.2019.109486

Taherzadeh, M. J., & Karimi, K. (2008). Pretreatment of lignocellulosic wastes to improve ethanol and biogas production: A review. *International Journal of Molecular Sciences, 9*(9), 1621–1651. https://doi.org/10.3390/ijms9091621

Tijmensen, M. J., Faaij, A. P., Hamelinck, C. N., & van Hardeveld, M. R. (2002). Exploration of the possibilities for production of Fischer Tropsch liquids and power via biomass gasification. *Biomass and Bioenergy, 23*(2), 129–152. https://doi.org/10.1016/S0961-9534(02)00037-5

Trninić, M., Stojiljković, D., Jovović, A., & Jankes, G. (2016). Biomass gasification technology: The state of the art overview. In *4th International symposium on environmental friendly energies and applications (EFEA), Belgrade, Serbia* (pp. 1–8). https://doi.org/10.1109/EFEA.2016.7748797

Ummalyma, S. B., Supriya, R. D., Sindhu, R., Binod, P., Nair, R. B., Pandey, A., & Gnansounou, E. (2019). Biological pretreatment of lignocellulosic biomass—Current trends and future perspectives. In *Second and third generation of feedstocks* (pp. 197–212). Elsevier. https://doi.org/10.1016/B978-0-12-815162-4.00007-0

Xiao, R., Yang, W., Cong, X., Dong, K., Xu, J., Wang, D., & Yang, X. (2020). Thermogravimetric analysis and reaction kinetics of lignocellulosic biomass pyrolysis. *Energy, 201*, 117537. https://doi.org/10.1016/j.energy.2020.117537

Yang, J., Xin, Z., He, Q., Corscadden, K., & Niu, H. (2019). An overview on performance characteristics of bio-jet fuels. *Fuel, 237*, 916. https://doi.org/10.1016/j.fuel.2018.10.079

Zhang, Q., Xiao, J., & Hao, J. (2023). Cumulative exergy analysis of lignocellulosic biomass to bio-jet fuel through aqueous-phase conversion with different lignin conversion pathways. *Energy, 265*, 126301. https://doi.org/10.1016/j.energy.2022.126301

Engineering Microbial Strains and Enzymes for Enhanced Productivity

George Dzorgbenya Ametefe, Oluwabusayo Odunola Oluyide, Babamotemi Olusola Itakorode, Ismail Muritala, Omotayo Oyedara, and Divine Senanu Ametefe

Abstract

The research mainly focused on a systematic review of open-access research articles published between 2019 and 2024 on the Springer link database (https://link.springer.com/). Of particular interests were the studies on the hydrolysis of lignocellulosic biomass (LCB) to produce valuable products and the genetic engineering of microorganisms to improve their potential. The study adhered to the Preferred Reporting Items for Systematic Reviews and Meta-Analyses guidelines. The engineering of microbial strains and enzymes for LCB hydrolysis has shown potential for enhancing productivity and creating value from underutilized biomass resources. Through genetic engineering, directed evolution, and rational design, it has been demonstrated that enzyme engineering can enhance the natural properties of enzymes for industrial applications. It has been demonstrated that recombinant DNA technologies can overcome these challenges by improving protein synthesis, stability of exogenous expression, control over gene expression, folding of proteins, and metabolic pathways. Prokaryotic chassis strains have been reportedly subjected to genome streamlining to improve exogenous expression stability, protein production, and fitness. Ethical studies were also reviewed. The study has the potential to significantly improve environmental remediation techniques, the pharmaceutical industry, and various biotechnological industries' productivity and efficacy.

G. D. Ametefe (✉) · O. O. Oluyide · B. O. Itakorode · I. Muritala · O. Oyedara
Department of Biotechnology, Faculty of Basic and Applied Sciences, College of Science, Engineering, and Technology, Osun State University, Osogbo, Osun State, Nigeria

D. S. Ametefe
Wireless Communication Technology Group (WiCOT), College of Engineering, School of Electrical Engineering, Universiti Teknologi MARA (UiTM), Shah Alam, Selangor, Malaysia

G. Mukherjee, S. Dhiman (eds.), *Value Addition and Utilization of Lignocellulosic Biomass*, https://doi.org/10.1007/978-981-96-2786-8_8

Keywords

Biomass · Biorefineries · Hydrolysis · Microbial Engineering · Strain development · CRISPR/Cas9 · Enzymes Engineering · Computational Engineering

1 Introduction

1.1 Value Addition Through Lignocellulosic Biomass (LCB) Hydrolysis

Plants or plant-based materials that are usually not used for food or feed are known as lignocellulosic biomass (LCBs), mostly composed of cellulose, hemicellulose, and lignin, and they commonly consist of garden trimmings, energy crops, forestry residues, and agricultural residues such as wheat straw (Wang et al., 2021). However, these LCBs need to be enhanced via value addition.

"Value addition" concerning biomass hydrolysis could refer to the extra worth added to the processed biomass being harnessed for beneficial use. Hence, the hydrolysis of lignocellulosic biomasses has been proven to contribute to the growth of the economy via enzyme production from engineered microbial strains and reduction in global warming (Afraz et al., 2024; Mohanakrishna & Modestra, 2023; Velvizhi et al., 2022).

For instance, the value addition process involved in the tract of animals (as regards biomass assimilation) for optimum growth is made possible by microorganisms, as the microorganisms, like *Klebsiella variicola* secret enzymes that promote biomass degradation for the body's needs (Al-Mamun et al., 2021; Jakeer et al., 2020).

1.1.1 Hydrolysis of Lignocellulose Biomass (LCB)

Hydrolysis is a typical type of chemical reaction in which water is utilized to break down the chemical interactions between molecules; enzymes have been used for beneficial purposes via the conversion of complex molecules into simpler ones (Ametefe et al., 2022, 2023; Ega, 2023; Akimova et al., 2024).

It should be however noted that chemicals such as acids have also been proven to hydrolyze LCB—that is, acid hydrolysis (Świątek et al., 2020; Zhou et al., 2023).

Hemicellulose can be hydrolyzed to produce a variety of different sugars, including the pentoses (5-carbon sugars) xylose and arabinose; and the hexoses (6-carbon sugars) glucose, galactose, and mannose, depending on the type of hemicellulose. The hydrolysis of cellulose will yield monomeric glucose because cellulose is a homopolysaccharide, meaning it only contains one type of sugar (Pramasari et al., 2023).

Lignocellulose is a sustainable feedstock of cellulose, hemicellulose, and lignin that help biorefineries produce commercial biomaterials including chemicals and biofuels (Ojo, 2023). Pretreatment of LCB is necessary because crystalline cellulose, which is found in large amounts, and the physical lignin barrier that encircles the cellulose fibers are thought to be the main causes of cellulose's recalcitrance (Jamaldheen et al., 2022). Furfural acid can be produced when the sugars hydrolyzed from hemicellulose are degraded by the more severe conditions needed for cellulose hydrolysis or ester group cleavage without the action of water by the process of pyrolysis (Djakaria et al., 2022; Montané et al., 2002).

Because of this, the majority of hydrolysis technologies use pretreatment procedures designed to reduce cellulose crystallinity, break up the matrix (especially the links between lignin and cellulose), and (in certain situations) hydrolyze hemicelluloses. This allows the hydrolysate to be separated from the cellulose, which can then receive more intensive or focused treatment (Xu et al., 2019; Zeng et al., 2021). https://www.celignis.com/bioprocess-hydrolysis.php#:~:text=The%20hydrolysis%20of%20cellulose%20will,the%20pentoses%20(5%2Dcarbon%20sugars.

A literature search of the Springer link database indicates studies concerning value addition via the hydrolysis of LCBs (Fig. 1; Charts 1, 2, and 3).

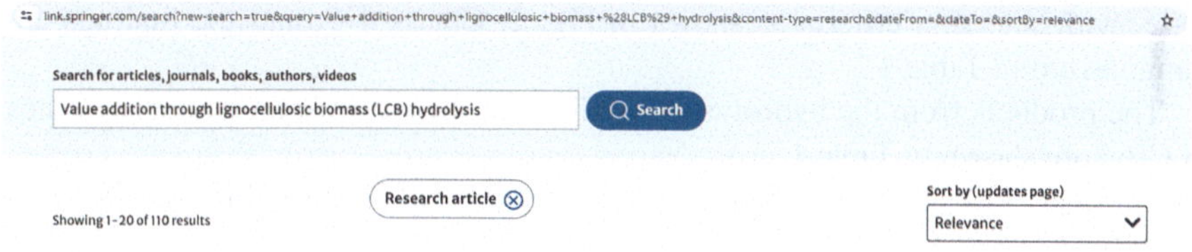

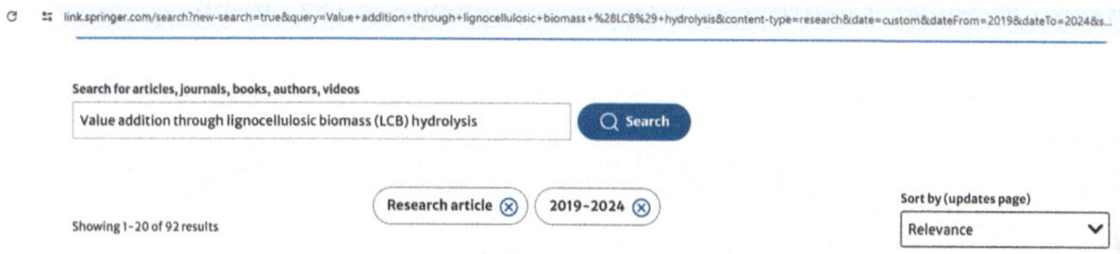

Chart 2: Value addition through lignocellulosic biomass (LCB) hydrolysis' *Research article & 2019-2024*

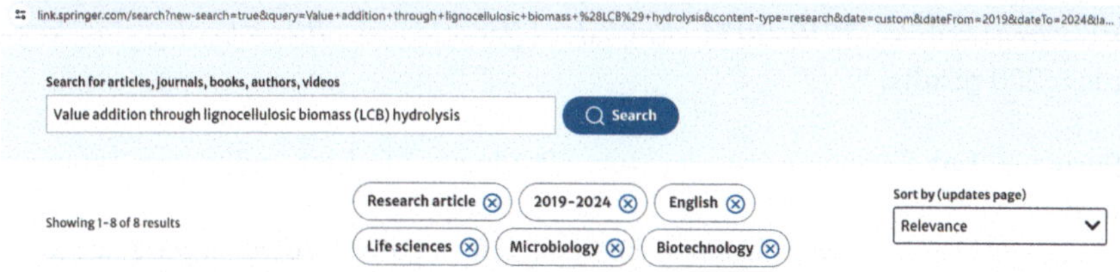

Chart 3: Value addition through lignocellulosic biomass (LCB) hydrolysis' *Research article, 2019-2024, English, Life sciences, Microbiology, & Biotechnology*

Fig. 1 Systematic Review Flow Chart. **Chart 1**: Value addition through lignocellulosic biomass (LCB) hydrolysis' *Research article*. **Chart 2**: Value addition through lignocellulosic biomass (LCB) hydrolysis' *Research article and 2019–2024*. **Chart 3**: Value addition through lignocellulosic biomass (LCB) hydrolysis' *Research article, 2019–2024, English, Life sciences, Microbiology, and Biotechnology*

Springer link database results for "value addition through lignocellulosic biomass (LCB) hydrolysis" *Research article* as keywords = 110 results.

↓

Springer link database results for "value addition through lignocellulosic biomass (LCB) hydrolysis" *Research article & 2019–2024* as keywords = 92 results.

↓

Springer link database results for "value addition through lignocellulosic biomass (LCB) hydrolysis" *Research article, 2019–2024, English, Life sciences, Microbiology, Biotechnology* as keywords = 8 results.

↓

Screened articles used for the review (open access) = 3 results.

1.1.2 Products Obtained from LCB Hydrolysis

Hydrolysis of LCB is expected to produce value in most cases for the benefit of man. Hence, the selected articles below show that some products produced from LCB in the last 3 months (April, May, and June 2024), on the Springer link database, with selection criteria as shown in Fig. 2, Charts 4–7, and the findings are summarized in Table 1.

The products from the hydrolysis of LCBs and their corresponding agro wastes (LCBs) are shown in Table 1.

Springer link database results for "Products of Lignocellulosic biomass hydrolysis."

Research article, English, Life Sciences as keywords = 1855 results.

↓

Springer link database results for "Products of Lignocellulosic biomass hydrolysis."

Research article, English, Life sciences, Biotechnology as keywords = 889 results.

↓

Springer link database results for "Products of Lignocellulosic biomass hydrolysis."

Research article, 2019–2024, English, Life sciences, Biotechnology as keywords = 220 results.

↓

Springer link database results for "Products of Lignocellulosic biomass hydrolysis."

Research article, Last 3 Months, English, Life sciences, Biotechnology as keywords = 18 results.

↓

Screened articles used for the review (showing LCB, microorganism used and products produced) *Research article, Last 3 Months, English, Life sciences, Biotechnology* as keywords = 10 results.

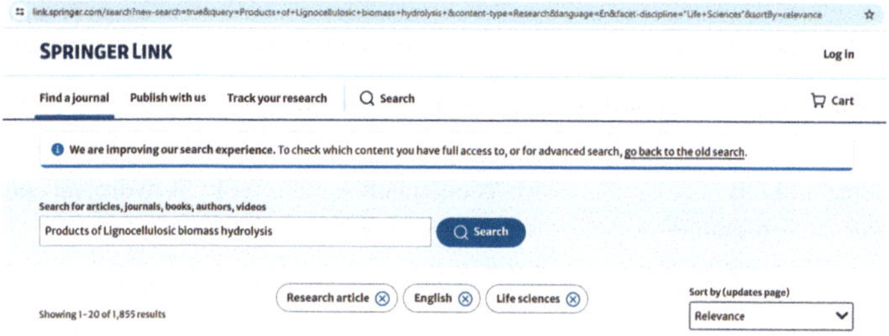

Chart 4: Products of Lignocellulosic hydrolysis with *Research article, English, and Life Sciences* as keywords

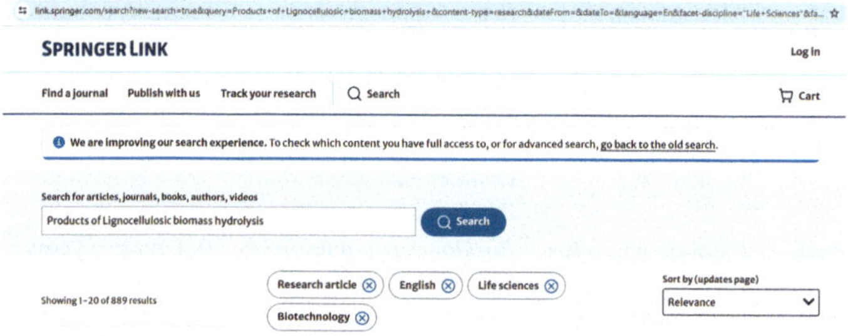

Chart 5: Products of Lignocellulosic hydrolysis with *Research article, English, Life Sciences & Biotechnology* as keywords

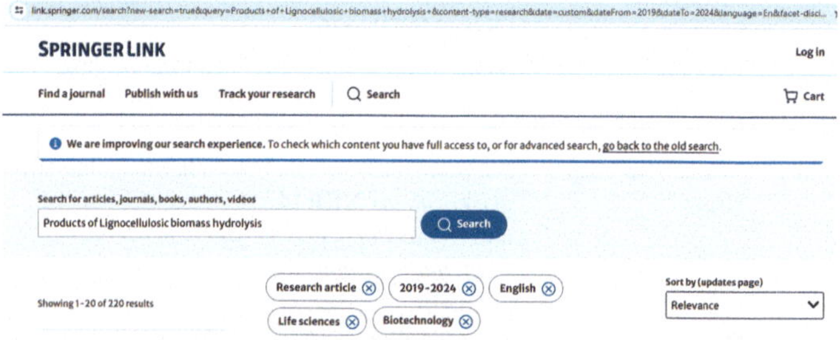

Chart 6: Products of Lignocellulosic hydrolysis with *Research article, 2019-2024, English, Life Sciences & Biotechnology* as keywords

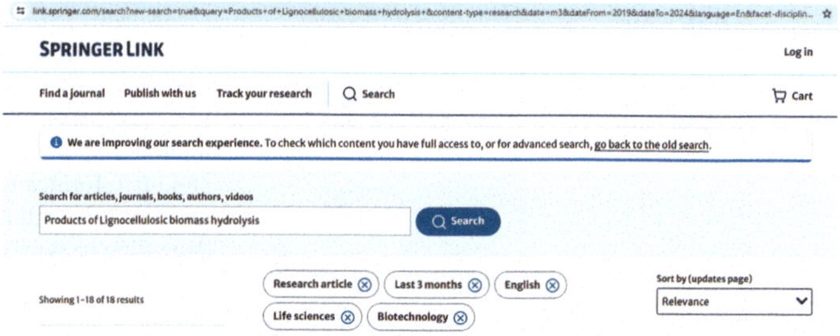

Chart 7: Products of Lignocellulosic hydrolysis with *Research article, Last 3 Months, English, Life Sciences & Biotechnology* as keywords

Fig. 2 Systematic Review Flow Chart. **Chart 4**: Products of lignocellulosic hydrolysis with *Research article, English, and Life Sciences* as keywords. **Chart 5**: Products of lignocellulosic hydrolysis with *Research article, English, Life Sciences, and Biotechnology* as keywords. **Chart 6**: Products of lignocellulosic hydrolysis with *Research article, 2019–2024, English, Life Sciences, and Biotechnology* as keywords. **Chart 7**: Products of lignocellulosic hydrolysis with *Research article, Last 3 Months, English, Life Sciences, and Biotechnology* as keywords

Table 1 Products from the hydrolysis of lignocellulosic biomass

S/no.	Reference	LCB	Microorganism	LCB hydrolysis products
1	Varriale et al. (2024)	*Perennial ryegrass* (*Lolium perenne*)	*Saccharomyces cerevisiae*	Bioethanol
2	Sharma et al. (2024)	Acid/alkali-treated bagasse	–	Cellulase
3	Chaoua et al. (2024)	Alfa stems (AS), dry palms (DP), olive pomace (OP), pinecones (PC), and tomato waste (TW)	*Reported for the first time for Bacillus mojavensis Bacillus safensis* *T. versicolor* K1, *Coniocheata hoffmannii* P4, *B. mojavensis* Y3	Laccase, cellulase, and xylanase
4	Arai et al. (2024)	Palm oil and other carbon sources (xylose, glucose, sucrose, glycerol, cellobiose)	*Burkholderia ambifaria* strain E5–3	Polyhydroxybutyrate (PHB), an eco-friendly biopolymer
5	Ali et al. (2024)	Sugarcane bagasse	*Clavispora lusitaniae* QG1 (MN592676)	Bioethanol
6	Yadav et al. (2024)	Rice husk	*Aspergillus protuberus*	β-Glucosidase
7	El-Sayed et al. (2024)	Wheat and bean straw	*Streptomyces griseorubens* NBR14 and *Nocardiopsis synnemataformans* NBRM9	Thermostable amylase
8	Duong et al. (2024)	Wheat straw	*Gloeophyllum trabeum* and other fungi	Hemicellulase
9	Gugel et al. (2024)	Olive leaves waste	*Lactobacillus casei*	Lactic acid
10	Xu et al. (2024)	Wheat straw	*Aspergillus niger*	Lytic polysaccharide monooxygenase and endo-β-1,4-glucanase activity (lignocellulose-degrading enzymes)

1.2 Common Microbial Strains Used for Lignocellulosic Biomass (LCB) Hydrolysis

Nature is interesting in that the previously learned life characteristics with every discovery appear real, unraveling mechanisms of feeding (nutrition) that cut across all kingdoms of living things. An instance is in the utilization of microbes for enzyme production. The simple concept in feeding is that, for humans to utilize the

Table 2 Some microorganisms for hydrolysis of LCB to valuable products and applications

		Microbial strains/ method of engineering	Lignocellulosic biomass	Enzyme produced	Applications
1	Chavan et al. (2024)	*Penicillium funiculosum* NCIM 1228 by Ethyl methanesulfonate (EMS) mutation	Bagasse	Cellulase and xylanase	The mutant strain was more active than the commercialized that was formulated
2	Prasad et al. (2024)	*Bacillus haynesii*	Lignin	Biofuel	Microbial pretreatment for biofuel production
3	Paul et al. (2024)	Isolated bacteria (*Bacillus albus* and *Pseudomonas mohnii*)	Rice straw	Cellulose and xylanase	Saccharification of rice straw using isolated cellulase–xylanase enzymes exhibited 60.33% production of total reducing sugar obtained by commercial cellulase–xylanase cocktail
4	Chetty et al. (2022)	Engineered (recombinant) *Saccharomyces cerevisiae*	No substrate as genes were inserted to enhance the amount of expression of the cellulases on cell-adhered surface	Cellulase	Bioethanol production

glucose from a meal of yam, the enzymes play a crucial role in the process; this mechanism is also replicated in microorganisms. However, for microorganisms, after the release of the microorganisms for catalysis, they leave the enzymes (which we need) in the medium.

Even though plants and animals produce these enzymes, microbial enzymes are relatively produced at a lower cost, and the production of these enzymes is "easier to control." The enzyme of interest to be secreted by the microbes is easier to predict with readily available raw material, producing the corresponding enzyme of interest (Raveendran et al., 2018). Many studies have utilized various LCBs for the production of microbial enzymes, bioethanol, and other value-added products (Tables 1 and 2).

1.3 Microbial Engineering Techniques for Enzyme Production

Enzymes are large biological molecules that accelerate biological processes; enzymes facilitate diverse biological reactions by selectively reacting with certain substrates (Saini et al., 2020). Enzymes derived from microbes have significant potential for utilization across several sectors due to their rapid efficacy and cost-efficient manufacturing capabilities (Yadav et al., 2018). These microbes are secreted by a myriad of microorganism strains as highlighted in Sect. 1.2.

1.3.1 Strain Development, Strain Improvement, or Engineering

The above terms in Sect. 1.3.1 could be used as the general term for improving the genetics of an organism so that it carries out biotechnological processes more effectively (Huo et al., 2022; Karn et al., 2023; Singh et al., 2023a). The aim is to create a strain of the organism that makes a particular desired product in large amounts and does not make much of anything else (so the product can be purified easily). The engineering of strain among other marketing options for an industry seems to be the most vital for an increase in profit. The ability of a microorganism to make any specific product is hugely hinged on its capability for the secretion of a particular set of enzymes; hence, enzyme production depends on the genetic makeup of the microorganism used (Cai et al., 2022).

Strain improvement depends on the metabolite of interest secreted by the organism, manipulation of the existing genetic apparatus in the organism, and in most cases the introduction of new properties into the organism by recombinant DNA technology or genetic engineering.

According to previously published studies, microbial engineering could have a big impact on the advancement of sustainable agriculture and a greener world (Nadarajah & Rahman, 2023; Singh et al., 2024). The genetic makeup of microorganisms is altered by microbial engineering techniques, such as gene editing with clustered regularly interspaced short palindromic repeats (CRISPR)/Cas9, RNA interference, and others, to enhance their advantageous functions in the metabolization of hazardous substances that support plant growth for increased yield and a clean environment (Eltarahony et al., 2023; Singh et al., 2024).

Before the CRISPR technique, tools like interference RNAs, ZFNs, and TALENs had already been developed for 'gene editing' (Ferreira & Choupina, 2022).

1.3.1.1 CRISPR/Cas9

Atsuo Nakata Group in 1987 at Osaka University, Japan first reported the presence of the clustered regularly interspaced short palindromic repeats (CRISPR) in *Escherichia coli* (Hatoum-Aslan & Howell, 2021). A palindrome is a series of letters that reads the same from left to right such as GAATTC (Fig. 3) and TAT. The palindrome has spaces filled with unidentical spacer DNAs; identified were also genes associated with CRISPR known as the Cas genes that make Cas proteins like helicase (proteins that unwind DNA) and nucleases (proteins that cut the DNA). Therefore, by translating the Cas 9 genes into proteins and converting CRISPR DNA into crRNa, which fits into the Cas protein to neutralize viral DNA, an *E. coli*

Fig. 3 Palindrome

equipped with the CRISPR-Cas system neutralizes the DNA of an incoming bacteriophage. If there are no spacer DNA matches, the Cas genes produce a class 1 Cas protein, which takes the invasive DNA, breaks it, copies it, and inserts it into the CRISPR system so that similar invasive DNA will be recognized by the spacer DNA the following time. Thus, the spacers represent the history of previous or old invasive DNAs (infections) that it came into contact with; this is an example of adaptive immunity, immunity gained through exposure to an antigen (Fig. 4) (Hatoum-Aslan & Howell, 2021; Newsom et al., 2021; Rath et al., 2015).

The distinguishing factor between the viral and the bacterial DNA is the protospacer adjacent motif (PAM) sequence. The system mimics the human immune system. PAM is a short piece of DNA sequence (typically ranging from two to six base pairs) that follows the DNA region the CRISPR system intends to cleave. The PAM, typically located three to four nucleotides downstream from the cut site, is necessary for a Cas nuclease to cut.

So, scientists imagined that the system could be used to inactivate or embed new genes as done in the laboratories of Jennifer Doudna and Emmanuelle Charpentier (Nobel Prize in Chemistry, 2020) working on *Streptococcus pyogenes*. Interestingly, they only had one Cas protein, the Cas 9. However, the system has both the spacer, which fits the invading DNA to inactivate it, and a tracer RNA, which pulls the crRNA in place. These scientists, after studying the operation of the system made a chimera of tracer and the spacer forming the trcrRNA-crRNA as well as fit in their DNA of interest as the spacer, with the chimera could be referred to as the guide RNA (gRNA) which cuts the DNA, but this cut DNA could be edited. In other words, a short DNA sequence, typically ranging from two to six base pairs, known as the protospacer adjacent motif, or PAM, comes after the DNA region that the CRISPR system, like CRISPR-Cas9, is designed to cleave. A Cas nuclease cannot cut without the PAM, typically located three to four nucleotides downstream from the cut site. So, in the absence of this system in the bacteria, the invading DNA of the bacteriophage hijacks the system makes copies of its DNA in the bacteria and kills the cell (Rath et al., 2015; https://cancerletter.com/the-cancer-letter/20201009_6/ (retrieved 29/07/2024)).

CRISPR/Cas9 is an RNA-mediated nuclease system that is a cost-effective and efficient gene editing method and is poised for numerous applications in functional genomics and epigenetics, including improving agricultural and environmental species, developing cell models for pharmaceutical and medical operations, and targeted therapy (Ferreira & Choupina, 2022).

1.3.1.2 RNA Interference (RNAi)

Most eukaryotic cells use a regulatory mechanism called RNA interference (RNAi) to control gene activity homology dependently. RNAi uses small double-stranded RNA (dsRNA) molecules as triggers (Almeida & Allshire, 2005). These roughly 21–22 bp long dsRNA molecules are known as small interfering RNAs (siRNA). Their distinctive 2 nt 3′ overhangs enable recognition by the RNAi enzymatic machinery, which in turn causes homology-dependent degradation of the target mRNA.

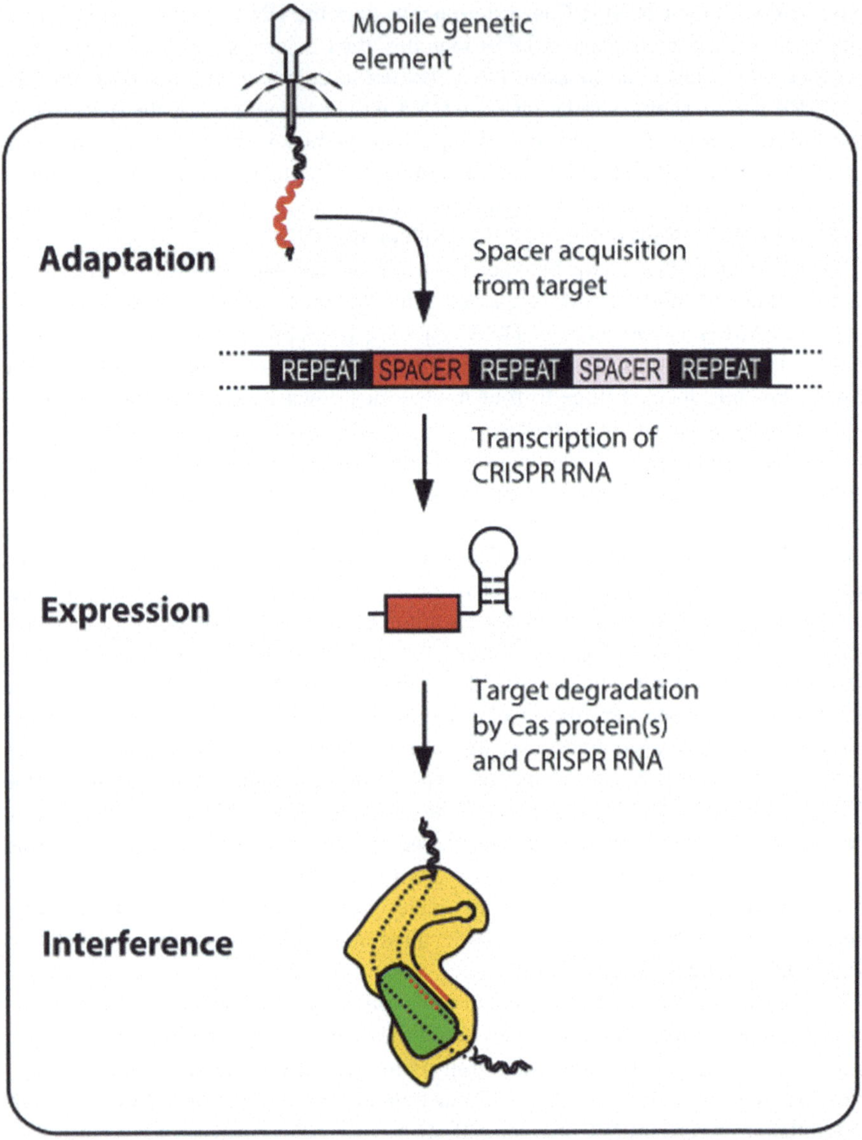

Fig. 4 Key steps of CRISPR-Cas immunity. (1) Adaptation: Insertion of fresh spacers into the CRISPR locus is an adaptation. (2) Expression: CRISPR RNA processing and transcription of the CRISPR locus. (3) Interference: CRISPR RNA and Cas protein(s) are used to detect and destroy mobile genetic elements (Rath et al., 2015)

Small double-stranded RNA (dsRNAs) plays a variety of roles in regulating gene expression. The endogenous micro-RNA (miRNA) pathway starts with Pol II-transcribed primary miRNAs, which are then processed in the nucleus to pre-miRNAs, exported to the cytoplasm, and processed once more into functional miRNAs. The main function of miRNAs is to hinder translation by incomplete WatsonCrick base pairing to the 3′ untranslated regions of targeted mRNAs. Alternatively, perfectly duplexed small interfering RNAs (siRNAs) can be produced intracellularly or given exogenously to cells. The guide strand is incorporated into the RNA-induced silencing complex (RISC), where it directs sequence-specific degradation of the target transcript regardless of where the base pairing takes place. There is no difference between the miRNA and siRNA pathways; what matters are the locations of the messages within the target transcripts and the degree of base pairing. By interacting with chromatin and directing histone and DNA methylation, which results in inactive chromatin, siRNAs can also cause transcriptional gene silencing (Daniel & John, 2008).

1.4 Enzymes Engineering for Enhanced Biomass Degradation

The majority of biotransformation relies on enzymes for essential operations. However, natural enzymes often require modification to enhance stability, specificity, and activity, as they may not be optimal for biotechnological applications. Lignocellulosic biomass, a readily available organic carbon source, holds great potential for green synthesis of biofuels and other bio-based compounds with minimal carbon and sulfur emissions (Omayma et al., 2023; Izydorczyk et al., 2024). Despite its abundant supply, the recalcitrance of lignocellulosic biomass poses a significant challenge to its industrial application. Efforts have been made to seek new enzymes with improved kinetic characteristics and to modify existing enzymes to overcome the challenges associated with lignocellulosic material degradation (Sethupathy et al., 2021; Sardar et al., 2024).

Enzyme engineering is required to ensure their compatibility with industrial settings. The engineering technology usually uses recombinant DNA to determine the specific genes responsible for enzyme synthesis. Subsequently, these genes are modified to enhance the stability and functionality of the current enzyme (Sinha & Shukla, 2019; Zhang et al., 2016). Enzyme engineering encompasses the process of genetically modifying enzymes by inserting or deleting nucleotides necessary to improve enzymatic properties, eliminate allosteric controls, and boost substrate selectivity, temperature stability, ideal pH, and other related factors (Kumar et al., 2017, 2018). Enzyme engineering, as an alternative to environmentally harmful petroleum-based chemicals, holds promise for developing resilient and efficient biocatalysts for the bio-based economy (Ribeiro et al., 2019). Techniques such as protein rational design and in vitro evolution can be utilized to modify the catalytic properties of enzymes, leading to improved production line yields through enhanced enzymatic parameters such as specificity, activity, efficiency, secret ability, and stability.

Biocatalysis has seen significant growth in recent decades, with advancements falling into two main categories. The first category involves enzyme immobilization as a method for conducting chemical reactions, which remains effective in enhancing protein characteristics today (Itakorode et al., 2024). This method entails enclosing an enzyme within a matrix or enzyme nanoparticle through molecular forces, resulting in the creation of a highly active, stable, and reusable catalytic species. Chemical-based methods have already successfully enhanced bioremediation and biomass conversion (Itakorode et al., 2022; Rosenboom et al., 2022).

The field of enzyme engineering encompasses a range of methodologies, with the use of technology to genetically engineer enzymes being the most recent approach. Two primary methods for enzyme engineering are directed evolution and rational design (Nirantar, 2021; Sellés et al., 2023). Directed evolution accelerates the process of natural selection by inducing changes in the protein-coding sequence through insertions, deletions, recombination, duplications, and mutations, leading to a diverse set of proteins. Subsequently, these variants are selected based on their enhanced characteristics (Nirantar, 2021). In contrast, rational design utilizes computer resources and an understanding of protein structure to intentionally create unique proteins (Vidal et al., 2023). These methods have proven useful in enhancing the characteristics of enzymes involved in the fermentation and depolymerization of lignocellulosic materials to produce valuable industrial compounds.

1.4.1 Rational Design Approach of Enzyme Engineering

A deep understanding of the structural and biochemical properties of proteins is essential for rational design to enhance protein function through precise modifications to achieve desired protein alterations based on prior knowledge (Gupta et al., 2017; Ribeiro et al., 2019; Reetz, 2022). Protein structure may be determined using techniques such as X-ray crystallography and nuclear magnetic resonance (NMR). The Protein Data Bank (PDB) has extensive structural information obtained from these methods. This information provides insights into how ligands or substrates interact with proteins. The study utilizes structural observation and site-saturation mutagenesis tests to identify the hot spot residues (Moroz et al., 2015). Hence, the rational approach involves specific point mutations, insertions, or deletions in the coding sequence, determined by the structure and function of the target biomolecule. Site-directed mutagenesis (SDM) is widely utilized for rational diversification, particularly to assess the impact of one or more amino acids on specific protein traits (Sardar et al., 2024). However, accurately predicting sequence–structure–function relationships, especially at the level of individual residues, remains challenging. Additionally, reliable structural data for the protein of interest is often lacking. Computational tools such as molecular dynamics simulations and structure homology modeling assist in predicting the significance of residues and suggesting potential modification sites (Bamezai et al., 2023). Nevertheless, their applicability is somewhat limited, especially with larger proteins and macromolecular complexes (Callaway, 2020). Consequently, carefully planned mutations may lead to undesired variants.

Additionally, the catalytic spot (site) must have a form that complements the reaction transition state to get the desired product (Raymond et al., 2015).

1.4.2 Direct Evolution Approach for Enzyme Engineering

A highly effective approach to enzyme engineering is directed evolution, which leverages the natural evolutionary process without requiring in-depth knowledge of the protein's tertiary structure. This technique facilitates the rapid identification of biomolecule variants with characteristics better suited for specific applications (Vidal et al., 2023). The procedure involves creating a collection of target enzyme mutants that explore the sequence space, aiming to identify an optimal solution during the screening process. Through iterative rounds of diversification, screening, and selection, these variants can be further refined. Given the vastness of the protein sequence space, comprehensive exploration is unfeasible, necessitating careful consideration of screening and diversification techniques.

Notably, libraries are broadly categorized as targeted and random. Targeted libraries modify specific gene regions or amino acid locations to enhance the likelihood of identifying desirable mutants, particularly for characteristics like substrate specificity, which disproportionately rely on a few locations. These libraries can be created using site-directed mutagenesis (Meister et al., 2019; Chua et al., 2020). Conversely, random libraries focus on the entire gene of interest and are more suitable for enhancing globally determined properties such as thermal stability or when detailed structure–function information is lacking, making it challenging to identify specific residues for diversification. Random libraries can be established both in vitro and in vivo.

1.4.2.1 In Vitro Methods

The utilization of physical or chemical mutagens to induce alterations in the target sequence formed the basis of some of the initial mutagenesis techniques. However, due to their inherent bias toward specific mutations, these methods have been surpassed by alternative approaches for controlled evolution. Among the most frequently utilized methods for random mutagenesis is error-prone PCR (epPCR). This technique involves the use of low-fidelity polymerases, such as Taq polymerase, under modified reaction conditions, including elevated concentrations of Mg^{2+} and Mn^{2+} ions, as well as varied deoxyribonucleotide concentrations, to augment the error rate. It is noteworthy that a specific codon is likely to undergo only a single mutation, limiting the mutagenesis of a given amino acid to a few alternatives rather than all 19, which is a drawback of error-prone PCR. Additionally, this method is susceptible to generating variants with frameshift mutations, often resulting in loss of functionality (Wenfa, 2023).

1.4.2.2 In Vivo Methods

It is worthy to note that in vitro mutagenesis techniques offer a level of control over mutation rate and spectrum, yet they have inherent limitations. These include the requirement for genetic material manipulation and restrictions on library size due to transformation efficiency. In contrast, in vivo methods bypass these limitations by leveraging the biological machinery of host cells to directly carry out mutagenesis. This approach enables the integration of screening or selection cycles with mutation, resulting in more efficient automated processes, such as contemporary continuous directed evolution (Tizei et al., 2021). In the realm of in vivo mutagenesis, there is a growing exploration of using enzymes other than DNA-dependent DNA polymerases as the primary source of mutagenesis. For example, the in vivo mutagenesis system for yeasts, based on the high error propensity of its encoded retrotranscriptase, showcases this expansion of the concept (Crook et al., 2016). Similarly, the MutaT7 system, which relies on a chimeric protein fusing T7 RNA polymerase with a nucleobase deaminase, introduces mutations (Moore et al., 2018).

CRISPR-based targeted mutation has recently seen significant advancements, illustrated by EvolvR, which allows editing windows of up to 350 nucleotides and offers focused mutation rates up to eight million times higher than those observed in wild-type cells (Halperin et al., 2018; Eric et al., 2024). These methods hold great potential for revolutionizing in vivo-directed evolution. However, the practical application of in vivo mutagenesis to expedite the evolution of biocatalysts acting on small-molecule substrates remains scarce, despite these advancements. This is primarily due to the absence of a standardized approach for assessing the fitness of created variations in vivo, even though successful directed evolution of novel enzymes working on small-molecule substrates through in vivo mutagenesis coupled to selection has been demonstrated.

1.5 Genetic Recombination

To enhance protein fitness, genetic recombination can be employed alongside mutagenesis techniques. Recombination methods enable the amalgamation of desired traits from homologous or nonhomologous genes. Random homologous recombination is achieved through well-established techniques such as DNA shuffling and synthetic shuffling (Yao et al., 2022; Bo et al., 2024), while SCHEMA, a semirational approach, utilizes computational algorithms and three-dimensional structures for guiding homologous recombination (Voigt et al., 2002). Postmutagenesis, homologous recombination is commonly used to blend mutations from different variants. Nevertheless, there are instances where recombining genes or DNA fragments with minimal similarity is intriguing. For instance, random nonhomologous recombination can be utilized to develop multifunctional chimeric proteins by combining distinct domains (Liu et al., 2023). Popular techniques for achieving random domain insertion in a homology-independent manner include multiplex inverted PCR, DNaseI or S1 nuclease, and transposon (Weiwu et al., 2021; Juma et al., 2023).

Hence, recombination could be viewed as

(a) Homologous recombination: the gene sequences undergoing recombination are very similar or closed-related DNA sequences.
(b) Nonhomologous recombination: the gene sequences undergoing recombination have little to no similarity.
(c) Reciprocal recombination: symmetrical exchange of genetic material occurs between two DNA strands.

(d) Site-specific recombination: Specialized nucleotide sequences exhibiting some degree of target site specificity are moved between nonhomologous sites within a genome.

(Didelot et al., 2012; Hallet & Sherratt, 1997; Pachauri, 2022).

Stemmer introduced DNA shuffling, the first in vitro homologous recombination method in 1994. Since then, many other homologous recombination methods have been developed, and methods for recombination of genes lacking sequence identity.

The aspiration to create state-of-the-art enzymes for bioprocesses is gradually becoming a reality. Enzyme evolution techniques have made significant strides, surpassing a billion mutations screened since early attempts over 40 years ago, with current approaches capable of high-throughput and quantitative readouts. However, there remains considerable scope for exploration and discovery of valuable enzymes, given the vast diversity present even in small enzymes.

It is also vital to note that enzyme engineering encompasses several techniques such as rational and semirational enzyme design, directed evolution, molecular dynamics and homology modeling, random mutagenesis, DNA shuffling, cell surface display approach, peptidomimetics, and de novo enzyme design.

1.6 Some Other Techniques Utilized for Enzyme Engineering Today

In the present day, computational tools and artificial intelligence (AI) approaches are being employed to develop optimal models for rapid enzyme modeling methodologies (Dhankhar et al., 2019).

1.6.1 Computational Engineering for Rapid Design

The main objective of the computational engineering of enzymes is to generate the three-dimensional structure of the active site in the transition state of a reaction, which is anticipated to provide the desired catalytic activity. Computational engineering has generated many enzyme virtual libraries in silico. The designs are subsequently evaluated and classified automatically using energy-scoring methods or geometric restrictions (Wijma et al., 2014).

1.6.2 Application of Artificial Intelligence Methodology

Artificial intelligence (AI) techniques for industrial processes are being developed to fulfill requirements such as increased cell productivity and selectivity, enhanced enzyme activity, and reduced operating expenses. Artificial intelligence encompasses several domains, including natural language processing, knowledge-based expert systems, robotics, and computer vision. As an illustration, the SB-913 ZEN genome editing therapy tools have been authorized to identify individuals with Hunter syndrome (Lim et al., 2019). The use of an artificial intelligence-generated biomaterial platform, based on nano-diamonds, has been shown as an effective method for treating and preventing reinfection following root canal therapy (Rawson et al., 2019). One additional biomaterial, polycaprolactone (PCL), has been utilized in the custom-made 3D printing process for breast rebuilding following tumor excision (Ho et al., 2020).

1.6.3 Enzyme Design from Scratch

De novo design approaches may be used to generate and design new enzymes with unique functions and structures to produce the desired result. De novo design approaches can also provide significant alterations in the protein structure. These strategies include designing active sites that do not possess the necessary catalytic site of the protein. The first de novo design was conducted using a quantum mechanical approach, together with the required catalyzed reaction (Kiss et al., 2013). The de novo enzyme design approaches are responsible for generating the majority of catalysts used in a wide range of catalytic processes. This design is referred to as "ozyme." For instance, the Diels–Alder reaction is a type of chemical reaction that often forms carbon–carbon (C–C) bonds by generating additional carbon atoms in the center. This reaction is commonly employed in synthetic chemistry. The enzyme is produced for this process by computational design methodologies. Nevertheless, several proteins are acquired using de novo creation techniques, although they function as sluggish catalysts. Nevertheless, the enzymes generated by this approach are known to usually have much lower catalytic activity compared to typical enzymes (Smith et al., 2014). The program Dezymer was initially used for de novo design to create an active site called His3-Fe-O_2 metal.

1.7 Pros and Cons of Microbial Engineering: Ethical Considerations

1.7.1 Microbial Engineering

This is an area of molecular biology that alters the genome sequence of microbes such as bacteria, algae, and fungi to develop novel products, applications, and routines as discussed previously. Alteration of the genetic sequence associated with traits of interest in wild microbes for medical, agricultural, industrial, and environmental applications is a key strategy for creating resourceful engineered microbes. These could find applications in plant and animal growth, health, quality and yield promotion, ecological clean-up, and others (Singh et al., 2024). Microbial engineering offers a range of potential benefits and drawbacks.

1.7.2 Pros of Microbial Engineering

1.7.2.1 Agriculture

Microbes could be engineered for the improvement of animal or crop yields, enhance nutrient uptake, and protect animals or plants from diseases, reducing the need for chemical fertilizers and pesticides. Research has demonstrated that microbial engineering can be of importance in developing a greener environment and agriculture sustainably (Nadarajah & Rahman, 2023). Microbial engineering techniques, which include gene editing, CRISPR/Cas9, RNAi, TALEN technologies, etc., are used to manipulate the genetic composition of microbes in an attempt to advance their advantageous effects in the metabolization of lethal substances that enhance growth in plants for better output and to clean the environment (Eltarahony et al., 2023). The engineering of microorganisms to create effective microbial inoculants could influence better crop growth and yield, tolerance against stresses (biotic and abiotic), biological control, increased soil fertility, and nutrient uptake (Kaul et al., 2021; Singh et al., 2023b; Salwan et al., 2023). Engineered microorganisms are pruning down the application of pesticides and inorganic fertilizers and can play roles as biopesticides and biofertilizers (Bharagava et al., 2020).

1.7.2.2 Bioremediation

Microbial engineering can facilitate the cleanup of environmental pollutants by enhancing the ability of microbes to break down contaminants such as oil spills or industrial waste. Different species of organisms on earth experience gross challenges in a bid to survive, as a result of many environmental contaminations. Microbial engineering applications could be explored to improve bioremediation quality, in replacement of failing conventional procedures to get rid of hazardous compounds from the environment. It is relatively safer and more economical to use engineered microbes rather than other methods for cleaning the environment (if such engineered microbe adheres to the ethical code of conduct). Biodegradation of xylenes, halobenzoates naphthalenes, octanes, toluenes, oil spills, trichloroethylene, etc. has been achieved using engineered microbes. They are more efficient than naturally occurring microbes and may degrade contaminants faster because they can rapidly adjust to new pollutants they come across or cometabolize. Thus, microbial engineering is a worthy development that will make the environment safe and eventually improve human health (Rafeeq et al., 2023).

1.7.2.3 Bioproduction

Microbes can be engineered to create valuable substances such as pharmaceuticals, biofuels, and industrial chemicals more cost-effectively and sustainably compared to conventional methods. Biodiesel production derived from alcohols and fatty acids may be sourced from engineered microbes. Many microorganisms such as algae, bacteria, and fungi could accrue intracellular fats, e.g., triacylglycerol into bulky amounts of biofuel (Al-Makishah, 2017). Oil from these engineered microorganisms can play roles of lactic biomaterial for biological fuel production during transesterification. Engineered microbes can potentially employ varieties of raw materials such as sugarcane with surprisingly more products per/ ha compared to other plant seeds and, thus, capable of producing biofuel that can utilize an insignificant proportion of farmland (Raj et al., 2022; Keerthana et al., 2022).

Recently, the production of high-energy biofuels is increasingly gaining attention. Scientists have been able to manipulate major pathways in *E. coli* to produce varieties of chained alcohols from sugars, additionally to *iso*-butanol, 1-butanol, 3-methyl-1-butanol, 2-methyl-1-butanol, and 2-phenylethanol (Sundarsingh et al., 2024).

1.7.2.4 Medical Applications

Microbial engineering could pave the way for the development of more effective and new treatments for various diseases, such as therapeutic proteins, vaccines, and targeted drug delivery systems. The human body naturally serves as a home to millions of microbes, with bacteria acknowledged to be the most common (>90%) among the microbes. Bacteria colonize different anatomical parts, which provide an enabling environment for them to thrive and reproduce. Many bacterial species (beneficial and disease-causing) interrelate with their hosts using biochemical signaling. Exploring such characteristics, beneficial and attenuated pathogenic bacteria have been modified to produce therapeutic molecules against a particular disease. New technology in microbial engineering has enabled scientists to execute complex genetic modifications in live bacteria and their products, which act out micron or submicron lipid-based vectors, for the production of therapeutic agents that are used in the control of many ailments (Shen et al., 2022).

1.7.2.5 Research Tools

Engineered microbes serve as valuable tools in scientific research, allowing researchers to study biological processes, gene functions, and metabolic pathways in detail.

1.7.3 Cons of Microbial Engineering

1.7.3.1 Environmental Risks

There is a concern that genetically modified microbes could escape into the environment and potentially disrupt ecosystems or introduce unintended consequences. Microorganisms can be engineered to create many valuable substances, including biofuels, plastics, and pharmaceuticals. Despite this importance, on many occasions, some of these engineered microbes compete with the metabolic pathways that the cells need to power for survival and reproduction (Trafton, 2019). A report by Snow et al. (2005) has it that when the Ecological Society of America evaluated the effects of potential uses of field-released genetically engineered organisms (GEOs) on ecology, some GEOs were found to play a positive role in sustainable agriculture, bioremediation, aquaculture, and environmental management, both in developed and developing countries. However, the indiscriminate use of GEOs in the environment may have negative ecological impacts. Furthermore, GEOs may lead to the creation of variant forms of pests and pathogens, make worse the effects of existing pests through hybridization with related transgenic organisms, and disrupt the ecosystem and irreversible damage or changes in genetic diversity within species.

1.7.3.2 Biosecurity Concerns

Engineered microbes could be used for nefarious purposes, such as bioterrorism or biowarfare if they *were to fall into the wrong hands*. Advanced knowledge in molecular biology, genetics, and biochemistry has made it possible to engineer living organisms. They could explore these developments to create effective and efficient microbes to treat ailments, intensification of food production, and enhance the quality of life of several people. However, such development could also be employed by state and nonstate actors to create engineered biological weapons. The threat of bioengineered microbes is analogous to the COVID-19 pandemic, and this has shown the massive effect that a single biological agent could have on society. Terrorist organizations may be monitoring these discoveries and then building on these to launch biological attacks with engineered agents (Wickiser et al., 2020).

1.7.3.3 Ethical Considerations

The manipulation of microbial life raises ethical questions regarding the potential for unintended consequences, the rights of organisms, and the implications of playing "creator" with living organisms.

1.8 Optimization Strategies for LCB Hydrolysis Through Enzymatic Processes

The need for optimization to improve microorganisms' function cannot be overemphasized. This is because for the production of enzymes to add value to biomass, there is a need for microbial strain improvement. Most approaches for strain improvement use the mutation concept as in recombinant yeasts strain to further increase the production of value-added products like bioethanol from lignocellulosic biomass (Table 3) (Chavan et al., 2024; Chetty et al., 2022). The criteria for the selection of the articles employed are shown in Fig. 5 and Charts 8–10.

Other traditional biotechnological optimization strategies used are the conditions for fermentation, like pH, temperature, and incubation periods (Ametefe & Orji, 2024; Paul et al., 2024; Prasad et al., 2024).

So, optimization strategies for LCB hydrolysis employed could exclusively target the microorganism and/or the conditions needed for fermentation for the production of value-added products (Table 3).

Springer link database results for "Optimization of lignocellulosic biomass (LCB) hydrolysis" *Research article* as keywords = 89 results.

↓

Springer link database results for "Optimization of lignocellulosic biomass (LCB) hydrolysis" *Research article and 2019–2024* as keywords = 75 results.

↓

Springer link database results for "Optimization of lignocellulosic biomass (LCB) hydrolysis" *Research article, 2019–2024, English, Life sciences, Microbiology, Biotechnology* as keywords = 5 results.

↓

Screened articles used for the review (as they were all not open access), the abstracts were = 5 results.

Table 3 Some common optimization strategies for LCB hydrolysis

S. no.	Reference	Microbial strains/ alternative process	Lignocellulosic biomass	Enzyme produced	Optimization (conditions)
1	Chavan et al. (2024)	*Penicillium funiculosum* NCIM 1228	Agro-waste	Cellulase	Random mutagenesis (derived by ethyl methanesulfonate (EMS) mutation)
2	Prasad et al. (2024)	*Bacillus haynesii* and *Paenibacillus alvei*	Rice straw and sugarcane bagasse	Cellulase	Temperature, pH, and incubation period
3	Govindaraj et al. (2023)	Novel Enzolv pretreatment	Banana fiber	Not stated	Not stated
4	Paul et al. (2024)	*Saccharomyces cerevisiae*	Rice straw	Cellulase and xylanase	Temperature, incubation time, and pH
5	Chetty et al. (2022)	*Saccharomyces cerevisiae*	Not stated (in the abstract); though crystalline cellulose was used for determination of the activity of the enzyme	Cellulase	Recombinant *Saccharomyces cerevisiae* strains

1.9 Challenges and Potential Remedies/Solutions for Engineering Microbial Strains and Enzymes for Enhanced Productivity

The majority of proteins were previously extracted from the living things that produced them. Low amounts of the necessary proteins were produced by this labor-intensive technique. This issue was resolved with the advancement of science, specifically biotechnology, as well as the overproduction of heterologous proteins in model organisms. In contrast to the native organism, the gene encoding the desired protein is overexpressed in other organisms, such as human cell lines, bacteria, yeasts, and insects; because *Escherichia coli* requires less time and frequently costs less than the alternative systems, it is a commonly utilized microbe (Liponska et al., 2019).

Microbe-based expression systems are used in the creation of a significant amount of recombinant proteins nowadays since they are comparatively easy to use and affordable. However, there are still impending challenges in making the most of this system. Some of the challenges include several microbial protein expression challenges that are difficult to optimize, like

- The creation of insoluble aggregates (Kane, 1995).
- DNA repeats that influence the translation of proteins and the ensuing folding of proteins (Kane, 1995).
- The creation of harmful endotoxins (*E. coli*) (Paddon et al., 2013).

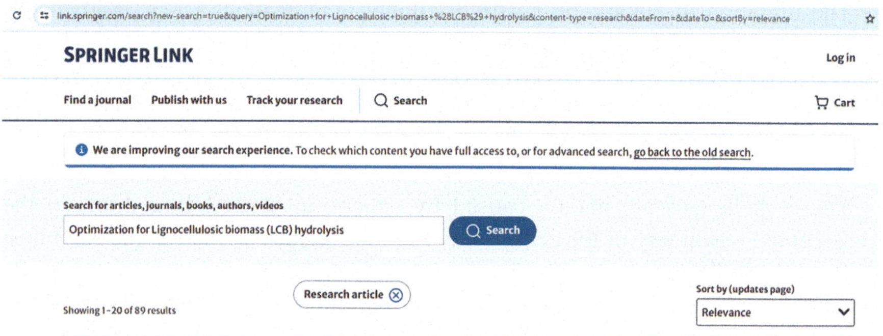

Chart 8: 'Optimization of lignocellulosic biomass (LCB) hydrolysis' *Research articles*

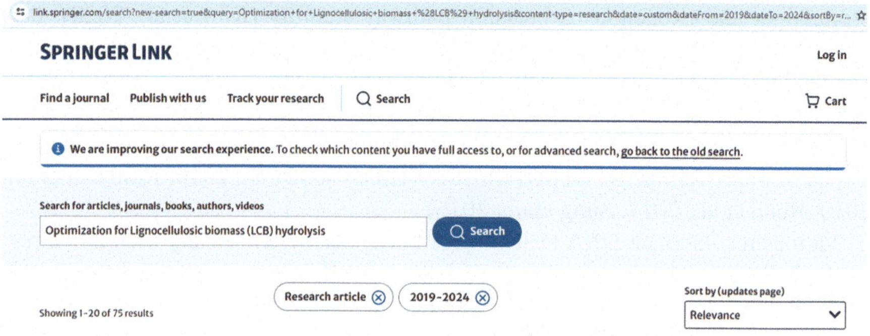

Chart 9: 'Optimization of lignocellulosic biomass (LCB) hydrolysis' *Research article & 2019-2024* as keywords

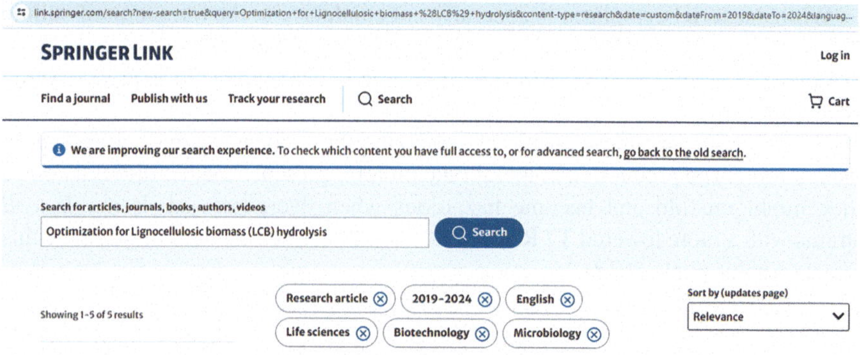

Chart 10: 'Optimization of lignocellulosic biomass (LCB) hydrolysis' *Research article, 2019-2024, English, Life sciences, Microbiology, Biotechnology* as keywords

Fig. 5 Systematic Review Flow Chart. **Chart 8**: "Optimization of lignocellulosic biomass (LCB) hydrolysis" *Research articles*. **Chart 9**: "Optimization of lignocellulosic biomass (LCB) hydrolysis" *Research article and 2019–2024* as keywords. **Chart 10**: "Optimization of lignocellulosic biomass (LCB) hydrolysis" *Research article, 2019–2024, English, Life sciences, Microbiology, Biotechnology* as keywords

- The imitation of eukaryotic posttranslational modifications (such as glycosylation) (Jenkins & Curling, 1994).
- Low protein yields (Li et al., 2019).
- Decreased cell survival brought on by undesired byproducts and cell death (Luli & Strohl, 1990; Kane, 1995).

These challenges are mainly caused by genetic instability, limitations in the translational machinery of the cell, or the metabolic capacity of the expression host (Kim et al., 2020).

Therefore, several microbial strains with carefully engineered genomes have been proposed as a solution to the persistent issues with the expression of recombinant proteins. Genome streamlining has already been applied to an increasing number of prokaryotic chassis strains to improve cellular fitness, recombinant protein production, and exogenous expression pathway stability (Kim et al., 2020). It is vital to note that genome streamlining is a natural evolutionary process of removing nonbeneficial genes since a smaller genome reduces the metabolic burden on the cell and improves fitness (Giovannoni et al., 2014; Sela et al., 2016; Stelzer et al., 2023). It has been successfully used as an engineering strategy in model heterotrophs to improve fitness and performance (Hutchison et al., 2016; Komatsu et al., 2010; Reuß et al., 2017; Sung et al., 2016).

Various recombinant DNA technologies have been systematically used to address these challenges. These technologies include controlling gene expression (Studier & Moffatt, 1986; Guzman et al., 1995; Wang & Kushner, 1991) directing proper bond formation and protein folding (Baneyx & Mujacic, 2004; Nallamsetty & Waugh, 2007), interfering with host metabolic pathways (Aristidou et al., 1995; Dittrich et al., 2005), directing or randomly evolving bacterial strains or enzymes (Dragosits & Mattanovich, 2013), and a host of other techniques (Liponska et al., 2019).

While bacterial cytoplasm is generally not conducive to the formation of sulfide bonds, strains of *E. coli* have been successfully designed to oxidize cysteines in the cytoplasm, making proteins with disulfide bonds challenging to express (Anton et al., 2016).

Heterologous enzymes capable of catalyzing those changes are coexpressed by these strains. The production of membrane proteins presents another difficulty since they might misfold and become hazardous when overexpressed. Using *E. coli* strains with a more lowered T7 RNA expression than standard ones may reduce this effect (Angius et al., 2018).

References

Afraz, M., Muhammad, F., Nisar, J., Shah, A., Munir, S., Ali, G., & Ahmad, A. (2024). Production of value added products from biomass waste by pyrolysis: An updated review. *Waste Management Bulletin, 1*(4), 30–40. https://doi.org/10.1016/j.wmb.2023.08.004

Akimova, D., Kakimov, A., Suychinov, A., Urazbayev, Z., Zharykbasov, Y., Ibragimov, N., Bauyrzhanova, A., & Utegenova, A. (2024). Enzymatic hydrolysis in food processing: Biotechnological advancements, applications, and future perspectives. *Potravinarstvo Slovak Journal of Food Sciences, 18*, 347–365.

Ali, S., Rana, Q. A., Riaz, F., Haq, A., Sajjad, W., Gauttam, R., Ali, M., & Badshah, M. (2024). Agricultural waste management by production of second-generation bioethanol from sugarcane bagasse using indigenous yeast strain. *Current Microbiology, 81*, 161. https://doi.org/10.1007/s00284-024-03668-y

Al-Makishah, N. H. (2017). Microbial biofuel production advancement. *International Journal of Pharmaceutical Research and Allied Sciences, 6*(3), 93–106.

Al-Mamun, A., Hossain, S., Debnath, G. C., Sultana, S., Rahman, A., Hasan, Z., Das, S. R., Ashik, A., Prodhan, Y., Aktar, S., Cho, K. M., & Haque, A. (2021). Unveiling lignocellulolytic trait of goat omasum inhabitant *Klebsiella variicola* strain HSTU-AAM51 in light of biochemical and genome analyses. *Brazilian Journal of Microbiology, 53*, 99–130. https://doi.org/10.1007/s42770-021-00660-7

Almeida, R., & Allshire, R. C. (2005). RNA silencing and genome regulation. *Trends in Cell Biology, 15*, 251–258.

Ametefe, G. D., & Orji, F. A. (2024). Pectin classifications, mode of action, and factors influencing pectinase production. *Journal of Agricultural, Food Science & Biotechnology, 2*(1), 36–53.

Ametefe, G. D., Iheagwam, F. N., Fashola, F., Ibidapo, O. I., Iweala, E. E. J., & Chinedu, S. N. (2022). Comparison of pectinase activity in the Flavedo and albedo of citrus and *Thaumatococcus daniellii* fruits. In A. O. Ayeni, S. E. Sanni, & S. U. Oranusi (Eds.), *Bioenergy and biochemical processing technologies. Green Energy and Technology*. Springer.

Ametefe, G. D., Ajani, O. O., Iweala, E. E. J., & Chinedu, S. N. (2023). Introduction to enzymes and Organocatalysis. In S. N. Meena, V. Nandre, K. Kodam, & R. S. Meena (Eds.), *New horizons in natural compound research. Progress in biochemistry and biotechnology* (1st ed.). Elsevier.

Angius, F., Ilioaia, O., Amrani, A., Suisse, A., Rosset, L., Legrand, A., et al. (2018). A novel regulation mechanism of the T7 RNA polymerase-based expression system improves overproduction and folding of membrane proteins. *Scientific Reports, 8*, 8572.

Anton, B. P., Fomenkov, A., Raleigh, E. A., & Berkmen, M. (2016). Complete genome sequence of the engineered Escherichia coli SHuffle strains and their wild-type parents. *Genome Announcements, 4*, e00230–e00216.

Arai, T., Aikawa, S., Sudesh, K., Arai, W., Rawi, N. F. M., Leh, C. P. P., Kassim, M. H. M., Tay, G. S., & Kosugi, A. (2024). Efficient production of polyhydroxybutyrate using lignocellulosic biomass derived from oil palm trunks by the inhibitor-tolerant strain *Burkholderia ambifaria* E5-3. *World Journal of Microbiology and Biotechnology, 40*, 242. https://doi.org/10.1007/s11274-024-04041-8

Aristidou, A. A., San, K. Y., & Bennett, G. N. (1995). Metabolic engineering of Escherichia coli to enhance recombinant protein production through acetate reduction. *Biotechnology Progress, 11*, 475–478.

Bamezai, S., Maresca, S. G., Morris, F., Hildebrandt, R., Amil, M. A. S., Sporadicate iGEM Team, & Ledesma-Amaro, R. (2023). Protein engineering in the computational age: An open source framework for exploring mutational landscapes *in silico*. *Engineering Biology, 7*(1–4), 29–38. https://doi.org/10.1049/enb2.12028

Baneyx, F., & Mujacic, M. (2004). Recombinant protein folding and misfolding in *Escherichia coli*. *Nature Biotechnology, 22*, 1399–1408.

Bharagava, R. N., Saxena, G., & Mulla, S. I. (2020). Introduction to industrial wastes containing organic and inorganic pollutants and bioremediation approaches for environmental management. In *Bioremediation of industrial waste for environmental safety* (pp. 1–18). Springer.

Bo, L., Yasin, E., Mikkel, N. S., Tommy, S. A., & Sebastian, U. S. (2024). Synthetic data shuffling accelerates the convergence of federated learning under data heterogeneity. *Transactions on Machine Learning Research*. https://openreview.net/pdf?id=c5o4HUypqm

Cai, P., Han, M., Zhang, R., Ding, S., Zhang, D., Liu, D., Liu, S., & Qian-Nan Hu, Q. N. (2022). SynBioStrainFinder: A microbial strain database of manually curated CRISPR/Cas genetic manipulation system information for biomanufacturing. *Microbial Cell Factories, 21*, 87.

Callaway, E. (2020). It will change everything: DeepMind's AI makes gigantic leap in solving protein structures. *Nature, 588*(7837), 203–204.

Chaoua, S., Flahaut, S., Cornu, B., Hiligsmann, S., & Chaouche, N. K. (2024). Unlocking the potential of Algerian lignocellulosic biomass: Exploring indigenous microbial diversity for enhanced enzyme and sugar production. *Archives of Microbiology, 206*, 277. https://doi.org/10.1007/s00203-024-04011-6

Chavan, S., Shete, A., & Dharne, M. S. (2024). Strain improvement for cellulolytic enzymes for effective saccharification of lignocellulosic biomass by mutant of *Penicillium funiculosum* NCIM 1228. *Systems Microbiology and Biomanufacturing, 4*, 716–730. https://doi.org/10.1007/s43393-024-00247-x

Chetty, B. J., Inokuma, K., Hasunuma, T., van Zyl, W. H., & den Haan, R. (2022). Improvement of cell-tethered cellulase activity in recombinant strains of *Saccharomyces cerevisiae*. *Applied Microbiology and Biotechnology, 106*, 6347–6361. https://doi.org/10.1007/s00253-022-12114-7

Chua, J. P. S., Go, M. K., Osothprarop, T., Mcdonald, S., Karabadzhak, A. G., Yew, W. S., Peisajovich, S., & Nirantar, S. (2020). Evolving a thermostable terminal Deoxynucleotidyl transferase. *ACS Synthetic Biology, 9*, 1725–1735.

Crook, N., Abatemarco, J. J., Sun, J., Wagner, M., Schmitz, A., & Alper, H. S. (2016). *In vivo* continuous evolution of genes and pathways in yeast. *Nature Communications, 7*, 13051.

Daniel, H. K., & John, J. R. (2008). RNAi mechanisms and applications. *BioTechniques, 44*(5), 613–616.

Dhankhar, R., Kumar, A., Kumar, S., Chhabra, D., Shukla, P., & Gulati, P. (2019). Multilevel algorithms and evolutionary hybrid tools for enhanced production of arginine deiminase from pseudomonas furukawaii RS3. *Bioresource Technology, 290*, 121789.

Didelot, X., Méric, G., Falush, D., & Darling, A. E. (2012). Impact of homologous and non-homologous recombination in the genomic evolution of Escherichia coli. *BMC Genomics, 13*, 256.

Dittrich, C. R., Vadali, R. V., Bennett, G. N., & San, K. Y. (2005). Redistribution of metabolic fluxes in the central aerobic metabolic pathway of E. coli mutant strains with deletion of the ackA-pta and poxB pathways for the synthesis of isoamyl acetate. *Biotechnology Progress, 21*, 627–631.

Djakaria, K., Tang, Z., Shao, J., Chen, X., Xiao, H., Smith, J., Yang, H., & Chen, H. (2022). Improving the production of fufural from cellulose catalytic pyrolysis using $WO_3/\gamma\text{-}Al_2O_3$ composites oxides. *Journal of Analytical and Applied Pyrolysis, 167*, 105648.

Dragosits, M., & Mattanovich, D. (2013). Adaptive laboratory evolution principles and applications for biotechnology. *Microbial Cell Factories, 12*, 64.

Duong, H. L., Paufler, S., Harms, H., Maskow, T., & Schlosser, D. (2024). Biocalorimetry-aided monitoring of fungal pretreatment of lignocellulosic agricultural residues. *Applied Microbiology and Biotechnology, 108*, 394. https://doi.org/10.1007/s00253-024-13234-y

Ega, L. (2023). Characterization and determination of hydrolysis process parameters of superior sweet starch CIP-types enzymatically. *Food Research, 7*(2), 316–322.

El-Sayed, M. H., Gomaa, A. E. R. F., Atta, O. M., & Hassane, A. M. A. (2024). Characteristics and kinetics of thermophilic actinomycetes' amylase production on agro-wastes and its application for ethanol fermentation. *World Journal of Microbiology and Biotechnology, 40*, 255. https://doi.org/10.1007/s11274-024-04009-8

Eltarahony, M., Ibrahim, E., Hegazy, G., & Sabry, A. (2023). Microbial remediation of mercury: An overview. In *Mercury toxicity: Challenges and solutions* (pp. 201–234).

Eric, A., Pierce, T. S., Aleman, K. T., Jayasundera, B. S., Ashimatey, K. K., Alia, R., Michael, C., Jaskolka, R. L., Myers, B. L., Lam, S. T., & Bailey, J. I. (2024). Gene editing for CEP290 - associated retinal degeneration. *New England Journal of Medicine, 390*, 1972. https://doi.org/10.1056/NEJMoa2309915

Ferreira, P., & Choupina, A. B. (2022). CRISPR/Cas9 a simple, inexpensive, and effective technique for gene editing. *Molecular Biology Reports, 49*, 7079–7086.

Giovannoni, S. J., Cameron, T. J., & Temperton, B. (2014). Implications of streamlining theory for microbial ecology. *The ISME Journal, 8*, 1553–1565.

Govindaraj, O., Gopal, N. O., Raja, A. S. M., & Uthandi, S. (2023). Influence of novel EnZolv pretreatment on the release of reducing sugar and proximate content of Banana fiber. *Indian Journal of Microbiology, 63*, 693–701. https://doi.org/10.1007/s12088-023-01130-4

Gugel, I., Marchetti, F., Costa, S., Gugel, I., Baldini, E., Vertuani, S., & Manfredini, S. (2024). 2G-lactic acid from olive oil supply chain waste: Olive leaves upcycling via *Lactobacillus casei* fermentation. *Applied Microbiology and Biotechnology, 108*, 379. https://doi.org/10.1007/s00253-024-13217-z

Gupta, S. K., Srivastava, S. K., Sharma, A., Nalage, V. H., et al. (2017). Metabolic engineering of CHO cells for the development of a robust protein production platform. *PLoS One, 12*(8), e0181455.

Guzman, L. M., Belin, D., Carson, M. J., & Beckwith, J. (1995). Tight regulation, modulation, and high-level expression by vectors containing the arabinose PBAD promoter. *Journal of Bacteriology, 177*, 4121–4130.

Hallet, B., & Sherratt, D. J. (1997). Transposition and site-specific recombination: Adapting DNA cut-and-paste mechanisms to a variety of genetic rearrangements. *FEMS Microbiology Reviews, 21*(2), 157–178.

Halperin, S. O., Tou, C. J., Wong, E. B., Modavi, C., Schaffer, D. V., & Dueber, J. E. (2018). CRISPR-guided DNA polymerases enable diversification of all nucleotides in a tunable window. *Nature, 560*(7717), 248–252.

Hatoum-Aslan, A., & Howell, O. G. (2021). CRISPR-Cas systems and anti-CRISPR proteins: Adaptive defense and counter-defense in prokaryotes and their viruses. In *Encyclopedia of virology* (Vol. 4, 4th ed., pp. 242–251).

Ho, D., Quake, S. R., McCabe, E. R., Chng, W. J., et al. (2020). Enabling technologies for personalized and precision medicine. *Trends in Biotechnology, 38*, 497–518. https://doi.org/10.1016/j.tibtech.2019.12.021

Huo, G., Foulquié-Moreno, M. R., & Thevelein, J. M. (2022). Development of an industrial yeast strain for efficient production of 2,3-butanediol. *Microbial Cell Factories, 21*, 199.

Hutchison, C. A. I. I. I., Chuang, R. Y., Noskov, V. N., Assad-Garcia, N., Deerinck, T. J., Ellisman, M. H., Gill, J., Kannan, K., Karas, B. J., Ma, L., Pelletier, J. F., Qi, Z. Q., Richter, R. A., Strychalski, E. A., Sun, L., Suzuki, Y., Tsvetanova, B., Wise, K. S., Smith, H. O., Glass, J. I., Merryman, C., Gibson, D. G., & Venter, J. C. (2016). Design and synthesis of a minimal bacterial genome. *Science, 351*, aad6253.

Itakorode, B. O., Okonji, R. E., & Torimiro, N. (2022). Cyanide bioremediation potential of *Klebsiella oxytoca* JCM 1665 rhodanese immobilized on alginate-glutaraldehyde beads. *Biocatalysis and Biotransformation, 41*, 299–308. https://doi.org/10.1080/10242422.2022.2087510

Itakorode, B. O., Itakorode, D. I., Torimiro, N., & Okonji, R. E. (2024). Kinetic and thermodynamic investigation of Rhodanese synthesized by enhanced *Klebsiella oxytoca* JCM 1665 strain: a comparison between the free and immobilized enzyme entrapped in alginate beads. *Preparative Biochemistry & Biotechnology, 54*(10), 1–10. https://doi.org/10.1080/10826068.2024.2347407

Izydorczyk, G., Skrzypczak, D., Mironiuk, M., Mikula, K., Samoraj, M., Gil, F., Taf, R., Moustakas, K., & Chojnacka, K. (2024). Lignocellulosic biomass fertilizers: production, characterization, and agri-applications. *The Science of the Total Environment, 923* (171343), 171343.

Jakeer, S., Verma, M., Sharma, J., Mattoo, F., Gupta, D., Singh, J., Kumar, M., & Gaur, N. A. (2020). Metagenomic analysis of the fecal microbiome of an adult elephant reveals the diversity of CAZYmes related to lignocellulosic biomass degradation. *Symbiosis, 81*, 209–222. https://doi.org/10.1007/s13199-020-00695-8

Jamaldheen, S. B., Kurade, M. B., Basak, B., Yoo, C. G., Oh, K. K., Jeon, B. H., & Kim, T. H. (2022). A review of physic-chemical delignification as a pretreatment of lignocellulosic biomass for enhanced bioconversion. *Bioresource Technology, 346*, 126591.

Jenkins, N., & Curling, E. M. (1994). Glycosylation of recombinant proteins: Problems and prospects. *Enzyme and Microbial Technology, 16*, 354–364.

Juma, J., Konongoi, S. L., Nsengimana, I., Mwangi, R., Akoko, J., Nyamota, R., Muli, C., Dobi, P. O., Kiritu, E., Osiany, S., Onwong'a, A. A., Gachogo, R. W., Sang, R., Christoffels, A., Roesel, K., Bett, B., & Oyola, S. O. (2023). Using multiplex amplicon PCR technology to efficiently and timely generate Rift Valley fever virus sequence data for genomic surveillance. *Viruses, 15*(2), 477. https://doi.org/10.3390/v15020477

Kane, J. F. (1995). Effects of rare codon clusters on high-level expression of heterologous proteins in *Escherichia coli. Current Opinion in Biotechnology, 6*, 494–500.

Karn, S. K., Bhambri, A., & Rawat, D. (2023). Development of lead (Pb) tolerant strain by protoplast technology and their remediation. *World Journal of Microbiology and Biotechnology, 39*, 27.

Kaul, S., Choudhary, M., Gupta, S., & Dhar, M. K. (2021). Engineering host microbiome for crop improvement and sustainable agriculture. *Frontiers in Microbiology, 12*, 635917.

Keerthana, D. M., Manikandan, S., Oviyapriya, M., Manickam, S., Assiri, M. A., Vickram, S., et al. (2022). Recent advances in biogas production using agro-industrial waste: A comprehensive review outlook of techno-economic analysis. *Bioresource Technology, 363*, 127871.

Kim, H.K., Lee, S., Kim, Y., Park, J., Min, S., Choi, J.W., Huang, T.P., Yoon, S., Liu, D.R., & Kim, H.H. (2020). High-throughput analysis of the activities of xCas9, SpCas9-NG, and SpCas9 at matched and mismatched target sequences in human cells. *Nat Biomedical Engineering, 4*, 111–124.

Kiss, G., Çelebi-Ölçüm, N., Moretti, R., Baker, D., & Houk, K. N. (2013). Computational enzyme design. *Angewandte Chemie, International Edition, 52*(22), 5700–5725.

Komatsu, M., Uchiyama, T., Omura, S., Cane, D. E., & Ikeda, H. (2010). Genome-minimized Streptomyces host for the heterologous expression of secondary metabolism. *Proceedings of the National Academy of Sciences of the United States of America, 107*, 2646–2651.

Kumar, V., Baweja, M., Liu, H., & Shukla, P. (2017). Microbial enzyme engineering: Applications and perspectives. In *Recent advances in applied microbiology* (pp. 259–273). Springer.

Kumar, V., Dangi, A. K., & Shukla, P. (2018). Engineering thermostable microbial xylanases toward its industrial applications. *Molecular Biotechnology, 60*(3), 226–235.

Li, D., Fu, G., Tu, R., Jin, Z., & Zhang, D. (2019). High-efficiency expression and secretion of human FGF21 in Bacillus subtilis by intercalation of a mini-cistron cassette and combinatorial optimization of cell regulatory components. *Microbial Cell Factories, 18*, 17.

Lim, J. J., et al. (2019). Maximizing efficiency of artificial intelligence-driven drug combination optimization through minimal resolution experimental design. *Advances in Therapy, 3*, 1900122. https://doi.org/10.1002/adtp.201900122

Liponska, A., Ousalem, F., Aalberts, D. P., Hunt, J. F., & Boel, G. (2019). The new strategies to overcome challenges in protein production in bacteria. *Microbial Biotechnology, 12*, 44–47.

Liu, C., Yue, Y., Xue, Y., Zhou, C., & Ma, Y. (2023). CRISPR-Cas9 assisted non-homologous end joining genome editing system of Halomonas bluephagenesis for large DNA fragment deletion. *Microbial Cell Factories, 22*(1), 211. https://doi.org/10.1186/s12934-023-02214-y

Luli, G. W., & Strohl, W. R. (1990). Comparison of growth, acetate production, and acetate inhibition of *Escherichia coli* strains in batch and fed-batch fermentations. *Applied and Environmental Microbiology, 56*, 1004–1011.

Meister, S. W., Hendrikse, N. M., & Löfblom, J. (2019). Directed evolution of the 3C protease from coxsackievirus using a novel fluorescence-assisted intracellular method. *Biological Chemistry, 400*, 405–415.

Mohanakrishna, G., & Modestra, J. A. (2023). Value addition through biohydrogen production and integrated process from hydrothermal pretreatment of lignocellulosic biomass. *Bioresource Technology, 369*, 128386. https://doi.org/10.1016/j.biortech.2022.128386

Montané, D., Salvadó, J., Torras, C., & Farriol, X. (2002). High-temperature dilute-acid hydrolysis of olive stones for fufural production. *Biomass and Bioenergy, 22*(4), 295–304.

Moore, C. L., Papa, L. J., & Shoulders, M. D. (2018). A processive protein chimera introduces mutations across defined DNA regions *in vivo*. *Journal of the American Chemical Society, 140*(37), 11560–11564.

Moroz, Y. S., Dunston, T. T., Makhlynets, O. V., Moroz, O. V., & Wu, Y. (2015). New tricks for old proteins: Single mutations in a nonenzymatic protein give rise to various enzymatic activities. *Journal of the American Chemical Society, 137*, 14905–14911.

Nadarajah, K., & Rahman, N. S. N. A. (2023). The microbial connection to sustainable agriculture. *Plants, 12*(12), 2307.

Nallamsetty, S., & Waugh, D. S. (2007). A generic protocol for the expression and purification of recombinant proteins in Escherichia coli using a combinatorial His6-maltose binding protein fusion tag. *Nature Protocols, 2*, 383–391.

Newsom, S., Parameshwaran, H. P., Martin, L., & Rajan, R. (2021). The CRISPR-Cas mechanism for adaptive immunity and alternative bacterial functions fuels diverse biotechnologies. *Frontiers in Cellular and Infection Microbiology, 10*, 619763.

Nirantar, S. R. (2021). Directed evolution methods for enzyme engineering. *Molecules, 26*(18), 5599. PMID: 34577070; PMCID: PMC8470892. https://doi.org/10.3390/molecules26185599

Ojo, A. O. (2023). An overview of lignocellulose and its biotechnological importance in high-value product production. *Fermentation, 9*, 990. https://doi.org/10.3390/fermentation9110990

Omayma, A., Wadalla, W. A., Atawy, M. Y., Bedaiwy, S. S., Ali, Y. A., & Mahmoud, G. (2023). Anaerobic digestion of lignocellulosic waste for enhanced methane production and biogas-digestate utilization. *Industrial Crops and Products, 195*, 116420. https://doi.org/10.1016/j.indcrop.2023.116420

Pachauri, S. (2022). Recombination. In J. Vonk & T. K. Shackelford (Eds.), *Encyclopedia of animal cognition and behavior*. Springer.

Paddon, C. J., Westfall, P. J., Pitera, D. J., Benjamin, K., Fisher, K., McPhee, D., Leavell, M. D., Tai, A., Main, A., Eng, D., et al. (2013). High-level semi-synthetic production of the potent antimalarial artemisinin. *Nature, 496*, 528–532.

Paul, M., Shroti, G. K., Mohapatra, S., Mohapatra, P. K. D., & Thatoi, H. (2024). A comparative study on pretreatment of rice straw and saccharification by commercial and isolated cellulase–xylanase cocktails towards enhanced bioethanol production. *Systems Microbiology and Biomanufacturing, 4*, 731–749. https://doi.org/10.1007/s43393-023-00228-6

Pramasari, D. A., Otaviani, M., Thontowi, A., Purnawan, A., Ermawar, R. A., Sandari, D., Ningrum, R. S., Laksana, R. P. B., Lianawati, A., Fahrezi, M. Z. M., Salsabila, Q., & Hermiati, E. (2023). The use of hemicellulose acid hydrolysate for hydrolysis of sugarcane trash and its fermentation for producing xylitol. *Industrial Crops and Products, 193*, 116163.

Prasad, R., Polaki, S., & Padhi, R. K. (2024). Isolation of delignifying bacteria and optimization of microbial pretreatment of biomass for bioenergy. *Biotechnology Letters, 46*, 183–199. https://doi.org/10.1007/s10529-023-03463-y

Rafeeq, P. H., Afsheen, N., Rafique, S., Arshad, A., Intisar, M., Hussain, A., Bilal, M., & Iqbal, H. M. N. (2023). Genetically engineered microorganisms for environmental remediation. *Chemosphere, 310*, 136751.

Raj, R.A., Murugesan, S., Sarathkumar, D., & Venkatachary, S.K. (2022). A practical method for restoring waste cooking and motor oil as possible biofuel energy. *In 2022 2nd International Conference on Innovative Sustainable Computational Technologies (CISCT)*. IEEE. https://doi.org/10.1109/cisct55310.2022.10046608

Rath, D., Amlinger, L., Rath, A., & Lundgren, M. (2015). The CRISPR-Cas immune system: Biology, mechanisms and applications. *Biochimie, 117*, 119–128.

Raveendran, S., Parameswaran, B., Ummalyma, S. B., Abraham, A., Mathew, A. K., Madhaven, A., et al. (2018). Applications of microbial enzymes in food industry. *Food Technology and Biotechnology, 56*(1), 16–30.

Rawson, T. M., et al. (2019). Microneedle biosensors for real-time, minimally invasive drug monitoring of phenoxymethylpenicillin: A first-in-human evaluation in healthy volunteers. *Lancet Digital Health, 1*, e335–e343.

Raymond, E.A., Mack, K.L., Yoon, J.H., Moroz, O.V., Moroz, Y.S., & Korendovych, I.V. (2015). Design of an allosterically regulated retroaldolase. *Protein Science: A Publication of the Protein Society 24* (4): 561–70.

Reetz, M. T. (2022). Witnessing the birth of directed evolution of stereoselective enzymes as catalysts in organic chemistry. *Advanced Synthesis and Catalysis, 364*, 3326–3335. https://doi.org/10.1002/adsc.2022004466

Reuß, D. R., Altenbuchner, J., Mäder, U., Rath, H., Ischebeck, T., Sappa, P. K., Thürmer, A., Guérin, C., Nicolas, P., Steil, L., Zhu, B., Feussner, I., Klumpp, S., Daniel, R., Commichau, F. M., Völker, U., & Stülke, J. (2017). Large-scale reduction of the Bacillus subtilis genome: Consequences for the transcriptional network, resource allocation, and metabolism. *Genome Research, 27*, 289–299.

Ribeiro, L. F., Amarelle, V., Alves, L. F., Viana de Siqueira, G. M., Lovate, G. L., Borelli, T. C., & Guazzaroni, M. E. (2019). Genetically engineered proteins to improve biomass conversion: New advances and challenges for tailoring biocatalysts. *Molecules, 24*(16), 2879. https://doi.org/10.3390/molecules24162879

Rosenboom, J. G., Langer, R., & Traverso, G. (2022). Bioplastics for a circular economy. *Nature Reviews Materials, 7*, 117–137. https://doi.org/10.1038/s41578-021-00407-8

Saini, D. K., Chakdar, H., Pabbi, S., & Shukla, P. (2020). Enhancing production of microalgal biopigments through metabolic and genetic engineering. *Critical Reviews in Food Science and Nutrition, 60*(3), 391–405.

Salwan, R., Sharma, M., Sharma, A., & Sharma, V. (2023). Insights into plant beneficial microorganism-triggered induced systemic resistance. *Plant Stress, 7*, 100140.

Sardar, S., Caferri, R., Camargo, F. V. A., Capaldi, S., Ghezzi, A., Dall'Osto, L., D'Andrea, C., Cerullo, G., & Bassi, R. (2024). Site-directed mutagenesis of the chlorophyll-binding sites modulates excited-state lifetime and chlorophyll-xanthophyll energy transfer in the monomeric light-harvesting complex CP29. *Journal of Physical Chemistry Letters, 15*(11), 3149–3158. https://doi.org/10.1021/acs.jpclett.3c02900

Sela, I., Wolf, Y. I., & Koonin, E. V. (2016). Theory of prokaryotic genome evolution. *Proceedings of the National Academy of Sciences of the United States of America, 113*, 11399–11407.

Sellés, V. L., Isalan, M., Heap, J. T., & Ledesma Amaro, R. (2023). A primer to directed evolution: Current methodologies and future directions. *RSC Chemistry & Biology, 4*(4), 271–291. https://doi.org/10.1039/d2cb00231k

Sethupathy, S., Morales, G. M., & Li, Y. (2021). Harnessing microbial wealth for lignocellulose biomass valorization through secretomics: A review. *Biotechnology for Biofuels, 14*, 154. https://doi.org/10.1186/s13068-021-02006-9

Sharma, G., Kaur, B., Singh, V., Raheja, Y., Di-Falco, M., Tsang, A., & Chadha, B. S. (2024). Genome and secretome insights: Unraveling the lignocellulolytic potential of *Myceliophthora verrucosa* for enhanced hydrolysis of lignocellulosic biomass. *Archives of Microbiology, 206*, 236. https://doi.org/10.1007/s00203-024-03974-w

Shen, H., Aggarwal, N., Wun, K. S., Lee, Y. S., Hwang, I. Y., & Chang, M. W. (2022). Engineered microbial systems for advanced drug delivery. *Advanced Drug Delivery Reviews, 187*, 114364.

Singh, A. K., Deeba, F., Kumar, M., Kumari, S., Wani, S. A., Paul, T., Naseem, A., & Gaur, N. A. (2023a). Development of engineered *Candida tropicalis* strain for efficient corncob-based xylitol-ethanol biorefinery. *Microbial Cell Factories, 22*, 201.

Singh, P., Rayeen, F., Singh, R., Tiwari, V., Tripathi, M., & Singh, P. K. (2023b). Enhancing agricultural sustainability through microbial-mediated abiotic stress tolerance. *Journal of Ecophysiology and Occupational Health, 23*(4), 1–15.

Singh, P., Singh, R., Singh, S., Chauhan, R. S., Bala, S., Pathak, N., Singh, P. K., & Tripathi, M. (2024). Microbial engineering for greener ecosystem and agriculture: Recent advances and challenges. *Journal of Pure and Applied Microbiology, 18*(2), 797–807.

Sinha, R., & Shukla, P. (2019). Protein engineering for improved health: Technological perspectives. *Current Protein & Peptide Science, 20*(9), 856–860.

Smith, M.D., Zanghellini, A., & Grabs-Röthlisberger, D. (2014). Computational Design of Novel Enzymes without Cofactors. *Methods in Molecular Biology (Clifton, N.J.), 1216*, 197–210.

Snow, A. A., Andow, D. A., Gepts, P., Hallerman, E. M., Power, A., Tiedje, M., & Wolfenbarger, L. L. (2005). Genetically engineered organisms and the environment: Current status and recommendations. *Ecological Applications, 15*(2), 377–404.

Stelzer, C. P., Pichler, M., & Stadler, P. (2023). Genome streamlining and clonal erosion in nutrient-limited environments: A test using genome-size variable populations. *Evolution, 77*, 2378–2391.

Studier, F. W., & Moffatt, B. A. (1986). Use of bacteriophage T7 RNA polymerase to direct selective high-level expression of cloned genes. *Journal of Molecular Biology, 189*, 113–130.

Sundarsingh, J. A., Fuad, A. F., Ranjitha, J., Raghavan, S., & Shankar, V. (2024). Engineering microbes for sustainable biofuel production and extraction of lipids-current research and future perspectives. *Fuel, 355*, 129532.

Sung, B. H., Choe, D., Kim, S. C., & Cho, B. K. (2016). Construction of a minimal genome as a chassis for synthetic biology. *Essays in Biochemistry, 60*, 337–346.

Świątek, K., Gaag, S., Klier, A., Kruse, A., Sauer, J., & Steinbach, D. (2020). Acid hydrolysis of lignocellulosic biomass: Sugars and furfurals formation. *Catalysts, 10*(4), 437. https://doi.org/10.3390/catal10040437

Tizei, P. A. G., Harris, E., Withanage, S., Renders, M., & Inheiro, V. B. (2021). A novel framework for engineering protein loops exploring length and compositional variation. *Scientific Reports, 11*(1), 9134.

Trafton, A. (2019). *Chemical engineers program bacteria to switch between different metabolic pathways, boosting their yield of desirable products*. Massachusetts Institute of Technology News Office.

Varriale, L., Geib, D., & Ulber, R. (2024). Short-term adaptation as a tool to improve bioethanol production using grass press-juice as fermentation medium. *Applied Microbiology and Biotechnology, 108*, 393. https://doi.org/10.1007/s00253-024-13224-0

Velvizhi, G., Goswami, C., Shetti, N. P., Ahmad, E., Pant, K. K., Tejraj, M., & Aminabhavi, T. M. (2022). Valorization of lignocellulosic biomass to value-added products: Paving the pathway towards low-carbon footprint. *Fuel, 313*, 122678.

Vidal, D.F., Silva, C.F., Borges, M.M.C., Nascimento, C.S., & Borges, K.B. (2023). Magnetic and mesoporous molecularly imprinted polymer synthesized by rational computation design: sample preparation and analysis of ractopamine. *Journal of Chromatography. B, Analytical Technologies in the Biomedical and Life Sciences, 1227* (123843), 123843.

Voigt, C. A., Martinez, C., Wang, Z. G., Mayo, S. L., & Arnold, F. H. (2002). Protein building blocks preserved by recombination. *Nature Structural Biology, 9*, 553.

Wang, R. F., & Kushner, S. R. (1991). Construction of versatile low-copy-number vectors for cloning, sequencing, and gene expression in Escherichia coli. *Gene, 100*, 195–199.

Wang, Y., Elzenga, T., & Elsas, J. D. (2021). Effect of culture conditions on the performance of lignocellulose-degrading synthetic microbial consortia. *Applied Microbiology and Biotechnology, 105*, 7981–7995.

Weiwu, G., Wai, L. K., Lixia, P., Jonathan, P., Tingting, Z., Gangqing, H., Yuzhang, W., Jun, Z., Bing, N., & Keji, Z. (2021). Multiplex indexing approach for the detection of DNase I hypersensitive sites in single cells. *Nucleic Acids Research, 49*(10), e56. https://doi.org/10.1093/nar/gkab102

Wenfa, N. (2023). *Development of a MATLAB Error-Prone PCR simulator for random mutation of target gene*. Elsevier. https://doi.org/10.2139/ssrn.4443798

Wickiser, J. K., Kevin, J. O., Washington, M., Hummel, S., & Burpo, F. J. (2020). *Engineered pathogens and unnatural biological weapons: The future threat of synthetic biology*. https://ctc.westpoint.edu/engineered-pathogens-and-unnatural-biological-weapons-the-future-threat-of-synthetic-biology/

Wijma, H. J., Floor, R. J., Jekel, P. A., et al. (2014). Computationally designed libraries for rapid enzyme stabilization. *Protein Engineering, Design & Selection, 27*, 49–58.

Xu, H., Che, X., Ding, Y., Kong, Y., Li, B., & Tian, W. (2019). Effect of crystallinity on pretreatment and enzymatic hydrolysis of lignocellulosic biomass based on multivariate analysis. *Bioresource Technology, 279*, 271–280.

Xu, Y., Dong, F., Wang, R., Ajmal, M., Liu, X., Lin, H., & Chen, H. (2024). Alternative splicing analysis of lignocellulose-degrading enzyme genes and enzyme variants in *Aspergillus*

niger. *Applied Microbiology and Biotechnology, 108*, 302. https://doi.org/10.1007/s00253-024-13137-y
Yadav, R., Singh, P., & Shukla, P. (2018). Metabolic engineering for probiotics and their genome-wide expression profiling. *Current Protein & Peptide Science, 19*(1), 68–74.
Yadav, P. S., Prasad, B. V. S., Chandra, M. S., Maddela, N. R., & Prasad, R. (2024). Leaching approach for β-glucosidase extraction from fermented Rice husk in solid state cultivation by *Aspergillus protuberus*. *Current Microbiology, 81*, 140. https://doi.org/10.1007/s00284-024-03641-9
Yao, Z., Jeon, H. S., Yoo, J. Y., Kang, Y. J., Kim, M. J., Kim, T. J., & Kim, J. H. (2022). DNA shuffling of *aprE* genes to increase fibrinolytic activity and thermostability. *Journal of Microbiology and Biotechnology, 32*(6), 800–807. https://doi.org/10.4014/jmb.2202.02017
Zeng, G., He, S., Li, Y., Sun, D., Li, H., Wen, X., & Wang, J. (2021). Pretreatment technology of lignocellulose. *E3S Web of Conferences, 271*, 04010.
Zhang, X. F., Yang, G. Y., Zhang, Y., Xie, Y., Withers, S. G., & Feng, Y. (2016). A general and efficient strategy for generating the stable enzymes. *Scientific Reports, 6*, 33797.
Zhou, S., Zhang, M., Zhu, L., Zhao, X., Chen, J., Chen, W., & Chang, C. (2023). Hydrolysis of lignocellulose to succinic acid: A review of treatment methods and succinic acid applications. *Biotechnology for Biofuels, 16*, 1. https://doi.org/10.1186/s13068-022-02244-5

Smart Recycling and Sustainable Lignocellulosic Waste Management

Nihar Gupta, W. P. T. Wijenayake, Deblina Roy, Rahul Kumar, Meenakshi Rangot, Priya Chugh, Gustavo Santoyo, Sourav Chattaraj, Sawant Shraddha Bhaskar, Snežana Andjelković, Beatriz Elena Guerra-Sierra, and Debasis Mitra

Abstract

Lignocellulosic waste management and smart recycling strategies are critical components of a sustainable bioeconomy and environmental stewardship. The increasing generation of lignocellulosic waste from agricultural, forestry, and industrial activities poses significant environmental and economic concerns. Conventional disposal methods such as landfilling and incineration contribute to pollution and resource depletion. In response, smart recycling techniques have emerged as viable solutions for converting lignocellulosic waste into valuable resources while minimizing environmental impact. Mechanical, chemical, and biological processes are integral to smart recycling, enabling the conversion of waste biomass into biofuels, biochemicals, biopolymers, and other high-value products. Advanced technologies such as pyrolysis, hydrothermal processing, enzymatic hydrolysis, and fermentation offer efficient pathways for lignocellulosic waste valorization, contributing to energy security and reducing greenhouse

N. Gupta
Department of Soil Science, Punjab Agricultural University, Ludhiana, Punjab, India

W. P. T. Wijenayake
Department of Agricultural Biology, Faculty of Agriculture, University of Peradeniya, Peradeniya, Sri Lanka

D. Roy
Department of Agronomy, Bidhan Chandra Krishi Viswavidyalaya, Nadia, West Bengal, India

R. Kumar · D. Mitra (✉)
Department of Microbiology Graphic Era (Deemed to be University), Dehradun, Uttarakhand, India
e-mail: debasismitra.mb@geu.ac.in; debasismitra.mb@geu.ac.in

G. Mukherjee, S. Dhiman (eds.), *Value Addition and Utilization of Lignocellulosic Biomass*, https://doi.org/10.1007/978-981-96-2786-8_9

gas emissions. The integration of circular economic principles into lignocellulosic waste management promotes material reuse, recycling, and resource efficiency. This approach emphasizes the transformation of waste streams into secondary raw materials, thereby reducing dependence on finite resources and mitigating environmental degradation. Case studies and real-world applications have underscored the feasibility and scalability of smart recycling techniques across diverse waste streams and regions. However, challenges persist in terms of technological maturity, economic viability, and regulatory frameworks, necessitating interdisciplinary collaboration and policy interventions to facilitate its widespread adoption and implementation. Hence, keeping in view the necessity of the synthesis of innovative technologies, sustainable practices, and policy initiatives is imperative for advancing lignocellulosic waste management.

Keywords

Waste utilization · Recycling techniques · Bioenergy production · Sustainable practices · Biomass conversion · Waste to energy

1 Introduction

A significant portion of woodlots, crop residues, animal manure, municipal solid trash, forest leftovers, and crops grown specifically for energy production are composed of lignocellulosic materials (Sims, 2003). Lignocelluloses consist of various inorganic components, including cellulose, hemicellulose, lignin, and extractives

M. Rangot
Basic Science (Microbiology), Dr. Yashwant Singh Parmar University of Horticulture & Forestry, Nauni, Himachal Pradesh, India

P. Chugh
School of Agriculture, Graphic Era Hill University, Dehradun, Uttarakhand, India

G. Santoyo
Instituto de Investigaciones Químico Biológicas, Universidad Michoacana de San Nicolás de Hidalgo, Ciudad Universitaria, Morelia, Michoacán, Mexico
e-mail: gustavo.santoyo@umich.mx

S. Chattaraj
Centre for Industrial Biotechnology Research, School of Pharmaceutical Sciences, Siksha 'O' Anusandhan, Deemed to be University, Bhubaneswar, Odisha, India

S. S. Bhaskar
Department of Plant Pathology, Bihar Agricultural University, Bhagalpur, Bihar, India

S. Andjelković
Institute for Forage Crops, Kruševac, Serbia

B. E. Guerra-Sierra
Grupo Microbiota, Facultad de Ciencias Naturales, Universidad de Santander, Bucaramanga, Colombia
e-mail: bguerra@udes.edu.co

(Sjöström, 1993). Cellulose chains are joined by hydrogen bonds to create elementary microfibrils (Ha et al., 1998). These fibrils are covered with lignin and joined to one another by hemicelluloses, which are amorphous polymers of various sugars and other polymers such as pectin. Microfibrils are frequently linked to the formation of bundles or macrofibrils (Delmer & Amor, 1995; Morohoshi, 1991). Cellulose is a unique and complex material resistant to biological and pharmacological interventions. It can be found as lignocelluloses in waste streams, partially purified as pure cellulose like cotton, or mixed with wastes from citrus fruits. Hemicelluloses found in softwood and hardwood contain mannose and xylose (Persson et al., 2006; Lavarack et al., 2002; Emmel et al., 2003). Hemicelluloses, consisting of galactose, glucose, and arabinose, are more easily hydrolyzed by acids than cellulose, which has a more crystalline structure. They are viable substitutes for cellulose because of their random branching structures and low susceptibility to hydrolysis (Sjöström, 1993; Tables 1 and 2).

Lignin, a complex molecule composed of phenyl propane, is the most resistant component of plant cell walls, with more lignin resulting in greater resistance to enzymatic and chemical breakdowns, making biodegradation particularly challenging. Softwood often contains more lignin than hardwood and agricultural waste. Lignin, hemicellulose, and cellulose are chemically bound to each other (Taherzadeh, 1999; Palmqvist & Hahn-Hägerdal, 2000). Lignocellulosic biomass, also known as lignocellulose, is a renewable and abundant raw material with fermentable components that makes it an ideal feedstock for bioethanol production. They can be divided into three main categories: waste biomass, energy crops, and virgin biomass. Virgin biomass is made up of naturally growing plants; energy crops are grown to produce high-yield lignocellulosic biomass, and waste biomass is a low-value byproduct of

Table 1 Common lignocellulosic materials (cellulose, hemicellulose, and lignin components)

Percentage of lignin (%)	Percentage of cellulose (%)	Percentage of hemicellulose (%)	Lignocellulosic materials
3.55	18.4	49.2	Water-hyacinth
5–10	60–70	10–20	Chemical pulps waste papers
12	45	31.4	Grass switch
20	60	20	Sorted refuse
2.7–5.7	1.6–4.7	1.4–3.3	Cattle manure (solid)
25–35	45–50	25–35	Steams softwood
0–15	85–99	0	Paper
30–40	25–30	25–30	Shells nut
18–30	40–55	25–40	Newspaper
0	15–20	80–85	Leaves
18–25	40–55	24–40	Steams hardwoods
10–3	25–40	35–50	Grasses
0	85–95	5–20	Hair seed (cotton)
15	45	35	Corn cobs
6.4	25	35.7	Coastal Bermuda grass

Table 2 The major chemical composition of most ordinary agro-industrial waste

Composition (% dry wt)	Barley straw	Soybean straws	Sugarcane bagasse	Maize stalks	Wheat straw	Rice husk	Rice straw
Lignin	14.50 ± 0.50	9.50 ± 4.50	17.65 ± 7.65	12.70 ± 5.70	12.60 ± 7.38	16.00 ± 8.00	17.00 ± 9.00
Cellulose	32.50 ± 1.50	63.50 ± 19.50	43.60 ± 11.60	37.30 ± 2.30	41.45 ± 8.55	37.50 ± 7.50	46.60 ± 10.40
Hemicellulose	26.50 ± 2.50	22.43	27.15 ± 17.15	25.90 ± 9.10	25.25 ± 10.25	22.00 ± 3.00	26.00 ± 7.00

Source: Blasi et al. (2023)

several industrial processes (Zabed et al., 2017; Arefin et al., 2021; Ge et al., 2011). Every category of lignocellulose source can function as an input for bioethanol manufacturing. The usefulness of biomass, plant species, and sections is determined by their polysaccharide contents, with lignocellulosic biomass being the most abundant feedstock for the manufacture of ethanol worldwide. It has many benefits, including affordability, ease of accessibility, non-interference with the production of food and feed, the requirement of no additional land, and a steady and consistent supply. Furthermore, its use to produce bioethanol reduces the issue of managing waste biomass and aligns with a zero-waste, sustainable, and ecologically friendly circular economy (Chen et al., 2021).

Most plant cell walls consist of polysaccharides such as cellulose, hemicelluloses, and lignocellulose, which are complex aromatic polymers made of lignin (Table 1). The concentration of these polymers varies depending on the plant species and parts; however, typically two-thirds of the total dry mass is composed of cellulose and hemicellulose (Zhang et al., 2021). The primary structural polysaccharide that provides the cell walls of plants with remarkable tensile strength and stiffness is cellulose. Its percentage typically falls between 30% and 50% of the dry weight of the lignocellulosic biomass. Cellulose, a highly crystalline polymer with high molecular weight, is a long-chain, straight, linear homopolysaccharide composed of cellobiose, a disaccharide, and β-D-glucopyranose units linked by β-1,4-glycosidic linkages, exhibiting a high tendency to form highly crystalline fibers (Heinze, 2016).

The formation of a fibrous state occurs because of the formation of a network of hydrogen bonds by the reactive hydroxyl groups at the C6, C2, and C3 glucopyranose units. Van der Waals interactions stabilize the individual cellulose chains arranged in microfibrils in this structure, which also induces a crystalline structure. In cellulose, there are patches of disorganized amorphous material between the highly structured crystalline sections. It is resistant to depolymerization and is insoluble in water owing to its explicit cellulose structure. The lignocellulosic dry mass of the plant cell wall primarily consists of hemicelluloses, complex polysaccharides with an average molecular weight of 30,000 Da containing α-L-rhamnose and α-L-fructose, and various sugar units such as pentoses, uronic acids, and hexoses. The most prevalent heteropolymers in hemicelluloses are xylan, glucomannan, glucuronoxylan, arabinoxylan, and xyloglucan. Hemicelluloses have an amorphous or partially crystalline structure. The cellulose and hemicellulose network, formed through hydrogen bonds and van der Waals interactions, supports and strengthens the cell wall, enabling it to bear weight (Abdel-Hamid et al., 2013).

All vascular plants contain lignin, a cross-linked, amorphous, and hyperbranched polymer that constitutes the third major component of lignocellulosic biomass. The product is formed through the enzymatically driven free-radical polymerization of three distinct phenylpropane units or monolignols: p-coumaryl, coniferyl, and sinapyl alcohols. Alcohols combine to form lignin, p-hydroxyphenyl, guaiacyl, and syringyl, with plant, cell, and cell wall structures affecting the monomer ratios. By acting as a cement between the fibers, lignin preserves the mechanical properties of the cell wall and prevents microorganisms from destroying structural polysaccharides. It is covalently bound to hemicellulose and a portion of cellulose (Sheng et al., 2021). The primary structural polymers of the cell walls have distinct chemical structures and, consequently, varied behaviors. The composite matrix, formed through covalent cross-linkages and non-covalent forces, such as hydrogen bonds and van der Waals interactions, is highly resilient and organized. Lignin is complicated, extremely refractory, and difficult to separate and depolymerize because of all these factors, as well as the crystalline form of cellulose and the minimal surface area available for enzymes. The conversion of lignocellulosic biomass into bioethanol is a complex process that requires expertise and technologies that combine heat, chemicals, enzymes, and microbes (Haghighi Mood et al., 2013). Bioethanol, derived from lignocellulose structural polymers such as cellulose and hemicelluloses, requires pretreatment to overcome recalcitrance in lignocellulosic biomass for optimal efficiency and profitability. Enzymatic hydrolysis and the conversion of lignocellulosic substrates to ethanol, cellulose, and hemicelluloses must be released from a complex matrix (Haghighi Mood et al., 2013; Table 1).

2 Environmental Challenges Associated with Lignocellulosic Waste

This study explored the effects of lignocellulosic waste on the environment by referencing important research that emphasizes the degree of air pollution caused by burning agricultural residues, the negative health effects of fine particulate matter, and the role of burning biomass in increasing greenhouse gas emissions. Researchers have found that particulate matter ($PM_{2.5}$), carbon monoxide (CO), methane (CH_4), and nitrous oxide (N_2O) are released into the atmosphere and cause respiratory and cardiovascular issues. These emissions have contributed to climate change and global warming. Methane and nitrous oxide emissions are particularly harmful because of their potential to cause global warming. Burning releases volatile organic compounds (VOCs) that can combine with nitrogen oxides to create ground-level ozone, which can worsen asthma, impair lung function, and increase the risk of respiratory infection. To mitigate the negative effects of open-field burning, this study suggests implementing sustainable agricultural methods, such as adding rice straw to the soil or using it as feedstock for bioenergy production (Gadde et al., 2009). Pope and Dockery's analysis of PM2.5 health effects highlights the significant correlation between exposure to small particles and various health outcomes. $PM_{2.5}$, which has a diameter of 2.5 micrometers or less, can penetrate the bloodstream and cause significant lung damage (Pope & Dockery, 2006). This review provides an overview of toxicological and epidemiological studies that show that exposure to $PM_{2.5}$ has negative health impacts. These consequences include decreased lung function, cardiovascular and respiratory disorders, and a high mortality rate. The authors emphasized that exposure to fine particulate matter, whether short-term (acute) or long-term (chronic), can have detrimental effects on health.

Long-term exposure to $PM_{2.5}$ has been linked to lung cancer and other chronic respiratory diseases, such as Chronic obstructive pulmonary disease (COPD). The study by Pope & Dockery focuses on studies showing how long-term exposure to fine particulate matter accelerates lung function degradation in adults and retards lung function development in children. Chronic exposure also increases the risk of heart failure, atherosclerosis, hypertension, and other cardiovascular disorders (Pope & Dockery, 2006).

The methods by which $PM_{2.5}$ causes harm are also highlighted in this review. Toxic materials, including organic chemicals and heavy metals, are carried by fine particles deep in the respiratory system, resulting in oxidative stress and inflammation. This inflammation may have systemic repercussions that affect the heart and other organs. Pope and Dockery concluded that it is critical for public health to lower $PM_{2.5}$ levels in the air. Public awareness efforts and regulatory actions to restrict emissions from transportation, industrial sources, and agricultural activities can help reduce health hazards related to fine particle air pollution (Pope & Dockery, 2006). Andreae and Merlet offered a thorough summary of aerosol and trace gas emissions from the burning of biomass, including lignocellulosic waste. Burning biomass significantly contributes to air pollution and greenhouse gas emissions, impacting climate change and altering atmospheric composition (Andreae & Merlet, 2001). This study assessed the emissions of various trace gases, including methane (CH_4), carbon dioxide (CO_2), nitrous oxide (N_2O), carbon monoxide (CO), and volatile organic compounds (VOCs). As the authors emphasize, biomass burning is a major source of CO_2, the primary greenhouse gas responsible for global warming. A significant portion of the global carbon budget is released during the burning of biomass. Despite being released in lesser amounts than CO_2, methane and nitrous oxide have a far greater potential to cause global warming. Despite their lower quantities, methane, whose global warming potential (GWP) is approximately 25 times greater than that of CO_2 over a 100-year period, and nitrous oxide, whose GWP is approximately 298 times greater than that of CO_2, make major contributions to climate change (Andreae & Merlet, 2001). This study also covered the emission parameters associated with various biomass sources such as grasslands, forest biomass, and agricultural leftovers. The type of biomass, circumstances surrounding combustion, and location of combustion affect these variables. For example, owing to differences in moisture content, combustion efficiency, and chemical composition, emissions from burning agricultural residues such as rice and wheat straw differ from those from burning biomass from forests (Andreae & Merlet, 2001).

Burning biomass releases large amounts of volatile organic compounds and carbon monoxide in addition to greenhouse gases. Volatile organic compounds (VOCs) participate in atmospheric chemical reactions that generate ground-level ozone and secondary organic aerosols. Carbon monoxide (CO) is a byproduct of incomplete combustion and presents direct health risks. Burning biomass releases other aerosols, such as organic carbon and black carbon. By collecting and dispersing sunlight, these particles can alter the Earth's radiation balance and consequently, climate patterns. Because black carbon can absorb sunlight and lower the albedo (reflectivity) of surfaces such as snow and ice, it has a particularly powerful warming effect (Andreae & Merlet, 2001; Figs. 1 and 2).

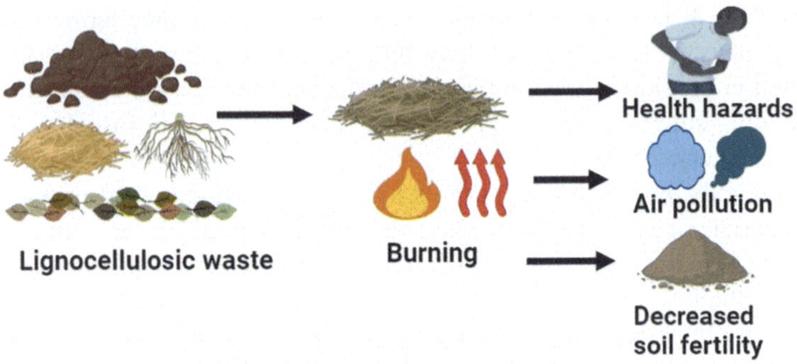

Fig. 1 Risks associated with the burning of lignocellulosic wastes

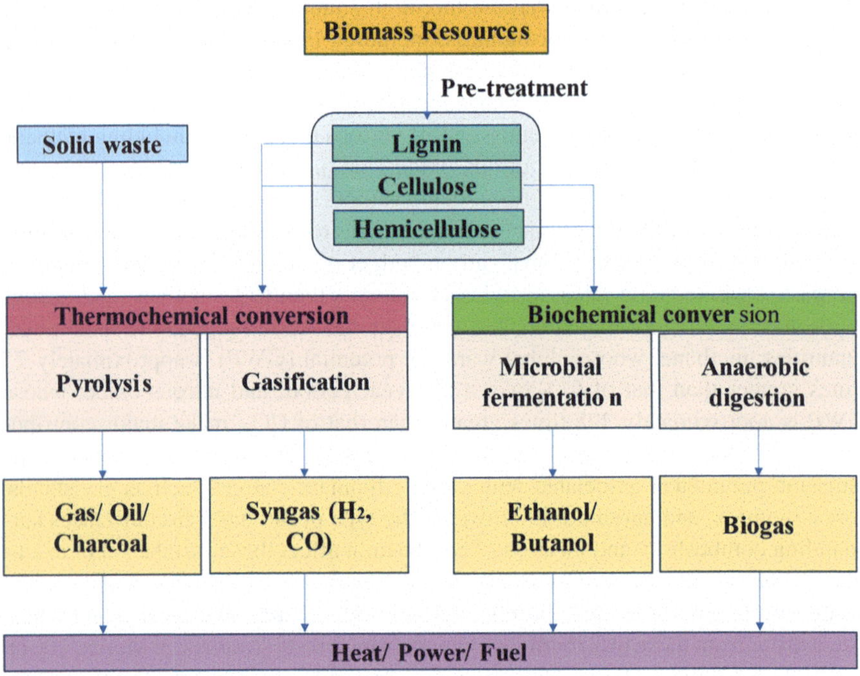

Fig. 2 Biomass Conversion Pathways: Schematic representation of thermochemical and biochemical processes for converting biomass into heat, power, and fuel

3 Smart Recycling Techniques

Better waste management helps prevent negative local effects, such as littering, local water and air pollution, landscape degradation from landfilling, and greenhouse gas emissions. Therefore, recycling is the only option to create sustainable ecosystems. Strategies for reuse and recycling that extend useful life offer long-term

Fig. 3 Smart recycling techniques of lignocellulosic wastes

benefits (Perkumienė et al., 2023). Since 2015, the EU has implemented rules centered on its recycling goals. The EU's average municipal garbage recycling rate was 25.10% of all waste generated in 2000 and 46.40% in 2017 (Eurostat, 2019). Waste management has been revolutionized by clever recycling using mechanical processes, which increases its economic viability, sustainability, and efficiency. The complexities and benefits of the mechanical processes in smart recycling were examined in this study, which offers a thorough overview of the most recent developments and their effects (Fig. 3).

4 Mechanical Processes

Automated optical sorting devices are a significant advancement in the recycling sector and use advanced cameras and sensors to identify and separate materials based on their color, shape, and composition. These devices can distinguish between different plastic materials, paper grades, and metals, such as steel and aluminum, from nonmetallic garbage. Automated optical sorting ensures improved purity levels of sorted recyclables, which directly affects the quality and value of the recycled materials. Contaminated or incorrectly sorted recyclables can result in inferior goods or unusable batches, thereby increasing the waste and costs. Automated optical sorting expedites the sorting process and increases the precision, thereby reducing the possibility of contamination. This supports the sustainability of recycling operations by reducing the need for manual sorting and improving efficiency and material purity. Automation enhances worker safety and increases the throughput. Recycling facilities often incorporate automated optical sorting systems as part of a comprehensive waste management plan, offering an effective and efficient recycling process when combined with other sorting technologies (Hossain & Poon, 2018).

Magnetic separation is a crucial industrial method for separating ferrous and non-ferrous metals from mixed waste streams by using strong magnets to draw iron and steel based on their magnetic properties. This method is widely used in recycling facilities owing to its efficiency and low cost. It improves recycling quality, reduces waste processing, and protects waste management equipment from damage from ferrous materials. Jody et al. (2011) asserted that magnetic separation is an essential technique in the recycling sector, owing to its cost-effectiveness and efficiency.

Modern recycling operations include eddy current separators, which are sophisticated pieces of equipment that separate non-ferrous metals, such as copper and aluminum, from other materials. These separators use electric current to create a magnetic field, pushing the metals away from the rest of the material stream. This repulsive force efficiently separates the metals from other materials. This method is particularly beneficial for recycling metals from complex waste streams where conventional techniques may not be effective. Eddy current separators are highly regarded for their precision in extracting valuable non-ferrous metals and handling a wide range of materials, thereby increasing the efficiency and profitability of recycling operations (Wang et al., 2013). Recycling materials are thoroughly cleaned using washing systems, which remove impurities such as adhesives, labels, and food stains, resulting in higher quality recycled goods and simpler processing, requiring pre-washing, main washing, and rinsing steps (Cucchiella et al., 2017). Paper and plastics are examples of light materials that are separated from heavy materials, such as glass and metals, using the airflow-in-air classification. This procedure improves the quality of the separated elements and is particularly helpful for mixed waste streams. Pires, Martinho, and Chang (2011) highlight the importance of air classification in improving how effective and efficient recycling operations are. Recycling often begins with shredding, which cuts waste into smaller, easiertohandle pieces. This step simplifies sorting and processing. Metals, polymers, and electronic debris are only a few items that can be handled using sophisticated shredders. Additionally, shredding helps homogenize materials, which improves the efficiency of later procedures (Jody et al., 2011). Granulation is the process of disintegrating shredded materials into small, uniform particles. This procedure is required for the preparation of materials, particularly plastics, for further processing as new goods. Consistent quality and ease of handling of granulated materials are essential for producing high-grade recycled goods (Cucchiella et al., 2017). Extrusion is a popular technique for repurposing plastic. This entails the shaping of melted plastic materials from shredded and granulated plastic into new products, including sheets, films, and pipelines. Modern extruders are capable of handling a variety of polymers and adding additives to improve the quality of the finished products. According to Hossain and Poon (2018), extrusion broadens the application range of recycled plastics while simultaneously increasing their value. The blends of several plastic materials and additives were combined to create a homogenous mixture.

5 Chemical Processes

For chemical recycling of plastic waste, polymer chains must first be broken down into smaller components such as methane, carbon dioxide, carbon monoxide, and hydrogen. After further processing, these chemicals are converted into monomers. Since chemical recycling provides additional options beyond physical and mechanical recycling (e.g., dissolution), it plays a crucial role in linking the petrochemical sector with waste management. The two types of chemical recycling are thermochemical recycling (pyrolysis and gasification) and chemical depolymerization (solvolysis). Dissolution is a solvent-based purification method that falls under advanced recycling rather than traditional mechanical recycling. Because this technology is considerably more closely related to chemical recycling, it has also been briefly explored. Chemical recycling should be considered in addition to mechanical and physical recycling because it provides more options (dissolution). It connects the petrochemical and waste management domains. Depolymerization, also known as chemolysis or solvolysis, breaks down polymers into oligomers or monomers using various combinations of heat, solvents, and chemistry. Potential pollutants are separated from the monomers and oligomers in the next phase to eliminate them. Subsequently, monomers and oligomers are added as secondary raw materials for regular plastic production operations. The grade of the polymers created in this manner is comparable to that of plastics made using conventional fossil fuels. This technique works best with homogeneous waste and polymers polymerized via polycondensation, such as polyesters (particularly PET) and polyamides (Vollmer et al., 2020).

Pyrolysis is a thermal decomposition process that occurs in the absence of oxygen, leading to the production of fuel-type products (such as pyrolysis oil) and valuable short-chain olefins like ethylene and propylene. The final product is affected by process variables such as temperature, pressure, and residence time. One of the most crucial factors is temperature because it affects the composition of the final product and regulates the breaking reaction of the polymer chain. Temperatures above 500 °C produce more gaseous or char products, whereas temperatures between 300 and 500 °C generate oil in liquid form. Hydrocarbon mixtures can be used in chemical reactions to create new polymers by replacing feedstocks such as naphtha or natural gas (Vollmer et al., 2020). Fuels, such as kerosene or diesel, can also be substituted with pyrolysis oil (hydrocarbon mix). Pyrolysis is a comparatively easy technology that works well for recycling plastic waste, which is challenging to depolymerize. Because the process occurs without oxygen, it is appropriate for polymers with a low oxygen content, such as polystyrene (PS) and polyolefins. Pyrolysis is used to remove high-oxygen-content such as polyethylene terephthalate (PET) or halogen-containing materials from waste streams. The Agilyx technique, an exception to this, utilizes a pre-reactor to heat polyvinyl chloride (PVC) and release hydrogen chloride (HCl). However, it faces challenges such as high energy consumption, risk of feedstock contamination, lower-quality recycled materials that require upgrading, and difficulties in scaling up to a 100-kilotonne capacity (Vollmer et al., 2020). Gasification is another method of thermal cracking.

Nevertheless, this is not the same as pyrolysis, because it is carried out with a restricted but regulated quantity of oxygen. Hydrocarbons and syngas (CO and H_2) mixtures can produce byproducts, such as tar, char, energy, hydrogen, and various chemicals. In contrast to pyrolysis, it can theoretically absorb polymers that contain halogens or oxygen. It can handle almost any organic material, including biomass and scrap plastics. Gasification requires fewer steps than pyrolysis does. After the waste has been pretreated (including the removal of water), it is gasified and cleaned to remove impurities, such as tar, ammonia, H_2S, and alkali metals (Hrbek, 2015). Researchers are looking for new methods to work at the temperatures between pyrolysis and gasification with a limited oxygen supply. Waste plastics can now be more easily converted into monomers or a benzene, toluene, and xylene (BTX) mixture owing to technology. For example, Synova converts polyolefins and biowaste into monomers, whereas BioBTX converts plastic wastes into BTX. Both have a higher output added value than syngas and are capable of handling more polluted waste streams than pyrolysis. Although the Synova demoplant is currently in the financing stage, the BioBTX demoplant is expected to be completed by 2024 (Hrbek, 2015).

6 Biological Process

Biological waste management technologies include composting, incineration, landfilling, anaerobic digestion, disposal, and bioconversion to produce biofuel (Demirbas, 2011; Bhatia et al., 2018). Composting involves the aerobic breakdown of organic wastes, such as fruit and vegetable scraps, into nutrient-rich soil that can be used as a fertilizer for plants. (Ayilara et al., 2020). The waste material is burned during incineration. A landfill is a structure that resembles a container used to dispose of rubber. Using a process called anaerobic digestion, which is carried out in the absence of oxygen, organic material can be immediately converted into biogas. Efficient handling of biowaste is crucial for environmental preservation and the improvement of living standards. Education and awareness of global waste management are essential for effective resource management and pollution control. Young people should be introduced to the concept of environmental sustainability to help reduce waste generation and disposal, ensuring that waste production does not exceed waste management capacity. Proper disposal is crucial for managing large volumes of biological waste in an energy-efficient and environmentally friendly manner. For waste to be disposed of properly, a thorough examination of its properties from the point of origin is required, including the raw materials used in their production, the manufacturing process, their hazardous nature, possible product repurposing, and the overall proper disposal of wastes (Demirbas, 2011). A comprehensive waste management system is crucial for efficient biological waste handling, focusing on the three Rs: reduction, reuse, and recycling. This system also involves converting garbage to energy, landfills, and gas to energy, and disposing of liquid and solid waste in landfills. Landfill disposal is the most cost-effective waste management (Singh et al., 2011). Landfills are more likely to be completely enclosed, air- and water-tight storage facilities, where organic waste is progressively broken down by decaying bacteria and other microorganisms (Kan, 2009). Approximately 16% of all municipal solid waste (MSW) was thought to be burned in 2016, with the remainder being disposed of in landfills (Demirbas et al., 2016). The decomposition of organic waste in landfills occurs in several stages, each characterized by the synthesis and consumption of specific metabolic products, along with fluctuations in the population of decomposing bacteria. In the first stage, aerobic bacteria consume oxygen present in the waste, a process that typically lasts up to one week. The second stage, known as the anaerobic acid stage, involves a large population of fermentative and hydrolytic bacteria. These bacteria break down polymers such as cellulose, proteins, lipids, and hemicellulose into simple monomers like glycerol, amino acids, and soluble sugars through hydrolysis. The metabolic byproducts generated during these stages serve as essential precursors for biogas and biofuel production (Demirbas, 2011; Bhatia et al., 2018). Landfill locations and disposal facilities differ according to waste type. Classes 1 and 2 deal with waste in the soil, Class 2 with waste from construction and renovation, Class 3 with MSW, Class 4 with commercial and industrial waste, and Classes 5 and 6 with the underground disposal of hazardous waste, which includes, among other things, radioactive waste and medical waste (Kumar et al., 2011; Nanda & Berruti, 2021). Greenhouse gases, such as carbon dioxide (CO_2) and methane (CH_4), are released into the atmosphere when waste decomposes in landfills. These gases pollute the environment when discharged (Idehai & Akujieze, 2015). These gases are created in landfills during aerobic and anaerobic degradation of MSW through chemical, biological, and thermal interactions. Methane (CH_4) accounts for approximately 1–2% of total greenhouse gas emissions (Bogner et al., 2011). According to a 2020 assessment by the U.S. Environmental Protection Agency (USEPA), landfills are the third-largest source of methane emissions in the United States. Therefore, if landfill gases contain a higher amount of combustible CH_4 gas, there is a potential for inadvertent explosions and fires at the site (Kumar et al., 2011). Table 3 highlights various techniques for the pretreatment of lignocellulosic biomass.

Table 3 Techniques for pre-treating lignocellulose biomass

Biological	Physiochemical	Chemical	Electrical	Physical
Living microorganisms enzymes	Explosion of ammonia explosion Ammonia recycle percolation Supercritical fluids Explosion of steam Heated liquid Carbon dioxide explosion	DESs CELF Wet oxidation Ozonolysis Organosolv Hydrolysis Alkaline Acid hydrolysis	PEF	Extrusion Ultrasound Microwave Milling

Source: Blasi et al. (2023)

Note: *PEF*-pulsed electric field, *AFEX*-ammonia fiber explosion, *ARP*-ammonia recycle percolation, *DESs*-deep eutectic solvents, *SCFs*-supercritical fluids, *CELF*-co-solvent-enhanced lignocellulosic fractionation

7 Valorization of Lignocellulosic Waste

7.1 Biofuel Production

For the purpose of producing sustainable and renewable energy, lignocellulosic biomass, which is mainly made up of hemicellulose, cellulose, and lignin, is a readily available and accessible feedstock that can be effectively converted to biomethane by anaerobic digestion. Through the biological process of anaerobic digestion, which occurs in the absence of oxygen, microbes break down complex organic matter to produce biogas, which is largely composed of carbon dioxide (CO_2) and methane (CH_4). Lignocellulosic biomass requires effective pretreatment techniques, and methods such as steam explosion, alkali treatments, and enzymatic hydrolysis have been developed to enhance its component breakdown (Nanda & Berruti, 2021). Buitron et al. studied the waste generated from hydrothermal pretreatment (HPT). This waste contains a lot of organic material, which can be used to produce methane through anaerobic digestion. Producing methane this way is both eco-friendly and cost-effective. This study assessed the waste materials generated by HPT lignocellulose pretreatment in terms of content, structure, hydrolysis rate, and theoretical biochemical methane potential (BMP). Despite the presence of compounds that limit crop growth, such as furan and lignin derivatives, sugarcane exhibited the highest BMP and degradability of any crop owing to its high organic matter concentration per mass of biomass. Furfural, however, has an inhibitory action at lower concentrations that turns stimulating at higher concentrations; as a result, adding furfural maximizes the generation of biomethane (Idehai & Akujieze, 2015). Differences in furfural concentration of steam-exploded hydrolysates caused varying levels of AD inhibition. A study by Mwene-Mbeja et al. to optimize the production of biomethane and biofertilizer by enhancing enzymatic reactions that convert organic matter—such as lipids, carbohydrates, and proteins—while improving the efficiency of biomethanation in industrial settings. However, the presence of furfural was found to inhibit methane generation. The study examined the roles of various enzyme types involved in key anaerobic digestion processes, including hydrolysis, acidification, acetate synthesis, and methane production. Certain enzymes degrade proteins, carbohydrates, and lipids under anaerobic conditions. These intermediates are then used as substrates to produce fertilizers and methane (Bogner et al., 2011). Co-digestion techniques using lignocellulosic biomass as a co-feed with additional organic substrates, such as food waste, energy crops, and animal dung, have been studied (Sannigrahi et al., 2010). Co-digestion improves the carbon-to-nitrogen ratio, nutritional balance, and diversity of microbial populations, all of which contribute to more stable and efficient biomethane production. The anaerobic digestion process provides waste treatment and nutrient recycling in addition to the creation of biomethane. A sustainable and circular bioeconomy can be achieved using digestate or digestion waste as nutrient-rich fertilizers (Buitron et al., 2019). Furthermore, using lignocellulosic biomass to produce biomethane helps to reduce greenhouse gas emissions and the need for fossil fuels. Because oilseed crops require a lot of land for cultivation and because energy crops compete with traditional food crops for market share, biodiesel production is limited in many nations (Mwene-Mbeja et al., 2020). Consequently, studies have concentrated on creating biodiesel from waste products such as residual biomass, used oils, and food scraps (Truong et al., 2020). Lignocellulosic biomass is an abundant and easily accessible feedstock that can be effectively utilized in a variety of conversion methods to produce biodiesel. It is composed of cellulose, hemicellulose, and lignin (Mata-Alvarez et al., 2000). Because lignocellulosic biomass has a complicated structure and is refractory, it is necessary to use appropriate pretreatment techniques to increase the accessibility of the biomass's constituent parts and to make the conversion processes easier. Biodiesel can also be produced from lignocellulosic biomass, but not directly through fermentation. Instead, the process involves multiple steps to convert the biomass into lipid-rich feedstock, which is then used to produce biodiesel. Since lignocellulosic biomass (wood, agricultural residues, etc.) primarily contains cellulose, hemicellulose, and lignin, it does not naturally contain the triglycerides (oils/fats) needed for biodiesel production. Therefore, the biomass must first be converted into lipid-rich compounds through microbial lipid accumulation. Certain oleaginous (oil-producing) microorganisms (e.g., *Cryptococcus curvatus*, *Yarrowia lipolytica*, and some filamentous fungi) can convert sugars (from lignocellulosic hydrolysates) into lipids (triglycerides). These microbial lipids can then be extracted and used for biodiesel production through transesterification. Fast pyrolysis and hydrothermal liquefaction can also break down lignocellulosic biomass into bio-oil, which contains fatty acid precursors. These can be upgraded into lipids suitable for biodiesel synthesis. Factors, such as substrate composition, temperature, pH, and enzyme loading, influence biodiesel generation (Aguiar Severo et al., 2019; Viola et al., 2012). Table 4 highlights the role of various microorganisms in fermentation.

Table 4 Microbial cultures employed in the process of fermentation

Condition of culture applied	Description	Microbial strategy	Microbial cultures	Reference
pH 7 fermentation at 50 °C for 24 h	A specific matrix holds two different kinds of microorganisms	Co-implemented culture	*Pichia stipites* and *Zymomonas mobilis*	Wirawan et al., 2020
pH 3.8–7.65	A certain sort of microorganism is trapped in a matrix	Immobilized culture	*Zymomonas mobilis*	Braga et al., 2021
pH 5–6, time of fermentation is 20 h	Growths originating from several microorganisms	Blends of cultures	Four strains of *Zymomonas mobilis* and *Paenibacillus* sp.	He et al., 2009
pH 5.89 fermentation at 28 °C for 3.6 days	Growths from two different types of cells	Co-culture	*Candida shehatae* and *aspergillus Niger*	Wu, 2019
NA	From a single cell, one form of microbe emerged	Pure culture	*Saccharomyces cerevisiae*	Malik et al., 2020

A study investigated the fermentation of lignocellulosic carbohydrates in microbial consortia, including bacteria and archaea, to synthesize biodiesel (Chintagunta et al., 2021). The microorganisms selected, the circumstances of fermentation, and the subsequent phases of processing all have a substantial impact on biodiesel output and quality (Kuhad et al., 2011). Recently, emerging biofuels have garnered considerable interest, one of which is renewable diesel. Sustainable resources, including vegetable and single-cell oils, biodiesel, and biodiesel, are hydrodeoxygenated to create them. Renewable diesel is sometimes referred to as second-generation biodiesel in addition to green diesel. It is distinguished from the traditional biodiesel produced through transesterification by its greater cleanliness, oxidative stability, and cold compatibility (Arantes & Saddler, 2011). The production of jet fuel and hydrogen from lignocellulosic biomass generated from energy crops, forestry leftovers, and agricultural residues has significant potential as a renewable and sustainable energy source. Hydrogen production and extraction using contemporary conversion technologies, such as biochemical and thermochemical reactions, have demonstrated promise. Hydrogen and syngas are produced by thermochemical processes, such as pyrolysis and gasification, while alternate techniques are provided by biological and enzymatic processes. Biological processes, supported by enzymes, break down cellulose and hemicellulose into simple sugars that canbe fermented. This process also produces biohydrogen, a key part of renewable natural gas. Microorganisms are involved in two methods to produce hydrogen: dark fermentation (which occurs without light) and photofermentation (which uses light) (Knothe, 2010).

Lignocellulosic biomass is being explored as a sustainable and renewable feedstock for biokerosene in the aviation industry, despite concerns about food crop usage and high biokerosene selling prices. Biobutanol, a promising biofuel feedstock, can be generated through acetone-butanol-ethanol fermentation. However, obstacles such as poor output, elevated toxicity, and downstream recovery remain (Jara-Cobos et al., 2023). Several processes for producing biobutanol from lignocellulosic biomass have been investigated, including alkali- and acid-pretreated rice straw, *Clostridium sporogenes*, acidogenic fermentation, mild alkali pretreatment, and enzymatic hydrolysis. Future strategies should focus on reducing the processing costs (Kannah et al., 2021). Lignocellulosic biomass, which comprises agricultural and forest wastes, is a renewable resource for the generation of bioenergy. Promising thermochemical conversion methods for economical and environmentally friendly production of biofuels include gasification and pyrolysis. Biological processes focus on converting lignocellulosic waste into biofuels or biogas, rather than directly producing synthesis gas (Das and Veziroğlu, 2001). Pyrolysis is a thermochemical process that uses lignocellulosic biomass to produce charcoal, syngas, and bio-oils. Although biochar is utilized for soil amendment and carbon sequestration, bio-oil is also used in chemicals and transportation fuels. Catalysts and additives have been used to improve selectivity and quality, and many technologies have been developed to maximize bio-oil yield and attributes. Research on issues such as the instability of bio-oil and its high oxygen concentration is still required. The goal of recent developments in pyrolysis and gasification technologies is to enhance environmental performance, product quality, and process efficiency. Advanced reactors and integrated systems are also being investigated to enhance quality and stability (Jara-Cobos et al., 2023).

7.2 Organic Compounds and Biopolymers

Heavy metals, dyes, and pesticides are among the most hazardous pollutants due to their toxicity, persistence in the environment, and potential to bioaccumulate, leading to severe ecological and human health risks. They may exist in different amounts in the environment, such as in soil, water, and atmosphere. There is a strong need to eliminate these hazardous compounds because they accumulate over time and have created several negative consequences for humans, such as mental illnesses, genetic mutations, congenital defects, and a decrease in the amount of oxygen in the environment (Singh & Wahid, 2015). The development of new technologies for the removal of heavy metal contaminants is essential, as eliminating excess heavy metals from the environment is necessary to achieve satisfactory quality standards, despite the considerable challenges involved. Biopolymers, which are polymeric materials produced by living organisms, have gained widespread interest as viable alternatives for sustainable solutions. Their numerous advantages have contributed to their growing popularity (Sikora et al., 2017). The primary reason for this is that they are produced solely from biomass rather than fossil fuels, which helps reduce environmental pollution both on land and at sea by reducing emissions such as carbon dioxide. In addition, natural polymers are used as a sustainable method to increase the mechanical qualities of soil, such as soil strength, which increases grain crop production (Eswaran et al., 2021). They are fully bio-based, made from living things, and are suitable natural alternatives to synthetic and chemical polymers (Eswaran et al., 2021). They are a diverse range of materials that can be categorized into three primary classes, depending on their multiple origins. These materials are recyclable, biodegradable, and decomposable in the environment (Cheng et al., 2012). Starch, cellulose, and chitosan are popular biopolymers found in nature and are used in film production and pharmaceutical and medical applications because of their sustainability, adaptability, and ease of handling, making them vital in various industries (Sannigrahi et al., 2010). Biopolymers derived from lignocellulosic biomass have attracted considerable attention within this field. The primary components of lignocellulose biomass are the complex carbohydrate polymers cellulose and hemicellulose, as well as the phenolic polymer lignin. It is the most prevalent portion of biomass derived from plants in the natural world and offers a variety of applications. For instance, they can be used to replace many conventional polymers or remove harmful pollutants using bio-sorbents (Alengebawy et al., 2021; Gupta et al., 2022). The use of biopolymers provides several benefits. Based on market research on biopolymers, the most promising materials for creating a more sustainable environment are lignocellulose, starch, and chitosan. However, laboratory-scale experiments remain the only available method for real-world application. Although they are some of the most useful and pleasant materials for daily use, the buildup of petrochemical plastics has seriously harmed the habitats of flora and animals, putting human safety at risk (Thuo et al., 2021). One-third of plastic waste contributes to land and marine pollution, as most polymers take hundreds of years to degrade (Lisitsyn et al., 2021). Despite being some of the most practical and comfortable things for daily living, the buildup of petrochemical plastic materials has seriously harmed the environment of flora and animals and endangered human safety (Thuo et al., 2021). Because most polymers take hundreds of years to break down, one-third of the plastic trash results in pollution on land or in the ocean (Russo et al., 2021). The amount of plastic debris that ends up in the ocean annually is millions of tons. "Microplastics" are created when this garbage breaks down. Plankton are bound to consume microplastics, as evidenced by numerous studies. The latter is the foundation of the food chain in the ocean and is the primary means of absorbing atmospheric carbon. The presence of microplastics has a catastrophic impact on the capacity of the oceans to absorb carbon.

Finding a substitute that has less impact on ecology is essential for a society in which pollution is becoming a more serious concern. The industrial sector is becoming increasingly interested in biodegradable plastics, and there is a growing demand for environmentally friendly and cost-effective materials to minimize waste management. The "circular economy" idea arose to lessen environmental harm, and biodegradable plastics have been suggested as a possible means of establishing a more sustainable global community (Machado et al., 2020). A product is considered biodegradable if it breaks down spontaneously because of the enzymatic activity of bacteria and fungi. When this system is automatically triggered, it shuts down without human assistance and prevents environmental contamination. Biodegradable plastics can originate from fossil fuels, such as oil, or renewable sources, such as vegetables and animals. Alternatively, bioplastics are produced from non-fossil fuel resources and have the benefit of being renewable sources because they typically include the use of biomass, a plant-based raw material, instead of fossil fuels (Kumari et al., 2022). The use of non-petroleum-based alternative materials to enhance or replace traditional plastic products is rapidly expanding. Owing to their handy features, easy availability, abundance, and technical viability, the utilization of novel products and biopolymers has become commonplace. Additionally, the quality of bioplastic materials can be strengthened and improved using composite mixtures of various polymers with additional natural ingredients, such as lignocellulose, silica, or maize starch. Lignocellulosic raw materials are the most prevalent type of biomass that is not edible, and approximately 140 gigatons (Gt) of lignocellulosic waste are produced each year, which can pose a severe environmental hazard and lead to numerous illnesses (Abdelhedi et al., 2022; Abdelwahab et al., 2012). In recent decades, scientists have explored biocomposites and biofibers as viable substitutes for synthetic materials. Natural fibers can be categorized into vegetable and animal fibers, with vegetable fibers consisting of lignin, hemicellulose, and cellulose and animal fibers consisting of hair and secretions (Aisyah et al., 2021). Composite materials reinforced with biofibers, particularly lignocellulosic fibers, are increasingly used in bioplastics due to their strength, thermal stability, biodegradability, and bending stiffness, making them valuable natural bioresources (Hahladakis et al., 2018; Fig. 4).

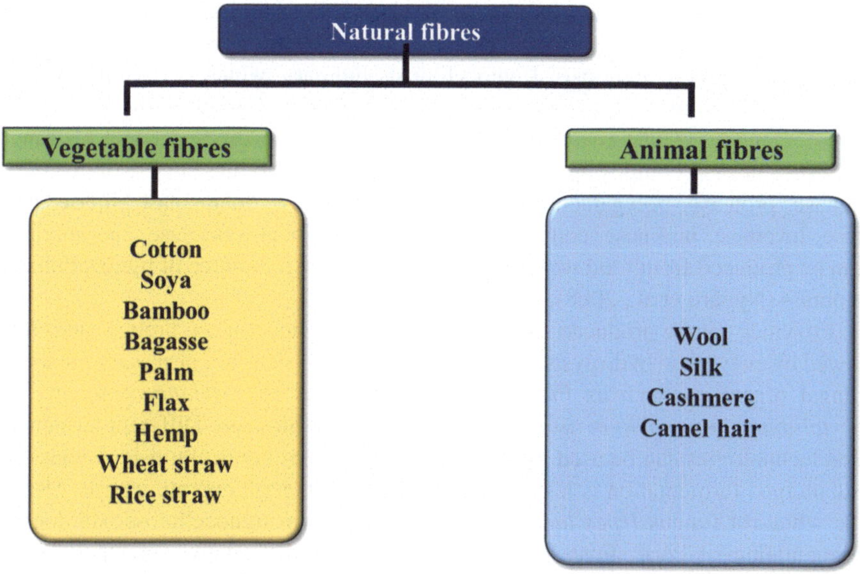

Fig. 4 Classification of biofibers

7.3 Bio-based Products

Fruit and vegetable biomass contain bioactive substances that are beneficial to human health. Phenolics, flavonoids, and carotenoids are some of these compounds. They can be extracted using a variety of techniques, including supercritical fluid, enzyme, microwave, ultrasound, liquid-liquid, solid-liquid, and solvent-based extraction (Alessandroni et al., 2022). Furthermore, lignocellulosic residues can be used to create novel bioactive compounds, pharmaceuticals, and nutraceuticals such as physiologically active phenol derivatives. Bioactive polyphenolic chemicals were recovered from banana crop byproducts using an environmentally friendly method based on aqueous ethanol and optimum operating conditions (Kainthola et al., 2019). Studies have demonstrated that sustainable wheat straws can be used to extract and separate valuable phenolic compounds using pretreatment and fractionation of biomass, examination of the extract using polymeric resins, and purification using supercritical CO_2 extraction. *Aspergillus niger* has been utilized to extract antioxidant chemicals and produce enzymes from various agro-industrial wastes, increasing antioxidant activity by 2.3-fold after 1 day (Batstone et al., 2002). The white-rot fungus *Inonotus obliquus* successfully synthesized bioactive exopolysaccharides using corn straw as lignocellulose, demonstrating the economic and technical feasibility of microbial synthesis from forestry and agricultural waste. This process also leads to the conversion of olive pruning and eucalyptus residues into lactic acid and other useful products, such as ferulic acid, which can be recovered through solid-phase extraction, unlike supercritical fluid extraction (Vinod et al., 2020).

Enzymes, which are natural proteins found in all living systems, are essential for biocatalysis in various reactions. However, their use in industrial processes is limited by high production costs. Lignocellulosic biomass, which is rich in polysaccharides, can be used as a source of enzymes to degrade lignocellulose by various microorganisms. *Obba rivulosa, Gelatoporia subvermispora*, and white-rot fungi are efficient in decomposing lignocellulosic biomass. Enzymes such as transglutaminase, protease, polygalacturonase, phytase, pectinase, mannanase, lipase, laccase, invertase, inulinase, cellulose, beta-xylosidase, beta-glucanase, and amylase can be produced from food waste, which is an inexpensive source of lignocellulosic biomass (Nygård et al., 2008).

Enzymes can be produced from food waste via lignocellulose pretreatment followed by enzymatic hydrolysis. The enzymatic hydrolysis is not necessary for some fungal organisms such as *Pleurotus* sp., *Aspergillus* sp., *Melanocarpus* sp., or *Scytalidium thermophilum* naturally growing on plant biomass. Different fermentation technologies can be used to produce enzymes from lignocellulosic biomasses. Submerged fermentation is less popular because of its high operating costs. Using the white-rot fungus *Irpex lacteus* CD2 to break down lignocellulose, solid-state fermentation of corn stover produced hydrolytic and oxidative enzymes. After 15 days, 80% of the acid-insoluble lignin, 17.2% of the cellulose, and 63% of the hemicellulose were degraded. On day five, overall cellulase activity peaked and continued to increase for 40–60 days. Enzymes must be extracted and refined from the fermentation media during the manufacturing stage. These downstream operations have a significant impact on the total production cost and severely restrict large-scale applications, owing to their low enzyme yield, concentration, and productivity. Microbes that produce enzymes through genetic or metabolic modifications and medium optimization can increase enzyme activity. Furthermore, enzyme recovery is simpler and less expensive when the purification and immobilization of enzymes are performed in a single action (Espro et al., 2021).

One such product with additional value that can be produced by microbial fermentation from a variety of waste materials is single-cell protein. Microbial biomass proteins, also known as single-cell proteins (SCPs), are a class of proteins isolated from single-cell microorganisms, including bacteria, fungi, yeast, and algae, and they are cultured through fermentation using various carbon sources, because the term "single-cell protein" was first used in 1968. SCPs have been studied extensively. SCPs are utilized in human nutrition for emulsification, vitamin delivery, and flavor enhancement. They are also used as animal feeds for fish, pigs, chickens, and calves. SCPs contain not only a high percentage of protein but also minerals, vitamins, enzymes, amino acids, and nucleic acids (da Costa Lopes et al., 2016).

However, because the waste products utilized as raw materials are inexpensive, easily accessible, and environmentally acceptable, they are still reasonably priced compared to other plant and animal sources. Waste lignocellulosic materials can be appropriately processed to convert cellulose into fermentable sugars, thereby allowing the production of SCPs. Biomass from cactus pear plants is a potential raw material for the synthesis of SCP in arid and semi-arid areas (Leite et al., 2021). The synthesis of single-cell proteins (SCPs) from lignocellulosic materials involves four key processes: cellulase production, enzymatic hydrolysis, physical and chemical pretreatment, and the assimilation or fermentation of cellulose and hemicellulose (Teles et al., 2019; Arias et al., 2023).

Lignocellulose is used as a substrate for microbial cell growth through fermentation. Submerged fermentation is the most popular method for producing soy proteins (SCPs). *Yarrowia lipolytica* can convert food waste into SCPs through one-stage and two-stage fermentation, with protein contents of 38.8% w/w and 24.2% w/w, respectively. Wheat bran, a high-carbohydrate agricultural residue, can be fermented with the yeast *Candida utilis* and fungus *Rhizopus oligosporus*, resulting in a maximum crude protein production of 41.02% (Ravindran & Jaiswal, 2016). As the need for antibiotics increases worldwide, lignocellulose provides a dependable supply of carbon for microorganisms to produce secondary metabolites. Antibiotics are biosynthesized when bacterial growth rate falls below a certain threshold during nutrient deprivation. Both solid-state fermentation (SSF) and submerged fermentation (SmF) have been used to produce antibiotics from easily accessible lignocellulosic agricultural residues using various microorganisms. However, SSF offers several advantages over SmF, including lower production costs, higher productivity, and reduced excess liquid waste (Garcia-Galan et al., 2011).

The manufacture of β-lactam antibiotics or cephalosporins via fermentation has been extensively studied. Using barley, Bombay rawa, rice bran, wheat rawa, and wheat bran as substrates, *Acremonium chrysogenum* ATCC 48272 produced cephalosporin C via solid-state fermentation. For wheat rawa, the optimal growth conditions were pH 6.5, an incubation period of 5 days at 30 °C, a salt solution-to-wheat bran ratio of 1.5:10 (v/w), 80% moisture content for the solid substrate, with yeast extract and 1% (w/w) soluble starch as additives. Under these conditions, the maximum antibiotic yield (22,281 μg/g) was obtained using wheat rawa. Additionally, when bran was boiled in distilled water, *Tolypocladium* sp. produced 16 ± 8 mg/100 g of cephalosporin C (Sharif et al., 2021).

8 Circular Economy Principles in Lignocellulosic Waste Management

Climate change and the desire to reduce the dependence on fossil fuels have accelerated the search for renewable energy sources. Countries such as India, the United States, China, Brazil, and the European Union have played a significant role in this process. The EU aims to achieve 32% renewable energy in its Member States' final gross consumption by 2030. Certain industries, such as the transportation system in Europe, rely mostly on fuels generated from petroleum. By 2050, Europeans will want to completely phase out fossil fuels and switch to renewable energy. Brazil's Renovabio Law intends to reduce greenhouse gas emissions by 10% by the year 2028. Because they are renewable, more first-generation biofuels such as ethanol, biodiesel, and biogas are produced. Global biofuel production in 2019 was

159 billion L; the US and Brazil produced 75% and 50% of the ethanol and biodiesel, respectively. By 2020, the EU wanted the transportation sector to use 7% biofuels made from grains, starches, sugars, and oilseeds. This is amid the competition for food production and energy. The EU also plans to use low-carbon biomass (LCB) from agro-waste for 2G biofuels, with the objective of using second-generation biofuels for 25% of its energy needs for transportation (Banu et al., 2021). Lignocellulosic biomass (LCB) is a promising substrate for biofuel production owing to its abundance, low cost, and waste recovery potential. Countries such as Brazil, China, India, and the US generate significant amounts of agricultural and forestry waste. The global market for cellulose fiber is projected to grow from $21.5 billion in 2017 to $41.5 billion by 2025 (Liu et al., 2018). Bioconversion involves the pretreatment, hydrolysis of cellulose and hemicellulose into fermentable sugars, fermentation, and distillation of the fermented broth. LFB can be transformed into biofuels in two ways: thermochemical transformation, which produces energy products such as pyrolysis and gasification, and biological conversion, which produces liquid biofuels (Kapoor et al., 2020).

Ethanol can be produced most effectively by biological conversion, with high selectivity compared to thermochemical conversion. This process involves pretreatment, enzymatic hydrolysis, and sugar fermentation using yeast and bacteria. Biobutanol offers advantages such as reduced volatility, increased energy content, and increased net heat capacity when added to gasoline or biofuels. Grass, corn straw, wheat straw, and sugarcane bagasse are examples of agricultural leftovers that can be used as feedstocks to produce biobutanol. *Clostridium* species are fermenting organisms, but lignocellulosic biomass must be rich in cellulose and hemicellulose with low lignin content. In the transportation sector, biodiesel, a renewable and biodegradable fuel, is gradually replacing petroleum-derived diesel (Koul et al., 2022). Due to the limitations of first-generation biodiesel production, there is now a greater focus on second- and third-generation biodiesel, which is derived from non-edible oils, unconventional plant oils, and microbial oils. Livestock waste (LCB) is an agricultural waste that is used as a feedstock for the production of biodiesel and as a substrate for oleaginous microorganisms (OMs) to create lipids. To facilitate the enzymatic hydrolysis of cellulose, fermentation with OMs, and eventually transesterification into biodiesel, lignin must be removed from the LCB before further processing can begin. Anaerobic digestion can produce gaseous biofuels that generate biomethane for transportation or in natural gas distribution networks. Physical, thermochemical, chemical, oxidative, biological, and hybrid pretreatment methods are promising for enhancing the anaerobic digestion of lignocellulosic biomass (LCB). Biohydrogen, a green biofuel produced through bioconversion, is a clean alternative to fossil fuels because of its high energy content and low greenhouse gas emissions. Primarily derived from corn stalks, pretreatment methods have been developed to enhance the bioconversion process, following the trend of biomethane production (Khan et al., 2022).

9 Case Studies and Applications of Smart Recycling Techniques

The possibilities of Internet of Things (IoT) technologies for garbage collection procedures are examined in this study. By embedding smart devices in glass containers to track the fill levels and streamline the collection process, the inefficiencies of conventional approaches can be overcome. Interconnecting physical items via the Internet for data collection and process optimization is known as the Internet of Things (IoT), which was first proposed by Ashton in 1999. Designed to improve resource efficiency, manufacturing efficiency, and urban living standards, it is a key technology in smart industries and cities. However, issues with energy usage, dependability, and data-transfer effectiveness are frequently encountered by IoT devices. The study highlights that to reduce energy consumption and extend the gadget's lifetime, it is crucial to choose the right wireless data-transfer technology. The study used smart devices in glass containers to measure filling levels, collect data, and optimize waste collection processes. These devices were deployed across multiple locations and collected data over several months by using low-power transmission technology. The collected data were analyzed in an IoT cloud to generate waste collection routes and schedules. The challenges include reliable wireless data transfer, battery life management, and infrastructure limitations. Low-power requirements also pose difficulties for device updates (Kamm et al., 2020). This study revealed that smart waste collection systems can enhance efficiency by transitioning from fixed schedules to data-driven, on-demand collection, reducing costs, lowering the environmental impact, and improving service quality. These insights can be applied to other smart-city and factory projects to serve as models for other sectors to adopt smart technology. The study also underscores the potential of IoT to transform urban infrastructure and improve quality of life (Kamm et al., 2020). The European Green Digital Coalition (EGDC) case study on smart waste management developed by Nokia and the Municipality of Nicosia, Cyprus, examines Nokia's Integrated Operations Centre (IOC) platform for smart waste management in Nicosia. The study, developed by Carbon Trust, assesses the platform's net carbon impact and identifies areas for improvement. The IOC platform is part of a smart-city project aimed at increasing productivity and efficiency through digital operational technologies. Sensoneo Quatro sensors measure the fill levels in the waste bins. These sensors transmit real-time data to the IOC platform via existing IoT networks or GPRS, which are then analyzed and displayed on a dashboard or a smartphone application. Based on this data, waste collection routes are optimized to include bins that are more than 50% full, reducing fuel consumption and emissions from garbage trucks. The study methodology includes first-order effects, which capture the lifecycle emissions of the sensors and related hardware, and second-order effects, which calculate carbon savings from reduced truck idling during optimized collection routes. The results indicate that the smart waste management solution achieves significant carbon savings: 1.96 tCO_2e per year saved for underground bins, −0.04 tCO_2e per year for glass recycling bins, and a total net carbon savings of 1.92 tCO_2e per year. However, this study has several limitations, including the quality and granularity of the data, the exclusion of water consumption for bin washing due to a lack of data, and the specificity of the implementation context. Higher order effects, such as potential economic rebounds from cost savings, were considered but were found to have low impact and likelihood. In conclusion, the EGDC case study on smart waste management highlights the potential of digital solutions such as Nokia's IOC platform to achieve significant carbon savings through optimized waste management. However, continuous improvement and adaptation of the methodology are essential to enhance the accuracy and applicability of these assessments in different contexts (Carbon Trust, 2023).

10 Future Directions and Challenges in Sustainable Lignocellulosic Waste Management

Future directions in sustainable lignocellulosic waste management are geared toward improving the efficiency and scalability of bioconversion processes, developing integrated biorefineries, and leveraging biotechnological innovations. The key to these advancements is the optimization of pretreatment methods, which is essential for increasing the accessibility of cellulose and hemicellulose for enzymatic hydrolysis and thereby improving the yield of fermentable sugars. Research has focused on enhancing pretreatment techniques, such as enzymatic reactions, acid hydrolysis, and steam explosion treatments, to minimize energy and chemical use while maximizing sugar recovery and tailoring these methods to different biomass types for cost reduction. Integrated biorefineries are a promising approach for managing lignocellulosic waste by utilizing every component of biomass, such as lignin, hemicellulose, and cellulose, to produce bio-based products, such as biofuels, chemicals, and materials. Biorefineries can achieve greater economic viability and sustainability by developing processes that can produce multiple products. Advances in catalytic processes and new chemical pathways are crucial for converting lignin to valuable chemicals and materials. Through the development of resilient microbial strains that can effectively convert feedstocks into desirable products in industrial settings, biotechnological advancements have revolutionized the management of lignocellulosic waste. These strains have better yields of chemicals and biofuels, and are tolerant to pretreatment inhibitors. They can also simultaneously use several sugars. Metabolic engineering approaches optimize metabolic pathways to enhance performance and stability.

Despite advancements in lignocellulosic waste management, challenges persist, including economic viability for large-scale operations, the need for cost-effective and environmentally friendly pretreatment technologies, optimization of production processes to compete with fossil fuel-based products, and the formation of inhibitory compounds during pretreatment, which can hinder microbial fermentation. Strategies to minimize or detoxify these inhibitors are crucial for improving the process efficiency. The transition to sustainable bio-based economies requires government funding, tax incentives, and policies promoting bio-based product adoption, along with public-private partnerships and cooperation on technical, financial, and regulatory issues between researchers, industry participants, and legislators. Sustainable lignocellulosic waste management requires a multidisciplinary approach, with researchers developing innovative technologies, industry stakeholders investing in infrastructure, and policymakers supporting regulations. By integrating advanced pretreatment methods, developing biorefineries, and leveraging biotechnological innovations, lignocellulosic waste can be transformed into valuable products, thereby reducing reliance on fossil resources and mitigating environmental impacts (Kumar & Sharma, 2017).

Conflicts of Interest None.

References

Abdel-Hamid, A. M., Solbiati, J. O., & Cann, I. K. O. (2013). Chapter One—Insights into lignin degradation and its potential industrial applications. In S. Sariaslani & G. M. Gadd (Eds.), *Advances in applied microbiology* (Vol. 82, pp. 1–28). Academic Press.

Abdelhedi, O., Salem, A., Nasri, R., Nasri, M., & Jridi, M. (2022). Food applications of bioactive marine gelatin films. *Current Opinion in Food Science, 43*, 206–215.

Abdelwahab, M. A., Flynn, A., Chiou, B. S., Imam, S., Orts, W., & Chiellini, E. (2012). Thermal, mechanical and morphological characterization of plasticized PLA–PHB blends. *Polymer Degradation and Stability, 97*, 1822–1828.

Aguiar Severo, I., Fortes Siqueira, S., Costa Deprá, M., Manzoni Maroneze, M., Queiroz Zepka, L., & Jacob-Lopes, E. (2019). Biodiesel facilities: What can we address to make biorefineries commercially competitive? *Renewable and Sustainable Energy Reviews, 112*, 686–705.

Aisyah, H. A., Paridah, M. T., Sapuan, S. M., Ilyas, R. A., Khalina, A., Nurazzi, N. M., Lee, S. H., & Lee, C. H. (2021). A comprehensive review on advanced sustainable woven natural fibre polymer composites. *Polymers, 13*, 471.

Alengebawy, A., Abdelkhalek, S. T., Qureshi, S. R., & Wang, M. Q. (2021). Heavy metals and pesticides toxicity in agricultural soil and plants: Ecological risks and human health implications. *Toxics, 9*, 42.

Alessandroni, L., Caprioli, G., Faiella, F., Fiorini, D., Galli, R., Huang, X., Marinelli, G., Nzekoue, F., Ricciutelli, M., Scortichini, S., et al. (2022). A shelf-life study for the evaluation of a new biopackaging to preserve the quality of organic chicken meat. *Food Chemistry, 371*, 131134.

Andreae, M. O., & Merlet, P. (2001). Emission of trace gases and aerosols from biomass burning. *Global Biogeochemical Cycles, 15*(4), 955–966. https://doi.org/10.1029/2000GB001382

Arantes, V., & Saddler, J. N. (2011). Cellulose accessibility limits the effectiveness of minimum cellulase loading on the efficient hydrolysis of pretreated lignocellulosic substrates. *Biotechnology for Biofuels, 4*, 3.

Arefin, M. A., Rashid, F., & Islam, A. (2021). A review of biofuel production from floating aquatic plants: An emerging source of bio-renewable energy. *Biofuels, Bioproducts and Biorefining, 15*, 574–591. [CrossRef].

Arias, A., Feijoo, G., & Moreira, M. T. (2023). Process modelling and environmental assessment on the valorization of lignocellulosic waste to antimicrobials. *Food and Bioproducts Processing, 137*, 113–123.

Ayilara, M. S., et al. (2020). Waste management through composting: Challenges and potentials. *Sustainability, 12*(11), 4456.

Banu, J. R., Kavitha, S., Tyagi, V. K., Gunasekaran, M., Karthikeyan, O. P., & Kumar, G. (2021). Lignocellulosic biomass based biorefinery: A successful platform towards circular bioeconomy. *Fuel, 302*, 121086.

Batstone, D. J., Keller, J., Angelidaki, I., Kalyuzhnyi, S. V., Pavlostathis, S. G., Rozzi, A., Sanders, W. T., Siegrist, H., & Vavilin, V. A. (2002). The IWA Anaerobic Digestion Model No 1 (ADM1). *Water Science and Technology, 45*, 65–73.

Bhatia, S. K., et al. (2018). Biowaste-to-bioenergy using biological methods–a mini-review. *Energy Conversion and Management, 177*, 640–660.

Blasi, A., Verardi, A., Lopresto, C. G., Siciliano, S., & Sangiorgio, P. (2023). Lignocellulosic agricultural waste valorization to obtain valuable products: An overview. *Recycling, 8*(4), 61.

Bogner, J. E., Spokas, K. A., & Chanton, J. P. (2011). Seasonal greenhouse gas emissions (methane, carbon dioxide, nitrous oxide) from engineered landfills: Daily, intermediate, and final California cover soils. *Journal of Environmental Quality, 40*(3), 1010–1020.

Braga, A., Gomes, D., Rainha, J., Amorim, C., Cardoso, B. B., Gudiña, E. J., Silvério, S. C., Rodrigues, J. L., & Rodrigues, L. R. (2021). Zymomonas Mobilis as an emerging biotechnological chassis for the production of industrially relevant compounds. *Bioresources and Bioprocessing, 8*, 128.

Buitron, G., Hernandez-Juarez, A., Hernandez-Ramírez, M. D., & Sanchez, A. (2019). Biochemical methane potential from lignocellulosic wastes hydrothermally pretreated. *Industrial Crops and Products, 139*, 111555.

Carbon Trust. (2023). *European Green Digital Coalition (EGDC) case study: Nokia Smart Waste Management*. Retrieved from https://www.carbontrust.com

Chen, J., Zhang, B., Luo, L., Zhang, F., Yi, Y., Shan, Y., Liu, B., Zhou, Y., Wang, X., & Lü, X. (2021). A review on recycling techniques for bioethanol production from lignocellulosic biomass. *Renewable and Sustainable Energy Reviews, 149*, 111370.

Cheng, C. L., Che, P. Y., Chen, B. Y., Lee, W. J., Lin, C. Y., & Chang, J. S. (2012). Biobutanol production from agricultural waste by an acclimated mixed bacterial microflora. *Applied Energy, 100*, 3–9.

Chintagunta, A. D., Zuccaro, G., Kumar, M., Kumar, S. P. J., Garlapati, V. K., Postemsky, P. D., Kumar, N. S. S., Chandel, A. K., & Simal-Gandara, J. (2021). Biodiesel production from lignocellulosic biomass using oleaginous microbes: Prospects for integrated biofuel production. *Frontiers in Microbiology, 12*, 658284.

Clark L. H.-D., J. H., Farmer, T. J., & Sherwooda, J. (2016). Green chem. *IEEE Wireless Communications, 18*.

Cucchiella, F., D'Adamo, I., & Gastaldi, M. (2017). Sustainable management of waste-to-energy facilities. *Renewable and Sustainable Energy Reviews, 80*, 62–78.

da Costa Lopes, A. M., Brenner, M., Falé, P., Roseiro, L. B., & Bogel-Łukasik, R. (2016). Extraction and purification of phenolic compounds from lignocellulosic biomass assisted by ionic liquid, polymeric resins, and supercritical CO2. *ACS Sustainable Chemistry & Engineering, 4*, 3357–3367.

Das, D., & Veziroğlu, T.N. (2001). Hydrogen production by biological processes: A survey of literature. *International Journal of Hydrogen Energy, 26*, 13–28.

Delmer, D. P., & Amor, Y. (1995). Cellulose biosynthesis. *Plant Cell, 7*, 987–1000. International Journal of Molecular Sciences 2008, 9 1642.

Demirbas, A. (2011). Waste management, waste resource facilities and waste conversion processes. *Energy Conversion and Management., 52*(2), 1280–1287.

Demirbas, A., et al. (2016). Optimization of municipal solid waste (MSW) disposal in Saudi Arabia. *Energy Sources, Part A: Recovery, Utilization, and Environmental Effects., 38*(13), 1929–1937.

Emmel, A., Mathias, A. L., Wypych, F., & Ramos, L. P. (2003). Fractionation of Eucalyptus grandis chips by dilute acid- catalysed steam explosion. *Bioresource Technology, 86*, 105–115.

Espro, C., Paone, E., Mauriello, F., Gotti, R., Uliassi, E., Bolognesi, M. L., Rodríguez-Padrón, D., & Luque, R. (2021). Sustainable production of pharmaceutical, nutraceutical and bioactive compounds from biomass and waste. *Chemical Society Reviews, 50*, 11191–11207.

Eswaran, S., Subramaniam, S., Geleynse, S., Brandt, K., Wolcott, M., & Zhang, X. (2021). Techno-economic analysis of catalytic hydrothermolysis pathway for jet fuel production. *Renewable and Sustainable Energy Reviews, 151*, 111516.

Eurostat. *Recycling rate of municipal waste*. https://ec.europa.eu/eurostat/. Online; accessed 04 September 2019.

Fenila, F., & Shastri, Y. (2016). Optimal control of enzymatic hydrolysis of lignocellulosic biomass. *Resource-Efficient Technologies, 2*(Suppl. S1), S96–S104.

Gadde, B., Bonnet, S., Menke, C., & Garivait, S. (2009). Air pollutant emissions from rice straw open field burning in India, Thailand, and The Philippines. *Environmental Pollution, 157*(5), 1554–1558. https://doi.org/10.1016/j.envpol.2009.01.004

Garcia-Galan, C., Berenguer-Murcia, A., Fernandez-Lafuente, R., & Rodrigues, R. C. (2011). Potential of different enzyme immobilization strategies to improve enzyme performance. *Advanced Synthesis and Catalysis, 353*, 2885–2904.

Ge, X., Burner, D. M., Xu, J., Phillips, G. C., & Sivakumar, G. (2011). Bioethanol production from dedicated energy crops and residues in Arkansas, USA. *Biotechnology Journal, 6*, 66–73.

Gupta, R. K., Guha, P., & Srivastav, P. P. (2022). Natural polymers in bio-degradable/edible film: A review on environmental concerns, cold plasma technology and nanotechnology application on food packaging—A recent trends. *Food Chemistry Advances, 1*, 100135.

Ha, M. A., Apperley, D. C., Evans, B. W., Huxham, I. M., Jardine, W. G., Vietor, R. J., Reis, D., Vian, B., & Jarvis, M. C. (1998). Fine structure in cellulose microfibrils: NMR evidence from onion and quince. *The Plant Journal, 16*, 183–190.

Haghighi Mood, S., Hossein Golfeshan, A., Tabatabaei, M., Salehi Jouzani, G., Najafi, G. H., Gholami, M., & Ardjmand, M. (2013). Lignocellulosic biomass to bioethanol, a comprehensive review with a focus on pretreatment. *Renewable and Sustainable Energy Reviews, 27*, 77–93.

Hahladakis, J. N., Velis, C. A., Weber, R., Iacovidou, E., & Purnell, P. (2018). An overview of chemical additives present in plastics: Migration, release, fate and environmental impact during their use, disposal and recycling. *Journal of Hazardous Materials, 344*, 179–199.

He, M.-X., Li, Y., Liu, X., Bai, F., Feng, H., & Zhang, Y.-Z. (2009). Ethanol production by mixed-cultures of Paenibacillus Sp. and Zymomonas Mobilis using the raw starchy material from sweet potato. *Annales de Microbiologie, 59*, 749–754.

Heinze, T. (2016). Cellulose: Structure and properties. In O. J. Rojas (Ed.), *Cellulose chemistry and properties: Fibers, nanocelluloses and advanced materials; Advances in polymer science* (pp. 1–52). Springer. isbn:978-3-319-26015-0.

Hossain, U., & Poon, C. S. (2018). Comparative LCA of recycled and conventional concrete mix. *Resources, Conservation and Recycling, 133*, 78–89.

Hrbek, J. IEA Bioenenergy. (2015). Thermal gasification based hybrid systems [online]. https://www.ieabioenergy.com/wpcontent/uploads/2018/10/Thermal_gasif_based_hybrids.pdf. Accessed 16 August 2023.

Idehai, I. M., & Akujieze, C. N. (2015). Estimation of landfill gas and its renewable energy potential in Lagos, Nigeria. *International Journal of Energy and Environmental Engineering, 6*, 329–343.

Jara-Cobos, L., Abril-González, M., & Pinos-Vélez, V. (2023). Production of hydrogen from lignocellulosic biomass: A review of technologies. *Catalysts, 13*, 766.

Jody, B. J., Daniels, E. J., & Pomykala, J. A., Jr. (2011). *End-of-life vehicle recycling: State of the art of resource recovery from shredder residue*. Argonne National Laboratory.

Kainthola, J., Kalamdhad, A. S., & Goud, V. V. (2019). A review on enhanced biogas production from anaerobic digestion of lignocellulosic biomass by different enhancement techniques. *Process Biochemistry, 84*, 81–90. [CrossRef].

Kamm, M., Gau, M., Schneider, J., & vom Brocke, J. (2020). *Smart waste collection processes—A case study about smart device implementation*. https://doi.org/10.24251/HICSS.2020.810

Kan, A. (2009). General characteristics of waste management: A review. *Energy Education Science and Technology Part A: Energy Science and Research., 23*, 55–69.

Kannah, R. Y., Kavitha, S., Preethi, Karthikeyan, O. P., Kumar, G., Dai-Viet, N. V., & Rajesh Banu, J. (2021). Techno-economic assessment of various hydrogen production methods—A review. *Bioresource Technology, 319*, 124175.

Kapoor, R., Ghosh, P., Kumar, M., Sengupta, S., Gupta, A., Kumar, S. S., Vijay, V., Kumar, V., Vijay, V. K., & Pant, D. (2020). Valorization of agricultural waste for biogas based circular economy in India: A research outlook. *Bioresource Technology, 304*, 123036.

Khan, M. U., Usman, M., Ashraf, M. A., Dutta, N., Luo, G., & Zhang, S. (2022). A review of recent advancements in pretreatment techniques of lignocellulosic materials for biogas production: Opportunities and limitations. *Chemical Engineering Journal Advances, 100263*.

Knothe, G. (2010). Biodiesel and renewable diesel: A comparison. *Progress in Energy and Combustion Science, 36*, 364–373.

Koul, B., Yakoob, M., & Shah, M. P. (2022). Agricultural waste management strategies for environmental sustainability. *Environmental Research, 206*, 112285.

Koutinas, A. A., Vlysidis, A., Pleissner, D., Kopsahelis, N., Lopez Garcia, I., Kookos, I. K., & Papanikolaou, S. (2014). Valorization of industrial waste and by-product streams via fermentation for the production of chemicals and biopolymers. *Chemical Society Reviews, 43*, 2587–2627.

Kuhad, R. C., Gupta, R., & Singh, A. (2011). Microbial cellulases and their industrial applications. *Enzyme Research, 11*, 280696.

Kumar, A. K., & Sharma, S. (2017). Recent updates on different methods of pretreatment of lignocellulosic feedstocks: A review. *Bioresources and Bioprocessing, 4*(1), 1–19.

Kumar, S., Chiemchaisri, C., & Mudhoo, A. (2011). Bioreactor landfill technology in municipal solid waste treatment: An overview. *Critical Reviews in Biotechnology, 31*(1), 77–97.

Kumari, G., Guleria, A., Singh, K., Kumar, N., Guleria, A., Kumar, D., & Lima, E. (2022). Chapter 15—Lignocellulosic biopolymers as potential biosorbents. In H. Li, S. Saravanamurugan, A. Pandey, & S. Elumalai (Eds.), *Biomass, biofuels, biochemicals* (pp. 391–429). Elsevier. ISBN 9780128244197.

Lavarack, B. P., Giffin, G. J., & Rodman, D. (2002). The acid hydrolysis of sugarcane bagasse hemicellulose to produce xylose, arabinose, glucose, and other products. *Biomass and Bioenergy, 23*, 367–380.

Leite, P., Belo, I., & Salgado, J. M. (2021). Co-management of agro-industrial wastes by solid-state fermentation for the production of bioactive compounds. *Industrial Crops and Products, 172*, 113990.

Lisitsyn, A., Semenova, A., Nasonova, V., Polishchuk, E., Revutskaya, N., Kozyrev, I., & Kotenkova, E. (2021). Approaches in animal proteins and natural polysaccharides application for food packaging: Edible film production and quality estimation. *Polymers, 13*, 1592.

Liu, W., Chen, W., Hou, Q., Wang, S., & Liu, F. (2018). Effects of combined pretreatment of dilute acid pre-extraction and chemical-assisted mechanical refining on enzymatic hydrolysis of lignocellulosic biomass. *RSC Advances, 8*(19), 10207–10210.

Machado, G., Santos, F., Lourega, R., Mattia, J., Faria, D., Eichler, P., & Auler, A. (2020). Biopolymers from lignocellulosic biomass. In A. P. Ingle, A. K. Chandel, & S. S. da Silva (Eds.), *Lignocellulosic biorefining technologies*. Wiley-Blackwell.

Malik, K., Salama, E.-S., Kim, T. H., & Li, X. (2020). Enhanced ethanol production by saccharomyces cerevisiae fermentation post acidic and alkali chemical pretreatments of cotton stalk lignocellulose. *International Biodeterioration and Biodegradation, 147*, 104869.

Mata-Alvarez, J., Macé, S., & Llabrés, P. (2000). Anaerobic digestion of organic solid wastes. An overview of research achievements and perspectives. *Bioresource Technology, 74*, 3–16.

Mittermeier, F., Bäumler, M., Arulrajah, P., de Jesús García Lima, J., Hauke, S., Stock, A., & Weuster-Botz, D. (2022). Artificial microbial consortia for bioproduction processes. *Engineering in Life Sciences, 23*, e2100152.

Morohoshi, N. (1991). Chemical characterization of wood and its components. In D. N. S. Hon & N. Shiraishi (Eds.), *Wood and cellulosic chemistry* (pp. 331–392). Marcel Dekker, Inc..

Mwene-Mbeja, T. M., Dufour, A., Leckad, J., Kaurd, B. S., & Vaneeckhaute, C. (2020). Enzymatic reactions in the production of biomethane from organic waste. *Enzyme and Microbial Technology, 132*, 109410.

Nanda, S., & Berruti, F. (2021). Municipal solid waste management and landfilling technologies: A review. *Environmental Chemistry Letters, 19*, 1433–1456.

Nygård, P., Tanem, B. S., Karlsen, T., Brachet, P., & Leinsvang, B. (2008). Extrusion-based wood fibre–PP composites: Wood powder and pelletized wood fibres—A comparative study. *Composites Science and Technology, 68*, 3418–3424.

Palmqvist, E., & Hahn-Hägerdal, B. (2000). Fermentation of lignocellulosic hydrolysates. II: Inhibitors and mechanisms of inhibition. *Bioresource Technology, 74*, 25–33.

Perkumienė, D., Atalay A., Safaa L., & Grigienė J. (2023). Sustainable waste management for clean and safe environments in the recreation and tourism sector: A case study of Lithuania, Turkey and Morocco. *Recycling, 8(4), 56.*

Persson, T., Matusiak, M., Zacchi, G., & Jonsson, A.-S. (2006). Extraction of hemicelluloses from process water from the production of masonite. *Desalination, 199*, 411–412.

Pires, A., Martinho, G., & Chang, N. B. (2011). Solid waste management in European countries: A review of systems analysis techniques. *Journal of Environmental Management, 92*(4), 1033–1050.

Pope, C. A., & Dockery, D. W. (2006). Health effects of fine particulate air pollution: Lines that connect. *Journal of the Air & Waste Management Association, 56*(6), 709–742. https://doi.org/10.1080/10473289.2006.10464485

Ravindran, R., & Jaiswal, A. K. (2016). Microbial enzyme production using lignocellulosic food industry wastes as feedstock: A review. *Bioengineering, 3*, 30.

Russo, T., Fucile, P., Giacometti, R., & Sannino, F. (2021). Sustainable removal of contaminants by biopolymers: A novel approach for wastewater treatment. Current state and future perspectives. *Processes, 9*, 719.

Sannigrahi, P., Ragauskas, A. J., & Tuskan, G. A. (2010). Poplar as a feedstock for biofuels: A review of compositional characteristics. *Biofuels, Bioproducts and Biorefining, 4*, 209–226.

Sharif, M., Zafar, M. H., Aqib, A. I., Saeed, M., Farag, M. R., & Alagawany, M. (2021). Single cell protein: Sources, mechanism of production, nutritional value and its uses in aquaculture nutrition. *Aquaculture, 531*, 735885.

Sheng, Y., Lam, S. S., Wu, Y., Ge, S., Wu, J., Cai, L., Huang, Z., Le, Q. V., Sonne, C., & Xia, C. (2021). Enzymatic conversion of pretreated lignocellulosic biomass: A review on influence of structural changes of lignin. *Bioresource Technology, 324*, 124631.

Sikora, A., Detman, A., Chojnacka, A., & Błaszczyk, M. K. (2017). Anaerobic digestion: I. A common process ensuring energy flow and the circulation of matter in ecosystems. II. A tool for the production of gaseous biofuels. In *Ferment processes* (Vol. 14, p. 271). InTech.

Sims, R. (2003). *Biomass and resources bioenergy options for a cleaner environment in developed and developing countries*. Elsevier Science.

Singh, L., & Wahid, Z. A. (2015). Methods for enhancing bio-hydrogen production from biological process: A review. *Journal of Industrial and Engineering Chemistry, 21*, 70–80.

Singh, R. P., et al. (2011). An overview for exploring the possibilities of energy generation from municipal solid waste (MSW) in Indian scenario. *Renewable and Sustainable Energy Reviews, 15*(9), 4797–4808.

Sjöström, E. (1993). *Wood chemistry: Fundamentals and applications*. Academic Press.

Taherzadeh, M. J. (1999). *Ethanol from lignocellulose: Physiological effects of inhibitors and fermentation strategies*. Ph.D. thesis in Biotechnology, Chemical Reaction Engineering, Chalmers University of Technology.

Talebnia, F., Bafrani, M. P., Lundin, M., & Taherzadeh, M. J. (2008). Optimization study of citrus wastes saccharification by dilute acid hydrolysis. *BioResources, 3*, 108–122.

Teles, A. S. C., Chávez, D. W. H., Oliveira, R. A., Bon, E. P. S., Terzi, S. C., Souza, E. F., Gottschalk, L. M. F., & Tonon, R. V. (2019). Use of grape pomace for the production of hydrolytic enzymes by solid-state fermentation and recovery of its bioactive compounds. *Food Research International, 120*, 441–448.

Thuo, M., Gregory, P., Banerjee, S., & Du, C. (2021). 1 Introduction: Biopolymers and biocomposites. In S. A. Madbouly & C. Zhang (Eds.), *Biopolymers and composites: processing and characterization* (pp. 1–26). De Gruyter.

Towards the circular economy, Vol. 1: Economic and business rationale for accelerated transition. Transition to a circular economy report, n.d.. Ellenmacarthur Foundation.org.

Truong, M. V., Nguyen, L. N., Li, K., Fu, Q., Hasan Johir, M. A., Fontana, A., & Nghiem, L. D. (2020). Biomethane production from anaerobic co-digestion and steel-making slag: A new waste-to-resource pathway. *Science of the Total Environment, 738*, 139764.

Vinod, A., Sanjay, M. R., Suchart, S., & Jyotishkumar, P. (2020). Renewable and sustainable biobased materials: An assessment on biofibers, biofilms, biopolymers and biocomposites. *Journal of Cleaner Production, 258*, 120978.

Viola, E., Blasi, A., Valerio, V., Guidi, I., Zimbardi, F., Braccio, G., & Giordano, G. (2012). Biodiesel from fried vegetable oils via transesterification by heterogeneous catalysis. *Catalysis Today, 179*, 185–190.

Vollmer, I., Jenks, M. J., Roelands, M. C., White, R. J., van Harmelen, T., de Wild, P., & Weckhuysen, B. M. (2020). Beyond mechanical recycling: Giving new life to plastic waste. *Angewandte Chemie International Edition, 59*(36), 15402–15423.

Wang, F., Li, J., & Xu, M. (2013). Recycling of waste printed circuit boards: A review of current technologies and treatment status in China. *Journal of Hazardous Materials, 236*, 245–256.

Wirawan, F., Cheng, C.-L., Lo, Y.-C., Chen, C.-Y., Chang, J.-S., Leu, S.-Y., & Lee, D.-J. (2020). Continuous cellulosic bioethanol co-fermentation by immobilized Zymomonas mobilis and suspended Pichia sipitis in a two-stage process. *Applied Energy, 266*, 114871.

Wu, Z. (2019). Mixed fermentation of Aspergillus niger and Candida shehatae to produce bioethanol with ionic-liquid-pretreated bagasse. *3 Biotech, 9*, 41.

Zabed, H., Sahu, J. N., Suely, A., Boyce, A. N., & Faruq, G. (2017). Bioethanol production from renewable sources: Current perspectives and technological Progress. *Renewable and Sustainable Energy Reviews, 71*, 475–501.

Zhang, B., Gao, Y., Zhang, L., & Zhou, Y. (2021). The plant cell wall: Biosynthesis, construction, and functions. *Journal of Integrative Plant Biology, 63*, 251–272.

Application of Artificial Intelligence Techniques in Biorefinery Processes

Protibha Nath Banerjee

Abstract

Lignocellulosic biomasses have emerged as one of the cheap and sustainable raw materials for the production of platform chemicals and fuels through various biorefinery processes. Depending on the biorefinery processing techniques, the conversion processes are influenced by various parameters like temperature, concentration of acid or enzymes, time, solid-to-liquid ratio and composition of the feedstocks, among others, which in turn influences the final output in terms of yield and quality. Mathematical modelling using response surface methodology and artificial intelligence techniques like artificial neural network (ANN), fuzzy inference system (FIS) and genetic algorithm (GA) are very promising predictive modelling tools for determining the interaction between the parameters, involved in the conversion processes, for the purpose of optimisation of reaction conditions to obtain desired yield and purity of the output. These techniques reduce the need for time and resource consuming experimentation and have shown outstanding accuracy in predicting the optimal reactions conditions by analysing the non-linear interactions between the different parameters. In the recent years, these artificial intelligence-based modelling tools have revolutionised the process designing of biorefinery techniques by improving the process intensification, which has significant impact on process economies. The artificial neural network and fuzzy inference system techniques provide an excellent advantage in pattern recognition when interpreting the influence of various parameters involved in complex biorefinery processes compared to the traditional mathematical modelling using response surface methodology. These artificial intelligence-based techniques are going to be an integral part of biorefinery process

P. N. Banerjee (✉)
Department of Chemistry, Pimpri Chinchwad University, Pune, India
e-mail: protibha.banerjee@pcu.edu.in

G. Mukherjee, S. Dhiman (eds.), *Value Addition and Utilization of Lignocellulosic Biomass*, https://doi.org/10.1007/978-981-96-2786-8_10

designing in the future years to come as computational techniques continue to grow. Therefore, this chapter deals with the various studies that have been attempted to use the power of ANN, FIS, GA and other emerging computational techniques in biorefinery process designing for circular bioeconomy.

Keywords

Artificial Intelligence (AI) · Machine Learning (ML) · Process Optimisation · Biomass Conversion · Artificial Neural Network · Mathematical modelling · Predictive Modelling · Sustainability · Process Simulation

1 Introduction to Biorefinery and the Current Technical and Economic Challenges

Biorefinery is a concept of converting the lignocellulosic biomasses (LCBs) into chemical, fuels and energy. It is promising and gaining interest among the different stakeholders (government agencies; fuel and chemical producing companies; research organisations, etc.) as it is going to reduce the dependence on fossil fuels and more sustainable in nature. Although fossil-based fuels have played a significant role in the industrial and economic development through years, they are non-sustainable and going to deplete very soon (Bentley, 2002). Therefore, there is a pressing need to shift from fossil-based economy to bio-based economy.

Despite being a sustainable and eco-friendly concept, biorefinery processes are still at its infancy stage due to various techno-economic challenges, compared to their fossil-based counterparts, which limits its market acceptance. The current techno-economic challenges associated with the biorefineries include feedstock composition, process efficiency, technological development, environmental impact, cost competitiveness, market acceptance, regulatory norms and sociological factors (Francisco et al., 2022; Kwang et al., 2020). This chapter, in particular, will deal with the technical challenges related to valorisation processes in biorefinery and how mathematical modelling using ANN-based techniques can play a vital role in addressing these challenges. The previous studies have shown that the valorisation processes require combination of different processing techniques in order to achieve the required yield and purity of the final products (Pradhan et al., 2022). The different techniques that have been used by various research groups are pretreatment (Tu & Hallett, 2019), thermal processing (Sarker et al., 2019) and enzymatic and fermentation processes (Østby et al., 2020; Mussatto & Teixeira, 2010). All these processes face severe challenges due to the structural complexity and recalcitrant nature of lignocellulosic biomasses. These challenges increase the cost of production and thereby affect the process economics to a large extent.

2 Mathematical Modelling as a Tool for Designing Cost-Effective Biorefining Processes

The LCB valorisation processes are complex and non-linear in nature due to the structural complexity and recalcitrant nature of lignocellulosic biomasses. In order to get optimum yield with desired purity the combination of different parameters like pH, temperature, concentration of acid or enzymes, time, solid-to-liquid ratio and composition of the feedstocks had been explored for the valorisation processes of lignocellulosic biomass and has been found to have significant impact on the yield and purity of the final products (Banerjee et al., 2014). Therefore, the determination of the optimal value of these parameters, for any of the above-mentioned valorisation processes, becomes crucial to determine the optimum conditions and improve the process efficiency. The manual studies for the optimisation of LCB valorisation processes are time consuming and a very costly affair. Mathematical modelling, on the other hand, can assist in determining the optimum values of these parameters by analysing the non-linear interactions between the different parameters and thereby reduces the expensive and time-consuming experimentation Pomeroy et al., 2022). The statistical modelling using response surface methodology (RSM) is a conventional approach, which has been explored extensively by research groups in the optimisation studies of LCB valorisation processes (Kim & Han, 2012; Manmai et al. 2021; Qi et al. 2009). RSM is a combination of experimental designs and a strategy to develop a new dataset by first- or second-order polynomial equations in a systematic test method (Ramaraj & Unpaprom, 2019). RSM analysis is predominantly done with symmetrical experimental designs, which are central composite design (CCD), the Box–Behnken design (BBD), and a three-level full factorial design (Bezerra et al., 2008; Chollom et al., 2019). Deepanraj et al. (2021) have studied the effect of solid concentration, pH, temperature and co-digestion on biogas production and chemical oxygen demand degradation using anaerobic digestion method. Nababan et al. (2022) carried out the optimisation studies on the enzymatic hydrolysis of jabon wood pulp for ethanol production. The effect of three independent variable, pulp freeness, enzyme loading and surfactant concentration on ethanol production, was optimised using Box-Behnken design. Lingru Ruan et al. (2024) studied the optimisation of NaOH pretreatment conditions of sugarcane bagasse by RSM using Box-Behnken Design. The three factors including NaOH concentration, time and solid-to-liquid ratio were optimised to obtain the optimal lignin removal and solid yield (Ruan et al. 2024). In the same study, they also performed the optimisation of the enzymolysis conditions of cellulase for pretreated sugarcane bagasse by RSM using BBD. Varilla-Mazaba et al. (2024) used RSM and desirability criterion to optimise the yield of xylose during acid hydrolysis of sugarcane bagasse. They evaluated the combined effect of the concentration of sulfuric acid, the solid–liquid ratio and reaction time on the maximisation of xylose yield and the minimum production of acetic acid as degraded impurity. Marcelina et al. (2022) did the optimisation studies on the production of levulinic acid from sugarcane bagasse. They evaluated the combined effect of three parameters—time, temperature and concentration of sulphuric acid on the yield of levulinic acid using central composite design. All the above-mentioned studies have shown the predictive modelling capabilities of RSM in the valorisation processes of LCB within biorefinery concept. The RSM approach of predictive modelling, over the period of time, has shown to improve the process efficiency of pretreatment methods by reducing the cost and time. However, despite the conventional mathematical modelling, using response surface methodology can explain the influence of one factor or interaction of multiple factors on the final output and can establish a non-linear relationship between the factors, the assumption of polynomial relationship between the independent variable and the output is one of the major limitations of this approach (Chakraborty & Shirivastava, 2019). The non-conventional approach, on the other hand, using artificial intelligence techniques like ANN, ANFIS and GA is gaining interest among the different research groups as they are emerging as very promising predictive modelling tools for determining the interaction between the parameters for the optimisation studies of LCB valorisation processes (Marcelina et al., 2022). These techniques exhibit higher precision on the prediction and provide compelling advantages for pattern recognition when interpreting the influence of ever-changing feedstock compositions for complex biomass conversion processes as they do not require any prior knowledge of reaction mechanisms or thermodynamic phenomena (Abdelkader et al., 2019).

3 Artificial Intelligence Techniques as Emerging Predictive Modelling Tools for Circular Bioeconomy

The conversion of renewable biological resources (agricultural, municipal and forestry waste, phytoplankton in water bodies, etc.) into value-added products in a closed-loop system to minimise waste formation, pollution and carbon emission is referred as circular bioeconomy. Circular bioeconomy falls in line with the United Nation's sustainable development goal no.12, which tells about sustainable production and consumption (United Nation, 2015). The shift to circular bioeconomy from linear economy is need of the hour for a sustainable world, and it reflects a systematic shift that builds long-term resilience and creates economic opportunities in a sustainable manner together with protecting the environment at the same time. Although circular bioeconomy is an excellent idea for a sustainable world, the actual realisation is still at its infancy stage due to the complexity and dynamic alteration of bioprocesses which in turn is due to the structural diversity of the LCB grown in different geographical regions. The governing key factors (e.g. organic loading rate, dissolved oxygen level, nutrients, temperature, pH, solids concentration, toxicity, etc.), as well as microbial community structure, intracellular metabolic interaction, and enzymatic activity, collectively influence the overall performance (Khanal et al., 2023). Therefore, bioprocesses require precise control and real time monitoring in order to maximise the process economics. The precise control over the reaction parameters, key factors and real time monitoring requires a proper process design and mathematical modelling can assist in doing so (we have seen the power of RSM in predictive modelling in Sect. 2). Artificial intelligence techniques are gaining interest in process designing and optimisation studies of bioprocesses as they are very precise and work effectively with non-linear and complex bioprocess system. The artificial intelligence techniques that have been used extensively by the various research groups are ANN, ANFIS and GA among others. In this section, we will discuss these three techniques used in bioprocess designing and optimisation studies. Artificial neural networks are mathematical data-processing techniques that mimic a model of biological neurons (Vani et al., 2015). The input, hidden and output layers are the three layers of an ANN. The input layer comprises neurons designating input variables, whereas the output layer comprises neurons representing output variables and the hidden layer, like the two above layers, is made up of a sequence of distinct neurons (Moodley et al., 2015). In ANN, the construction and mapping of non-linear correlation between the input variables and output is key factor where it executes the non-linear operation on combined information and finally gives the outputs. While doing so it learns and recognises the pattern from the input data without making assumptions about the interrelationship between the input factors (Khajeh et al., 2012). The application of ANN has been adopted by various research group for bioprocess design and optimisation studies (Rodriguez-Granrose et al., 2021). The technique of ANN had been used by Moodley et al. (2015) for predicting sugar yields from sugarcane leaf waste using inorganic salt-based pretreatment. Two ANN models were developed, one for steam-assisted and another for microwave-assisted. Both the models gave coefficients of determination (R^2) of 0.97 indicating higher accuracy at predicting the reducing sugar yield under the given conditions. Marcelina et al. (2022) carried out the optimisation studies on the production of levulinic acid from sugarcane bagasse. They took three input variables—temperature, time and concentration of sulphuric acid, and output as the yield of levulinic acid. The datasets were classified into three parts—training (80%), testing (10%) and validation (10%). The feedforward backpropagation network with 3–10-1 architecture was employed to model the production of levulinic acid where '3' is the input layer, '10' is the hidden layer and '1' is the output.

The highest coefficient of determination value of 0.96 indicates that the model-predicted values are in agreement with the experimental values. The optimisation studies using ANN in the production of bioethanol from sorghum starch by the process of hydrolysis followed by fermentation were carried out by Sebayang et al. (2017). In this study, two ANN models were developed—one for prediction of reducing sugar with 4-5-1 architecture and another for prediction of ethanol with 3-6-1 architecture. The four input variables for reducing sugar prediction were substrate loading, amylase enzyme concentration, amyloglucosidase enzyme concentration and strokes speed, while three input variable, yeast concentration, reaction temperature and agitation, were used for the prediction of ethanol yield. The R^2 value was found to be 0.994, which again shows the high predictive accuracy of the ANN techniques. Chouaibi et al. (2020) studied the optimisation of bioethanol production from pumpkin peel waste. The feedforward backpropagation network with 4-15-1 architecture was employed to model the production of reducing sugar while 4-6-1 was employed to model the bioethanol yield. The model showed the R^2 value of 0.99 indicating again the high predictive ability of the ANN techniques. Elsayed et al. (2021) designed an ANN model for conversion of date palm fronds to citric acid by New Trichoderma species PWN6 using 7-3-1 architecture. The input layer comprised of seven neurons (steam exploded biomass, date palm frond size (mm), incubation time (day), incubation temperature (°C), inoculation (spore/g), tricalcium phosphate (mg/g), and ammonium sulphate concentration (mg/g)) that is determined by the number of the examined independent factors (Elsayed et al., 2021). The output layer has one neuron (OA, μmol/g), representing the response factor. The best model obtained had three neurons in the hidden layer and showed an R^2 value of 0.960 for the training and 0.967 for the validation dataset, indicating the high predictive ability of the model (Elsayed et al., 2021).

ANFIS is another mathematical modelling tool under the domain of AI techniques, which are gaining interest among the different stakeholders related to lignocellulosic bioprocesses. Fuzzy inference system (FIS) is a process of mapping from a given input to an output using fuzzy logic. When FIS is combined with ANN, it is called ANFIS techniques. It is an approximate method for modelling process behaviour using incomplete and ambiguous data. An ANFIS integrates the reasoning capabilities of FIS with the learning capabilities of ANN. FIS basically consists of three parts—fuzzification where input variables are compared with the membership functions on the antecedent part to obtain the membership values of each label, fuzzy inference mechanism, which combines the membership values on the premise part to get the firing strength and generate the qualified consequents, and defuzzification in which the qualified consequents are aggregated to produce the output. FIS employs a rule-based structure with the following main elements: (1) an IF-THEN fuzzy rule base; (2) a repository of membership functions utilised in fuzzy rules and (3) an inference procedure that applies the rules to relate outputs and inputs (Ruan et al., 2024). Fuzzification divides data into subcategories based on linguistic classification (low, medium and high) as well as membership degrees. The fuzzy outputs generated by the inference procedure are defuzzified to produce an unambiguous number value (Ruan et al., 2024). Fuzzy systems are classified into three types: Mamdani's system, Tsukamoto's system and Sugeno's system. Mamdani's system is the most popular tool for fuzzy modelling (Takagi & Sugeno, 1985; Cobaner et al., 2009). Many research groups have used ANFIS in diverse range of bioprocess modelling where other tools are inadequate, too complex or inapplicable (Al-Mahasneh et al., 2016). Ahmad Jahanbakhshi and Rouhalla (Jahanbahkshi & Salehi, 2019) used both ANN and ANFIS for modelling the bioethanol production from watermelon waste using Saccharomyces cerevisiae yeast. Two independent variables, viz., agitator speed and amount of yeast, were taken as input variables and bioethanol yield as output. The performance of ANFIS was evaluated for six models with Bell, Gaussian and triangular functions with two and three test membership functions. The Gaussian model with two membership functions having an MSE of 0.3129 and R^2 of 0.9993 was the best ANFIS. The high R^2 and low MSE values confirmed that the ANFIS and ANN models were efficient in bioethanol yield predictions. Uzuner and Cekmecelioglu (2016) compared ANN and ANFIS model for the production of polygalacturonase by *Bacillus subtilis* in submerged fermentation.

The architecture of the ANN was 5–7-1 with five neurons in the input layer, one hidden layer with 7 neurons and one neuron in the output layer. Five fermentation variables (pH, temperature, time, yeast extract concentration and K_2HPO_4 concentration) served as the input of the ANN and ANFIS models, and the polygalacturonase was the output. The R^2 value for ANFIS and ANN was found to be 0.978 and 1.0, respectively, which proved that both models performed similarly in terms of predictive ability. Kushwaha et al. (2023) developed an ANFIS model for forecasting the carbon dioxide fixation rate of various microalgal strain. The temperature, pH, CO2% and amount of nitrogen and phosphorous (mg/L) were taken as the input variables, while the CO_2 fixation rate was taken as the output variable. Sugeno inference system was used for ANFIS modelling where 70% of the dataset is employed to train the model, and the remaining 30% is used for testing the models. The R^2 value was found to be 0.839 showing the less predictive ability of ANFIS model in this case. In the same study, the authors carried out the optimisation studies using genetic algorithm–ANFIS (GA-ANFIS) where the R^2 value was found to be 0.978. These results show the high predictive accuracy of GA-ANFIS model compared to the ANFIS model. Genetic algorithm (GA) will be discussed later in this section. Zareei and Khodai (2017) used ANFIS modelling to predict and optimise biogas production from cow manure with maize straw from anaerobic digestion. Three input variables—carbon to nitrogen (C/N) ratio, total solid content and stirring intensity—were used, and biogas production was taken as output. Sugeno system was used for ANFIS modelling where 80% of dataset was used for training and 20% dataset was used to test the model. The R^2 between observed and predicted biogas production values was 0.99 showing that the predicted value is in agreement with the experimental results. Highest biogas production was achieved from carbon to nitrogen ratio 26.76, total solid content 9% and moderate stirring. Biogas production increased about 8% with optimal conditions suggested by the ANFIS model. Rego et al. (2019) developed ANN and NFIS model for biogas production in biodigester from mixture of swine manure and rice husk. The temperature (°C), pH, FOS/TAC ratio and type of biodigesters were taken as input neurons and biogas production was taken as output neurons. The 67% of the data was used for training and 37% of the data was used for testing the model. The R^2 value of ANN was found to be 0.777 at its best configuration while from ANFIS the R^2 value was found to be 0.812 at its best configuration, showing better predictive ability of ANFIS over the ANN model.

Genetic algorithm (GA) is a computational technique, which uses the concept of biological evolution in solving the myriads of real-world optimisation problems. J H Hollan first proposed the concept of GA in the year 1992. It is an optimisation algorithm that is inspired from the natural selection. It is a population-based search algorithm, which utilises the concept of Darwin's survival of fittest (Michalewicz et al., 2013). The basic elements of GA are chromosome representation, fitness selection and biological-inspired operators. Typically, the chromosomes take the binary string format. In chromosomes, each locus (specific position on chromosome) has two possible alleles (variant forms of genes)—0 and 1 (Michalewicz et al., 2013). Chromosomes are considered as points in the solution space. These are processed using genetic operators by iteratively replacing its population. The fitness function is used to assign a value for all the chromosomes in the population (Michalewicz et al, 1996). The biological-inspired operators are selection, mutation and crossover. In selection, the chromosomes are selected on the basis of its fitness value for further processing. In crossover operator, a random locus is chosen, and it changes the subsequences between chromosomes to create off-springs (Michalewicz et al, 1996). In mutation, some bits of the chromosomes will be randomly flipped on the basis of probability (Michalewicz et al., 2013; Michalewicz, 1996; Holland, 1975). GA is gaining tremendous interest among various research groups for optimisation of different bioprocesses as its predictive accuracy and precision are very high. Jacob et al. (2016) employed ANN coupled GA for the optimisation of anaerobic co-digestion of potato waste and aquatic weed. The datasets were divided into 70% training, 15% testing and 15% to validate the model. Different network topologies were examined for their predictability. The developed ANN model was then used as fitness function in GA to select the optimal set of input variables that can result in maximum methane yield. The effective network topology was found to be 3–12-1, where 3 represents the input 401 variables, 12 as hidden neurons and 1 as the output. The MSE and R values were found to be 0.136 and 0.9972, respectively, which shows the high predictive accuracy of ANN-GA modelling. Kumar et al. (2017) employed ANN-GA model to optimise the xylanase production from *Thermomyces lanuginosus* VAPS-24 using low cost agro-industrial residue. In their study, the artificial neural network mechanisms have three layers with inputs, outputs and neurons to generate the best fit model. The input data and output data were explored from the RSM design matrix. Every input has three variables and three replications. The various parameters such as input, output, net and fitness obtained from ANN were used in genetic algorithm (GA) to predict the optimum reaction for the maximum production of xylanase. The overall correlation coefficient was found to be 0.99172, which again reveals the high predictive accuracy of ANN-GA model. Varbanov et al. (2023) studied the optimisation of environmental impact of microalgal biorefinery having multiple products using GA. There are seven variables in this equation as a function of environmental impact. These variables are the input parameters in the processes in the biorefinery such as power used in cultivation, methanol in transesterification, heat in transesterification, solid residue input in biochar, liquid residue input in anaerobic digestion, methane input in CHP and the power used in transesterification. The variables were selected and used for input variables due to the different attributions to the environmental impact. The results showed that ozone formation had the highest value of R^2, while aquatic eutrophication had the lowest R^2 value. Najila et al. (2020) applied ANN-GA to maximise exoglucanase production by Penicillium roqueforti ATCC 10110 under solid-state fermentation, using a blend of agro-industrial wastes as substrate. The time, temperature, humidity and pH were used as input neurons while exoglucanase yield was taken as output neurons. The R^2 value was found to be greater than 0.99 showing the generalisation capability of the model and the optimisation process increased the enzyme activity by up to 1263% compared with the preliminary analysis using individual substrates, demonstrating the high efficiency of the algorithms on predicting and optimising enzyme production (Najila et al., 2020).

4 Future Prospects and Conclusion

The above discussion reveals that artificial intelligence techniques have superior predictive capability in terms of accuracy and precision in the optimisation of bioprocesses. These AI techniques can be an accelerator of transition to circular bioeconomy from the present petroleum-based economy. With the continuous development of computational techniques, AI can be trained with the more accurate data to help tune and optimise the existing biorefinery processes in a more efficient way to reduce production cost and increase market acceptability. The ANN technique has been extensively employed by different research groups, while the use of ANFIS and GA has not been explored much. However, all the techniques discussed in this chapter have shown the higher predictive ability than the conventional RSM methods. The utilisation of these AI techniques in valorisation processes of LCB will not only lead to the development of sustainable production process but will also help to generate employment, reduce environmental pollution and boost the socio-economic security of the people. The artificial intelligence techniques will be an integral part and parcel of circular bioeconomy for sustainable production of chemicals, fuels and energy. Although lot of work related to the development of AI techniques for biorefinery processes has been carried out and published by different research groups, these techniques are still at its infancy stage, and more precise and robust results are required to fully integrate the artificial intelligence and machine learning techniques into the existing biorefinery processes.

References

Abdelkader, H., Karim, M., Cherif, B., Farid, D. (2019). Comparison of artificial neural network (ANN) and response surface methodology (RSM) prediction in compressive strength of recycled concrete aggregates. *Construction and Building Materials, 209*, 425-436.

Al-Mahasneh, M., Aljarrah, M., Rababah, T., & Alu'datt. (2016). Application of hybrid neural fuzzy system (ANFIS) in food processing and technology. *Food Engineering Reviews, 8*(3), 351–366.

Banerjee, P. N., Pranovich, A., Dax, D., & Willfor, S. (2014). Non-cellulosic heteropolysaccharides from sugarcane bagasse - sequential extraction with pressurized hot water and alkaline peroxide at different temperatures. *Bioresource Technology, 155*, 446–450.

Bentley, R. W. (2002). Global oil and gas depletion: an overview. *Energy Policy., 30*(3), 189–205.

Bezerra, M. A., Santelli, R. E., Oliveira, E. P., Villar, L. S., & Escaleira, L. A. (2008). Response surface methodology (RSM) as a tool for optimization in analytical chemistry. *Talanta, 76*(5), 965–977.

Chakraborty, S., & Shrivastava, C. (2019). Comparison between multiresponse-robust process design and numerical optimization: A case study on baking of fermented chickpea flour-based wheat bread. *Journal of Food Process Engineering, 42*(3), 13008.

Chollom, M. N., Rathilal, S., Swalaha, F. M., Bakare, B. F., & Tetteh, E. K. (2019). Comparison of response surface methods for the optimization of an upflow anaerobic sludge blanket for the treatment of slaughterhouse wastewater. *Environmental Engineering Research, 25*(1), 114–122.

Chouaibi, M., Ben Daoued, K., Riguane, K., Rouissi, T., & Ferrari, G. (2020). Production of bioethanol from pumpkin peel wastes: Comparison between response surface methodology and artificial neural networks. *Industrial Crops and Products, 155*, 112822.

Cobaner, M., Unal, B., & Kisi, O. (2009). Suspended sediment concentration estimation by an adaptive neuro-fuzzy and neural network approaches using hydro-meteorological data. *Journal of Hydrology, 367*, 52–61.

da Silva, N., Nunes, L. L., Carneiro, L. H., de Menezes, S., Silva, M., de Carvalho, A., Pimentel, B., Silva, T. P., Pacheco, C. S. V., de Carvalho Tavares, I. M., Santos, P. H., das Chagas, T. P., da Silva, E. G. P., de Oliveira, J. R., Bilal, M., & Franco, M. (2020). Simplex-centroid design and artificial neural network-genetic algorithm for the optimization of Exoglucanase production by Penicillium Roqueforti ATCC 10110 through solid-state fermentation using a blend of agro-industrial wastes. *Bioenergy Research, 13*(4), 1130–1143.

Deepanraj, B., Senthilkumar, N., Ranjitha, J., et al. (2021). Biogas from food waste through anaerobic digestion: Optimization with response surface methodology. *Biomass Conversion and Biorefinery, 11*, 227–239.

Elsayed, M. S., Eldadamony, N. M., Alrdahe, S. S. T., & Saber, W. I. A. (2021). Definitive screening design and artificial neural network for modeling a rapid biodegradation of date palm fronds by a new Trichoderma sp. PWN6 into citric acid. *Molecules, 26*(16), 5048. extraction. *Journal of Supercritical Fluids, 69*, 91–96.

Francisco, G., Calvo-Flores Francisco, J., & Martin-Martinez. (2022). *Frontiers in Chemistry, 10*, 973417.

Hammoudi, A., Moussaceb, K., Belebchouche, C., & Dahmoune, F. (2019). Comparison of artificial neural network (ANN) and response surface methodology (RSM) prediction in compressive strength of recycled concrete aggregates. *Construction and Building Materials, 209*, 425–436.

Holland, J. H. (1975). *Adaptation in natural and artificial systems*. The Univ of Michigan Press.

Jacob, S., & Banerjee, R. (2016). Modelling and optimization of anaerobic codigestion of potato waste and aquatic weed by response surface methodology and artificial neural network coupled genetic algorithm. *Bioresource Technology, 214*, 386–395.

Jahanbakhshi, A., & Salehi, R. (2019). Processing watermelon waste using Saccharomyces cerevisiae yeast and the fermentation method for bioethanol production. *Journal of Food Process Engineering, 42*(7), e13283.

Khajeh, M., Moghaddam, M. G., & Shakeri, M. (2012). *Application of artificial neural network in predicting the extraction yield of essential oils of Diplotaenia cachrydifolia by supercritical fluid* (Vol. 69, p. 91).

Khanal, S. K., Tarafdar, A., & You, S. (2023). Artificial intelligence and machine learning for smart bioprocesses. *Bioresource Technology, 375*, 128826.

Kim, I., & Han, J.-I. (2012). Optimization of alkaline pretreatment conditions for enhancing glucose yield of rice straw by response surface methodology. *Biomass and Bioenergy, 46*, 210–217.

Kwang H. K., & Chang G. Y. (2020). *Biorefinery: Current status, challenges, and new strategies. Applied science*. MDPI.

Kumar, V., Chhabra, D., & Shukla, P. (2017). Xylanase production from Thermomyces lanuginosus VAPS-24 using low cost agro-industrial residues via hybrid optimization tools and its potential use for saccharification. *Bioresource Technology, 243*, 1009–1019.

Kushwaha, O., Uthayakumar, H., & Kumaresan, K. (2023). Modeling of carbon dioxide fixation by microalgae using hybrid artificial intelligence (AI) and fuzzy logic (FL) methods and optimization by genetic algorithm (GA). *Environmental Science and Pollution Research, 30*, 24927–24948.

Manmai, N., Unpaprom, Y., & Ramaraj, R. (2021). Bioethanol production from sunflower stalk: Application of chemical and biological pretreatments by response surface methodology. *Biomass Conversion and Biorefinery, 11*, 1759–1773.

Marcelina, O., Ashish Kapoor, P., Kumar, S., Rangasamy, G., Ponnuchamy, M., Rajagopal, M., & Banerjee, P. N. (2022). Modelling of sugarcane bagasse conversion to levulinic acid using response surface methodology (RSM), artificial neural networks (ANN), and fuzzy inference system (FIS): A comparative evaluation. *Fuel, 329*(8), 125409.

Michalewicz, Z. (2013). *Genetic algorithms + data structures = evolution programs* (3rd ed.). Springer.

Michalewicz, Z., & Schoenauer, M. (1996). Evolutionary algorithms for constrained parameter optimization problems. *Evolutionary Computation, 4*(1), 1–32.

Moodley, P., Rorke, D. C. S., & Gueguim Kana, E. B. (2015). Development of artificial neural network tools for predicting sugar yields from inorganic salt-based pretreatment of lignocellulosic biomass. *Bioresource Technology, 273*, 682–686.

Mussatto, S. I., & Teixeira, J. A. (2010). Lignocellulose as raw material in fermentation processes. In *Current research, technology and education topics in applied microbiology and microbial biotechnology*.

Nababan, M. Y. S., Fatriasari, W., & Wistara, N. J. (2022). Response surface methodology for enzymatic hydrolysis optimization of jabon alkaline pulp with Tween 80 surfactant addition. *Biomass Conversion and Biorefinery, 12*, 2165–2174.

Østby, H., Hansen, L. D., Horn, S. J., Eijsink, V. G. H., & Várnai, A. (2020). Enzymatic processing of lignocellulosic biomass: Principles, recent advances and perspectives. *Journal of Industrial Microbiology and Biotechnology, 47*(9–10), 623–657.

Pomeroy, B., Grilc, M., & Likozar, B. (2022). Artificial neural networks for bio-based chemical production or biorefining: A review. *Renewable and Sustainable Energy Reviews, 153*, 111748.

Pradhan, D., Jaiswal, S., & Jaiswal, A. K. (2022). Artificial neural networks in valorization process modelling of lignocellulosic biomass. *Biofuel, Bioproduct & Biorefining, 16*(6), 1849–1868.

Qi, B., Chen, X., Shen, F., Su, Y., & Wan, Y. (2009). Optimization of enzymatic hydrolysis of wheat straw pretreated by alkaline peroxide using response surface methodology. *Industrial and Engineering Chemistry Research, 48*(15), 7346–7353.

Ramaraj, R., & Unpaprom, Y. (2019). Optimization of pretreatment condition for ethanol production from cyperus difformis by response surface methodology. *3 Biotech, 9*(6), 1–9.

Rodriguez-Granrose, D., Jones, A., Loftus, H., Tandeski, T., Heaton, W., & Foley, K. T. (2021). Design of experiment (DOE) applied to artificial neural network architecture enables rapid bioprocess improvement. *Bioprocess and Biosystems Engineering, 44*(6), 1301–1308.

Ruan, L., Haifeng, W., Shiya, W., Zhou, L., Shangxin, W., & Shang, C. (2024). Optimizing the conditions of pretreatment and enzymatic hydrolysis of sugarcane bagasse for bioethanol production. *ACS Omega, 9*(27), 29566–29575.

Sarker, T. R., Nanda, S., Dalai, A. K., & Meda, V. (2019). A review of Torrefaction Technology for Upgrading Lignocellulosic Biomass to solid biofuels. *Bioenergy Research, 14*, 645–669.

Sebayang, A. H., Masjuki, H. H., Ong, H. C., Dharma, S., Silitonga, A. S., & Kusumo, J. M. (2017). Optimization of bioethanol production from sorghum grains using artificial neural networks integrated with ant colony. *Industrial Crops and Products, 97*, 146–155.

Takagi, T., & Sugeno, M. (1985). Fuzzy identification of system and its applications to modeling and control. *IEEE Transactions on Systems, Man, and Cybernetics, 15*, 116–132.

Tu, W.-C., & Hallett, J. P. (2019). Recent advances in the pretreatment of lignocellulosic biomass. *Current Opinion in Green and Sustainable Chemistry, 20*, 11–17.

United Nations—Department of Economic and Social Affairs: Sustainable Development. (2015).

Uzuner, S., & Cekmecelioglu, D. (2016). Comparison of artificial neural networks (ANN) and adaptive neuro-fuzzy inference system (ANFIS) models in simulating polygalacturonase production. *BioResources, 11*(4), 8676–8685.

Vani, S., Sukumaran, R. K., & Savithri, S. (2015). Prediction of sugar yields during hydrolysis of lignocellulosic biomass using artificial neural network modelling. *Bioresource Technology, 188*, 128–135.

Varbanov, P. S., Seferlis, P., & Van Fan, Y. (2023). Athanasios Papadopoulos, environmental impact optimization of microalgae biorefinery using life cycle assessment and evolutionary computing. *Chemical Engineering Transactions, 103*, 379–384.

Varilla-Mazaba, A., Raggazo-Sánchez, J. A., Calderón-Santoyo, M., et al. (2024). Multi-response optimization of acid hydrolysis in sugarcane bagasse to obtain high xylose concentration. *Biomass Conversion and Biorefinery, 14*, 173–181.

Zareei, S., & Khodaei, J. (2017). Modeling and optimization of biogas production from cow manure and maize straw using an adaptive neuro-fuzzy inference system. *Renewable Energy, 114*, 423–427.

Harnessing Bioproducts for a Sustainable Circular Economy

Prasann Kumar and Joginder Singh

Abstract

The ideas of circular economy and bioproducts mark a fundamental shift in our attitude to resource management and sustainable development. The fundamental character of bioproducts, their relevance in advancing a circular economy, and their effects on several sectors and businesses are investigated in this chapter. Bioproducts, made using environmentally friendly techniques and derived from sustainable biological resources like plants, animals, and microorganisms, encompass a range of items including biofuels, bioplastics, biochemicals, and biomaterials. These products serve as practical alternatives to traditional petroleum-based counterparts. By leveraging the natural properties of organic materials and applying innovative technologies, bioproducts offer significant potential to mitigate environmental degradation, reduce carbon emissions, and decrease reliance on finite fossil resources. Achieving the goals of sustainable production and consumption depends on integrating bioproducts within a circular economy paradigm. The ideas of the circular economy help close the material flow loop, reducing waste generation and optimizing resource use. Remanufacturing, reuse, and recycling help one to accomplish this. By converting organic waste streams into valuable resources, enhancing natural biodegradation, and enabling cascading bioreactor systems, bioproducts contribute to improved circularity. Furthermore, using bioproducts supports a regenerative approach to economic development whereby natural resources are preserved, and the benefits of

P. Kumar (✉)
Department of Agronomy, School of Agriculture, Lovely Professional University, Phagwara, Punjab, India

J. Singh
Department of Botany, Nagaland University, Nagaland, India

G. Mukherjee, S. Dhiman (eds.), *Value Addition and Utilization of Lignocellulosic Biomass*, https://doi.org/10.1007/978-981-96-2786-8_11

ecosystems are enhanced. This chapter provides an overview of the key aspects of bioproducts and their role in the circular economy.

Keywords

Agriculture · Biotic · Circular · Biological · Economic · Bioproducts · Microorganism · Environmental · Fossils · Resources

1 Introduction

In many sectors, bioproducts derived from renewable biological resources represent a significant step toward sustainability and environmental stewardship. The mentioned products biofuels, bioplastics, biochemicals, and bio-based materials are derived from biomass, which includes organic matter such as plants, agricultural waste, and microbial cultures (Ab Abdullah et al., 2024). Bioproducts, derived from biological sources, can occur in place of goods produced from fossils. This helps to lower the environmental impact usually connected with conventional manufacturing processes and the carbon footprint. Bioproducts are important since they help support a circular economy—reusing, recycling, and regenerating resources, lowering waste, and improving resource efficiency. Among the worldwide issues of resource depletion, pollution, and climate change, bioproducts offer a viable answer in line with world sustainability objectives. They inspire innovation in agriculture, energy, and materials science by helping grow environmentally friendly technologies and supporting sustainable business practices. Moreover, bioproducts are essential in reducing reliance on limited resources, supporting energy stability, and enhancing financial resilience. Their significance is further emphasized by their ability to create new markets and job opportunities, thus driving economic growth while ensuring environmental preservation. Additionally, incorporating bioproducts across various industries fosters the transition to a low-carbon economy, helping to reduce greenhouse gas emissions and preserve biodiversity. Promoting sustainable development and addressing pressing environmental problems depend much on bioproducts. They help to build a future that is more resilient and sustainable (Chen et al., 2023). Using circular economy ideas marks a significant change in how waste management, consumption, and manufacturing are handled in societies. Unlike the traditional linear economy that follows a "take-make-disposal" strategy, the circular economy seeks to maximize resource efficiency by extending the use of materials, tools, and resources. Using recycling, reusing, refurbishing, and remanufacturing techniques, this method seeks to lower waste and environmental damage by completing the cycle of product lifetimes (Di Fraia et al., 2024). As the planet deals with growing problems, including resource depletion, environmental damage, and climate change, the concept of a circular economy has grown more appealing. It guarantees the preservation or enhancement of the value of goods and resources over time, enabling a rejuvenation system. Several basic ideas drive the circular economy. At first, it gives the design of products with an eye toward durability, repairability, and recycling as top priorities. This guarantees that materials can be efficiently recovered and applied once more when the product ends its lifetime. Moreover, it emphasizes the need for sustainable sources by encouraging the use of biodegradable and renewable materials. Furthermore, it supports implementing innovative business models that minimize the need for ownership and promote the efficient use of resources, such as product-as-a-service and sharing systems. The possibilities of technology, which include digital platforms allowing tracking of goods, material recovery, and resource flow optimization, support these values even more in their evolution. This technology has great possibilities for developing the circular economy (Saravanan et al., 2023). Beyond only improving the surroundings, circular economy ideas have great social and financial value. By reducing reliance on limited resources and minimizing waste, particularly during supply chain interruptions and resource shortages, the circular economy helps to build economic resilience. It also creates opportunities for companies and employment markets, especially in the recycling, repair, and remanufacturing sectors. Promoting resource efficiency and future resilience helps the circular economy to be a complete approach to attaining sustainability (Ashokkumar et al., 2024). By using the possibilities of renewable biological resources, the strategic integration of bioproducts into the circular economy efficiently addresses the urgent demand for sustainable development. The basis for this integration is the mutually beneficial link between bioproducts and circular economy ideas, which aim to minimize environmental damage, enhance resource use, and strengthen economic stability. Sustainable replacements for fossil-based goods are biomass-based bioproducts, including plants, agricultural wastes, and microorganisms. Industries can guarantee resource efficiency by including them in the circular economy and ensuring that materials are regularly recycled instead of dumped as waste. Bioproducts provide a strong justification for their inclusion in the circular economy, as they help reduce dependence on limited resources and mitigate the environmental damage associated with fossil fuel extraction and processing. Since they are derived from renewable resources that can be replenished, bioproducts are inherently more sustainable and promote a fairer and more innovative approach to resource management (Swaminaathan et al., 2024). Moreover, bioproducts often show a lower carbon footprint than their traditional counterparts, helping lower greenhouse gas emissions and slow climate change. The ability of bioproducts to support the creation of new and sustainable value chains is another important justification. Industries can create creative business models that give efficient use of resources top priority by including bioproducts in the circular economy, such as closed-loop manufacturing systems whereby waste from one process is used as the input for another. Creating new markets and innovative ideas lowers waste and increases the financial viability of bioproducts by supporting economic growth and adaptation. Moreover, the use of bioproducts supports the aim of the circular economy to create a regenerative system benefiting society and the surroundings. Creating jobs in the local economy supports sustainable agricultural practices, advocates biodiversity, and helps build strong local businesses. Pursuing a sustainable future whereby economic development is separated from environmental damage requires including bioproducts in the circular economy (Nibbi et al., 2019).

2 Bioproducts: Concepts and Production

2.1 Classification of Bioproducts

Bioproducts can be categorized according to their source, use, and manufacturing methods, demonstrating how biological resources can be harnessed to produce environmentally friendly alternatives to traditional products (Rapp et al., 2021). The source of raw materials serves as a primary basis for categorizing bioproducts into plant-based, microbial-based, and animal-based types. Plant-based bioproducts, derived from crops, agricultural residues, or forest biomass, include biofuels such as ethanol and biodiesel, bioplastics like polylactic acid (PLA), and natural fibers used in textiles and construction. Microbial-based bioproducts are generated via fermentation using bacteria, yeast, or fungi (Sarsaiya et al., 2019). These processes result in the production of biochemicals, antibiotics, enzymes, and bioactive compounds. These bioproducts have applications in the food, manufacturing, and pharmaceutical industries. Bioproducts derived from animals come from animal tissues, fats, and other auxiliary materials. Manufacturing biopharmaceuticals, leathers, and materials based on collagen makes excellent use of them. Furthermore, classifying bioproducts, depending on their use in different sectors, is possible. These cover biogas, renewable energy sources, and biofuels, among other bioenergy products. Among other industrial chemicals, plastics, and solvents, they also include biochemicals. This category also contains biomaterials, which range in products from biodegradable packaging materials (Mabate & Pletschke, 2024), bioplastics, and bio-based composites. Highly flexible and capable of replacing traditional materials in various sectors, including energy, transportation, packaging, healthcare, and agriculture, bioproducts are derived through direct extraction from biomass, chemical conversion, and biotechnological processes such as fermentation and enzymatic reactions. These production techniques help classify bioproducts based on their manufacturing methods. Direct extraction involves obtaining natural compounds from plants or animal tissues, whereas chemical conversion transforms biomass into valuable products through chemical reactions (Ashokkumar et al., 2022). Similarly, biotechnological techniques, including fermentation, utilize microorganisms to produce bioproducts with diverse applications. This classification highlights the promising range of bioproducts and their capacity to enable a sustainable circular economy. By substituting renewable and environmentally friendly alternatives for conventional goods, these bioproducts efficiently satisfy the needs of many sectors (Table 1).

Table 1 Classification of bioproducts and their manufacturing process

Classification of bioproducts	Manufacturing process	Uses
1. Biofuels	Fermentation of biomass, transesterification of oils	Transportation fuels, energy generation
2. Bioplastics	Polymerization of bio-based monomers (e.g., PLA, PHA)	Packaging, consumer goods, automotive parts
3. Biolubricants	Chemical modification of vegetable oils	Industrial machinery, automotive engines
4. Bio-based adhesives	Extraction of natural polymers (e.g., starch, soy protein)	Wood products, packaging, construction
5. Bio-based solvents	Fermentation or extraction of natural compounds	Cleaning agents, paints, coatings
6. Bio-based surfactants	Fermentation of sugars and oils	Detergents, personal care products
7. Bio-based chemicals	Fermentation, enzymatic conversion of biomass	Industrial chemicals, pharmaceuticals
8. Biocomposites	Combination of natural fibers with bio-based resins	Construction materials, automotive parts
9. Biodegradable packaging	Production of materials like PLA, PHA, cellulose	Food packaging, disposable cutlery, bags
10. Bio-based textiles	Processing of natural fibers (e.g., cotton, hemp)	Clothing, home furnishings, industrial fabrics
11. Bio-based insulation	Production of cellulose, wool, or cork-based materials	Building insulation, soundproofing
12. Bio-based paints and coatings	Processing of natural oils, resins, and pigments	Protective coatings, decorative paints
13. Bio-based detergents	Synthesis of enzymes and natural surfactants	Household cleaning, industrial cleaning
14. Bio-based fertilizers	Composting, anaerobic digestion of organic waste	Agriculture, horticulture
15. Biopesticides	Extraction or fermentation of natural substances	Pest control in agriculture
16. Bio-based pharmaceuticals	Fermentation, extraction, and purification of bioactive compounds	Medicinal products, nutraceuticals
17. Bio-based enzymes	Fermentation and extraction from microorganisms	Industrial processes, food production, detergents
18. Bio-based flavors and fragrances	Extraction or fermentation of natural compounds	Food industry, cosmetics, perfumes
19. Bio-based energy storage	Development of bio-derived electrolytes and anodes	Batteries, capacitors
20. Bio-based construction materials	Processing of wood, bamboo, straw, and other biomass	Building structures, interior finishes
21. Bio-based polymers	Fermentation or extraction of monomers like lactic acid	Plastics, textiles, adhesives

(continued)

Table 1 (continued)

Classification of bioproducts	Manufacturing process	Uses
22. Bio-based activated carbon	Pyrolysis of biomass, chemical activation	Water purification, air filtration
23. Bio-based lubricant additives	Chemical modification of vegetable oils or animal fats	Enhancing lubricant properties in machinery
24. Bio-based oils and waxes	Cold pressing, extraction, and refining of plant oils	Cosmetics, candles, industrial applications
25. Bio-based sweeteners	Fermentation or extraction of natural sweet compounds	Food and beverage industry
26. Bio-based biopolymers	Polymerization of sugars, proteins, or other biomolecules	Packaging, medical devices, textiles
27. Bio-based membranes	Processing of cellulose, chitosan, or protein-based materials	Filtration, separation processes, medical applications
28. Bio-based resins	Extraction and processing of natural resins like rosin	Adhesives, coatings, composites
29. Bio-based nanomaterials	Synthesis from biomass-derived precursors	Electronics, medical devices, packaging
30. Bio-based rubber	Processing of natural rubber latex or guayule	Tires, seals, belts, medical devices
31. Bio-based cleaning agents	Extraction of natural surfactants and enzymes	Household cleaning, industrial applications

2.2 Manufacturing Processes and Technologies

Manufacturing processes and technologies are the central components of industrial production, involving the techniques and instruments needed to convert raw materials into finalized goods. These processes are essential for diverse industries, from automotive and aerospace to electronics and consumer goods. The selection of the manufacturing process is contingent upon factors such as the material's characteristics, the product's design, the production volume, and cost considerations (Oliveira et al., 2021). Conventional manufacturing methods encompass casting, machining, forming, and joining. Casting is pouring molten material into a mold to create a specific shape. At the same time, machining involves removing material from a workpiece to achieve the desired geometry. Forming processes, such as forging and extrusion, manipulate materials by exerting force upon them, while joining processes, such as welding and soldering, are employed to assemble components into a finished product (Kee et al., 2023; Acebes-Huerta et al., 2024; Heitzman et al., 2024). The development in manufacturing technologies has produced more accurate, flexible, and efficient production techniques. Additive manufacturing, sometimes known as 3D printing, is a significant technological development whereby objects are created by layer additions depending on digital designs. This technology is excellent for prototyping and manufacturing small quantities since it allows the construction of complex forms that would be difficult or impossible with traditional techniques. One significant advancement in using robotics and computer-operated machinery to perform highly accurate and consistent tasks is automation. This includes faster manufacturing and lower labor costs (Kumar et al., 2024). The modern output includes digital manufacturing technologies, computer-aided design (CAD), and computer-aided manufacturing (CAM). While CAM systems translate designs into manufacturing instructions, guaranteeing extraordinary quality, CAD software helps engineers create exact product designs. Furthermore, the arrival of Industry 4.0 technologies, including artificial intelligence (AI) and the Internet of Things (IoT), is transforming the manufacturing sector. Smart factories made possible by these technologies allow machines to interact with one another and continuously optimize output in real time, thus improving the quality of the products (Alazaiza et al., 2022). By lowering waste, reducing energy consumption, and enabling the use of renewable materials, these technologies enhance the efficiency and quality of manufacturing processes and support sustainable production methods. Satisfying the needs of a fast-changing global market, encouraging innovation, and ensuring the competitiveness of businesses worldwide depend on constant improvements in manufacturing technologies and processes.

2.3 Environmental and Economic Benefits of Bioproducts

Bioproducts provide significant environmental and economic advantages, making them an increasingly appealing substitute for conventional, fossil fuel-based products (Jain et al., 2022). Bioproducts are environmentally friendly because they are made from renewable biological resources like plants, agricultural residues, and microorganisms. This reduces our dependence on limited fossil fuels and helps to lower greenhouse gas emissions. Bioproducts typically employ less energy-intensive procedures and produce fewer pollutants than traditional manufacturing techniques. For example, using biofuels can aid in reducing carbon dioxide emissions, thereby contributing to climate change mitigation (Ahmed et al., 2023). Moreover, bioplastics, characterized by their ability to biodegrade or compost, can mitigate the escalating issue of plastic pollution, particularly in marine ecosystems and waste disposal sites, by undergoing a more natural decomposition process in the environment. From an economic perspective, bioproducts play a crucial role in driving the growth of a bioeconomy by promoting innovation and generating new industries and employment prospects (Costa et al., 2023). Producing many living organisms to create bioproducts, such as biofuels or biochemicals, can enhance agricultural sectors and rural economies by generating extra sources of income for farmers and landowners. In addition, the processing and manufacturing of bioproducts can result in establishing new manufacturing facilities, creating jobs, and promoting regional economic growth. The economy also encourages the expansion of energy sources, decreasing reliance on imported fossil fuels and improving energy security (Kumar et al., 2023; Mujtaba et al., 2023). Bioproducts can enhance the development of self-sufficient economies that are less susceptible to the impact of global market fluctuations by promoting local production and consumption. Bioproducts adhere to the principles of the circular economy by promoting resource efficiency and waste reduction. Bioproducts play a crucial role in production by transforming agricultural residues, waste materials, or non-food crops into valuable resources. This helps to minimize waste and maximize economic benefits (Priya et al., 2023). This practice preserves resources and diminishes disposal expenses and environmental repercussions. Incorporating bioproducts into diverse industries, ranging from energy to materials science, signifies a substantial advancement toward a more sustainable and economically feasible future, where economic expansion is aligned with environmental conservation (Table 2).

3 Circular Economy: Principles and Practices

3.1 Fundamentals of Circular Economy

The core principles of a circular economy center on redefining economic expansion through sustainable, restorative, and regenerative methods that reduce waste, prolong the product lifespan, and optimize resource utilization. The circular economy seeks to close the loop of product lifecycles by advocating for the reuse, refurbishment, recycling, and remanufacturing of materials, in contrast to the linear economy's "take-make-dispose" model. This approach decreases the negative effects on the environment related to production and consumption and promotes economic stability by establishing sustainable value chains and decreasing reliance on limited resources. The circular economy's central tenet is that design can help to eradicate waste and pollution. Product and process design seeks to maximize material lifetime and minimize waste from the outset (Nibbi et al., 2019). This means selecting long-lasting, reusable, or decomposable materials with great care and designing products that fit easily removed, fixed, or improved upon (Oliveira et al., 2021). By prioritizing sustainability during the design process, businesses can significantly reduce the environmental impact of their products and help the circular economy run free from waste. Maintaining goods and materials in active use is another fundamental concept (Chen et al., 2023). This is achieved by using maintenance, repair, reuse, and remanufacturing, which extend the lifetime of goods and help them not to become wasted early on. For example, choosing remanufactured parts instead of new ones in the automotive industry will help save resources, lower energy consumption, and reduce greenhouse gas emissions (Saravanan et al., 2023). Similarly, encouraging the reusing of goods via second-hand markets or sharing platforms helps to maximize resource value and reduce the need for new manufacturing. The circular economy restores natural systems (Ashokkumar et al., 2024). This approach seeks to revive and improve ecosystems by replenishing valuable nutrients in the soil and establishing conditions that enable natural systems to rejuvenate (Mabate & Pletschke, 2024) rather than only minimizing damage. Practices like regenerative agriculture, which aims to restore soil health, improve biodiversity, and capture carbon, support the mitigation of climate change and guarantee food security using their clear evidence. By matching economic activities with natural cycles, the circular economy helps to sustain environmental systems as well as human ones over the

Table 2 Environmental and economic benefits of bioproducts

Benefit category	Benefit	Description
1. Reduced greenhouse gas emissions	Lower carbon footprint	Bioproducts often have a lower carbon footprint than fossil-based products, reducing GHG emissions.
2. Renewable resources	Sustainable raw materials	Bioproducts use renewable resources, reducing dependency on finite fossil fuels.
3. Waste reduction	Waste valorization	Converts waste materials into valuable products, minimizing landfill use and pollution.
4. Energy efficiency	Reduced energy consumption	Production of bioproducts often requires less energy compared to conventional materials.
5. Biodegradability	Reduced landfill waste	Many bioproducts are biodegradable, reducing the accumulation of persistent waste in landfills.
6. Soil health improvement	Organic fertilizers	Bio-based fertilizers enhance soil health and fertility, reducing the need for synthetic chemicals.
7. Water conservation	Reduced water usage	Some bioproducts are produced with less water compared to conventional products.
8. Pollution reduction	Lower toxic emissions	Bioproducts often emit fewer toxins and pollutants during production and disposal.
9. Biodiversity protection	Conservation of ecosystems	Sustainable sourcing of biomass helps protect natural habitats and biodiversity.
10. Carbon sequestration	Carbon storage	Plants used for bioproducts capture and store carbon dioxide, mitigating climate change.
11. Rural development	Job creation	The bioproducts industry creates jobs in rural areas, boosting local economies.
12. Economic diversification	New market opportunities	Expands market opportunities for agricultural and forestry products.
13. Lower resource dependency	Reduced reliance on imports	Uses locally sourced materials, reducing dependency on imported resources.
14. Industrial efficiency	Optimized production processes	Bioproducts often lead to more efficient industrial processes and reduced waste.
15. Cost reduction	Lower long-term costs	In many cases, bioproducts can reduce long-term costs due to lower environmental compliance expenses.
16. Enhanced product performance	Improved material properties	Bioproducts can offer enhanced performance characteristics such as strength and flexibility.
17. Innovation promotion	Stimulates technological advancement	Drives innovation and development of new technologies and processes in various sectors.

(continued)

Table 2 (continued)

Benefit category	Benefit	Description
18. Waste-to-energy	Renewable energy production	Converts waste materials into bioenergy, providing a sustainable energy source.
19. Local economy boost	Support for small and medium enterprises	Provides opportunities for SMEs in the bioproducts supply chain.
20. Reduced air pollution	Cleaner production processes	Bioproducts production often results in lower air pollution compared to traditional processes.
21. Sustainable agriculture	Promotes organic farming practices	Encourages organic farming methods, reducing the need for synthetic chemicals.
22. Green building materials	Sustainable construction	Provides eco-friendly materials for green building and construction projects.
23. Resource efficiency	Maximized utilization of biomass	Improves the efficiency of resource use by converting biomass into multiple products.
24. Eco-friendly packaging	Reduces plastic waste	Offers biodegradable or compostable packaging alternatives, reducing plastic pollution.
25. Community engagement	Promotes community involvement	Involves local communities in bioproduct production and decision-making processes.
26. Climate resilience	Adaptation to climate change	Bio-based systems and materials can adapt to changing climatic conditions.
27. Health benefits	Reduced chemical exposure	Uses natural ingredients, reducing exposure to harmful synthetic chemicals.
28. Environmental remediation	Soil and water remediation	Some bioproducts help in cleaning contaminated soil and water through bioremediation processes.
29. Energy security	Diversified energy sources	Reduces reliance on fossil fuels and enhances energy security through renewable bioenergy sources.
30. Reduced greenhouse gas emissions in agriculture	Lower methane and nitrous oxide emissions	Sustainable agricultural practices associated with bioproducts reduce methane and nitrous oxide emissions.
31. Increased agricultural productivity	Enhanced soil fertility and crop yields	Bio-based fertilizers and soil conditioners can boost crop productivity and soil health.
32. Enhanced waste management	Efficient waste processing and recycling	Bioproducts contribute to better waste management and recycling systems.
33. Eco-friendly personal care	Natural ingredients	Offers personal care products made from natural and sustainable sources.
34. Energy cost savings	Reduced energy bills	Bio-based energy sources can lower energy costs for consumers and businesses.

(continued)

Table 2 (continued)

Benefit category	Benefit	Description
35. Sustainable transportation	Biofuels for vehicles	Provides renewable fuels for transportation, reducing reliance on fossil fuels.
36. Long-term economic viability	Sustainable business models	Supports long-term economic viability through sustainable practices and resource management.
37. Conservation of resources	Efficient use of non-food biomass	Utilizes non-food biomass for various applications, conserving food resources.
38. Increased recycling rates	Improved material recovery	Enhances recycling processes by providing alternative materials that are easier to process.
39. Global competitiveness	Strengthened market position	Enhances global competitiveness by offering sustainable and innovative products.
40. Reduced dependency on synthetic inputs	Lower use of chemical inputs	Reduces reliance on synthetic chemicals in agriculture, industry, and consumer products.
41. Educational opportunities	Research and development careers	Creates educational and career opportunities in the fields of biotechnology and environmental science.

long term. Moreover, the circular economy depends on creative business models that separate the use of resources from economic growth. This covers business models such as product-as-a-service, whereby businesses keep ownership of items and present them to consumers as a service. This strategy motivates the production of more efficient and durable goods. Another example is the sharing economy, where several users share resources to lower the demand for new goods and promote better use of present assets (Kumar et al., 2024). These models provide companies with fresh opportunities to create income and sustainably give value to consumers, thus improving resource efficiency. The circular economy cannot be facilitated without technology, mainly via digital tools that increase tracking, management, and resource optimization ability. Big data, artificial intelligence (AI), and the Internet of Things (IoT), among other technologies, help to monitor product consumption, project maintenance needs, and streamline resource flow. Real-time product state monitoring, made possible by intelligent sensors, helps to enable quick maintenance. It lowers the possibility of early disposal by digital platforms. It can also help consumers to link with second-hand markets, repair companies, or sharing networks, thus extending the lifetime of goods (Swaminaathan et al., 2024). The success of the circular economy is influenced by present technological developments that allow companies and consumers to make more environmentally friendly choices. The circular economy essentially marks a basic transformation that challenges accepted ideas of manufacturing and consumption. Establishing a sustainable economy within the planetary constraints requires cooperation among many sectors, governments, and consumers. Following ideas, including waste reduction, extending product and material use, restoring natural systems, and implementing creative business models, helps the circular economy offer a more sustainable, resilient, and fair future. This strategy reveals fresh economic opportunities, addresses the pressing environmental challenges of our day, and promotes long-term prosperity. Every person and organization is accountable for helping this cooperative effort; the circular economy is defined as a genuinely inclusive and powerful movement.

3.2 Strategies for Resource Optimization and Waste Minimization

Essential elements of sustainable development are minimizing waste and using resources best. Their main goals are to maximize resource efficiency and reduce waste in manufacturing and consumption processes. These approaches seek to increase economic efficiency, lower environmental impact, and prolong material lifetime by optimizing resource value (Nibbi et al., 2019; Sarsaiya et al., 2019). The following vital ideas show how companies and sectors might maximize resource use and reduce waste, thus supporting a more circular economy. One quite effective way to maximize resources is to apply lean manufacturing ideas. Lean manufacturing aims to maximize operations, increase efficiency, and lower unnecessary resource consumption by eradicating inefficiencies in production processes. This method entails always observing and enhancing procedures to find where resources might be better used or saved. Companies can use optimal production schedules and inventory control to increase operational efficiency and sustainability by lowering material waste and energy consumption and preserving production levels (Rapp et al., 2021; Costa et al., 2023). Lean manufacturing prioritizes quality control to guarantee the exact production of goods. This method aims to minimize scrap generated and lower the frequency of rework. Using eco-design ideas—creating goods with a specific eye toward their environmental impact over their lifetime—is another key tactic. Once their lifetime is reached, this entails choosing renewable, recyclable, or biodegradable materials and building sturdy, readily repairable, recyclable goods. By including environmental elements throughout the design process, companies can lower waste generation and resource consumption and encourage product recycling or repurposing (Saravanan et al., 2023; Ashokkumar et al., 2024). Circular business models prioritizing reusing and recycling materials instead of new resource extraction will help optimize resources. Product-as-a-service refers to a business model where businesses retain ownership of goods and provide them to consumers as a service. The sharing economy—which promotes the shared use of products and resources among several users—is one example of these models. These models help reduce waste and increase the lifetime of materials by encouraging the effective use of current resources and lessening the demand for new products. Especially when the lifetime of a product ends, reverse logistics systems are essential for efficiently retrieving and reinserting goods into the production cycle. Maximizing resource use and lowering waste depend on technological innovation. The IoT, big data, and AI have made it possible for more accurate monitoring, management, and resource optimization. IoT sensors can track the condition and use of materials in real time, which lets you plan maintenance needs and lowers the chance of equipment breaking down and material waste (Heitzman et al., 2024; Mabate & Pletschke, 2024). Using big data analytics allows businesses to spot trends in resource use, improve processes, and reduce inefficiencies. AI-driven systems guarantee the most effective use of resources by automating waste management and resource distribution, thus guaranteeing efficiency. Moreover, strategies for waste minimization sometimes call for using sensible waste management techniques, including energy recovery, composting, and recycling. By separating waste streams and guaranteeing the correct processing of recyclable materials, companies can drastically lower their environmental impact and help divert significant volumes of waste away from landfills. Organic waste composting helps the soil to restore vital nutrients (Swaminaathan et al., 2024). Concurrent with this, waste-to-energy and other energy recovery methods help to transform waste products into usable energy, enhancing resource use efficiency. Achieving sustainable development and moving toward a circular economy depends on putting plans to maximize resources and minimize waste into effect. Companies can significantly lower resource consumption and waste generation by using lean manufacturing techniques, including eco-design ideas, applying circular business models, and using technology developments. These approaches promote environmental sustainability, economic efficiency, resilience, and long-term profitability, fostering a fairer and more sustainable future.

3.3 Role of Bioproducts in Circularity Enhancement

Promoting circularity depends on bioproducts since they help minimize waste and optimize resources by including renewable resources in production cycles and reducing dependency on fossil-based materials (Oliveira et al., 2021). Including bioproducts in many sectors offers a way to control material consumption properly, promote environmental sustainability, and increase economic resilience, given the worldwide trend toward sustainability. Maximizing the lifetime of goods and materials helps preserve their value, thus defining circularity. Bioproducts naturally follow this idea from sustainable biological sources, including plants, agricultural waste, and microorganisms (Alazaiza et al., 2022). Using renewable resources, bioproducts minimize the environmental effects of their extraction and processing, thus reducing reliance on finite fuels. This habit helps protect natural resources and reduces greenhouse gas emissions, helping slow global warming. Because they can biodegrade and compost, bioproducts greatly help to achieve circularity. Unlike conventional plastics and materials that remain in the environment for a long time, many bioproducts are designed primarily to break down organically once in their lifetime. This process not only replenishes vital nutrients for the soil but also significantly reduces waste accumulation, offering a hopeful future view (Ashokkumar et al., 2022). For example, bioplastics made from plant-based materials like polylactic acid (PLA) can be composted in factories, thus completing the cycle of material use and lessening the demand in landfills. Furthermore, bioproducts improve circularity by advocating using waste and byproducts from different sectors as primary sources. Industrial symbiosis is developing new products using waste or byproducts exchanged between different sectors. This system maximizes resource use and helps to reduce waste (Abdullah et al., 2024). For instance, agricultural wastes, including straw, husks, and other plant materials, can be converted into biofuels, biochemicals, and biomaterials, turning what would have been thrown away into valuable resources. In addition to reducing the environmental effects of waste management, this creates new economic opportunities and motivates and involves bioeconomy participants. Furthermore, helping to create circular business models stressing the effective use of resources and waste reduction are bioproducts. For example, the concept of a bio-based circular economy calls for bioproducts inside closed-loop systems whereby materials are repeatedly recycled and reintegrated into the manufacturing process. Biorefineries provide one instance where biomass is converted into several bioproducts, including biofuels, bioplastics, and biochemicals (Ashokkumar et al., 2024). These biorefineries also recover and rework waste streams and byproducts concurrently. Through the best use of biomass and waste minimization, these systems encourage circularity. As much as in the environment, bioproducts encourage circularity in social and economic spheres. By creating new markets for renewable materials, bioproducts help to promote economic diversification and job generation. In rural areas with plenty of biomass resources, this is especially helpful. Lowering dependence on imported fossil fuels and materials increases resilience and helps local economies to flourish. Because bioproducts encourage creativity in manufacturing and product design, their economic possibilities are just beginning. This encourages the development of more environmentally friendly products that satisfy consumer tastes for greebles. Thanks to bioproducts, circularity can also be included in many sectors, including agriculture, energy, packaging, and building. By lowering the reliance on synthetic chemicals and enhancing soil conditions in agriculture, bioproducts—such as fertilizers and biopesticides—are vital in supporting sustainable farming methods. In the energy sector, biofuels provide a sustainable replacement for fossil fuels, helping to lower carbon emissions in generating electricity and transportation. A sustainable packaging solution that helps to solve plastic waste is bioplastics. Likewise, sustainable building materials, including straw, hemp, and wood, are used in the building sector. Improving circularity using renewable resources, promoting waste reduction, and enabling the expansion of circular business models depends on bioproducts. Their ability to substitute materials derived from fossils, break down organically following their life cycle, and use waste as a valuable resource makes them essential in changing a circular economy. Promoting environmental sustainability, economic resilience, and innovation depends significantly on biomass products. They help to create a more sustainable and circular future whereby resources are used effectively, waste is minimized, and the value of materials is preserved for as long as possible (Kumar et al., 2024).

4 Applications of Bioproducts

4.1 Bioproducts in Agriculture and Forestry

Bioproducts from biological sources primarily target agriculture and forestry due to their potential to enhance sustainability, reduce environmental damage, and promote healthier ecosystems. The product range encompasses fertilizers, biopesticides, biostimulants, and additional inputs sourced from biological materials that improve soil health, facilitate plant growth, and assist in pest control (Nibbi et al., 2019). The increasing global demand for food and timber creates a promising opportunity to integrate bioproducts in agriculture and forestry, thereby improving productivity and minimizing environmental impact. Biofertilizers are organic materials composed of living microorganisms. These microorganisms enhance plant nutrient availability upon exposure to seeds, plants, or soil (Ashokkumar et al., 2022). Unlike chemical fertilizers, biofertilizers provide advantages, including improved soil structure, enhanced nutrient availability, and augmented plant growth without detrimental environmental effects. Conversely, chemical fertilizers may lead to soil degradation and water contamination. *Rhizobium, Azospirillum*, and *mycorrhizal* fungi are commonly utilized microorganisms in biofertilizer production (Kee et al., 2023). These microorganisms facilitate nitrogen fixation, phosphorus solubilization, and nutrient cycling, enhancing soil quality and promoting plant growth. Biopesticides derived from organic sources, including plants, bacteria, and particular minerals, provide a sustainable substitute for synthetic pesticides (Oliveira et al., 2021). They play a vital role in integrated pest management (IPM) systems by explicitly targeting certain pests while minimizing adverse effects on beneficial organisms and the environment. *Bacillus thuringiensis* (Bt) is a bacterium widely utilized in agriculture as a biopesticide because of its capacity to generate toxins harmful to insects. Biopesticides include plant-derived substances like neem oil, which exhibits insecticidal properties against various pests, and pheromones that interfere with pest mating (Priya et al., 2023). Their utilization reduces chemical pesticide residues in food and diminishes the risk of pest resistance. Biostimulants are substances or microorganisms that enhance plant growth, augment nutrient absorption, and bolster the plant's resilience to stress. Biostimulants improve the efficacy of plant resource utilization by promoting natural processes, whereas fertilizers provide nutrients (Acebes-Huerta et al., 2024). Biostimulants include seaweed extracts, humic acids, and advantageous bacteria such as *Bacillus* spp. Biostimulants are utilized in forestry to improve the growth and durability of seedlings, particularly in reforestation efforts where young trees are vulnerable to environmental stressors (Jain et al., 2022). Bioproducts are crucial for the sustainable management of forests in forestry. Biofertilizers and biostimulants enhance tree growth in nurseries and during reforestation, improving seedling establishment and survival rates. Biopesticides efficiently manage forest pests and diseases, reducing dependence on chemical treatments that may adversely affect non-target species and disrupt ecosystems (Kumar et al., 2023). Moreover, bioproducts are increasingly employed in forest restoration efforts, as they enhance soil fertility and promote the growth of native plant species. Adopting bioproducts in agriculture and forestry offers numerous environmental and economic benefits. Bioproducts diminish reliance on synthetic chemicals, consequently lowering the risk of soil and water contamination while promoting biodiversity. Furthermore, they improve carbon sequestration by enhancing soil health and increasing organic matter content. From an economic standpoint, bioproducts can reduce costs for farmers and foresters, improve crop yields, and provide access to markets with significant demand for sustainably produced goods. Despite the benefits of bioproducts, their widespread adoption faces challenges, including variable efficacy, regulatory impediments, and insufficient awareness among agricultural and forestry practitioners. Ongoing research and development are crucial for tackling these challenges by improving the efficiency of bioproducts, creating robust regulatory frameworks, and promoting education and outreach. As the global focus on sustainable agriculture and forestry intensifies, bioproducts are poised to significantly influence these sectors' futures. They offer a means to implement more sustainable and eco-friendly practices (Swaminaathan et al., 2024).

4.2 Bioproducts in Energy Generation and Storage

Bioproducts are increasingly recognized for their potential to fundamentally revolutionize energy production and storage, offering eco-friendly alternatives to conventional fossil fuels and chemical batteries. Bioproducts in the energy sector are derived from renewable biological sources, including plants, algae, and microorganisms (Nibbi et al., 2019). These bioproducts encompass biofuels, biogas, biochemicals, and bio-based materials utilized for energy storage. These products utilize waste streams and renewable resources to reduce greenhouse gas emissions, enhance energy security, and promote a circular economy (Oliveira et al., 2021). A notable application of bioproducts in energy generation is the production of biofuels derived from biomass that serve as substitutes for petroleum-based fuels. Bioethanol, produced through the enzymatic degradation of sugars in crops like corn and sugarcane, is widely utilized as an alternative to petrol. Biodiesel, derived from vegetable oils or animal fats, is a significant biofuel that can be employed in diesel engines with minimal modifications (Alazaiza et al., 2022). Cellulosic ethanol and algae-derived biofuels exemplify advanced biofuels, representing the next generation of renewable fuels. These advanced biofuels are beneficial as they are derived from non-food biomass sources, including agricultural residues, forest waste, and algae (Kee et al., 2023). This mitigates the food-versus-fuel controversy and enhances sustainability. Biogas, an important bioproduct, is produced through the anaerobic decomposition of organic materials, including agricultural residues, manure, and food waste (Abdullah et al., 2024). The process generates biogas with a high methane concentration, which is suitable for heating, electricity generation, or vehicle fuel following purification (Ashokkumar et al., 2024). Biogas production provides a sustainable energy source and facilitates waste management by reducing the volume of organic waste sent to landfills. Moreover, the byproduct of anaerobic digestion, known as digestate, can function as a nutrient-rich fertilizer, thereby improving the closed-loop system of the circular economy. Bioproducts, biofuels, and biogas are advancing notably in energy storage. Bio-based batteries and supercapacitors represent an emerging area of research aimed at creating sustainable alternatives to traditional energy storage technologies (Heitzman et al., 2024). These technologies often rely on scarce and environmentally detrimental resources like lithium and cobalt. Lignin, a byproduct from the paper and biofuel industries, is presently being explored as a potential material for producing bio-based anodes in lithium-ion batteries. Additionally, cellulose, an abundant natural polymer, is presently being investigated for its prospective use in bio-based electrolytes and separators. Biologically sourced materials can offer energy storage solutions that are environmentally sustainable, biodegradable, and economically viable. A promising domain is the application of bioelectrochemical systems, particularly microbial fuel cells (MFCs), which generate electricity via the metabolic activities of microorganisms. Microbial fuel cells (MFCs) utilize bacteria to transform organic matter into electrical energy, offering a distinctive approach for harnessing energy from wastewater, organic waste, and marine ecosystems. Despite being in the nascent stages of development, MFCs exhibit significant potential for decentralized energy generation, particularly in remote or off-grid areas. Integrating bioproducts into energy generation and storage systems enhances the transition to a low-carbon economy by reducing reliance on fossil fuels and lowering carbon emissions. The rising global demand for clean energy is anticipated to enhance the utilization of bioproducts. Progress in biotechnology, process optimization, and the development of novel materials will drive this expansion. Nonetheless, challenges persist, including the capacity to enhance production processes, competition for biomass resources, and the necessity for supportive policies and infrastructure. Bioproducts are crucial to the future energy paradigm, offering sustainable energy generation and storage solutions (Kumar et al., 2024). By employing biological processes, these products significantly enhance the development of a more resilient and environmentally sustainable energy system, effectively tackling the intertwined challenges of energy security and climate change. As research and innovation advance, bioproducts are poised to play an increasingly important role in the global transition to sustainable energy (Priya et al., 2023; Ashokkumar et al., 2024).

4.3 Bioproducts in Materials and Manufacturing

Bioproducts are revolutionizing the field of materials and manufacturing by providing renewable and sustainable substitutes for conventional materials obtained from petrochemicals and other finite resources. These bioproducts, namely biopolymers, bio-composites, and bio-based chemicals, are obtained from living organisms such as plants, algae, and microorganisms. By incorporating them into different manufacturing processes, their presence diminishes the negative effects on the environment and creates opportunities for advancements in material characteristics, product development, and industrial uses. Biopolymers are highly important bioproducts in materials and manufacturing, offering an environmentally friendly substitute for traditional plastics. Biopolymers from renewable sources like corn starch, sugarcane, or cellulose produce biodegradable plastics that break down more easily than petroleum-based plastics. Polylactic acid (PLA) and polyhydroxyalkanoates (PHA) are extensively utilized biopolymers in the industry. PLA, derived from fermented plant starch, is widely utilized in packaging, disposable utensils, and 3D printing. Polyhydroxyalkanoate (PHA), produced by bacteria using organic feedstocks, finds applications in various fields, including medical devices and agricultural films. Using biopolymers in the manufacturing industry tackles the escalating issue of plastic pollution and diminishes reliance on fossil fuels. Bioproducts are having a significant impact in the field of biocomposites. These composites combine natural fibers, such as hemp, flax, or jute, with bio-based or conventional polymers. This combination results in composites that have improved mechanical properties. Biocomposites are becoming more prevalent in the automotive, construction, and consumer goods sectors, providing a sustainable substitute for conventional composite materials composed of glass or carbon fibers. In the automotive industry, biocomposites produce lightweight components that help enhance fuel efficiency and decrease carbon emissions (Acebes-Huerta et al., 2024; Swaminaathan et al., 2024). Using natural fibers in these composites diminishes the environmental impact and capitalizes on the renewable and biodegradable characteristics of the raw materials. Bio-based chemicals obtained from renewable biological sources constitute an essential group of bioproducts in materials and manufacturing. These chemicals are fundamental components of various products, such as solvents, adhesives, and coatings. Lactic acid is a bio-based compound utilized to manufacture PLA and produce pharmaceuticals, food additives, and personal care products. Moreover, bio-based succinic acid, obtained through fermenting sugars, is utilized to produce biodegradable polyesters and manufacture resins, coatings, and plasticizers. The transition toward bio-based chemicals in manufacturing is motivated by the imperative to decrease greenhouse gas emissions, limit the utilization of hazardous substances, and advance the circular economy. Bioproducts in materials and manufacturing have environmental advantages and present novel prospects for innovation in material science and engineering. The distinctive characteristics of biopolymers and biocomposites, including their ability to decompose naturally, be replenished, and be modified for specific purposes, enable the creation of innovative materials with customized attributes. Researchers are currently investigating the utilization of bio-based nanomaterials, specifically cellulose nanocrystals, to produce high-performance composites that exhibit exceptional strength, transparency, and barrier properties. These technological advancements create opportunities to develop new applications in various fields, including packaging, electronics, and biomedical devices. Although bioproducts offer numerous benefits in materials and manufacturing, there are still obstacles to their widespread acceptance and use (Nibbi et al., 2019; Sarsaiya et al., 2019; Ahmed et al., 2023; Kee et al., 2023; Kumar et al., 2023; Priya et al., 2023). These factors encompass the ability to increase production processes efficiently, the cost-effectiveness of bioproducts compared to conventional materials, and the necessity of favorable policies and infrastructure to facilitate the shift toward bio-based manufacturing. Moreover, additional investigation and advancement are required to enhance the efficiency and longevity of bioproducts and investigate novel uses and markets. Bioproducts play an increasingly significant role in the materials and manufacturing industries, providing sustainable alternatives that decrease environmental harm and stimulate innovation. As the worldwide need for environmentally friendly products increases, incorporating bioproducts into manufacturing processes is expected to speed up, leading to a more sustainable and robust industrial future. Bioproducts can revolutionize the production and utilization of materials, leading to a more sustainable and circular economy thanks to continuous research, technological progress, and supportive policy frameworks (Table 3).

5 Policy and Market Dynamics

Regulatory frameworks and governmental incentives profoundly impact the adoption and advancement of bioproducts. These frameworks and incentives can either facilitate or obstruct the transition to a bio-based economy (Nibbi et al., 2019; Alazaiza et al., 2022). They significantly influence market dynamics, guarantee adherence to safety and environmental regulations, and promote innovation and investment within the industry. The regulatory frameworks for bioproducts are designed to guarantee safety, environmental sustainability, and rigorous compliance with quality standards for human consumption (Kee et al., 2023; Mujtaba et al., 2023). The regulations concerning bioproducts, including biofuels, bioplastics, and bio-based chemicals, differ significantly across various regions. Bioproduct regulation in the United States is administered by multiple federal agencies, including the Environmental Protection Agency (EPA), the Food and Drug Administration (FDA), and the Department of Agriculture (USDA). Biofuels such as ethanol and biodiesel are governed by the Renewable Fuel Standard (RFS), which requires incorporating renewable fuels in transportation fuel supplies. The primary objective of the RFS is to diminish greenhouse gas emissions and decrease reliance on fossil fuels by promoting and facilitating the utilization of biofuels. The USDA promotes bioproducts via its BioPreferred Program, which advocates using bio-based products in federal procurement. The European Union's regulatory framework is defined by the Renewable Energy Directive (RED), which establishes targets for utilizing renewable energy, including bioenergy, across member states. The amended RED II directive significantly underscores the incorporation of sustainability criteria for biofuels, bioliquids, and biomass fuels. This ensures that these products will substantially influence the climate goals established by the European Union. The European Union's Bioeconomy Strategy delineates a framework of regulations and policies designed to foster the sustainable production and utilization of bio-based products. Numerous governments globally have instituted diverse incentives to promote the advancement and utilization of bioproducts (Nibbi et al., 2019; Sarsaiya et al., 2019; Oliveira et al., 2021; Rapp et al., 2021; Alazaiza et al., 2022). These incentives reduce production costs, promote the development of innovative ideas and

Table 3 Different applications of bioproducts

Application	Benefits	Methods of use	Advantages	Efficiency
1. Biofuels (ethanol, biodiesel)	Reduces greenhouse gas emissions, renewable source	Blended with fossil fuels, used in dedicated engines	Lower emissions, energy security	High energy efficiency in modern engines
2. Bioplastics (PLA, PHA)	Biodegradable, reduces plastic pollution	Used in packaging, disposable items, and consumer goods	Reduces waste, compostable in industrial settings	Comparable to conventional plastics in many applications
3. Biolubricants	Biodegradable, reduces environmental impact	Used in machinery, automotive engines	Lower toxicity, less environmental harm	Effective lubrication, especially in extreme conditions
4. Bio-based adhesives	Non-toxic, derived from renewable resources	Used in wood products, packaging, and construction	Safer for health, less environmental impact	Strong bonding performance
5. Bio-based solvents	Reduced toxicity, renewable source	Used in paints, coatings, and cleaning products	Less harmful to health and environment	Effective for many industrial applications
6. Bio-based surfactants	Biodegradable, derived from natural sources	Used in detergents, personal care products	Lower environmental impact, skin-friendly	Effective cleaning and foaming agents
7. Bio-based chemicals	Sustainable production, reduced dependency on petroleum products	Used in industrial processes, pharmaceuticals	Lower environmental impact, renewability	Efficient for a variety of chemical applications
8. Biocomposites	Lightweight, renewable resources	Used in construction materials, automotive parts	Reduces reliance on non-renewable materials	High strength-to-weight ratio
9. Biodegradable packaging	Reduces landfill waste, compostable	Used in food packaging, disposable cutlery	Reduces plastic pollution, compostable in industrial facilities	Effective in reducing environmental impact
10. Bio-based textiles	Sustainable, reduces reliance on synthetic fibers	Used in clothing, home furnishings, and industrial fabrics	Environmentally friendly, biodegradable	Comparable durability to conventional textiles

(continued)

Table 3 (continued)

Application	Benefits	Methods of use	Advantages	Efficiency
11. Bio-based insulation	Renewable, improves energy efficiency	Used in building construction	Better thermal performance, eco-friendly	High thermal insulation performance
12. Bio-based paints and coatings	Non-toxic, derived from natural ingredients	Used in home and industrial applications	Safer for indoor air quality, less environmental impact	Effective coverage and durability
13. Bio-based detergents	Environmentally friendly, biodegradable	Used in household and industrial cleaning	Less harmful to aquatic life, effective cleaning	Comparable performance to conventional detergents
14. Bio-based fertilizers	Enhances soil health, reduces synthetic chemical use	Applied to crops and soil	Improves soil fertility, sustainable	Effective for increasing crop yields
15. Biopesticides	Natural pest control, reduces chemical pesticide use	Applied to crops as sprays or soil treatments	Safer for non-target species, less environmental impact	Effective pest management in agriculture
16. Bio-based pharmaceuticals	Sustainable production, reduced reliance on synthetic drugs	Used in medical treatments, nutraceuticals	Lower environmental impact, potentially fewer side effects	Effective in treating various health conditions
17. Bio-based enzymes	Biodegradable, effective in various industrial processes	Used in food processing, detergents, and pharmaceuticals	High specificity, reduces chemical use	Efficient in targeted applications
18. Bio-based flavors and fragrances	Natural origin, reduces synthetic chemical use	Used in food and beverage products, personal care items	Safer for consumers, environmentally friendly	Effective in providing desired flavors and scents
19. Bio-based energy storage	Renewable materials, enhances energy storage capabilities	Used in batteries, supercapacitors	Reduced reliance on non-renewable materials	High energy density and performance
20. Bio-based construction materials	Sustainable, reduces reliance on traditional materials	Used in building structures, interior finishes	Eco-friendly, often renewable	Comparable strength and durability to conventional materials

(continued)

Table 3 (continued)

Application	Benefits	Methods of use	Advantages	Efficiency
21. Bio-based polymers	Renewable resources, biodegradable	Used in various applications including packaging and textiles	Reduces environmental impact, sustainable	Effective for many applications including biodegradable options
22. Bio-based activated carbon	Renewable, effective in pollution control	Used in water and air purification systems	High adsorption capacity, less environmental impact	Effective for removing contaminants
23. Bio-based lubricant additives	Derived from renewable resources, reduces wear and tear	Added to conventional lubricants	Enhances performance, less toxic	Improves lubricant properties
24. Bio-based oils and waxes	Natural ingredients, renewable	Used in cosmetics, candles, and industrial applications	Safer for health, less environmental impact	Effective in various applications
25. Bio-based sweeteners	Natural origin, reduced calorie content	Used in food and beverages	Lower calorie content, natural alternative to synthetic sweeteners	Effective for sweetening with reduced health concerns
26. Bio-based biopolymers	Derived from renewable sources, biodegradable	Used in packaging, medical devices, and textiles	Reduces environmental impact, compostable	Effective for a wide range of applications
27. Bio-based membranes	Derived from natural materials, efficient filtration	Used in water purification, gas separation	Effective in filtering, less environmental impact	High filtration efficiency
28. Bio-based resins	Renewable, less toxic	Used in adhesives, coatings, and composites	Lower toxicity, sustainable production	Effective bonding and coating performance
29. Bio-based nanomaterials	Sustainable production, versatile applications	Used in electronics, medicine, and environmental applications	High performance, reduced environmental impact	Effective in advanced technological applications
30. Bio-based rubber	Derived from natural sources, renewable	Used in tires, seals, and industrial products	Reduces dependency on synthetic rubber, renewable	Effective performance in various applications

(continued)

Table 3 (continued)

Application	Benefits	Methods of use	Advantages	Efficiency
31. Bio-based cleaning agents	Natural ingredients, less harmful to the environment	Used in household and industrial cleaning	Reduces chemical exposure, biodegradable	Effective cleaning properties
32. Bio-based coatings	Environmentally friendly, less toxic	Used in industrial and consumer product coatings	Safer for health and environment, effective protection	Comparable performance to conventional coatings
33. Bio-based water treatment chemicals	Effective in removing contaminants, renewable	Used in water purification and treatment systems	Reduces reliance on synthetic chemicals, sustainable	High efficiency in purifying water
34. Bio-based fibers	Renewable, strong	Used in textiles, composites, and insulation materials	Environmentally friendly, high strength	Effective for various applications
35. Bio-based adhesive films	Sustainable, non-toxic	Used in packaging and labeling	Safer for health, reduces plastic use	Effective for various packaging needs
36. Bio-based pharmaceutical carriers	Biodegradable, controlled release	Used in drug delivery systems	Reduces environmental impact, effective drug delivery	High efficiency in drug administration
37. Bio-based soil conditioners	Enhances soil structure and fertility	Applied to soil for improving crop growth	Improves soil health, sustainable	Effective in enhancing agricultural productivity
38. Bio-based additives for foods	Natural origin, reduces synthetic chemicals	Used to enhance flavors, textures, and preservation	Healthier alternatives to synthetic additives	Effective in food processing and preservation
39. Bio-based polymers for electronics	Renewable, high performance	Used in electronic device components	Reduces reliance on conventional polymers, sustainable	Effective in electronic applications
40. Bio-based energy sources	Renewable, reduces fossil fuel use	Used in power generation (biomass, bioenergy)	Enhances energy security, reduces emissions	High efficiency in converting biomass to energy
41. Bio-based environmental sensors	Sustainable, effective monitoring	Used in environmental monitoring and analysis	Reduces reliance on synthetic materials, effective sensing	High sensitivity and accuracy in environmental monitoring

technologies, and foster a conducive market environment for products originating from biological sources. A prevalent strategy to incentivize individuals entails offering subsidies and tax credits to producers and consumers of bioproducts. The Biomass Crop Assistance Program (BCAP) in the United States offers financial support to biomass feedstock producers. The Biodiesel Tax Credit provides a tax incentive to biodiesel manufacturers for every gallon produced. The European Union offers financial assistance via the Horizon Europe initiative, which finances research and innovation projects within the bioeconomy sector, encompassing the development of advanced bioproducts. A different type of motivation entails the implementation of compulsory blending mandates, which stipulate a designated ratio of biofuels mixed with traditional fuels (Macias Aragonés & Arroyo Torralvo, 2024). This initiative establishes a reliable and assured biofuel market, fostering and incentivizing investments in biofuel production facilities. Brazil's National Alcohol Program (Proálcool) requires the incorporation of ethanol into petrol, thus fostering the development of the nation's bioethanol sector. Governments are progressively employing public procurement policies to enhance the demand for bioproducts. Public procurement can establish a reliable market for these products and stimulate private sector investment by prioritizing the acquisition of bio-based products in government contracts. The USDA's BioPreferred Program mandates that federal agencies procure bio-based products whenever feasible, thereby establishing a substantial market. The bioproduct market has steadily expanded over the last 10 years, propelled by environmental, economic, and technological influences. Comprehending these patterns and factors is crucial for individuals or groups seeking to exploit economic opportunities (Di Fraia et al., 2024; Mabate & Pletschke, 2024). The increasing focus on environmental sustainability is a major driver for adopting bioproducts. In response to escalating concerns about climate change, resource depletion, and pollution, there is a growing demand from consumers, businesses, and governments for sustainable alternatives to conventional products. Bioproducts from renewable sources can biodegrade, effectively mitigating environmental harm and fostering a circular economy. Bioplastics are increasingly recognized as a viable substitute for traditional fossil fuel-based plastics, contributing to the mitigation of plastic pollution. Bioplastics can be synthesized from renewable resources like corn starch or sugarcane, and many are specifically engineered to decompose naturally or be converted into compost. This attribute mitigates their adverse environmental impacts. Regulatory frameworks and government incentives are essential for fostering the adoption of bioproducts. Renewable energy policies mitigate greenhouse gas emissions and foster sustainable agriculture, cultivating a conducive environment for expanding the bioproducts market. The European Union's Renewable Energy Directive and the United States Renewable Fuel Standard have substantially facilitated the promotion and adoption of biofuels in their respective markets. The adoption and execution of bioproducts significantly depend on pivotal progress in biotechnology, chemistry, and materials science. Advancements in these domains have facilitated the creation of novel bioproducts exhibiting enhanced performance, cost-efficiency, and sustainability (Macias Aragonés & Arroyo Torralvo, 2024; Swaminaathan et al., 2024). Advancements in synthetic biology have resulted in the creation of engineered microorganisms capable of efficiently generating bio-based chemicals and materials. Advancements in processing technologies have enabled the conversion of a wider variety of biomass feedstocks into valuable bioproducts. Consumer preferences increasingly favor sustainable and eco-friendly products, resulting in heightened demand for bioproducts. Consumers are increasingly cognizant of their purchasing choices' environmental and social ramifications, and a significant segment is prepared to pay a premium for sustainable products. This phenomenon is especially pronounced in sectors like food packaging, personal care items, and household goods, where bioproducts are promoted as environmentally sustainable substitutes for traditional products. Numerous companies are incorporating sustainability initiatives by integrating bioproducts into their overarching corporate social responsibility (CSR) strategies. Companies acknowledge that integrating bioproducts into their supply chains can facilitate achieving sustainability objectives, diminish carbon emissions, and bolster their brand reputation. Numerous multinational corporations have committed to using bioplastics in packaging or procuring bio-based materials. Notwithstanding the increasing interest in bioproducts, numerous challenges must be confronted to enhance their production and facilitate widespread adoption. Nonetheless, these challenges offer prospects for advancement and development in the bioproducts sector. Bioproducts encounter a substantial obstacle due to their restricted capacity to compete on price. Currently, the cost of various bioproducts surpasses that of their traditional counterparts. The principal factors contributing to this are the escalating expenses of biomass feedstocks, the restricted economic advantages of large-scale production, and the necessity for specialized processing technologies. The cost disparity can hinder bioproducts from competing effectively in the market, especially in sectors that emphasize affordability. Nonetheless, opportunities exist to diminish expenses via innovation, economies of scale, and process optimization. Biotechnology and process engineering advancements may enhance biomass conversion efficiency and lower production expenses. Moreover, as the demand for bioproducts escalates, it becomes viable to establish larger production facilities, leading to cost reductions via economies of scale. The accessibility and sustainability of biomass feedstocks are critical considerations in evaluating the viability of expanding bioproducts. The competition for land, water, and other resources essential for biomass cultivation may restrict feedstock availability and raise concerns regarding the sustainability of bioproduct production. Furthermore, utilizing food crops for bioproduct production raises ethical concerns about ensuring food security. Utilizing non-food biomass sources, such as agricultural residues, forest waste, and algae, to manufacture bioproducts is a viable solution to these challenges. These unrefined resources can mitigate food security issues and diminish the environmental impact of escalating biomass. Furthermore, enhancements in the cultivation and collection of feedstock can augment the long-term sustainability and accessibility of biomass resources. Regulatory and market obstacles may impede the advancement of bioproducts. The presence of intricate and contradictory regulations across various domains can generate uncertainty and elevate the expenses related to the market introduction of bioproducts. Moreover, the absence of standardized labeling and bioproduct certification can pose challenges for consumers and businesses in recognizing and selecting bio-based alternatives. These challenges can be surmounted by instituting consistent regulations, enforcing standardized labeling, and creating certification programs that offer clear information regarding the sustainability and efficacy of bioproducts. Collaboration between governmental entities and industry stakeholders can create a regulatory framework that promotes the development and adoption of bioproducts. To fulfill the demand for market expansion, developing infrastructure and supply chains for bioproducts is essential. Bioproducts frequently necessitate specialized processing facilities, distribution networks, and storage solutions that may not be readily accessible. Moreover, the biomass supply chain segmentation may lead to logistical difficulties and elevated expenses. To address these challenges, allocating resources to enhance infrastructure and integrate supply chains for bioproducts is feasible. Collaboration between the public and private sectors can accelerate the development of essential infrastructure, including biorefineries and distribution networks, to foster the expansion of the bioproducts market. Furthermore, digital technologies and data analytics can be utilized to enhance supply chain operations and the efficiency of biomass sourcing and processing. The restricted uptake of bioproducts is due to inadequate market awareness and consumer education. Numerous consumers and enterprises do not broadly acknowledge the benefits and availability of these products, potentially obstructing market growth. Moreover, misconceptions regarding bioproducts' performance, cost, and sustainability may resist their adoption (Jain et al., 2022; Ahmed et al., 2023; Chen et al., 2023; Costa et al., 2023).

6 Future Perspectives

Research and development (R&D) in bioproducts fosters a more sustainable economy reliant on biological resources. The escalating environmental concerns and the demand for resource efficiency are fostering a heightened interest in bioproducts, leading to several trends shaping this domain's future. A significant trend in bioproduct research and development is the improved utilization of biomass. Researchers are progressively focusing on optimizing various biomass sources, including agricultural residues, forestry waste, algae, and non-food crops, to produce a wide range of bioproducts. This trend is driven by the need to reduce reliance on food crops, which may lead to competition for food resources and provoke ethical dilemmas. Researchers are examining the viability of lignocellulosic biomass, including straw, wood chips, and grasses, for producing biofuels, bioplastics, and bio-based chemicals. The present research predominantly centers on creating more efficient and economically viable technologies for converting lignocellulosic biomass into valuable products. A burgeoning trend involves the application of synthetic biology and metabolic engineering for the research and development of bioproducts. These domains involve modifying and enhancing biological systems, such as microorganisms, to increase their ability to produce specific bioproducts. By manipulating microbial metabolic pathways, scientists can increase the yield of desired substances, reduce production costs, and enable the development of novel bioproducts that were previously difficult or impossible to achieve. Genetically modified yeast and bacteria are presently utilized to produce bio-based chemicals, including lactic acid, succinic acid, and isoprene. These chemicals are utilized in various industrial sectors, such as bioplastics and synthetic rubber manufacturing. The concept of biorefineries, which integrate multiple processes to convert biomass into diverse bioproducts, is gaining prominence in research and development. Biorefineries, akin to petroleum refineries, aim to optimize biomass utilization by producing various products, including biofuels, biochemicals, and biomaterials, to enhance their value. This holistic strategy improves the economic viability of bioproducts and maximizes resource efficiency by minimizing waste. Researchers are exploring innovative biorefinery models that can effectively transform diverse biomass sources into a broader array of valuable products. There is a growing demand for the establishment of decentralized biorefineries that operate on smaller scales and are situated nearer to biomass sources, thus reducing transportation costs and environmental impact. The circular economy, emphasizing the recycling and repurposing of biological resources, is a crucial driver for continuous progress in bioproduct research and development. Waste valorization, the process of converting waste materials into valuable products, is essential to this trend. Researchers are developing sophisticated technologies to transform agricultural, industrial, and municipal waste into bioproducts, such as biofuels, bioplastics, and bio-based chemicals. Food waste is presently being examined as a feedstock for biogas production, whereas plastic waste is being converted into bio-based monomers to synthesize new materials. These efforts enhance the circular economy by reducing waste, conserving resources, and creating new economic opportunities. Technological innovations and pivotal discoveries are crucial for developing and commercializing bioproducts. Recent advancements in biotechnology, materials science, and process engineering have generated opportunities to develop more efficient, sustainable, and versatile bioproducts. CRISPR-Cas9 and other gene-editing technologies are revolutionizing bioproduct research and development by enabling precise alterations to the genomes of organisms utilized in bioproduction. These technologies empower scientists to enhance the metabolic capabilities of microorganisms, plants, and algae, leading to heightened bioproduct production and the creation of novel compounds. CRISPR has been employed to genetically alter yeast strains for enhanced biofuel production and bacteria that can generate bioplastics from renewable feedstocks. The ability to modify the genetic makeup of organisms utilized in production accelerated the development of next-generation bioproducts that demonstrate improved performance and sustainability. Fermentation is a crucial component in producing bioproducts, and recent innovations in fermentation technologies significantly improve the efficiency and scalability of the process. Continuous fermentation, high-density cell cultures, and innovative bioreactors enhance bioproducts' cost-effective and efficient production. Continuous fermentation facilitates the uninterrupted production of bioproducts, reducing downtime and improving overall productivity. Developing advanced bioreactors featuring improved mixing, oxygen transfer, and temperature control enhances the capacity for scaling up and sustaining uniform production of bioproducts. Nanotechnology is increasingly impactful in advancing bioproducts, offering innovative opportunities to enhance material properties and performance. Researchers are examining nanomaterials, such as cellulose nanocrystals, chitosan nanoparticles, and protein-based nanostructures, to create bio-based materials with superior strength, flexibility, and barrier characteristics. Cellulose nanocrystals, derived from plant fibers, are presently employed to manufacture advanced bio-composites that demonstrate outstanding performance attributes. These bio-composites are utilized across multiple industries, including packaging, electronics, and automotive. Nanotechnology improves the delivery and efficacy of bio-based pharmaceuticals and nutraceuticals. Artificial intelligence (AI) and machine learning (ML) are increasingly essential in bioproduct research and development (R&D) as they facilitate the analysis of extensive datasets, optimization of production processes, and accelerated discovery of innovative bioproducts. Artificial intelligence models are presently utilized to predict the behaviors of biological systems, optimize metabolic pathways, and engineer novel enzymes and microorganisms for bioproduction. Moreover, machine learning algorithms are employed to improve the efficiency of bioprocesses, including fermentation and downstream processing, by determining optimal conditions and minimizing resource usage. The integration of artificial intelligence (AI) and machine learning (ML) into the research and development (R&D) of bioproducts is promoting rapid and economical innovation. Bioproduct research, development, and technology advancements significantly impact sustainable development and the circular economy. Given the global environmental challenges of climate change, resource depletion, and pollution, bioproducts offer a promising pathway to a sustainable and resilient future. A significant outcome of bioproducts for sustainable development is their ability to reduce dependence on fossil fuels. Bioproducts reduce greenhouse gas emissions and conserve finite fossil resources by replacing petroleum-derived materials with bio-based substitutes. Biofuels offer a sustainable alternative to petrol and diesel, aiding in reducing carbon emissions within the transportation sector. Bioplastics and bio-based chemicals reduce the carbon footprint of industrial production by employing renewable feedstocks, which generally require less energy consumption. Bioproducts are crucial for enhancing resource efficiency, a core tenet of the circular economy. Bioproducts improve resource efficiency and minimize waste by utilizing renewable biological resources and converting waste into valuable goods. Optimizing biomass value is illustrated by establishing biorefineries that generate multiple products from a singular feedstock. Furthermore, employing non-food biomass sources, including agricultural residues and algae, for bioproduct production mitigates competition for land and water resources, thus enhancing resource efficiency. The production of bioproducts is closely linked to adopting sustainable practices in agriculture and forestry. Bioproducts can promote the implementation of sustainable agricultural and forestry practices that emphasize soil health, biodiversity, and carbon sequestration by generating demand for biomass feedstocks. For example, cultivating cover crops to generate biomass can improve soil fertility and reduce erosion. Sustainable forestry practices can concurrently guarantee a steady supply of timber for bioproducts while protecting forest ecosystems. Moreover, integrating bioproducts into agricultural systems can promote the development of bio-based fertilizers and pesticides, thereby alleviating the environmental impacts of agriculture. The bioproducts sector possesses the capacity to drive economic growth and create job opportunities, particularly in rural and marginalized regions. Establishing biorefineries, biomass processing facilities, and bio-based manufacturing plants can stimulate local economies by creating employment opportunities and supporting small and medium-sized enterprises (SMEs). Furthermore, developing new markets for bioproducts can foster innovation and entrepreneurship, attract investment, and enhance economic resilience. Bioproducts can facilitate economic diversification and reduce dependence on conventional industries vulnerable to resource constraints or environmental regulations, thus fostering economic growth. Bioproducts are essential in advancing the circular economy initiative, which seeks to minimize waste, extend product longevity, and create closed-loop systems that restore natural resources. Bioproducts promote the sustainability and resilience objectives of the circular economy by advocating for using renewable resources and enabling the recycling and repurposing of materials. Biodegradable bioplastics can be composted after their life cycle, enriching the soil with essential nutrients and reducing landfill pressure. Similarly, augmenting the worth of waste materials through the production of bioproducts finalizes the cycle of resource utilization, converting waste into a valuable resource.

7 Conclusion

Bioproducts facilitate the circular economy by offering a sustainable substitute for traditional linear production methods. Bioproducts substantially diminish our reliance on fossil fuels, mitigate greenhouse gas emissions, and minimize waste by harnessing renewable biological resources and enhancing material efficiency. Integrating bioproducts into various sectors, including agriculture, energy, and manufacturing, fosters a more resilient and efficient economy that optimizes resource utilization. Biorefineries, waste valorization, and advanced biotechnologies such as synthetic biology and nanotechnology illustrate the capacity of bioproducts to foster sustainable practices across diverse industries. These innovations enhance the environmental sustainability of production processes and create new economic opportunities, particularly in rural and underdeveloped regions, by fostering job creation and local entrepreneurship. Bioproducts exemplify sustainability by efficiently closing resource loops, extending product longevity, and revitalizing natural systems in the circular economy context. They enable the transition from a linear model of production and consumption to a circular paradigm, wherein resources and energy are perpetually reused and recycled within the economic framework. Given the increasing global focus on sustainability, bioproducts will be essential in promoting the circular economy. This will facilitate the alignment of economic advancement with environmental stewardship, ultimately ensuring a more sustainable future for subsequent generations. Consequently, the continuous progress and adoption of bioproducts are crucial for achieving the goals of the circular economy, positioning them as a vital element of sustainable development strategies worldwide.

References

Abdullah, M. A., Nazir, M. S., Hussein, H. A., Shah, S. M. U., Azra, N., Iftikhar, R., Iqbal, M. S., Qamar, Z., Ahmad, Z., Afzaal, M., Om, A. D., Shaharah, M. I., Rak, A. E., & Hung, Y.-T. (2024). New perspectives on biomass conversion and circular economy based on integrated algal-oil palm biorefinery framework for sustainable energy and bioproducts co-generation. *Industrial Crops and Products, 213*, 118452. https://doi.org/10.1016/j.indcrop.2024.118452

Acebes-Huerta, A., Martínez-Botía, P., Carbajo-Argüelles, G., Fernández-Fuertes, J., Muñoz-Turrillas, M. C., Ojea-Pérez, A. M., López-Vázquez, A., Eble, J. A., & Gutiérrez, L. (2024). Characterization of the molecular composition and in vitro regenerative capacity of platelet-based bioproducts and related subfractions. *Acta Biomaterialia, 177*, 132–147. https://doi.org/10.1016/j.actbio.2024.01.029

Ahmed, S. F., Kabir, M., Mehjabin, A., Oishi, F. T. Z., Ahmed, S., Mannan, S., Mofijur, M., Almomani, F., Badruddin, I. A., & Kamangar, S. (2023). Waste biorefinery to produce renewable energy: Bioconversion process and circular bioeconomy. *Energy Reports, 10*, 3073–3091. https://doi.org/10.1016/j.egyr.2023.09.137

Alazaiza, M. Y. D., Albahnasawi, A., Ahmad, Z., Bashir, M. J. K., Al-Wahaibi, T., Abujazar, M. S. S., Abu Amr, S. S., & Nassani, D. E. (2022). Potential use of algae for the bioremediation of different types of wastewater and contaminants: Production of bioproducts and biofuel for green circular economy. *Journal of Environmental Management, 324*, 116415. https://doi.org/10.1016/j.jenvman.2022.116415

Ashokkumar, V., Flora, G., Venkatkarthick, R., Senthil Kannan, K., Kuppam, C., Mary Stephy, G., Kamyab, H., Chen, W.-H., Thomas, J., & Ngamcharussrivichai, C. (2022). Advanced technologies on the sustainable approaches for conversion of organic waste to valuable bioproducts: Emerging circular bioeconomy perspective. *Fuel, 324*, 124313. https://doi.org/10.1016/j.fuel.2022.124313

Ashokkumar, V., Chandramughi, V. P., Kumar, G., Ngamcharussrivichai, C., Piechota, G., Igliński, B., Kothari, R., & Chen, W.-H. (2024). Advancements in lignocellulosic biomass: A critical appraisal of fourth-generation biofuels and value-added bioproduct. *Fuel, 365*, 130751. https://doi.org/10.1016/j.fuel.2023.130751

Chen, Y., Song, C., Cui, X., Han, J., Paithoonrangsarid, K., Gan, Q., & Lu, Y. (2023). Halophilic microalga-based circular economy producing functional food by reclaiming high-salinity seafood processing sewage. *Biomass and Bioenergy, 178*, 106952. https://doi.org/10.1016/j.biombioe.2023.106952

Costa, J. A. V., Zaparoli, M., Cassuriaga, A. P. A., Cardias, B. B., Vaz, B. d. S., de Morais, M. G., & Moreira, J. B. (2023). Biochar production from microalgae: A new sustainable approach to wastewater treatment based on a circular economy. *Enzyme and Microbial Technology, 169*, 110281. https://doi.org/10.1016/j.enzmictec.2023.110281

Di Fraia, S., Godvin Sharmila, V., Banu, J. R., & Massarotti, N. (2024). A comprehensive review on upcycling of food waste into value added products towards a circular economy: Holistic approaches and life cycle assessments. *Trends in Food Science and Technology, 143*, 104288. https://doi.org/10.1016/j.tifs.2023.104288

Heitzman, B. S., Bueno, G. W., Camargo, T. R., Proença, D. C., Yaekashi, C. T. O., da Silva, R. M. G., & Machado, L. P. (2024). Duckweed application in nature-based system for water phytoremediation and high-value coproducts at family agrisystem from a circular economy perspective. *Sci Total Environ, 919*, 170714. https://doi.org/10.1016/j.scitotenv.2024.170714

Jain, A., Sarsaiya, S., Kumar Awasthi, M., Singh, R., Rajput, R., Mishra, U. C., Chen, J., & Shi, J. (2022). Bioenergy and bio-products from bio-waste and its associated modern circular economy: Current research trends, challenges, and future outlooks. *Fuel, 307*, 121859. https://doi.org/10.1016/j.fuel.2021.121859

Kee, P. E., Cheng, Y.-S., Chang, J.-S., Yim, H. S., Tan, J. C. Y., Lam, S. S., Lan, J. C.-W., Ng, H. S., & Khoo, K. S. (2023). Insect biorefinery: A circular economy concept for biowaste conversion to value-added products. *Environmental Research, 221*, 115284. https://doi.org/10.1016/j.envres.2023.115284

Kumar, V., Vangnai, A. S., Sharma, N., Kaur, K., Chakraborty, P., Umesh, M., Singhal, B., Utreja, D., Carrasco, E. U., Andler, R., Awasthi, M. K., & Taherzadeh, M. J. (2023). Bioengineering of biowaste to recover bioproducts and bioenergy: A circular economy approach towards sustainable zero-waste environment. *Chemosphere, 319*, 138005. https://doi.org/10.1016/j.chemosphere.2023.138005

Kumar, R., Neha, Y., Ravishankar, G. A., & Srivatsan, V. (2024). Chapter 21- algae-based bioremediation bioproducts and biofuels for biobusiness. In P. MNVB (Ed.), *Bioremediation and bioeconomy* (2nd ed., pp. 583–635). Elsevier.

Mabate, B., & Pletschke, B. I. (2024). Sequential and enzyme-assisted extraction of algal bioproducts from Ecklonia maxima. *Enzyme and Microbial Technology, 173*, 110364. https://doi.org/10.1016/j.enzmictec.2023.110364

Macias Aragonés, M., & Arroyo Torralvo, F. (2024). How to select the best approach for circular economy assessment? 3D positioning framework, decision support tool and critical analysis for bio-based systems. *Environmental Impact Assessment Review, 106*, 107493. https://doi.org/10.1016/j.eiar.2024.107493

Mujtaba, M., Fernandes Fraceto, L., Fazeli, M., Mukherjee, S., Savassa, S. M., Araujo de Medeiros, G., do Espírito Santo Pereira, A., Mancini, S. D., Lipponen, J., & Vilaplana, F. (2023). Lignocellulosic biomass from agricultural waste to the circular economy: A review with focus on biofuels, biocomposites and bioplastics. *Journal of Cleaner Production, 402*, 136815. https://doi.org/10.1016/j.jclepro.2023.136815

Nibbi, L., Chiaramonti, D., & Palchetti, E. (2019). Project BBChina: A new master program in three Chinese universities on bio-based circular economy; from fields to bioenergy, biofuel and bioproducts. *Energy Procedia, 158*, 1261–1266. https://doi.org/10.1016/j.egypro.2019.01.416

Oliveira, A., Pereira, A. S., Lemos, P. C., Guerra, J. P., Silva, V., & Faria, P. (2021). Effect of innovative bioproducts on air lime mortars. *Journal of Building Engineering, 35*, 101985. https://doi.org/10.1016/j.jobe.2020.101985

Priya, A. K., Alagumalai, A., Balaji, D., & Song, H. (2023). Bio-based agricultural products: A sustainable alternative to agrochemicals for promoting a circular economy. *RSC Sustain, 1*(4), 746–762. https://doi.org/10.1039/d3su00075c

Rapp, G., Garcia-Montoto, V., Bouyssiere, B., Thiebaud-Roux, S., Montoya, A., Trethowan, R., Pratt, P., Mozet, K., Portha, J.-F., & Coniglio, L. (2021). Indian mustard bioproducts dry-purification with natural adsorbents—a biorefinery for a green circular economy. *Journal of Cleaner Production, 286*, 125411. https://doi.org/10.1016/j.jclepro.2020.125411

Saravanan, A., Karishma, S., Senthil Kumar, P., & Rangasamy, G. (2023). A review on regeneration of biowaste into bio-products and bioenergy: Life cycle assessment and circular economy. *Fuel, 338*, 127221. https://doi.org/10.1016/j.fuel.2022.127221

Sarsaiya, S., Jain, A., Kumar Awasthi, S., Duan, Y., Kumar Awasthi, M., & Shi, J. (2019). Microbial dynamics for lignocellulosic waste bioconversion and its importance with modern circular economy, challenges and future perspectives. *Bioresource Technology, 291*, 121905. https://doi.org/10.1016/j.biortech.2019.121905

Swaminaathan, P., Saravanan, A., & Thamarai, P. (2024). Utilization of bioresources for high-value bioproducts production: Sustainability and perspectives in circular bioeconomy. *Sustainable Energy Technologies and Assessments, 63*, 103672. https://doi.org/10.1016/j.seta.2024.103672

Recent Findings and Advances in Sustainable Conversion of Lignocellulosic Waste to Bioplastic Precursors for a Circular Economy

Shilajit Barua, Gourab Shome, Sayak Dolai, and Jisan Sarwar

Abstract

Urban and agricultural lignocellulosic waste poses challenges for waste management due to limited landfill capacity, inefficient disposal, environmental pollution, and greenhouse gas emissions. However, it offers resource recovery opportunities through composting, production of bioenergy, bioplastics manufacturing, and anaerobic digestion. These approaches of waste valorization enable waste reduction, support the circular economy, and promote the sustainable use of energy and materials. Resource recovery from lignocellulosic waste extracts valuable materials and energy, aiding waste diversion, sustainability, and resource conservation. Lignocellulosic waste is a renewable resource for bioplastic production, offering an eco-friendly alternative to fossil-based plastics. This approach encourages sustainable waste management by transforming agricultural and urban waste into valuable products. It also helps in decreasing environmental pollution, aligns with circular economy principles, and minimizes dependence on non-renewable resources. Bioplastics, being biodegradable or compostable, offer a sustainable alternative to traditional plastics, helping to mitigate plastic pollution, thus reducing the environmental impact associated with traditional plastic products. Lignocellulosic wastes, such as paper, cardboard, agricultural residues, yard waste, etc., are collected from households, businesses, agricultural fields, or industrial sources and serve as feedstocks for bioplastic production. They undergo microbial fermentation, in which particular bacteria or fungi transform cellulose and hemicellulose into biopolymers such as polylactic acid (PLA) or polyhydroxyalkanoates (PHAs). Alternatively, enzymatic hydrolysis can be used to decompose lignocellulosic biomass into sugars, which are subsequently polymerized to form bioplastics. Additionally, chemical

S. Barua (✉) · G. Shome · S. Dolai · J. Sarwar
Department of Microbiology, Vijaygarh Jyotish Ray College, Kolkata, India

G. Mukherjee, S. Dhiman (eds.), *Value Addition and Utilization of Lignocellulosic Biomass*, https://doi.org/10.1007/978-981-96-2786-8_12

processes can modify lignin, a major component of lignocellulose, to create bioplastics with enhanced properties. Several bacteria, viz., *Cupriavidus necator*, *Escherichia coli*, *Bacillus subtilis*, *Pseudomonas putida*, *Clostridium* spp., *Rhodococcus* spp., etc., have been reported to be capable of producing bioplastics using lignocellulosic substrate. This chapter provides a comprehensive overview of global lignocellulosic agricultural and municipal waste production and its valorization to support a circular economy. It highlights recent trends in bioplastic development from diverse lignocellulosic sources and examines the critical role of microbes in bioplastic production, including fermentation and enzymatic processes. Additionally, it addresses the current scenario of bioplastic production worldwide, showcasing innovative approaches and technologies. The chapter concludes with an exploration of the promising prospects for lignocellulosic bioplastics, emphasizing their potential to contribute to sustainable waste management and resource recovery within a circular economy framework.

Keywords

Lignocellulosic waste · Bioplastic · Bacteria · Sustainable resource recovery

1 Introduction

The concept of a circular economy, intertwined with the bioeconomy, embodies a sustainable framework for resource production and consumption that prioritizes reuse, refurbishment, and recycling (D'Amato & Korhonen, 2021; Jain et al., 2022). This approach aims to minimize waste and maximize resource efficiency by utilizing renewable biological resources (Tan & Lamers, 2021) and converting waste streams into value-added products (Kover et al., 2022). The circular bioeconomy fosters a system that repurposes biomass to produce new materials, enhancing efficiency and conserving resources (Gatto, 2024). It aligns with the principles of industrial ecology, biomimicry, and cradle-to-cradle design, emphasizing the importance of closed-loop systems and holistic product life cycles (Chen Austin & Beermann, 2023; de Keyser & Mathijs, 2023; Hakovirta, 2023; Tsakalova et al., 2024). A sustainable bioeconomy necessitates integrating all three pillars of sustainability—environmental protection, social equity, and economic viability—right from the beginning (Liobikiene et al., 2019; Sharma & Malaviya, 2023). Although full circularity may be idealistic, improving circularity is vital for tackling resource scarcity and enhancing economic sustainability (Venkatesh, 2022). The bio-based economy, tied to renewable sources, can reduce waste and establish a circular economy, contributing to sustainability goals (SDGs) (Yaashikaa et al., 2020). Integrating waste management with the bioeconomy supports multiple SDGs by converting waste into valuable resources. Transforming organic waste into bioenergy, bioplastics, and biochemicals promotes SDG 7 (Affordable and Clean Energy) and SDG 12 (Responsible Consumption and Production) (Banerjee & Arora, 2021). This approach minimizes landfill use (Gibson, 2021), reduces greenhouse gas emissions (Virgin et al., 2022), and encourages sustainable farming practices (Valle et al.,

2024), contributing to SDG 13 (Climate Action) and SDG 15 (Life on Land). Additionally, it fosters green job creation and sustainable economic growth (Belmonte-Ureña et al., 2021), supporting SDG 8 (Decent Work and Economic Growth). This highlights the interconnectedness between waste management, the bioeconomy, and global sustainability goals.

Lignocellulosic waste, from agriculture, forestry, and kitchen sources like fruit and vegetable peels, has become an increasing environmental concern due to its potential negative impact. Primarily composed of cellulose, hemicellulose, and lignin, improper disposal of these waste materials can lead to the contamination of soil and water resources, contributing to air, water, and land pollution (Eugenia Eugenio et al., 2019). These wastes pose significant environmental risks, impacting soil health, wildlife habitats, and ecosystem stability. Allelochemicals from crop residues harm soil health (Fu et al., 2023), while traditional practices like incineration of lignocellulosic waste diminish soil fertility by destroying microflora and can cause forest fires, resulting in further destruction of wildlife habitats (Porichha et al., 2021; Poskart et al., 2021). The recalcitrant nature of these wastes leads to their accumulation in soil, obstructing root development and aeration, disrupting microbial communities, and ultimately compromising soil fertility, plant growth, and ecosystem stability (Alokika & Singh, 2023). Combusting these organic materials can release toxic compounds, including particulate matter, volatile organic compounds, and greenhouse gases, exacerbating air pollution and climate change (Haller et al., 2023). Improper disposal of lignocellulosic kitchen waste, like fruit and vegetable peels, through landfilling can lead to the release of methane and carbon dioxide, which also contributes to global warming (Pandey & Kumari, 2022).

Globally, 140 gigatons of lignocellulosic waste generated annually pose serious management challenges, often leading to burning or being left unmanaged (Blasi et al., 2023). However, valorizing this waste into sustainable chemicals and polymers presents a viable solution, with studies exploring its conversion into bioethanol, bioplastics, and bio-based chemicals (Mujtaba et al., 2023). These initiatives not only minimize waste disposal but also offer a renewable source of materials, decreasing dependence on fossil resources and fostering a circular economy. For example, extensive research has focused on converting lignin from lignocellulosic waste into carbon fibers (Kanhere et al., 2022). Additionally, extracting valuable chemicals like furfural and hydroxymethylfurfural has gained attention, as these platform chemicals serve as building blocks for a wide range of products, including fuels, solvents, and polymers (Zhao et al., 2021). Researchers have investigated the utilization of lignocellulosic waste as a renewable resource for biofuel (bioethanol/biogas) production, thus providing a sustainable alternative to fossil fuels and reducing greenhouse gas emissions from waste disposal. Cellulose and hemicellulose are converted into fermentable sugars, then fermented microbiologically to produce bioethanol or microbiologically carry out anaerobic digestion to generate biogas, a renewable methane-CO_2 mix (Bhatia et al., 2018; Castro Garcia et al., 2023).

Lignocellulosic biomass from agricultural, forestry, and kitchen residues has also become an important resource for producing bioplastics too. These waste streams can be transformed into various biopolymers, such as polylactic acid (PLA), polyhydroxyalkanoates (PHAs), and cellulose-based plastics, via different bioconversion processes (Khodaei et al., 2021). Integrating bioplastic production from lignocellulosic waste into a circular economy reduces reliance on finite fossil resources and mitigates plastic pollution, as bioplastics are generally biodegradable and recyclable. This approach also improves resource efficiency and generates new revenue from underutilized agricultural and forestry residues (Rosenboom et al., 2022).

This chapter explores recent findings regarding the conversion of lignocellulosic waste into bioplastics within a circular bioeconomy. It highlights the recently reported environmental and economic benefits of turning agricultural, forestry, and kitchen residues into biodegradable plastics, reducing reliance on fossil resources, and addressing plastic pollution. The chapter also emphasizes innovative technologies that transform waste into valuable products, promoting sustainability and resource efficiency.

2 Lignocellulosic Waste: An Overview

2.1 Types and Sources

Lignocellulosic materials are valuable, renewable resources crucial to modern industries (Esrafili & Martínez-Carrera, 2015). Significant waste from agriculture, kitchens, forestry, and food industries produces lignocellulosic by-products, often burned or left to decompose, causing environmental issues (Esrafili & Martínez-Carrera, 2015; Broda et al., 2022). Table 1 outlines various types and sources of these waste materials.

Though often burned or left unused, contributing to environmental issues, lignocellulosic wastes are becoming renewable, abundant, and valuable feedstocks for various applications, like bioethanol production (Brienza et al., 2022). Lignocellulosic biomass also offers strong potential for bioplastic production, providing a renewable, abundant, and cost-effective alternative to fossil-fuel-based plastics (Govil et al., 2020; Maraveas, 2020).

2.2 Global Production of Lignocellulosic Waste and Their Sources

A substantial quantity of lignocellulosic waste is generated globally, which is estimated at around 181.5 billion tons annually, though only 8.2 billion tons are utilized (Singh et al., 2022). This waste largely comes from agriculture, forestry, and industry, with agriculture contributing the most (Table 2), including residues like rice straw, corn stover, sugarcane bagasse, and wheat straw (Cazier et al., 2024).

Asia generates about 667.6 million tons of rice straw annually (Aziz et al., 2023), underscoring its leading role in lignocellulosic waste production. Major food-producing nations like China, the USA, India, Brazil, and Turkey also contribute significantly to agricultural waste. Lignocellulosic biomass has the potential to

Table 1 Types and sources of lignocellulosic wastes

Type of lignocellulosic waste	Source	References
Agricultural waste		
Crop residues	Straw, corn stover, sugarcane bagasse	Khan et al. (2021)
Fruit and vegetable residues	Peels, skins, seeds	Jena and Singh (2022)
Nut shells	Almond shells, walnut shells	Kali et al. (2024)
Husks	Rice husks, wheat husks	Lou et al. (2023)
Forest waste		
Sawdust	Sawmill operations	Mohammad (2023)
Wood chips	Lumber mills, forest thinning	Abu Bakar and Salim (2023), Chang et al. (2023)
Bark	Logging operations	Benouadah et al. (2022)
Leaves and branches	Forest management practices	Udali et al. (2024)
Kitchen waste		
Vegetable peelings	Household vegetable preparation	Coulibaly et al. (2022)
Fruit peels	Household fruit preparation	Kumar et al. (2023)
Coffee grounds	Household and commercial use	Lee et al. (2023)
Tea bags	Household and commercial use	Purkait et al. (2023)

produce approximately 442 billion liters of bioethanol per year—far exceeding current production (Pirzadah et al., 2015). Despite challenges due to its complex structure, advancements in pretreatment and biotechnology are improving its conversion to biofuels and valuable products, fostering a sustainable circular economy (Sharma et al., 2023).

2.3 Valorization of Lignocellulosic Waste in a Circular Bioeconomy

Valorizing lignocellulosic biomass is essential for promoting the circular bioeconomy by transforming agricultural and forestry residues as well as organic fractions of household and municipality wastes into valuable products, thereby supporting waste reduction and enhancing sustainability in line with the sustainable development goals (SDGs) (Sharma et al., 2024).

In biorefineries, lignocellulosic biomass is efficiently utilized to minimize waste and optimize resource use, supporting sustainable consumption and production (SDG12) (Rajendran et al., 2024), while the integrated biorefinery model enables the simultaneous production of biofuels, biochemicals, and biomaterials, promoting a zero-waste philosophy and advancing clean energy access (SDG7) (Duque-Acevedo et al., 2021; Sarker et al., 2024). For example, bioethanol, a renewable

Table 2 Global annual production of different lignocellulosic wastes

Source	Type of waste	Annual global production (million tonS)	References
Agricultural waste	Wheat straw	776	Babar et al. (2022)
	Rice straw	800–1000	Harun et al. (2022)
	Rice husk	Approx. 150	Kordi et al. (2024)
	Wheat bran	150	Yan et al. (2022)
	Corn stover	>1000	Alavijeh et al. (2023)
	Sugarcane bagasse	700	Ungureanu et al. (2022)
	Cotton stalks	90.3–129	Pandirwar et al. (2023)
	Barley straw	300	Kuittinen et al. (2022)
Forest Waste	Sawdust	38,000	Zhou et al. (2023)
	Wood chips	44.3	GLOBAL BIOENERGY STATISTICS (2022)
	Tree branches and leaves	4640	Bożym et al. (2021)
	Bark	359	Pásztory et al. (2016)
Kitchen-based waste	Fruit and vegetable peels	1300	Gupta et al. (2023)
	Coffee grounds	>10	Lee et al. (2023)
	Tea leaves	0.18–0.2 (China) 0.19 (India) 0.03 (Turkey)	Miao et al. (2023)

fuel, is produced by fermenting sugars derived from lignocellulosic biomass and serves as a substitute for gasoline (Haq et al., 2024). Similarly, biobutanol, an alcohol used as a fuel additive, is produced via fermentation processes similar to those used for bioethanol (Chen et al., 2024). Biodiesel, in contrast, is generated through the transesterification of oils from lignocellulosic feedstocks, providing a sustainable alternative to traditional fossil-derived diesel (Uthandi et al., 2022). Biogas, commonly used for heating and electricity generation, mainly consists of methane, produced by anaerobic digestion of lignocellulosic materials (Chettri et al., 2024). Hydrogen, another clean energy option, can be obtained from lignocellulosic biomass through gasification or fermentation, serving as a carrier of clean energy (Sridevi et al., 2024). Furfural, an organic compound obtained through the dehydration of pentoses, plays a key role in producing solvents, resins, and other chemicals (Bao et al., 2024). Levulinic acid, produced by acid hydrolysis of cellulose, is widely used in the production of solvents, plasticizers, and pharmaceuticals (Khan et al., 2023). Xylitol, a sugar alcohol derived from hemicellulose, is frequently used as a sweetener in foods and dental care products (Jain & Ghosh, 2023). Lignin-derived products, including phenolic compounds, adhesives, and carbon fibers, add significant value to lignocellulosic waste (Martínková et al., 2023). Cellulose nanocrystals, extracted from cellulose, have versatile applications in composites, coatings, and as reinforcement agents (Rovera et al., 2023). Bioactive compounds with antioxidant properties have also been produced from lignocellulosic biomass, offering potential health benefits (Bala et al., 2023).

Bioplastics, produced by fermenting sugars from lignocellulosic biomass, present biodegradable alternatives to conventional plastics (Mujtaba et al., 2023; Ahmed et al., 2024). By-products of different lignocellulose waste valorization processes can be repurposed into high-nutrition animal feed, contributing to circular agriculture (Marak et al., 2024). Additionally, residual biomass can be converted into biofertilizers, supporting sustainable agriculture by enhancing soil health (Izydorczyk et al., 2024). Finally, lignocellulosic waste can undergo pyrolysis to produce biochar, which is beneficial for soil improvement and carbon sequestration (Wang et al., 2024).

Energy generation from agricultural waste reduces reliance on non-renewable resources and supports carbon neutrality, contributing to climate change mitigation (SDG 13) (Olugbemide et al., 2024). Efficient biomass processing in lignocellulosic biorefineries enables the co-production of value-added products, lowering costs, attracting investors, creating rural jobs, and promoting economic growth (SDG 8) (Correia et al., 2024).

2.4 Bioplastics from Lignocellulosic Waste

Bioplastics made from lignocellulosic wastes present a sustainable alternative to traditional plastics, addressing environmental and resource challenges (Mujtaba et al., 2023). Lignocellulosic biomass, rich in cellulose, hemicellulose, and lignin, is abundant in agricultural and forestry residues (Yousuf et al., 2019). Transforming this biomass into fermentable sugars enables the production of bioplastics like PLA and PHA, offering a sustainable alternative to fossil fuel-based plastics. Utilizing such waste materials minimizes environmental pollution, manages agricultural residues, and reduces greenhouse gas emissions, lowering the overall carbon footprint (Vigneswari et al., 2024). Additionally, valorizing lignocellulosic biomass for bioplastics enhances economic viability, creates rural job opportunities, and supports local economies (Machineni & Rao Anupoju, 2022). Educational initiatives on bioplastics have empowered communities to adopt sustainable practices, fostering a culture of environmental responsibility too (Anokye et al., 2024).

3 Bioplastics

3.1 An Introductory Remark

Unlike traditional plastics that are derived from petroleum, bioplastics are produced from renewable biomass sources such as vegetable oils, corn starch, straw, wood chips, and organic waste (Sabry, 2022). These materials are biodegradable and can decompose under suitable conditions into natural substances like water, carbon dioxide, and compost, reducing environmental impact (Singh & Verma, 2020).

Bioplastics vary in chemical composition depending on the type and feedstock used. The primary types of bioplastics include polylactic acid (PLA), polyhydroxyalkanoates (PHAs), starch-based plastics, cellulose-based plastics, bio-derived polyethylene (bio-PE), and bio-based polyethylene terephthalate (bio-PET) (Siracusa & Blanco, 2020). Production often involves modified traditional processes, such as ethanol dehydration used in the production of bio-PE (Soo et al., 2024). Microbially-derived polymers like PLA and PHAs also show promise as biodegradable alternatives (Swetha et al., 2024). Advancements in production techniques have led to bioplastics with enhanced properties and functionality compared to conventional plastics (Raj et al., 2022).

Bioplastics include both bio-based and biodegradable plastics. Bio-based plastics are made from renewable biological resources, such as sugarcane or corn, as defined in PR 17228:2018-03. According to EN ISO 472, "plastics" are materials with a high polymer content that can be molded during production. According to DIN EN 16575:2014-10, biomass refers to non-fossilized, biodegradable organic material of biological origin, including plants, animals, microorganisms, and their by-products, residues, or waste. It is considered renewable as long as its use does not exceed its natural replenishment (Vamsi, 2020). Bio-based plastics are derived partly or entirely from renewable sources rather than fossil fuels (Jiang et al., 2024). On the other hand, biodegradable plastics can be produced from fossil fuels, yet they break down into natural elements like water and carbon dioxide through the action of microbes (Che Ab Aziz et al., 2023).

3.2 Attributes of Bioplastic

Bioplastics include various types of polymers that are either compostable or made from renewable sources (Ali et al., 2023). They are valued for their reduced environmental impact, as they break down more readily into simpler, natural substances like water, carbon dioxide, and biomass after disposal (Khandeparkar et al., 2024). Typically sourced from plants, such as corn starch, sugarcane, and potato starch, bioplastics stand apart from conventional plastics made from fossil fuels (Mangal et al., 2023). The degradation rate depends on the type of bioplastic and environmental factors; some break down quickly under optimal conditions, while others take a longer time to decompose (Meereboer et al., 2020). Higher temperatures in composting or industrial settings accelerate their decomposition (Ruggero et al., 2023), while moisture can enhance the breakdown of certain types, though marine environments do not always hasten degradation (Phosri et al., 2022). Bioplastics vary in mechanical properties, such as flexibility and tensile strength, which can be tailored for specific uses (Nanda et al., 2022). Their appearance can also be adjusted to meet requirements for transparency, translucency, or color (Chauhan et al., 2024). In some instances, blending bioplastics with conventional plastics enhances environmental performance, but careful management of blend proportions is needed for optimal results (Taneepanichskul et al., 2022). Generally, biodegradable plastics are designed to degrade into non-toxic substances; however, it is important to assess the

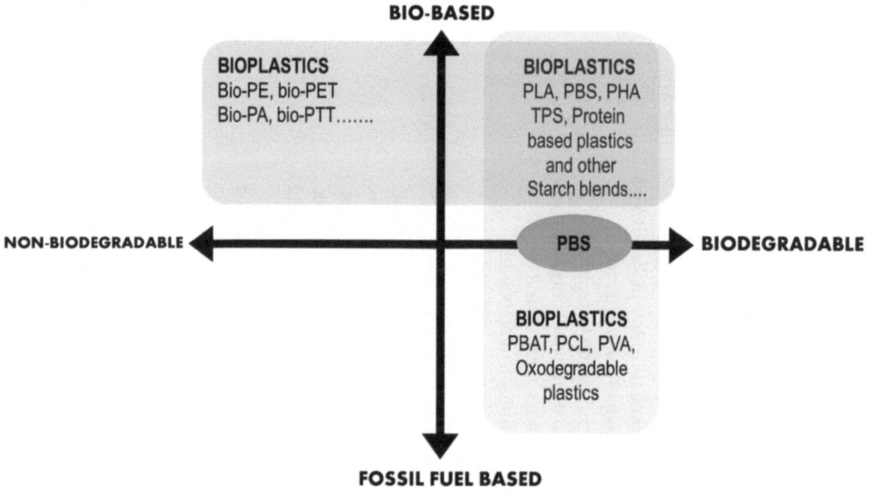

Fig. 1 Classification of bioplastics

by-products or additives released during degradation to ensure they do not pose environmental or health risks.

3.3 Types of Bioplastics

Bioplastics encompass biodegradable types like PLA, PHA, and polybutylene adipate terephthalate (PBAT), as well as bio-based variants such as bio-PE and bio-PET. Some, like PLA and PHA, are both bio-based and biodegradable. However, bio-based does not always mean biodegradable, nor does biodegradable imply bio-based (Dalton et al., 2022). Non-biodegradable bioplastics, including bio-PE, bio-PP, and bio-PET, are biomass-derived (Rahman & Bhoi, 2021) (Fig. 1). Conversely, fossil-based plastics like poly(butylene adipate-*co*-terephthalate; PBAT), polybutylene succinate (PBS), polycaprolactone (PCL), and polyvinyl alcohol (PVOH or PVA) are biodegradable (Havstad, 2020) (Table 3).

3.4 Renewable Resources for Bioplastics

Bioplastics originate from renewable resources like plants, algae, and microorganisms, with major sources including agricultural wastes and by-products like corn starch, sugarcane, and cellulose, which absorb CO_2 as they grow, potentially reducing emissions. Microbial fermentation of these wastes can yield biopolymers such as PHA and PLA, enhancing the eco-friendliness of bioplastics. However, resource selection depends on the desired properties and applications. For example, corn starch is often used for its ease of conversion into lactic acid, the precursor of PLA,

Table 3 Different biodegradable plastics

Name	Bio-based / fossil fuel based	Structure	Method of production	Distinguishing feature and usage	References
Polybutyrate adipate terephthalate (PBAT)	Fossil fuel based		Using a polycondensation reaction, adipic acid, terephthalic acid, and butanediol are combined to create biodegradable aliphatic-aromatic copolyester	Excellent toughness, biodegradability, and processability make PBAT a popular material for packaging (wrapping), biodegradable organic waste bags, agricultural mulch films, and disposable packaging and tableware	Burford et al. (2023)
Polybutylene succinate (PBS)	Fossil fuel based	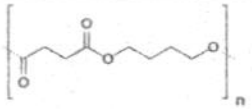	Synthesized by catalyzing the polymerization of succinic acid and 1,4-butanediol; uses bio-based or fossil monomers, usually fossil based	PBS is notable for mechanical strength, ductility, toughness, impact resistance, and thermal resistance over 90 °C. It is utilized in packaging films, shopping bags, agricultural mulch films, plant pots, hygiene products, and food contact applications	Cakmak (2024)
Polycaprolactone (PCL)	Fossil fuel based	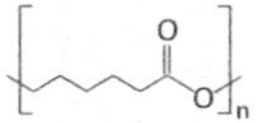	Derived from caprolactone by ring-opening polymerization	Due to its excellent biocompatibility, mechanical strength, processability, high hydrophobicity, and low in vivo degradability, it is widely used in medical applications and blended with bio-based biodegradable plastics like starch-based, PLA, or PHA	Archer et al. (2023), Pawar et al. (2023)

Polyvinyl alcohol (PVA)	Fossil fuel based	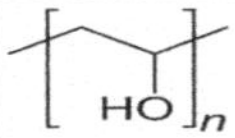	Manufactured by polymerizing vinyl acetate and partially hydrolyzing it, this process replaces ester groups with hydroxyl groups using aqueous sodium hydroxide, leading to the precipitation of polyvinyl alcohol	PVA is a water-soluble, biodegradable polymer known for high biocompatibility and self-cross-linking due to dense hydroxyl groups. However, it hydrolyzes quickly and is bioinert, limiting protein and cell adhesion. It is used in water-soluble films for various applications, including water treatment, dyes, detergents, agricultural chemicals, disinfectants, and industrial cleaning	Hesan et al. (2023)
Oxodegradable plastics	Fossil fuel based	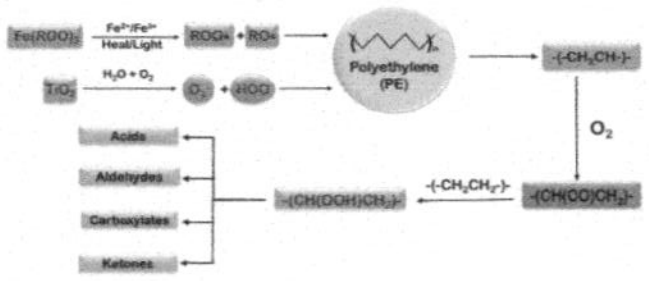	Oxodegradable plastics are traditional fossil-based polymers like PE, PP, or PS with added transition metals. These inorganic metal salts trigger degradation through oxygen, accelerated by exposure to light and/or heat	Utilization of oxodegradable plastics includes film applications such as shopping bags, agricultural mulch films, and, most recently, certain plastic bottles	Thapliyal (2022)

(continued)

Table 3 (continued)

Name	Bio-based / fossil fuel based	Structure	Method of production	Distinguishing feature and usage	References
Thermoplastic starch polymers (TPS)	Bio-based	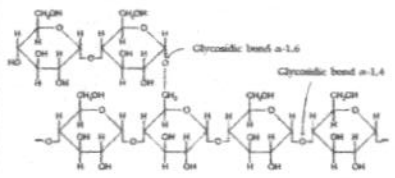	Created by heating native starch with plasticizers (water, urea, glycerol) at 70–90 °C, producing an amorphous, free-flowing material	Mixing plasticized native starch with polymers such as PLA and PCL under heat and shear produces biodegradable materials similar to petroleum-based plastics. These can be used in packaging (loose-fill, bags, sacks), agriculture (mulching film, plant pots), and hygiene and cosmetic products (nappies, sanitary items)	Abas and Attia (2023)
Polylactic acid-based plastics (PLA)	Bio-based	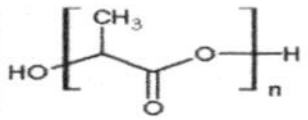	Direct polycondensation of lactic acid at high temperatures yields low molecular weight PLA in a viscous state; ring-opening polymerization of lactides (cyclic dimers of lactic acid) is needed for higher molecular weight materials	PLA is a biodegradable thermoplastic with a high modulus (~3.5 GPa) and strength (60–70 MPa). It is processed using standard methods and used in packaging, textiles, hygiene products, agricultural foils, cutlery, disposable tableware, and injection stretch blow molded bottles and jars	Mohapatra and Singh (2024)

Polyhydroxyalkanoates (PHAs)	Bio-based	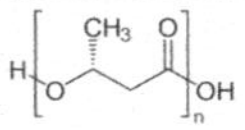	PHAs are natural polyesters synthesized by several hundreds of prokaryotic genera, including Gram-positive and Gram-negative eubacteria, cyanobacteria, and even archaea through sugar and lipid fermentation, creating diverse homopolymers and copolymers as intracellular carbon and energy reserves	PHA polymers, processable with conventional plastic equipment, are used in cosmetics, hygiene products, packaging, and golf tees. Unlike PLA, PHAs are UV stable, heat resistant to 180 °C, and water-resistant. PHBV [poly(3-hydroxy-butyrate-*co*-3--hydroxyvalerate)] is ideal for packaging and medical uses due to its slow degradation and biocompatibility. However, PHB's widespread use is hindered by cost, brittleness, and low thermal stability. Copolymers are preferred for better mechanical properties	Koller and Mukherjee (2022), Acharjee et al. (2024)
Protein-based plastics	Bio-based	Zein 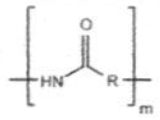	Soya protein plastics are primarily made through extrusion, compression, and injection molding. Zein, an alcohol-soluble protein from corn endosperm, is extracted with aqueous alcohol and dried into granular powder	Soya protein films provide effective barriers to oxygen and UV radiation, making them suitable for packaging and agricultural foils that biodegrade into soil nutrients. Soya protein can also be processed into foams and insulation materials. Zein, similar to beeswax, is tough, greaseproof, hydrophobic, and ideal for films and coatings, though brittle	Baloyi et al. (2023)

(continued)

Table 3 (continued)

Name	Bio-based / fossil fuel based	Structure	Method of production	Distinguishing feature and usage	References
Polybutylene succinate (PBS)	Bio and fossil based	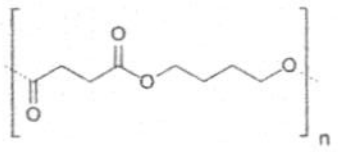	Combining succinic acid (SA) and 1,4-butanediol (1,4-BDO) through polycondensation creates PBS, currently 50% bio-based, since SA is produced by fermentation. Full bio-based PBS is possible via bio-based 1,4-BDO from SA hydrogenation or sugar fermentation	PBS offers excellent processability and mechanical properties, unlike brittle PLA and PHB. It resembles flexible plastics like PE and PP, with 350% elongation, 30–35 MPa tensile strength, and 700 MPa tensile modulus. PBS can be tailored through copolymerization with different dicarboxylic acids or diols, making it versatile for applications like packaging, disposable items, agricultural films, nonwoven fabrics, and biomedical devices	Aliotta et al. (2022)

while sugarcane is favored for producing durable bio-based plastics like polyethylene. Cellulose from wood or cotton provides a strong, flexible material ideal for packaging. Algae grow quickly and require minimal land, while bacteria can produce biodegradable PHAs. Each resource contributes specific qualities to the bioplastic, influencing factors such as strength, flexibility, degradation rate, and environmental impact.

3.5 Key Feedstocks in the Bioplastic Production

Bioplastics are produced from renewable lignocellulosic sources like corn starch, sugarcane, cellulose, potato starch, and wheat bran, yielding biodegradable plastics like PLA, TPS, PBS, Bio-Pur, etc. Algae and bacteria produce PHA, while soybean and palm oil make bio-based polyurethane. Agricultural by-products, including cane bagasse, rice husks, and tomato peels, as well as chitin from crustacean shells, serve as valuable feedstocks. These resources provide eco-friendly alternatives to conventional plastics, supporting sustainability and reducing environmental impact. Different feedstocks for bioplastics and the types of bioplastics they produce are represented in Table 4.

3.6 Lignocellulosic Waste as Cheaper Feedstock for Bioplastic Production

3.6.1 Corn Starch

Corn starch derived from lignocellulosic agricultural waste, such as husks, stalks, and leaves, is becoming a sustainable feedstock for bioplastic production, offering an eco-friendly alternative to conventional plastics (Moghaddam et al., 2024). The starch is first hydrolyzed into glucose using amylase, followed either by microbial fermentation to lactic acid for polymeric synthesis of polylactic acid (PLA) or biopolymers like polyhydroxyalkanoates (PHAs) (Kawaguchi et al., 2022). Thermoplastic starch (TPS) is formed by blending starch with plasticizers (Bangar et al., 2021). Starch-polyester blends combine corn starch with polyesters like polycaprolactone (PCL) or polybutylene succinate (PBS), resulting in the enhancement of both mechanical properties and the biodegradability of the composite (Ribba et al., 2021). Starch-polyvinyl alcohol (PVA) blends exhibit enhanced water solubility and flexibility (Song et al., 2024), while starch-based foams are biodegradable alternatives to polystyrene in protective packaging (Alban-Bolaños et al., 2024). Starch-nanocomposites incorporate nanoparticles to improve mechanical, thermal, and barrier properties for advanced packaging (Peidayesh et al., 2024).

Recent research focuses on enhancing the mechanical properties of corn starch-based bioplastics, improving strength, flexibility, and durability. Nandiyanto et al. (2020) found that glycerol boosts the flexibility and tensile strength of corn starch-based bioplastics, though the mechanical performance of these bioplastics still lags behind conventional plastics. Casiño et al. (2023) showed that adding bamboo

Table 4 Feedstocks for bioplastic production

Feedstock	Few examples of bioplastic produced	References
Corn starch	Polylactic acid (PLA), thermoplastic starch (TPS)	García-Cruz et al. (2020)
Sugarcane	PLA, polybutylene succinate (PBS), polyhydroxyalkanoates (PHAs) bio-based polyethylene (bio-PE)	Ratshoshi et al. (2021), Barbalho et al. (2023)
Cellulose (wood, cotton)	Cellulose acetate, cellophane	Nigam et al. (2022)
Potato starch	Starch-based foams, starch-based blends, starch-based films	Falua et al. (2022)
Wheat bran	PCL/cellulose composite, thermoplastic composites, PHA	Rossi et al. (2024)
Cassava	Thermoplastic starch (TPS), PLA	Gómez et al. (2023)
Algae	Polylactic acid (PLA), polyhydroxyalkanoates (PHAs)	Cheah et al. (2023), Narayanasamy et al. (2024)
Soybean oil	Bio-based polyurethane (bio-PUR), modified starch-based bioplastics	Xu et al. (2022)
Palm oil	Poly(dodecyl 2-tetradecylmalonte) (PDTDM) and poly(hexyl 2-tetradecylmalonte) (PHTDM)	Japir et al. (2021)
Cane bagasse	Cellulose-based bioplastics, lignin-based bioplastics	Shabbirahmed et al. (2022)
Bacterial cultures	PHA, PLA, PHB, bio-PU	Chawla et al. (2023)
Sugar beet pulp	Polylactic acid (PLA), PHA, cellulose nanofiber	Marzo-Gago et al. (2023)
Rice husks	TPS, cellulose-based composites, chitosan-based bioplastics	Hayatun et al. (2020)
Vegetable wastes	PHA, PLA, protein nanofibrils	Li et al. (2022)
Chitin (crustacean shells)	Chitosan-based bioplastics	Jin et al. (2024)

fibers, especially *Dendrocalamus asper*, improved the tensile strength and durability of starch-based bioplastics while maintaining biodegradability. It has also been reported that storage temperature influences the performance of starch-based bioplastics, with better resistance at low temperatures (4 °C) but moderate deterioration in strength and flexibility was observed at 25 °C. At higher temperatures (40 °C), significant deterioration and increased moisture absorption were reported, which compromised the material's structural integrity (Ayu et al., 2023).

3.6.2 Cellulose

Cellulose, a natural polymer abundant in plant cell walls, is an ideal feedstock for bioplastics due to its availability and renewability. Derived from wood and plant-based wastes, cellulose is chemically modified to produce bioplastics like cellulose acetate and cellophane, known for their flexibility and biodegradability. These bioplastics offer good mechanical strength, making them suitable for packaging and textiles, though they can be sensitive to moisture, potentially affecting their performance under certain conditions. Cellulose acetate is a cellulose-based bioplastic, produced by acetylating cellulose with acetic anhydride. It is a flexible, transparent bioplastic used in films, coatings, and textiles, offering good clarity and biodegradability (Nigam et al., 2022). Another cellulose-based bioplastic called cellophane is prepared by dissolving and regenerating cellulose and is primarily used for packaging and providing a biodegradable barrier to gases or/and moisture (Macnamara, 2024). Regenerated cellulose materials, like viscose and lyocell, are used in textiles, offering good strength and comfort (Huang & Wang, 2022). Yet another cellulose-based composite, nanocellulose, made from nanoscale cellulose fibers, is applied in high-strength films, coatings, and composites due to its exceptional mechanical properties and biodegradability (Rao et al., 2024).

Recent research on cellulose-based bioplastics focuses on improving durability, biodegradability, and functionality by combining biopolymers and optimizing composting methods to address scalability and sustainability challenges.

Cellulose-based bioplastic was developed from waste paper, using ionic liquid 1-butyl-3-methylimidazolium chloride to create membranes with tensile strength over 127 MPa—100 times stronger than the original paper. This film, with 80% optical transmittance in the 450–800 nm range, was found suitable for transparent applications. It exhibited excellent biodegradability, thus highlighting the sustainability benefits of using waste paper (Liu et al., 2023). Guzman-Puyol et al. (2022) created transparent bioplastics from renewable cellulose and naringin, a natural flavonoid, for food packaging. The film exhibited UV-blocking and antioxidant properties, maintained high tensile strength and flexibility while providing effective moisture and gas barriers, and confirmed biodegradability under composting conditions. In another study, life cycle assessment (LCA) was performed to evaluate the environmental impacts of different end-of-life management strategies for cellulose-based bioplastics in municipal solid waste systems, focusing on greenhouse gas emissions (Gadaleta et al., 2023). The study revealed that the environmental impact varied by disposal method. Combining cellulose-based bioplastics with organic waste resulted in the highest impact, releasing 5.8 kg CO_2 equivalent per functional unit, attributed to lower compost quality and reduced energy recovery. In contrast, mixing them with plastic or mixed waste produced negative impact values, with reductions of −9.8 kg CO_2 eq/FU and −7.7 kg CO_2 eq/FU, respectively. The authors concluded that integrating these bioplastics into plastic waste streams could enhance waste management and decrease carbon footprints.

3.6.3 Sugarcane

Sugarcane is an excellent renewable resource for bioplastic production due to its high sucrose content, which ferments into bio-based monomers. Fermenting sugarcane juice with microorganisms like *Lactobacillus* produces lactic acid, later polymerized into biodegradable PLA this approach, along with the use of sugarcane by-products, such as bagasse, boosts resource efficiency, reduces carbon footprints, supports waste reduction, and promotes biodegradability and circular economy.

Polylactic acid (PLA) is produced by fermenting sugarcane juice to produce lactic acid, which is then polymerized. Valued for its biodegradability, PLA is widely utilized in packaging, disposable cutlery, and 3D printing. (Agrawal & Kumar, 2023). Bio-based polyethylene (Bio-PE) is synthesized by converting sugarcane ethanol into ethylene, resulting in a plastic chemically identical to conventional PE, used in bottles and bags (Barbalho et al., 2023). Polyhydroxyalkanoates (PHAs) are derived from bacterial fermentation sugars from sugarcane and are employed in medical applications and packaging (Silverio et al., 2024). Succinic acid (SA) converted from cane sugar can be polymerized to synthesize polybutylene succinate (PBS), which is suitable for packaging and agriculture (Pei et al., 2023). Bio-based PET, partially derived from sugarcane-based ethanol, could serve as a biodegradable alternative to fossil-fuel-based PET used in beverage bottles and textiles (Hálfdánsdóttir & Ilievska, 2024).

Additionally, sugarcane bagasse, the fibrous residue left after juice extraction, can be used to produce different types of bioplastics. Bagasse rich in cellulose can be used to produce transparent and flexible cellulose acetate (Maryana et al., 2022) and bagasse-based biocomposites which are reinforced with other bioplastics like PLA or PHA, making them suitable for more durable applications, such as automotive parts and construction materials (Sai et al., 2024). The lignin in sugarcane bagasse has been extracted to produce high-strength and biodegradable lignin-based bioplastics, which find their application in packaging, coatings, and also as adhesives (Fazeli et al., 2023).

Recent research on sugarcane-derived bioplastics focuses on enhancing production and mechanical properties and broadening applications as sustainable alternatives to traditional plastics. Moungprayoon et al. (2022) optimized sugarcane juice for cultivating *Paracoccus* sp. KKU01, producing 55.5 g/L biomass with 36.2% polyhydroxybutyrate (PHB) content, improving PHB's economic feasibility. Barbalho et al. (2023) enhanced bio-polyethylene composites by incorporating sugarcane bagasse and treated Curauá fiber, boosting tensile strength, thermal stability, and adhesion. Yong et al. (2024) optimized PHA production from bagasse using *Bacillus cereus*, achieving an 86.4 g/L yield. Kyokusiima and Nyerere (2024) created a biodegradable film from bagasse and maize straw at a 75:25 cellulose-fiber ratio. Martins et al. (2024) examined how acetylating xylan from sugarcane bagasse influences bioplastic formulation, properties, and biodegradation. Testing different acetylation conditions produced water-soluble xylan with low and medium degrees of substitution (DS), enhancing the mechanical and thermal resistance of xylan-starch bioplastics. While low DS had no effect on disintegration rates, medium DS (0.9) reduced microbial activity. All samples provided oil barrier properties, indicating the benefits of acetylated xylan in bioplastic production.

3.6.4 Potato Starch

Potato starch-based bioplastics are developed from natural potato polymers, primarily amylose and amylopectin. Processed via gelatinization or combined with plasticizers like glycerol, they form flexible, biodegradable films with strong mechanical properties like tensile strength and elongation. Their eco-friendly, renewable nature makes them suitable for packaging, agricultural films, and disposables. Ongoing research aims to improve their durability, performance, and cost-effectiveness, supporting wider industrial applications for various types of bioplastics. Starch-based films are flexible, transparent materials made by gelatinizing potato starch with plasticizers like glycerol, which is majorly used in food packaging, agriculture, and making disposables (Li et al., 2024). Starch-based blends, prepared by combining potato starch with biopolymers like PLA or PHA, enhance mechanical properties and processability for versatile applications, like packaging and preparing molded products (Jayarathna et al., 2022). In addition, potato starch-based composites, reinforced with natural fibers like cellulose and wheat bran, provide increased strength and durability, making them suitable for automotive and construction uses (Muñoz-Gimena et al., 2023). Starch-based foams, produced by incorporating blowing agents into the starch matrix, are lightweight with good insulation and ideal for packaging and insulation applications (Dircio-Morales et al., 2023).

Olagundoye and Morayo (2022) have recently reported having developed films with 0, 5, 10, 20, and 30 wt% of banana pseudostem cellulose, observing improved physicochemical, mechanical, and thermal properties at higher cellulose levels. The 10% cellulose film had optimal properties, while the 30% was thickest and biodegraded in 12 days. Patel et al. (2023) developed ten films using potato starch with glycerol and sorbitol blends as plasticizers, varying their concentrations and ratios, which improved thermal stability, tensile strength, and morphological compatibility compared to films with only glycerol. Additionally, Masurkar (2024) tested films made from dried potato peels mixed with cornstarch, gelatin, and glycerol, showing a tensile strength of 17 MPa and significant flexibility, while biodegradability tests revealed an 80% weight loss after 6 months, with key microorganisms and enzymes aiding degradation. Another study investigated the effect of drying temperature on potato and rice starch bioplastics, finding that higher temperatures produced thinner films with optimal properties at 65 °C, including the best tensile and bursting strengths (Nikhilesh & Jethva, 2024).

3.6.5 Wheat Bran

Wheat bran, high in cellulose, hemicellulose, and lignin, is a sustainable feedstock for bioplastic production. The process for creating PCL/cellulose composite-based bioplastic from wheat bran involves cellulose extraction via enzymatic or chemical methods (Chen et al., 2023). Rahman et al. (2017) highlighted the potential of wheat bran in the production of thermoplastic composites with enhanced mechanical properties, while Van-Thuoc et al. (2008) explored its role in PHB production with *Halomonas boliviensis* LC1. Bioplastics derived from wheat bran are biodegradable, decomposing naturally through microbial action, which supports eco-friendly waste management. Wheat bran can be used to produce various types of bioplastics, including cellulose acetate, regenerated cellulose, cellulose nanofibers, and xylan-based plastics, thus broadening its applications (Martins et al., 2024).

3.6.6 Cassava

Cassava, due to its high starch content, is a renewable and biodegradable feedstock for bioplastic production. Cassava starch is affordable, plentiful, and has excellent film-forming properties, making it an ideal precursor of bioplastics (Matheus et al., 2023). Through gelatinization and plasticization, cassava starch could be converted to thermoplastic starch (TPS), which can be blended with biopolymers like PLA or PHA to improve durability and mechanical strength (Gómez et al., 2023). Cassava-based bioplastics are biodegradable and reduce environmental impact, while cassava farming uses minimal water and inputs, benefiting the environment and supporting rural economies (Hierro-Iglesias et al., 2022; Lilavanichakul & Yoksan, 2023).

3.6.7 Sugar Beet Pulp

Sugar beet pulp (SBP), a byproduct of sugar extraction, is gaining attention as a sustainable bioplastic resource. Rich in cellulose, hemicellulose, and pectin, it provides a renewable, biodegradable alternative to conventional plastics. SBP can be processed into films or combined with biopolymers to improve mechanical properties. Marzo-Gago et al. (2023) evaluated SBP's potential in bioplastics, optimizing sugar extraction through thermal and chemical pretreatments—sulfuric acid treatment notably boosting glucose yields. These sugars can be fermented by lactic acid bacteria to produce biodegradable polymers like PLA and PHA. Sarafidou et al. (2024) enhanced pectin yield and quality through enzymatic and chemical treatments, facilitating PHB production. Fadel et al. (2022) explored SBP-derived cellulose nanofibers (CNFs) for paper coatings, noting their antibacterial, antioxidant, moisture, and grease resistance. Incorporation of SBP in PLA composites improved its biodegradation significantly when exposed to soil due to microbial colonization and matrix disintegration, suggesting SBP's potential to enhance the environmental performance of PLA materials (Schirp et al., 2024).

3.6.8 Rice Husk

Rice husk, the outer layer of rice grains, due to its high cellulose, silica, and lignin content, is also a promising resource for durable and eco-friendly bioplastic production. Karaca et al. (2022) examined the simultaneous extraction of cellulose fibers and biogenic silica from rice husk to create cellulose-based composite films. Using $ZnCl_2$ as a solvent, the study found that increasing silica improved barrier and tensile strength, reaching up to 1347 kN/m^2 with 2% silica. The films biodegraded between 25.5% and 55.7% over 21 days in soil, making them sustainable packaging options. Brites et al. (2024) utilized hydrothermal extraction to isolate bioactive compounds like phenolics and antioxidants from rice husks, enhancing the mechanical and barrier properties of chitosan-based bioplastics, ideal for packaging. These bioplastics also displayed improved biodegradability, supporting environmental sustainability. Alonso-González et al. (2021) studied composites of rice husk bioplastics and thermoplastic starch, finding that increasing the mixing temperature from 100 to 140 °C improved starch plastification, enhancing tensile strength and elasticity. However, temperatures above 160 °C led to starch degradation, compromising mechanical properties, emphasizing the importance of temperature optimization in producing high-quality rice husk bioplastics.

3.6.9 Food Waste

Transforming food waste into bioplastics presents an innovative solution to environmental pollution and waste management challenges. Food waste, abundant in starch, cellulose, lipids, and proteins, can be converted into biodegradable polymers, offering a sustainable replacement for petroleum-based plastics. These bioplastics deliver comparable durability and functionality besides lowering the negative environmental impacts of conventional plastics. Giwa et al. (2024) reviewed advancements in polyhydroxyalkanoates (PHAs) production from food waste, focusing on microbial methods to convert residues like fruits, vegetables, and leftovers. They emphasized selecting optimal microbial strains for higher PHA yields and discussed various pretreatment methods to enhance fermentable sugars, essential for microbial growth. Mechanical and heat pretreatments improve conversion and degradability, while chemical methods suit lignocellulosic wastes. Enzymatic treatments enhance hydrolysis and reduce volatile emissions. The study also examined by-products like hydroxymethyl 5-hydroxymethylfurfural (HMF), furfural, and volatile fatty acids (VFA) that could influence microbial activity, suggesting biorefinery integration for greater sustainability.

Visco et al. (2022) studied agri-food wastes, such as brewer's spent grain and olive pomace, as resources for bioplastics, either by direct transformation or as biofillers. Their findings underscored the environmental and economic benefits, such as reducing fossil fuel reliance and carbon footprints. They also noted challenges with separating bioplastics from conventional plastics and highlighted the importance of EU-funded projects in supporting research and innovation in this field. PLA and PHA bioplastics were also reported to be produced from fermented food waste, with processes using lactic acid bacteria (Swetha et al., 2023) or specific substrates for PHB production by *Haloferax mediterranei* (Longo et al., 2024). Additionally, Jogi and Bhat (2020) explored converting food waste, like vegetable peels and dairy by-products, into bioplastics by direct extraction or through natural and genetically modified organisms. They reported that potato protein concentrate enables compression molding, while rapeseed oil by-products support injection molding. Genetically engineered microbes were reported to improve fermentation, boosting production rates and reducing costs, thus converting food waste into eco-friendly materials and mitigating environmental impacts. In another recent review, Peidayesh et al. (2024) reported on converting food protein waste into materials like protein nanofibrils and hydrogels, useful in food packaging, biomedical devices, and drug delivery. The review highlighted extraction techniques such as enzymatic hydrolysis and mechanical processing that enhance material properties. Protein-based biodegradable plastics were also explored as eco-friendly alternatives to fossil-fuel plastics, offering natural degradability that supports waste reduction and environmental sustainability.

3.7 Role of Microbes in Bioplastic Production

Microbes play a vital role in converting lignocellulosic waste, including agricultural and forestry by-products rich in cellulose (40–50%), hemicellulose (20–30%), and lignin (10–20%), into bioplastics. Chemical or microbial processes break down these complex materials, enabling microbial fermentation. Specific microbes release enzymes like cellulases and hemicellulases to break down cellulose and hemicellulose, producing sugars that bacteria and fungi ferment to produce bioplastics. Examples of a few recently reported microbes involved in bioplastic production are represented in Table 5.

Table 5 Microbes involved in bioplastic production

Bacterial species	Bioplastic produced	Lignocellulosic waste utilized	Key findings	Citation
Cupriavidus necator	Polyhydroxyalkanoates (PHAs)	Agricultural residues, corn stover	Efficiently converts glucose from cellulose into PHAs	Li and Wilkins (2020)
Bacillus megaterium	Polyhydroxybutyrate (PHB)	Sugarcane bagasse, wheat straw	High yield of PHB from pre-treated lignocellulosic biomass	Naitam et al. (2022)
Bacillus licheniformis	Polyhydroxybutyrate (PHB)	Rice straw, soybean residue	Effective in converting pre-treated lignocellulosic waste into PHB	Nalini and Sathiyamurthi (2023)
Ralstonia eutropha	Polyhydroxyalkanoates (PHAs)	Various lignocellulosic sources	Utilizes sugars derived from lignocellulosic wastes to produce PHAs	Jaffur et al. (2023)
Bacillus licheniformis	Poly lactic acid (PLA)	Corn stover, wheat straw	Utilizes enzymes to convert lignocellulose to sugars, which are then fermented to PLA	Chawla and Goyal (2024)
Lactobacillus brevis *Lactobacillus rhamnosus* *Lactobacillus delbrueckii*	Poly lactic acid (PLA)	Rice husks, sugarcane bagasse	Hydrolyzes lignocellulosic biomass to fermentable sugars; produces PLA through fermentation	Widjaja et al. (2022)

(continued)

Table 5 (continued)

Bacterial species	Bioplastic produced	Lignocellulosic waste utilized	Key findings	Citation
Bacillus sonorenesis strain DGS15	Poly lactic acid (PLA)	Wheat straw, corn stover	Enzymatic hydrolysis of lignocellulosic biomass followed by fermentation to PLA	Chawla and Goyal (2023)
Metabolically engineered *E. coli*	Poly(3-hydroxybutyrate-*co*-lactate) [P(3HB-*co*-LA)]	Corn stover hydrolysate	Decomposes lignocellulosic waste into sugars, which are converted to PLA	Wu et al. (2021)
Pediococcus acidilactici DQ2	Poly lactic acid (PLA)	Corn Stover hydrolysate	Simultaneous saccharification and high titer lactic acid fermentation	Zhao et al. (2013)
Clostridium butyricum NCIMB 8082	Bio-PU	Corn steep liquor (CSL)	Fermentative conversion of CSL to 1,3-propanediol, the precursor for bio-PU	Liberato et al. (2022)

Polyhydroxyalkanoates (PHAs) and polyhydroxybutyrate (PHB) are intracellular storage materials produced by bacteria under nutrient-limited conditions with an abundant carbon supply (Fu et al., 2023). Lignocellulosic waste offers an ample carbon source, encouraging bacteria to accumulate PHAs, like PHB, when grown with limited nitrogen, phosphorus, or oxygen (Chawla et al., 2023). Microbes like *Cupriavidus necator*, *Ralstonia eutropha*, and *Pseudomonas putida* can store up to 90% of their dry cell weight as PHA granules (Bellini et al., 2022; Chandra et al., 2023; Thiele et al., 2024), providing energy reserves (Fukala & Kučera, 2024). Through fermentation, microbes convert sugars or lipids into PHA or PHB, which are later extracted and purified for bioplastics. PHB is valued for its crystallinity and biodegradability. Polylactic acid (PLA) is produced by lactic acid bacteria like *Lactobacillus*, which ferments glucose, sucrose, or starch into lactic acid for polymerization into PLA (Widjaja et al., 2022). Bio-polyurethane (Bio-PU) uses polyols from sources like vegetable oils or microbial products. Microorganisms can produce polyols like glycerol or 1,3-propanediol, which can be reacted with isocyanates to form bio-PU (Liberato et al., 2022).

4 The Global Scenario of Bioplastic Production

The "Bioplastics Market Development Update, 2023," released by European Bioplastics provides a comprehensive overview of the current state and future projections of the bioplastics industry. The global production of bioplastics is rapidly expanding due to rising environmental concerns and the need for sustainable materials. Bioplastics represent about 0.5% of the 400 million tons of plastic produced annually. In 2023, global bioplastics production capacity reached 2.18 million tons and is projected to grow to 7.43 million tons by 2028, reflecting a compound annual growth rate (CAGR) of 17.43%. The industry operated at an average utilization rate of 82%, producing roughly 1.79 million tons out of the total capacity. Production of different bioplastic materials in 2023 (Bioplastics Market Development Update, 2023) has been summarized in Table 6.

The report projects that bioplastics production will increase from 2.67 million tons to 7.42 million tons between 2024 and 2028, driven by growth in both biodegradable and non-biodegradable bioplastics. Notably, the production of biodegradable bioplastics, such as polylactic acid (PLA), polyhydroxyalkanoates (PHAs), polybutylene adipate terephthalate (PBAT), starch-containing polymer compounds (SCPC), etc., is expected to rise due to technological advancements and growing global demand for bioplastics.

Asia is leading the global bioplastics production in 2022, accounting for 41.4% of global capacity, particularly in the production of polylactic acid (PLA) and polyhydroxyalkanoates (PHAs) in China (Oghazi et al., 2024; Soo et al., 2024). Chinese companies like BBCA Group are expected to boost PLA output, potentially surpassing demand by 2023 (Caicedo et al., 2021). Europe, the second-largest producer at 26.5%, emphasizes innovation in sustainable materials but may see its share decrease as Asia's production grows, potentially exceeding 60% by 2027. North America holds 19% of the market, focusing on PLA and bio-based PET, while South America, with around 13%, centers on bio-based polyethylene (PE) (Renewable Carbon News, 2024).

Table 6 Production of different bioplastic materials in 2023

Bioplastic material	Production capacity (in 1000 tons)
Polylactic acid (PLA)	1100
Polyhydroxyalkanoates (PHAs)	100
Polybutylene succinate (PBS)	100
Starch blends	600
Polyethylene (PE)	400
Polyamide (PA)	200
Polyethylene terephthalate (PET)	200
Polypropylene (PP)	300
Others	1200
Total	2180

Global bioplastics demand is rising due to stricter environmental regulations on single-use plastics (Barrowclough & Birkbeck, 2022), increasing consumer preference for eco-friendly products, and heightened awareness of plastic pollution (Findrik & Meixner, 2023). Technological advances in bioplastic production are improving durability, barrier properties, and cost-effectiveness, making bioplastics more competitive with conventional plastics (Siddiqui et al., 2024). Corporate sustainability goals have further driven investments, partnerships, and R&D for sustainable packaging solutions.

5 Recent Prospects of Lignocellulosic Bioplastics in a Circular Economy

Lignocellulosic biomass from plant materials like wood, straw, and agricultural residues is increasingly valued as a feedstock for the production of bioplastics. Advances in processing, microbial fermentation, and biopolymer development focus on optimizing pretreatment techniques to boost digestibility. Enzymatic hydrolysis and chemical treatments improve access to cellulose and hemicellulose, enabling bioplastic production (e.g., PHAs, PLA) from non-food sources, reducing food security concerns associated with traditional feedstocks like corn and sugarcane (Kawaguchi et al., 2022; Mujtaba et al., 2023). Integrating lignocellulosic bioplastics into a circular economy offers significant potential for converting agricultural waste into valuable products and reducing waste. Non-food sources—such as apple pomace, banana peels, coconut fiber, and sugarcane bagasse—are studied for bioplastic production, which could reduce costs and avoid additional land use, preserving agricultural land for food production (Lu et al., 2022). Bioplastics from lignocellulosic biomass also offer sustainability benefits by reducing plastic-related emissions and supporting renewable resources (Nisar et al., 2023). The European Bioeconomy strategy emphasizes these innovations for economic growth and environmental impact reduction (Tsakalova et al., 2024).

Valued at approximately USD 15.3 billion in 2024, the bioplastics market is expected to reach USD 45.2 billion by 2029, driven by a high CAGR of 31.4% in sectors utilizing renewable feedstocks like lignocellulose (Markets & Markets, 2024). Polylactic acid (PLA) alone is projected to generate around USD 1.31 billion by 2026, illustrating lignocellulosic sources' significant impact (Raj et al., 2022). This shift toward lignocellulosic feedstocks, such as agricultural residues, wood, and other non-food plant materials for bioplastic production, is fueled by their sustainability and environmental benefits (Gong et al., 2024). The market's growth is further supported by consumer awareness and regulatory pressures on plastic waste. Leading companies like Coca-Cola and PepsiCo are investing in bioplastics, enhancing market expansion (Censere News & Articles, 2015). The U.S. bioplastics market is projected to reach USD 1.2 billion by 2028, with companies like Biome Technologies reporting significant sales revenue increases, reflecting growing demand.

6 Conclusion

Utilizing lignocellulosic wastes for bioplastic development represents a significant advancement in achieving a circular bioeconomy and aligning with the United Nation's Sustainable Development Goals (SDGs). By converting agricultural residues, such as wheat bran, rice straw and husks, corn cobs, sugarcane juice and bagasse, forest wastes, and household kitchen wastes, into valuable bioplastics, this approach not only reduces reliance on fossil-based plastics but also addresses waste management challenges. The biodegradability of these bioplastics ensures minimal environmental impact, contributing to goals like responsible consumption and production (SDG 12) and climate action (SDG 13). Additionally, this process supports innovation and sustainable industrialization (SDG 9) by fostering new technologies and value chains in the bio-based economy. The integration of lignocellulosic wastes into bioplastic production exemplifies a closed-loop system, where waste is transformed into resources, reinforcing the principles of a circular economy and promoting long-term environmental sustainability. This approach is crucial for advancing global sustainability efforts and ensuring a balanced relationship between economic growth and ecological preservation.

References

Abas, K. M., & Attia, A. A. M. (2023). Thermoplastic starch (TPS)-based composite films for wastewater treatment: Synthesis and fundamental characterization. *BMC Chemistry, 17*(1), 84. https://doi.org/10.1186/s13065-023-00998-z

Abu Bakar, N. H., & Salim, N. (2023). Challenges and opportunities in wood waste utilization. In S. N. Sarmin, M. Jawaid, & R. Elias (Eds.), *Wood waste management and products. Sustainable materials and technology*. Springer. https://doi.org/10.1007/978-981-99-1905-5_1

Acharjee, S. A., Gogoi, B., Bharali, P., Sorhie, V., & Alemtoshi, B. W. (2024). Recent trends in the development of polyhydroxyalkanoates (PHAs) based biocomposites by blending with different bio-based polymers. *Journal of Polymer Research, 31*(4), 98. https://doi.org/10.1007/s10965-024-03947-z

Agrawal, D., & Kumar, V. (2023). Recent progress on sugarcane-bagasse based lactic acid production: Technical advancements, potential and limitations. *Industrial Crops and Products, 193*, 116132. https://doi.org/10.1016/j.indcrop.2022.116132

Ahmed, M. J., Ashfaq, J., Sohail, Z., Channa, I. A., Sanchez-Ferrer, A., Ali, S. N., & Chandio, A. D. (2024). Lignocellulosic bioplastics in sustainable packaging—Recent developments in materials design and processing: A comprehensive review. *Sustainable Materials and Technologies, 41*, e01077. https://doi.org/10.1016/j.susmat.2024.e01077

Alavijeh, R. S., Shahvandi, A., Okoro, O. V., Denayer, J. F. M., & Karimi, K. (2023). Biorefining of corn stover for efficient production of bioethanol, biodiesel, biomethane, and value-added byproducts. *Energy Conversion and Management, 283*, 116877. https://doi.org/10.1016/j.enconman.2023.116877

Alban-Bolaños, P., Ayala-Aponte, A. A., Villada-Castillo, H. S., Rodríguez-Herrera, R., Ávalos-Belmonte, F., & Solanilla-Duque, J. F. (2024). Biodegradable flexible foam: Novel material based on cassava TPS obtained by extrusion. *Colloid and Polymer Science, 302*(4), 597. https://doi.org/10.1007/s00396-023-05204-z

Ali, S. S., Abdelkarim, E. A., Elsamahy, T., Al-Tohamy, R., Li, F., Kornaros, M., Zuorro, A., Zhu, D., & Sun, J. (2023). Bioplastic production in terms of life cycle assessment: A state-of-the-art review. *Environmental Science and Ecotechnology, 15*, 100254. https://doi.org/10.1016/j.ese.2023.100254

Aliotta, L., Seggiani, M., Lazzeri, A., Gigante, V., & Cinelli, P. (2022). A brief review of poly (butylene succinate) (PBS) and its main copolymers: Synthesis, blends, composites, biodegradability, and applications. *Polymers, 14*(4), 844. https://doi.org/10.3390/polym14040844

Alokika, A., & Singh, B. (2023). Utilization of lignocellulosic plant residues for compost formation and its role in improving soil fertility. *Pedosphere, 33*(5), 700–716. https://doi.org/10.1016/j.pedsph.2022.11.008

Alonso-González, M., Felix, M., Guerrero, A., & Romero, A. (2021). Rice bran-based bioplastics: Effects of the mixing temperature on starch plastification and final properties. *International Journal of Biological Macromolecules, 188*, 932. https://doi.org/10.1016/j.ijbiomac.2021.08.043

Anokye, K., Mohammed, A. S., Agyemang, P., Agya, B. A., Amuah, E. E. Y., & Sodoke, S. (2024). Understanding the perception and awareness of senior high school teachers on the environmental impacts of plastic waste: Implications for sustainable waste education and management. *Social Sciences & Humanities Open, 10*, 100999. https://doi.org/10.1016/j.ssaho.2024.100999

Archer, E., Torretti, M., & Madbouly, S. (2023). Biodegradable polycaprolactone (PCL) based polymer and composites. *Physical Sciences Reviews, 8*(11), 4391–4414. https://doi.org/10.1515/psr-2020-0074

Ayu, G. E., Lubis, M., Ginting, M. H. S., & Hayat, N. (2023). Physical resistance of corn starch based bioplastic at various storage temperature. *Journal of Engineering Science and Technology, 18*(1), 235–243. https://jestec.taylors.edu.my/Vol%2018%20Issue%201%20February%20%202023/18_1_16.pdf

Aziz, T., Shah, Z., Sarwar, A., Ullah, N., Khan, A. A., Sameeh, M. Y., Haiying, C., & Lin, L. (2023). Production of bioethanol from pretreated rice straw, an integrated and mediated upstream fermentation process. *Biomass Conversion and Biorefinery*. https://doi.org/10.1007/s13399-023-04283-w

Babar, A., Lakho, H., Memon, M., Gul, H., Memon, F., & Siddiqui, F. H. (2022). The impact of wheat straw ash as a partial cement replacement material on concrete properties. *Resources, Conservation and Recycling, 3*.

Bala, S., Garg, D., Sridhar, K., Inbaraj, B. S., Singh, R., Kamma, S., Tripathi, M., & Sharma, M. (2023). Transformation of agro-waste into value-added bioproducts and bioactive compounds: Micro/nano formulations and application in the agri-food-pharma sector. *Bioengineering, 10*(2), 152. https://doi.org/10.3390/bioengineering10020152

Baloyi, J. T., Taylor, J., & Taylor, J. R. N. (2023). Bioplastic film making properties of quality protein maize (QPM) zein. *Cereal Chemistry, 100*(4), 805. https://doi.org/10.1002/cche.10665

Banerjee, S., & Arora, A. (2021). Sustainable bioprocess technologies for urban waste valorization. *Case Studies in Chemical and Environmental Engineering, 4*, 100166. https://doi.org/10.1016/j.cscee.2021.100166

Bangar, S. P., Whiteside, W. S., Ashogbon, A. O., & Kumar, M. (2021). Recent advances in thermoplastic starches for food packaging: A review. *Food Packaging and Shelf Life, 30*, 100743. https://doi.org/10.1016/j.fpsl.2021.100743

Bao, Y., Du, Z., Liu, X., Liu, H., Tang, J., Qin, C., et al. (2024). Furfural production from lignocellulosic biomass: One-step and two-step strategies and techno-economic evaluation. *Green Chemistry, 26*, 6318–6338. https://doi.org/10.1039/D4GC00883A

Barrowclough, D., & Birkbeck, C. D. (2022). Transforming the global plastics economy: The role of economic policies in the global governance of plastic pollution. *Social Sciences, 11*(1), 26. https://doi.org/10.3390/socsci11010026

Bellini, S., Tommasi, T., & Fino, D. (2022). Poly(3-hydroxybutyrate) biosynthesis by *Cupriavidus necator*: A review on waste substrates utilization for a circular economy approach. *Bioresource Technology Reports, 17*, 100985. https://doi.org/10.1016/j.biteb.2022.100985

Belmonte-Ureña, L. J., Plaza-Úbeda, J. A., Vazquez-Brust, D., & Yakovleva, N. (2021). Circular economy, degrowth and green growth as pathways for research on sustainable development goals: A global analysis and future agenda. *Ecological Economics, 185*, 107050. https://doi.org/10.1016/j.ecolecon.2021.107050

Benouadah, N., Pranovich, A., Labidi, J., & Willför, S. (2022). Valorization of waste bark for biorefineries: Chemical characterization of Eucalyptus camaldulensis inner and outer barks. *Holzforschung, 76*(3), 285. https://doi.org/10.1515/hf-2021-0173

Bhatia, S. K., Joo, H. S., & Yang, Y. H. (2018). Biowaste-to-bioenergy using biological methods—A mini-review. *Energy Conversion and Management, 177*, 640. https://doi.org/10.1016/j.enconman.2018.09.090

Bioplastics Market Development Update. (2023). Retrieved from https://www.european--bioplastics.org/bioplastics-market-development-update-2023-2/

Blasi, A., Verardi, A., Lopresto, C. G., Siciliano, S., & Sangiorgio, P. (2023). Lignocellulosic agricultural waste valorization to obtain valuable products: An overview. *Recycling, 8*(4), 61. https://doi.org/10.3390/recycling8040061

Bożym, M., Gendek, A., Siemiątkowski, G., Aniszewska, M., & Malaťák, J. (2021). Assessment of the composition of forest waste in terms of its further use. *Materials, 14*(4), 973. https://doi.org/10.3390/ma14040973

Brites, P., Aguiar, M. I., Gonçalves, J., Ferreira, P., & Nunes, C. (2024). Sustainable valorisation of bioactive molecules from rice husks through hydrothermal extraction for chitosan-based bioplastic production. *International Journal of Biological Macromolecules, 271*, 132489. https://doi.org/10.1016/j.ijbiomac.2024.132489

Brienza, F., Van Aelst, K., Devred, F., Magnin, D., Tschulkow, M., Nimmegeers, P., ... & Cybulska, I. (2022). Unleashing lignin potential through the dithionite-assisted organosolv fractionation of lignocellulosic biomass. Chemical Engineering Journal, 450, 138179. https://doi.org/10.1016/j.cej.2022.138179

Broda, M., Yelle, D. J., & Serwańska, K. (2022). Bioethanol production from lignocellulosic biomass—Challenges and solutions. *Molecules, 27*(24), 8717. https://doi.org/10.3390/molecules27248717

Burford, T., Rieg, W., & Madbouly, S. (2023). Biodegradable poly(butylene adipate-co-terephthalate) (PBAT). *Physical Sciences Reviews, 8*(8), 1127. https://doi.org/10.1515/psr-2020-0078

Caicedo, C., Jasso-Salcedo, A. B., de Abril Alexandra Soriano-Melgar, L., Díaz-Cruz, C. A., Jiménez-Regalado, E. J., & Aguirre-Loredo, R. Y. (2021). Biobased and biodegradable packaging plastics for food preservation. In *Handbook of bioplastics and biocomposites engineering applications*. Wiley. https://doi.org/10.1002/9781119160182.ch18

Cakmak, O. K. (2024). Biodegradable polymers—A review on properties, processing, and degradation mechanism. *Circular Economy and Sustainability, 4*(1), 339. https://doi.org/10.1007/s43615-023-00277-y

Casiño, V. J. A., Delos Reyes, N. Q., Jr., Casas, R. Y., Saldo, I. J. P., & Peñafiel-Dandoy, M. J. (2023). Characterization of cornstarch-based bioplastic reinforced with three different species of bamboo shoots. *American Journal of Environmental Protection, 11*(1), 25–33. https://doi.org/10.12691/env-11-1-4

Castro Garcia, A., Cheng, S., McGlynn, S. E., & Cross, J. S. (2023). Machine learning model insights into base-catalyzed hydrothermal lignin depolymerization. *ACS Omega, 8*(35), 32078. https://doi.org/10.1021/acsomega.3c04168

Cazier, E. A., Pham, T. N., Cossus, L., Abla, M., Ilc, T., & Lawrence, P. (2024). Exploring industrial lignocellulosic waste: Sources, types, and potential as high-value molecules. *Waste Management, 188*, 11–38. https://doi.org/10.1016/j.wasman.2024.07.029

Censere News & Articles. (2015). *Bioplastics set for significant growth through 2018*. Published on August, 2015. Retrieved from https://www.censere.com/index.php/en/articles-publications/articles/bioplastics-set-for-significant-growth-through-2018?format=amp

Chandra, R., Thakor, A., Mekonnen, T. H., Charles, T. C., & Lee, H. S. (2023). Production of polyhydroxyalkanoate (PHA) copolymer from food waste using mixed culture for carboxylate production and *Pseudomonas putida* for PHA synthesis. *Journal of Environmental Management, 336*, 117650. https://doi.org/10.1016/j.jenvman.2023.117650

Chang, H., Han, H.-S., Anderson, N., Kim, Y.-S., & Han, S.-K. (2023). The cost of forest thinning operations in the Western United States: A systematic literature review and new thinning cost model. *Journal of Forestry, 121*(2), 193. https://doi.org/10.1093/jofore/fvac037

Chauhan, K., Kaur, R., & Chauhan, I. (2024). Sustainable bioplastic: A comprehensive review on sources, methods, advantages, and applications of bioplastics. *Polymer-Plastics Technology and Materials, 63*(8), 913–938. https://doi.org/10.1080/25740881.2024.2307369

Chawla, S. K., & Goyal, D. (2023). Optimization of pretreatment of wheat straw using response surface methodology for production of lactic acid using *Bacillus sonorenesis* strain DGS15. *Bioenergy Research, 16*(2), 967. https://doi.org/10.1007/s12155-022-10439-9

Chawla, S. K., & Goyal, D. (2024). Enhanced production of lactic acid from pretreated rice straw using co-cultivation of *Bacillus licheniformis* and *Bacillus sonorenesis*. *3 Biotech, 14*(6), 1–17. Retrieved from https://link.springer.com/article/10.1007/s13205-024-04014-6

Chawla, M., Narwal, S., Dhankhar, R., & Kalshan, S. (2023). Microbial production of bioplastics: An eco-friendly alternative. *Ecology, Environment and Conservation, 29*. https://doi.org/10.53550/eec.2023.v29i03s.048

Che Ab Aziz, N. I., Zakaria, Y., Md Muslim, N. Z., & Nik Hassan, N. F. (2023). Emerging and advanced technologies in biodegradable plastics for sustainability. In *Handbook of biodegradable materials*. Springer. https://doi.org/10.1007/978-3-031-09710-2_21

Cheah, W. Y., Er, A. C., Aiyub, K., Yasin, N. H. M., Ngan, S. L., Chew, K. W., et al. (2023). Current status and perspectives of algae-based bioplastics: A reviewed potential for sustainability. *Algal Research, 71*, 103078. https://doi.org/10.1016/j.algal.2023.103078

Chen Austin, M., & Beermann, K. (2023). Including nature-based success measurement criteria in the life cycle assessment. In *Life cycle assessment—Recent advances and new perspectives*. IntechOpen. https://doi.org/10.5772/intechopen.110401

Chen, J., Huang, Y., Deng, L., Jiang, H., Yang, Z., Yang, R., & Wu, D. (2023). Preparation and research of PCL/cellulose composites: Cellulose derived from agricultural wastes. *International Journal of Biological Macromolecules, 235*, 123785. https://doi.org/10.1016/j.ijbiomac.2023.123785

Chen, X., Li, Y., Li, X., Shi, J., & Liu, L. (2024). Exploring the potential of multiple lignocellulosic biomass as a feedstock for biobutanol production. *Fuel, 357*, 129697. https://doi.org/10.1016/j.fuel.2023.129697

Chettri, D., Verma, A. K., Ghosh, S., & Verma, A. K. (2024). Biogas from lignocellulosic feedstock: Current status and challenges. *Environmental Science and Pollution Research, 31*, 1–26. https://doi.org/10.1007/s11356-023-29805-x

Correia, B., Matos, H. A., Lopes, T. F., Marques, S., & Gírio, F. (2024). Sustainability assessment of 2G bioethanol production from residual lignocellulosic biomass. *Processes, 12*(5), 987. https://doi.org/10.3390/pr12050987

Coulibaly, F., Ba, A., Sissoko, S., Maïga, M., Koné, A. K., Coulibaly, D., Diawara, M. O., & Diakité, H. S. (2022). Green kitchen waste: A feed resource for urban and peri-urban livestock. *Scholars Journal of Agriculture and Veterinary Sciences, 9*(9), 124–131. https://doi.org/10.36347/sjavs.2022.v09i09.002

D'Amato, D., & Korhonen, J. (2021). Integrating the green economy, circular economy and bioeconomy in a strategic sustainability framework. *Ecological Economics, 188*, 107143. https://doi.org/10.1016/j.ecolecon.2021.107143

Dalton, B., Bhagabati, P., de Micco, J., Padamati, R. B., & O'Connor, K. (2022). A review on biological synthesis of the biodegradable polymers polyhydroxyalkanoates and the development of multiple applications. *Catalysts, 12*(3), 319. https://doi.org/10.3390/catal12030319

de Almeida Barbalho, G. H., da Silva Nascimento, J. J., da Silva, L. B., Gomez, R. S., de Farias, D. O., Diniz, D. D. S., Santos, R. S., de Figueiredo, M. J., & de Lima, A. G. B. (2023). Biopolyethylene composites based on sugar cane and Curauá fiber: An experimental study. *Polymers, 15*(6), 1369. https://doi.org/10.3390/polym15061369

de Keyser, E., & Mathijs, E. (2023). A typology of sustainable circular business models with applications in the bioeconomy. *Frontiers in Sustainable Food Systems, 6*. https://doi.org/10.3389/fsufs.2022.1028877

Dircio-Morales, M. A., Fonseca-Florido, H. A., Velazquez, G., Ávila-Orta, C. A., Ramos-De Valle, L. F., Hernández-Gámez, F., Rivera-Salinas, J. E., & Soriano-Corral, F. (2023). Relationship among extrusion conditions, cell morphology, and properties of starch-based foams—A review. *Starch/Staerke, 75*(1–2). https://doi.org/10.1002/star.202200103

Duque-Acevedo, M., Belmonte-Ureña, L.J., Cortés-García, F.J., Camacho-Ferre, F. (2021). Recovery of Agricultural Waste Biomass: A Sustainability Strategy for Moving Towards a Circular Bioeconomy. In: Baskar, C., Ramakrishna, S., Baskar, S., Sharma, R., Chinnappan, A., Sehrawat, R. (eds) Handbook of Solid Waste Management. Springer, Singapore. https://doi.org/10.1007/978-981-15-7525-9_25-2

Esrafili, A., & Martínez-Carrera, D. (2015). Oyster mushrooms (Pleurotus) are useful for utilizing lignocellulosic biomass. *Academic Journals, 14*(1), 52–67. https://doi.org/10.5897/ajb2014.14249

Eugenia Eugenio, M., Ibarra, D., Martín-Sampedro, R., Espinosa, E., Bascón, I., & Rodríguez, A. (2019). Alternative raw materials for pulp and paper production in the concept of a lignocellulosic biorefinery. In *Cellulose*. IntechOpen. https://doi.org/10.5772/intechopen.90041

Fadel, S. M., Abou-Elseoud, W. S., Hassan, E. A., Ibrahim, S., & Hassan, M. L. (2022). Use of sugar beet cellulose nanofibers for paper coating. *Industrial Crops and Products, 180*, 114787. https://doi.org/10.1016/j.indcrop.2022.114787

Falua, K. J., Pokharel, A., Babaei-Ghazvini, A., Ai, Y., & Acharya, B. (2022). Valorization of starch to biobased materials: A review. *Polymers, 14*(11), 2215. https://doi.org/10.3390/polym14112215

Fazeli, M., Mukherjee, S., Baniasadi, H., Abidnejad, R., Mujtaba, M., Lipponen, J., Seppälä, J., Rojas, O. J., & Rojas, O. J. (2023). Lignin beyond the status quo: Recent and emerging composite applications. *Green Chemistry, 26*(2), 593. https://doi.org/10.1039/d3gc03154c

Findrik, E., & Meixner, O. (2023). Drivers and barriers for consumers purchasing bioplastics—A systematic literature review. *Journal of Cleaner Production, 410*, 137311. https://doi.org/10.1016/j.jclepro.2023.137311

Fu, X., Xu, H., Zhang, Q., Xi, J., Zhang, H., Zheng, M., Xi, B., & Hou, L. (2023). A review on polyhydroxyalkanoates production from various organic waste streams: Feedstocks, strains, and production strategy. *Resources, Conservation and Recycling, 198*, 107166. https://doi.org/10.1016/j.resconrec.2023.107166

Fukala, I., & Kučera, I. (2024). Natural polyhydroxyalkanoates—An overview of bacterial production methods. *Molecules, 29*(10), 2293. https://doi.org/10.3390/molecules29102293

Gadaleta, G., Ferrara, C., de Gisi, S., Notarnicola, M., & de Feo, G. (2023). Life cycle assessment of end-of-life options for cellulose-based bioplastics when introduced into a municipal solid waste management system. *Science of the Total Environment, 871*, 161958. https://doi.org/10.1016/j.scitotenv.2023.161958

García-Cruz, H. I., Jaime-Fonseca, M. R., Von Borries-Medrano, E., & Vieyra, H. (2020). Extrusion parameters to produce a PLA-starch derived thermoplastic polymer. *Revista Mexicana de Ingeniería Química, 19*(Suppl 1), 395–412. Retrieved from https://rmiq.org/iqfvp/Numbers/V19/S1/Poly1529.pdf

Gatto, A. (2024). *Towards sustainable global food and biomass systems: Interactions between food loss and waste reductions, dietary shifts, and transitioning to a circular bio-based economy* [Doctoral dissertation, Wageningen University]. Retrieved from https://edepot.wur.nl/648608

Gibson, J. (2021). *Proposal to reduce waste pollution and promote efficient waste materials lifecycle management through employment programs and recycling infrastructure investment.* Wright State University.

Giwa, A. S., Shafique, E., Ali, N., & Vakili, M. (2024). Recent advances in food waste transformations into essential bioplastic materials. *Molecules, 29*(16), 3838. https://doi.org/10.3390/molecules29163838

GLOBAL BIOENERGY STATISTICS. (2022). Retrieved from https://www.worldbioenergy.org/uploads/221223%20WBA%20GBS%202022.pdf

Gómez-López, R. A., Montilla-Buitrago, C. E., Villada-Castillo, H. S., Sáenz-Galindo, A., Avalos-Belmontes, F., & Serna-Cock, L. (2023). Co-plasticization of starch with glycerol and isosorbide: Effect on retrogradation in thermo-plastic cassava starch films. *Polymers, 15*(9), 2104. https://doi.org/10.3390/polym15092104

Gong, L., Passari, A. K., Yin, C., Kumar Thakur, V., Newbold, J., Clark, W., Jiang, Y., Kumar, S., & Gupta, V. K. (2024). Sustainable utilization of fruit and vegetable waste bioresources for bioplastics production. In. *Critical Reviews in Biotechnology, 44*(2), 236. https://doi.org/10.1080/07388551.2022.2157241

Govil, T., Wang, J., Samanta, D., David, A., Tripathi, A., Rauniyar, S., Salem, D. R., & Sani, R. K. (2020). Lignocellulosic feedstock: A review of a sustainable platform for cleaner production of nature's plastics. *Journal of Cleaner Production, 270*, 122521. https://doi.org/10.1016/j.jclepro.2020.122521

Gupta, P. J., Trivedi, M. J., & Soni, H. P. (2023). Proximate analysis of vegetable and fruit waste collected from the Kadi Vegetable Market, Gujarat. *Biosciences Biotechnology Research Asia, 20*(2), 673–679. https://doi.org/10.13005/bbra/3121

Guzman-Puyol, S., Hierrezuelo, J., Benítez, J. J., Tedeschi, G., Porras-Vázquez, J. M., Heredia, A., Athanassiou, A., Romero, D., & Heredia-Guerrero, J. A. (2022). Transparent, UV-blocking, and high barrier cellulose-based bioplastics with naringin as active food packaging materials. *International Journal of Biological Macromolecules, 209*, 1985. https://doi.org/10.1016/j.ijbiomac.2022.04.177

Hakovirta, M. (2023). Sustainable business models driving carbon neutrality. In *Springer climate* (Vol. Part F1826). https://doi.org/10.1007/978-3-031-45202-4_4.

Hálfdánsdóttir, D. Á. & Ilievska, S. (2024). *Analysis of sustainable polymers for environmentally friendly packaging solutions* [Doctoral dissertation]. Retrieved from https://skemman.is/bitstream/1946/47208/1/bsverkefnilokaskjal_%20dadey_sandra.pdf

Haller, H., Paladino, G., Dupaul, G., Gamage, S., Hadzhaoglu, B., Norström, S., Eivazi, A., Holm, S., Hedenström, E., & Jonsson, A. (2023). Polluted lignocellulose-bearing sediments as a resource for marketable goods—A review of potential technologies for biochemical and thermochemical processing and remediation. *Clean Technologies and Environmental Policy, 25*(2), 409–425. https://doi.org/10.1007/s10098-021-02147-3

Haq, I. U., Saleem, A., Chaudhary, R., Alessa, A. H., Nawaz, A., & Du, C. (2024). Role of microbial laccases in valorization of lignocellulosic biomass to bioethanol. *Frontiers in Bioengineering and Biotechnology, 12*, 1441075. https://doi.org/10.3389/fbioe.2024.1441075

Harun, S. N., Hanafiah, M. M., & Noor, N. M. (2022). Rice straw utilisation for bioenergy production: A brief overview. *Energies, 15*(15), 5542. https://doi.org/10.3390/en15155542

Havstad, M. R. (2020). Biodegradable plastics. In *Plastic waste and recycling* (pp. 97–129). Academic Press. https://doi.org/10.1016/B978-0-12-817880-5.00005-0

Hayatun, A., Jannah, M., Ahmad, A., & Taba, P. (2020). Synthetic bioplastic film from rice husk cellulose. *Journal of Physics: Conference Series, 1463*(1), 012009. https://doi.org/10.1088/1742-6596/1463/1/012009

Hesan, M., Lotfi, M., Gholipour-Kanani, A., & Shafiee, M. (2023). Surface activity characterization of synthesized polyvinyl alcohol-based nanogel at air-water interface under highly dynamic conditions. *Surfaces and Interfaces, 42*, 103314. https://doi.org/10.1016/j.surfin.2023.103314

Hierro-Iglesias, C., Chimphango, A., Thornley, P., & Fernández-Castané, A. (2022). Opportunities for the development of cassava waste biorefineries for the production of polyhydroxyalkanoates in Sub-Saharan Africa. *Biomass and Bioenergy, 166*, 106600. https://doi.org/10.1016/j.biombioe.2022.106600

Huang, K., & Wang, Y. (2022). Recent applications of regenerated cellulose films and hydrogels in food packaging. *Current Opinion in Food Science, 43*, 7. https://doi.org/10.1016/j.cofs.2021.09.003

Izydorczyk, G., Skrzypczak, D., Mironiuk, M., Mikula, K., Samoraj, M., Gil, F., Taf, R., Moustakas, K., & Chojnacka, K. (2024). Lignocellulosic biomass fertilizers: Production, characterization, and agri-applications. *Science of the Total Environment, 923*, 171343. https://doi.org/10.1016/j.scitotenv.2024.171343

Jaffur, N., Jeetah, P., & Kumar, G. (2023). Biotransformation of lignocellulosic biomass hydrolysate into polyhydroxybutyrate biopolymer via *Ralstonia eutropha*. *International*

Journal of Chemical Engineering and Applications, 14(3), 28–36. https://doi.org/10.18178/ijcea.2023.14.3.802

Jain, V., & Ghosh, S. (2023). Biotransformation of lignocellulosic biomass to xylitol: An overview. *Biomass Conversion and Biorefinery, 13*(11), 9643. https://doi.org/10.1007/s13399-021-01904-0

Jain, A., Sarsaiya, S., Kumar Awasthi, M., Singh, R., Rajput, R., Mishra, U. C., Chen, J., & Shi, J. (2022). Bioenergy and bio-products from bio-waste and its associated modern circular economy: Current research trends, challenges, and future outlooks. *Fuel, 307*, 121859. https://doi.org/10.1016/j.fuel.2021.121859

Japir, A. A. W., Salih, N., & Salimon, J. (2021). Synthesis and characterization of biodegradable palm palmitic acid based bioplastic. *Turkish Journal of Chemistry, 45*(3), 585. https://doi.org/10.3906/kim-2011-31

Jayarathna, S., Andersson, M., & Andersson, R. (2022). Recent advances in starch-based blends and composites for bioplastics applications. *Polymers, 14*(21), 4557. https://doi.org/10.3390/polym14214557

Jena, S., & Singh, R. (2022). Agricultural crop waste materials—A potential reservoir of molecules. *Environmental Research, 206*, 112284. https://doi.org/10.1016/j.envres.2021.112284

Jiang, L., Sabzi, M., & Zhang, J. (2024). Biodegradable and biobased polymers. In *Applied plastics engineering handbook* (pp. 133–165). Elsevier. https://doi.org/10.1016/B978-0-323-88667-3.00009-6

Jin, J., Luo, B., Xuan, S., Shen, P., Jin, P., Wu, Z., & Zheng, Y. (2024). Degradable chitosan-based bioplastic packaging: Design, preparation and applications. *International Journal of Biological Macromolecules, 266*, 131253. https://doi.org/10.1016/j.ijbiomac.2024.131253

Jogi, K., & Bhat, R. (2020). Valorization of food processing wastes and by-products for bioplastic production. Sustainable Chemistry and Pharmacy, 18, 100326. https://doi.org/10.1016/j.scp.2020.100326

Kali, A., Amar, A., Loulidi, I., Jabri, M., Hadey, C., Lgaz, H., et al. (2024). Characterization and adsorption capacity of four low-cost adsorbents based on coconut, almond, walnut, and peanut shells for copper removal. *Biomass Conversion and Biorefinery, 14*(3), 3655–3666. https://doi.org/10.1007/s13399-022-02564-4

Kanhere, S. V., Tindall, G. W., Ogale, A. A., & Thies, M. C. (2022). Carbon fibers derived from liquefied and fractionated poplar lignins: The effect of molecular weight. *IScience, 25*(12), 105449. https://doi.org/10.1016/j.isci.2022.105449

Karaca, A. E., Özel, C., Özarslan, A. C., & Yücel, S. (2022). The simultaneous extraction of cellulose fiber and crystal biogenic silica from the same rice husk and evaluation in cellulose-based composite bioplastic films. *Polymer Composites, 43*(10), 6838. https://doi.org/10.1002/pc.26729

Kawaguchi, H., Takada, K., Elkasaby, T., Pangestu, R., Toyoshima, M., Kahar, P., et al. (2022). Recent advances in lignocellulosic biomass white biotechnology for bioplastics. *Bioresource Technology, 344*, 126165. https://doi.org/10.1016/j.biortech.2021.126165

Khan, M. F. S., Akbar, M., Xu, Z., & Wang, H. (2021). A review on the role of pretreatment technologies in the hydrolysis of lignocellulosic biomass of corn stover. *Biomass and Bioenergy, 155*, 106276. https://doi.org/10.1016/j.biombioe.2021.106276

Khan, M. A., Dharmalingam, B., Chuetor, S., Cheng, Y. S., & Sriariyanun, M. (2023). Comprehensive review on effective conversion of lignocellulosic biomass to levulinic acid. *Biomass Conversion and Biorefinery*. https://doi.org/10.1007/s13399-023-04663-2

Khandeparkar, A. S., Paul, R., Sridhar, A., Lakshmaiah, V. V., & Nagella, P. (2024). Eco-friendly innovations in food packaging: A sustainable revolution. *Sustainable Chemistry and Pharmacy, 39*, 101579. https://doi.org/10.1016/j.scp.2024.101579

Khodaei, D., Álvarez, C., & Mullen, A. M. (2021). Biodegradable packaging materials from animal processing co-products and wastes: An overview. *Polymers, 13*(15), 2561. https://doi.org/10.3390/polym13152561

Koller, M., & Mukherjee, A. (2022). Polyhydroxyalkanoates (PHAs)—Production, properties, and biodegradation. In *Biodegradable polymers in the circular plastics economy* (pp. 145–204). Wiley. https://doi.org/10.1002/9783527827589.ch6

Kordi, M., Farrokhi, N., Pech-Canul, M. I., & Ahmadikhah, A. (2024). Rice husk at a glance: From agro-industrial to modern applications. *Rice Science, 31*(1), 14–32. https://doi.org/10.1016/j.rsci.2023.08.005

Kover, A., Kraljić, D., Marinaro, R., & Rene, E. R. (2022). Processes for the valorization of food and agricultural wastes to value-added products: Recent practices and perspectives. *Systems Microbiology and Biomanufacturing, 2*(1), 50–66. https://doi.org/10.1007/s43393-021-00042-y

Kuittinen, S., Hietaharju, J., Bhattarai, I., Hassan, M. K., Kupiainen, L., Kangas, J., Tanskanen, J., & Pappinen, A. (2022). Technoeconomic analysis and environmental sustainability estimation of bioalcohol production from barley straw. *Biocatalysis and Agricultural Biotechnology, 43*, 102427. https://doi.org/10.1016/j.bcab.2022.102427

Kumar, S., Konwar, J., Purkayastha, M., Purkayastha, M. D., Kalita, S., Mukherjee, A., & Dutta, J. (2023). Current progress in valorization of food processing waste and by-products for pectin extraction. *International Journal of Biological Macromolecules, 239*, 124332. https://doi.org/10.1016/j.ijbiomac.2023.124332

Kyokusiima, S., & Nyerere, G. (2024). *Development of packaging bio plastic film from sugar cane bagasse fibre reinforced with maize straw*. Retrieved from https://ir.busitema.ac.ug/bitstream/handle/20.500.12283/4262/Nyerere_%26_Kyokusiima_Abstract.pdf?sequence=2&isAllowed=y

Lee, Y. G., Cho, E. J., Maskey, S., Nguyen, D. T., & Bae, H. J. (2023). Value-added products from coffee waste: A review. *Molecules, 28*(8), 3562. https://doi.org/10.3390/molecules28083562

Li, M., & Wilkins, M. (2020). Fed-batch cultivation and adding supplements to increase yields of polyhydroxybutyrate production by *Cupriavidus necator* from corn stover alkaline pretreatment liquor. *Bioresource Technology, 299*, 122676. https://doi.org/10.1016/j.biortech.2019.122676

Li, H., Zhou, M., Mohammed, A. E. A. Y., Chen, L., & Zhou, C. (2022). From fruit and vegetable waste to degradable bioplastic films and advanced materials: A review. *Sustainable Chemistry and Pharmacy, 30*, 100859. https://doi.org/10.1016/B978-0-323-91677-6.00023-4

Li, H., Wang, J., Liu, Y., Chen, J., Wang, C., Hu, Y., & Hu, K. (2024). Production of biodegradable potato starch films containing Lycium barbarum polysaccharide and investigation of their physicochemical properties. *Food Packaging and Shelf Life, 44*, 101320. https://doi.org/10.1016/j.fpsl.2024.101320

Liberato, V. S., Martins, F. F., Ribeiro, C. M. S., Coelho, M. A. Z., & Ferreira, T. F. (2022). Two-waste culture medium to produce 1,3-propanediol through a wild *Clostridium butyricum* strain. *Fuel, 322*, 124202. https://doi.org/10.1016/j.fuel.2022.124202

Lilavanichakul, A., & Yoksan, R. (2023). Development of bioplastics from cassava toward the sustainability of cassava value chain in Thailand. *Sustainability (Switzerland), 15*(20), 14713. https://doi.org/10.3390/su152014713

Liobikiene, G., Balezentis, T., Streimikiene, D., & Chen, X. (2019). Evaluation of bioeconomy in the context of strong sustainability. *Sustainable Development, 27*(5), 955. https://doi.org/10.1002/sd.1984

Liu, M., Tong, S., Tong, Z., Guan, Y., & Sun, Y. (2023). A strong, biodegradable and transparent cellulose-based bioplastic stemmed from waste paper. *Journal of Applied Polymer Science, 140*(13), e53671. https://doi.org/10.1002/app.53671

Lou, J., Babadi, M. R., Otadi, M., Tarahomi, M., Van Le, Q., Khonakdar, H. A., & Li, C. (2023). Agricultural waste valorization towards (nano) catalysts for the production of chemicals and materials. *Fuel, 351*, 128935. https://doi.org/10.1016/j.fuel.2023.128935

Longo, A., Fanelli, F., Villano, M., Montemurro, M., & Rizzello, C. G. (2024). Bioplastic Production from Agri-Food Waste through the Use of Haloferax mediterranei: A Comprehensive Initial Overview. Microorganisms, 12(6), 1038. https://doi.org/10.3390/microorganisms12061038

Lu, H., Yadav, V., Bilal, M., & Iqbal, H. M. N. (2022). Bioprospecting microbial hosts to valorize lignocellulose biomass—Environmental perspectives and value-added bioproducts. *Chemosphere, 288*, 132574. https://doi.org/10.1016/j.chemosphere.2021.132574

Machineni, L., & Rao Anupoju, G. (2022). Review on valorization of lignocellulosic biomass for green plastics production: Sustainable and cleaner approaches. *Sustainable Energy Technologies and Assessments, 53*, 102698. https://doi.org/10.1016/j.seta.2022.102698

Macnamara, J. F., Jr. (2024). *High oxygen and moisture barrier biodegradable material* (Doctoral dissertation, Michigan State University). Retrieved from https://www.proquest.com/openview/0ff5f1111e2a4b9e45dc7ab8077e5f7c/1?pq-origsite=gscholar&cbl=18750&diss=y

Mangal, M., Rao, C. V., & Banerjee, T. (2023). Bioplastic: An eco-friendly alternative to non-biodegradable plastic. *Polymer International, 72*(11). https://doi.org/10.1002/pi.6555

Marak, T. B., Tiwari, A., & Roy, A. (2024). Conversion of agricultural residues into high-value animal feed. In *Transforming agriculture residues for sustainable development: From waste to wealth* (pp. 205–221). Springer Nature Switzerland. Retrieved from https://link.springer.com/chapter/10.1007/978-3-031-61133-9_9

Maraveas, C. (2020). Production of sustainable and biodegradable polymers from agricultural waste. *Polymers, 12*(5), 1127. https://doi.org/10.3390/POLYM12051127

Markets and Markets. (2024). Bioplastics & biopolymers market by product type (biodegradable, non-biodegradable/bio-based), end-use industry (packaging, consumer goods, textiles, agriculture & horticulture, AnT, coatings & adhesives), & region-global forecast to 2029. In *Webpage bioplastics and biopolymers market*. Published on April, 2024. Retrieved from https://www.marketsandmarkets.com/Market-Reports/biopolymers-bioplastics-market-88795240.html

Martínková, L., Grulich, M., Pátek, M., Křístková, B., & Winkler, M. (2023). Bio-based valorization of lignin-derived phenolic compounds: A review. *Biomolecules, 13*(5), 717. https://doi.org/10.3390/biom13050717

Martins, J. R., Llanos, J. H. R., Abe, M. M., Costa, M. L., & Brienzo, M. (2024). New blend of renewable bioplastic based on starch and acetylated xylan with high resistance to oil and water vapor. *Carbohydrate Research, 537*, 109068. https://doi.org/10.1016/j.carres.2024.109068

Maryana, R., Muryanto, Triwahyuni, E., Oktaviani, O., Prasetia, H., Das, A. K., & Sudiyani, Y. (2022). Extraction of cellulose acetate from Cajuput (Melaleuca leucadendron) twigs and sugarcane (Saccharum officinarum) bagasse by environmentally friendly approach. *Waste and Biomass Valorization, 13*, 1535. https://doi.org/10.1007/s12649-021-01610-y

Marzo-Gago, C., Díaz, A. B., & Blandino, A. (2023). Sugar beet pulp as raw material for the production of bioplastics. *Fermentation, 9*(7), 655. https://doi.org/10.3390/fermentation9070655

Masurkar, S. (2024). Production and biodegradability studies of edible bioplastics from potato peel waste. *Bulletin of Pure and Applied Sciences-Zoology, 43B*(1s), 110–121. https://www.bpasjournals.com/zoology/index.php/journal/article/view/58/11

Matheus, J. R. V., de Farias, P. M., Satoriva, J. M., de Andrade, C. J., & Fai, A. E. C. (2023). Cassava starch films for food packaging: Trends over the last decade and future research. *International Journal of Biological Macromolecules, 225*, 658. https://doi.org/10.1016/j.ijbiomac.2022.11.129

Meereboer, K. W., Misra, M., & Mohanty, A. K. (2020). Review of recent advances in the biodegradability of polyhydroxyalkanoate (PHA) bioplastics and their composites. *Green Chemistry, 22*(17), 5519–5558. https://doi.org/10.1039/d0gc01647k

Miao, S., Wei, Y., Chen, J., & Wei, X. (2023). Extraction methods, physiological activities and high value applications of tea residue and its active components: A review. *Critical Reviews in Food Science and Nutrition, 63*(33), 12150–12168. https://doi.org/10.1080/10408398.2022.2099343

Moghaddam, M. K., Gheshlagh, F. G., & Moezzi, M. (2024). Extraction and characterization of cellulose microfibers from cornhusk for application as reinforcing agent in biocomposite. *International Journal of Biological Macromolecules, 264*, 130669. https://doi.org/10.1016/j.ijbiomac.2024.130669

Mohammad, N. A. B. (2023). Concern on wood waste utilization: Environment and economic evaluation. In S. N. Sarmin, M. Jawaid, & R. Elias (Eds.), *Wood waste management and products. Sustainable materials and technology*. Springer. https://doi.org/10.1007/978-981-99-1905-5_3

Mohapatra, M., & Singh, S. (2024). Bioplastic: Unravelling the sustainable approach for petroleum plastic. In K. Bala, T. Ghosh, V. Kumar, & P. Sangwan (Eds.), *Harnessing microbial potential for multifarious applications. Energy, environment, and sustainability* (pp. 205–233). Springer. https://doi.org/10.1007/978-981-97-1152-9_8

Moungprayoon, A., Lunprom, S., Reungsang, A., & Salakkam, A. (2022). High cell density cultivation of Paracoccus sp. on sugarcane juice for poly(3-hydroxybutyrate) production. *Frontiers in Bioengineering and Biotechnology, 10*, 878688. https://doi.org/10.3389/fbioe.2022.878688

Mujtaba, M., Fernandes Fraceto, L., Fazeli, M., Mukherjee, S., Savassa, S. M., Araujo de Medeiros, G., do Espírito Santo Pereira, A., Mancini, S. D., Lipponen, J., & Vilaplana, F. (2023). Lignocellulosic biomass from agricultural waste to the circular economy: A review with focus on biofuels, biocomposites and bioplastics. *Journal of Cleaner Production, 402*, 136815. https://doi.org/10.1016/j.jclepro.2023.136815

Muñoz-Gimena, P. F., Oliver-Cuenca, V., Peponi, L., & López, D. (2023). A review on reinforcements and additives in starch-based composites for food packaging. *Polymers, 15*(13), 2972. https://doi.org/10.3390/polym15132972

Naitam, M. G., Tomar, G. S., Kaushik, R., Singh, S., & Nain, L. (2022). Agro-industrial waste as potential renewable feedstock for biopolymer poly-hydroxyalkanoates (PHA) production. *Enzyme Engineering, 11*(4). https://doi.org/10.35841/2329-6674.22.11.190

Nalini, S., & Sathiyamurthi, S. (2023). Bioplastic from microbial and agricultural biomass. In *Green sustainable process for chemical and environmental engineering and science: Natural materials-based green composites 2: Biomass*. Elsevier. https://doi.org/10.1016/B978-0-323-95183-8.00010-X

Nanda, S., Patra, B. R., Patel, R., Bakos, J., & Dalai, A. K. (2022). Innovations in applications and prospects of bioplastics and biopolymers: A review. *Environmental Chemistry Letters, 20*(1), 379. https://doi.org/10.1007/s10311-021-01334-4

Nandiyanto, A. B. D., Fiandini, M., Ragadhita, R., Sukmafitri, A., Salam, H., & Triawan, F. (2020). Mechanical and biodegradation properties of cornstarch-based bioplastic material. *Materials Physics and Mechanics, 44*(3), 380–391. https://doi.org/10.18720/MPM.4432020_9

Narayanasamy, A., Patel, S. K., Singh, N., Rohit, M. V., & Lee, J. K. (2024). Valorization of algal biomass to produce microbial polyhydroxyalkanoates: Recent updates, challenges, and perspectives. *Polymers, 16*(15), 2227. https://doi.org/10.3390/polym16152227

Nigam, S., Das, A. K., Matkawala, F., & Patidar, M. K. (2022). An insight overview of bioplastics produced from cellulose extracted from plant material, its applications and degradation. *Environmental Sustainability, 5*(4), 423. https://doi.org/10.1007/s42398-022-00248-3

Nikhilesh, A., & Jethva, K. R. (2024). Optimization of drying temperature for enhanced physical and mechanical properties of potato and rice starch-based bioplastic. *International Journal of Advanced Biochemistry Research, 8*(6), 440–446. Retrieved from https://www.biochemjournal.com/articles/1346/8-6-20-765.pdf

Nisar, B., Pahalvi, H. N., Gulzar, A., Rashid, S., Majeed, L. R., & Kamili, A. N. (2023). Bioplastics: Solution to a green environment and sustainability. In *Role of green chemistry in ecosystem restoration to achieve environmental sustainability*. Elsevier. https://doi.org/10.1016/B978-0-443-15291-7.00021-3

Oghazi, P., Mostaghel, R., & Hultman, M. (2024). International industrial manufacturers: Mastering the era of digital innovation and circular economy. *Technological Forecasting and Social Change, 201*, 123160. https://doi.org/10.1016/j.techfore.2023.123160

Olagundoye, A. A., & Morayo, A. O. (2022). Characterization of potato peel starch-based bioplastic reinforced with banana Pseudostem cellulose for packaging applications. *International Journal of Innovative Science and Research Technology, 7*(4).

Olugbemide, A. D., Ifijen, I. H., Mahata, C., Vicente, F. A., & Likozar, B. (2024). Valorization of deep eutectic solvent pretreated lignocellulosic biomass for improved biogas production. In *Solid waste management in delta region for SDGs fulfillment: Delta sustainability by waste management* (pp. 127–153). Springer. Retrieved from https://link.springer.com/chapter/10.1007/978-3-031-58253-0_6

Pandey, S., & Kumari, N. (2022). Sustainable utilization and management of agricultural and kitchen waste. In *Agricultural and kitchen waste*. Routledge. https://doi.org/10.1201/9781003245773-7

Pandirwar, A. P., Khadatkar, A., Mehta, C. R., Majumdar, G., Idapuganti, R., Mageshwaran, V., & Shirale, A. O. (2023). Technological advancement in harvesting of cotton stalks to establish sustainable raw material supply chain for industrial applications: A review. *Bioenergy Research, 16*(2), 741. https://doi.org/10.1007/s12155-022-10520-3

Pásztory, Z., Mohácsiné, I. R., Gorbacheva, G., & Börcsök, Z. (2016). The utilization of tree bark. *BioResources, 11*(3), 7859–7888. https://doi.org/10.15376/biores.11.3.Pasztory

Patel, M., Islam, S., Kallem, P., Patel, R., Banat, F., & Patel, A. (2023). Potato starch-based bioplastics synthesized using glycerol–sorbitol blend as a plasticizer: Characterization and performance analysis. *International journal of Environmental Science and Technology, 20*(7), 7843. https://doi.org/10.1007/s13762-022-04492-2

Pawar, R., Pathan, A., Nagaraj, S., Kapare, H., Giram, P., & Wavhale, R. (2023). Polycaprolactone and its derivatives for drug delivery. *Polymers for Advanced Technologies, 34*(10), 3296–3316. https://doi.org/10.1002/pat.6140

Pei, P., Zou, R., Wang, X., Liu, J., Liu, L., Deng, X., Li, X., Yu, M., Tan, J., & Li, S. (2023). Biocomposite optimization with NaOH-modified bagasse fiber, polybutylene succinate, and poly(lactic acid) using RSM approach. *BioResources, 18*(3), 5683–5702. https://doi.org/10.15376/biores.18.3.5683-5702

Peidayesh, H., Ondriš, L., Saparová, S., Kovaľaková, M., Fričová, O., & Chodák, I. (2024). Biodegradable nanocomposites based on blends of poly(butylene adipate-co-terephthalate) (PBAT) and thermoplastic starch filled with montmorillonite (MMT): Physico-mechanical properties. *Materials, 17*(3), 540. https://doi.org/10.3390/ma17030540

Phosri, S., Kunjiek, T., Mukkhakang, C., Suebthep, S., Sinsup, W., Phornsirigarn, S., & Charoeythornkhajhornchai, P. (2022). Biodegradability of bioplastic blown film in a marine environment. *Frontiers in Marine Science, 9*, 917397. https://doi.org/10.3389/fmars.2022.917397

Pirzadah, T. B., Malik, B., Kumar, M., & Rehman, R. U. (2015). Lignocellulosic biomass: As future alternative for bioethanol production. In *Biomass and bioenergy: Applications*. Springer. https://doi.org/10.1007/978-3-319-07578-5_8

Porichha, G. K., Hu, Y., Rao, K. T. V., & Xu, C. C. (2021). Crop residue management in India: Stubble burning vs. other utilizations including bioenergy. *Energies, 14*(14), 4281. https://doi.org/10.3390/en14144281

Poskart, A., Skrzyniarz, M., Sajdak, M., Zajemska, M., & Skibiński, A. (2021). Management of lignocellulosic waste towards energy recovery by pyrolysis in the framework of circular economy strategy. *Energies, 14*(18), 5864. https://doi.org/10.3390/en14185864

Purkait, M. K., Haldar, D., & Debnath, B. (2023). *Technological advancements in product valorization of tea waste*. Elsevier. https://doi.org/10.1016/C2021-0-03501-X

Rahman, M. H., & Bhoi, P. R. (2021). An overview of non-biodegradable bioplastics. *Journal of Cleaner Production, 294*, 126218. https://doi.org/10.1016/j.jclepro.2021.126218

Rahman, A., Ulven, C. A., Johnson, M. A., Durant, C., & Hossain, K. G. (2017). Pretreatment of wheat bran for suitable reinforcement in biocomposites. *Journal of Renewable Materials, 5*, 62. https://doi.org/10.7569/JRM.2017.634133

Raj, T., Chandrasekhar, K., Naresh Kumar, A., & Kim, S. H. (2022). Lignocellulosic biomass as renewable feedstock for biodegradable and recyclable plastics production: A sustainable approach. *Renewable and Sustainable Energy Reviews, 158*, 112130. https://doi.org/10.1016/j.rser.2022.112130

Rajendran, D. S., Venkatraman, S., Rahul, R., Afrrin, M., Karthik, P., & Vaidyanathan, V. K. (2024). Manifesting sustainability toward food waste into bioenergy: Biorefinery in a circular economic approach. In *Circular bioeconomy perspectives in sustainable bioenergy production* (pp. 431–460). Springer Nature Singapore. Retrieved from https://link.springer.com/chapter/10.1007/978-981-97-2523-6_18

Rao, H. J., Singh, S., Ramulu, P. J., Singh, N., Santos, T. F., Santos, C. M., et al. (2024). Nature-inspired nano cellulose materials, advancements in nano cellulose preparation and versatile applications. In *Nanocellulose-sources, preparations, and applications*. IntechOpen. Retrieved from https://www.intechopen.com/chapters/89068

Ratshoshi, B. K., Farzad, S., & Görgens, J. F. (2021). Techno-economic assessment of polylactic acid and polybutylene succinate production in an integrated sugarcane biorefinery. *Biofuels, Bioproducts and Biorefining, 15*(6), 1871–1887. https://doi.org/10.1002/bbb.2287

Renewable Carbon News. (2024). Current market study forecasts annual growth of 17% for bio-based polymers between 2023 and 2028. Demand from Asia and the USA in particular is driving growth, Europe is lagging behind. *New report released on the global bio-based polymer market 2023—A deep and comprehensive insight into a dynamically growing market.* Published on 11 March, 2024. Retrieved from https://renewable-carbon.eu/news/current-market-study-forecasts-annual-growth-of-17-for-bio-based-polymers-between-2023-and-2028-demand-from-asia-and-the-usa-in-particular-is-driving-growth-europe-is-laggi ng-behind/

Ribba, L., Lorenzo, M. C., Tupa, M., Melaj, M., Eisenberg, P., & Goyanes, S. (2021). Processing and properties of starch-based thermoplastic matrix for green composites. In *Materials horizons: From nature to nanomaterials*. Springer. https://doi.org/10.1007/978-981-15-9643-8_4

Rosenboom, J. G., Langer, R., & Traverso, G. (2022). Bioplastics for a circular economy. *Nature Reviews Materials, 7*(2), 117. https://doi.org/10.1038/s41578-021-00407-8

Rossi, D., Rossi, S., Cinelli, P., & Seggiani, M. (2024). Emerging opportunities in the valorisation of wheat bran byproduct as additive in polymer composite materials. *Materials Today Sustainability, 27*, 100832. https://doi.org/10.1016/j.mtsust.2024.100832

Rovera, C., Carullo, D., Bellesia, T., Büyüktaş, D., Ghaani, M., Caneva, E., & Farris, S. (2023). Extraction of high-quality grade cellulose and cellulose nanocrystals from different lignocellulosic agri-food wastes. *Frontiers in Sustainable Food Systems, 6*, 1087867. https://doi.org/10.3389/fsufs.2022.1087867

Ruggero, F., Roosa, S., Onderwater, R., Delacuvellerie, A., Lotti, T., Gori, R., Lubello, C., & Wattiez, R. (2023). Characterization of bacterial communities responsible for bioplastics degradation during the thermophilic and the maturation phases of composting. *Journal of Material Cycles and Waste Management, 25*(6), 3270. https://doi.org/10.1007/s10163-023-01751-3

Sabry, F. (2022). Bioplastic: Life in bioplastic is more fantastic. Is it biobased or biodegradable plastics? Is it victory or pure fiction? (Vol. 3). *One billion knowledgeable*. Retrieved from https://www.barnesandnoble.com/w/bioplastic-fouad-sabry/1140927917

Sai, D. S., Singh, S. K., Chattaraj, A., Chandhu, K. V., Durga, D. S. L., Samudram, L. P., Kumar, D. H., & Singh, J. I. P. (2024). Development of bagasse/PLA composites, it's mechanical & chemical properties: A review. *AIP Conference Proceedings, 2962*(1), 020018. https://doi.org/10.1063/5.0192258

Sarafidou, M., Vittou, O., Psaki, O., Filippi, K., Tsouko, E., Vardaxi, A., et al. (2024). Evaluation of alternative sugar beet pulp refining strategies for efficient pectin extraction and poly (3-hydroxybutyrate) production. *Biochemical Engineering Journal, 208*, 109368. https://doi.org/10.1016/j.bej.2024.109368

Sarker, A., Ahmmed, R., Ahsan, S. M., Rana, J., Ghosh, M. K., & Nandi, R. (2024). A comprehensive review of food waste valorization for the sustainable management of global food waste. Sustainable Food Technology. https://doi.org/10.1039/D3FB00156C

Schirp, A., Deetz, R., Kopitzky, R., & Schirp, C. (2024). Composite sheets based on polylactic acid and sugar beet pulp: A solution to accelerate biological degradation of PLA on soil under outdoor exposure. *Polymer Composites, 45*, 15318. https://doi.org/10.1002/pc.28840

Shabbirahmed, A. M., Haldar, D., Dey, P., Patel, A. K., Singhania, R. R., Dong, C. D., & Purkait, M. K. (2022). Sugarcane bagasse into value-added products: A review. *Environmental Science and Pollution Research, 29*(42), 62785–62806. https://doi.org/10.1007/s11356-022-21889-1

Sharma, R., & Malaviya, P. (2023). Ecosystem services and climate action from a circular bioeconomy perspective. *Renewable and Sustainable Energy Reviews, 175*, 113164. https://doi.org/10.1016/j.rser.2023.113164

Sharma, S., Tsai, M. L., Sharma, V., Sun, P. P., Nargotra, P., Bajaj, B. K., Chen, C. W., & di Dong, C. (2023). Environment friendly pretreatment approaches for the bioconversion of lignocellulosic biomass into biofuels and value-added products. *Environments, 10*(1), 6. https://doi.org/10.3390/environments10010006

Sharma, P., Bano, A., Singh, S. P., Varjani, S., & Tong, Y. W. (2024). Sustainable organic waste management and future directions for environmental protection and techno-economic perspectives. *Current Pollution Reports, 10*, 459–477. Retrieved from https://link.springer.com/article/10.1007/s40726-024-00317-7

Siddiqui, S. A., Yang, X., Deshmukh, R. K., Gaikwad, K. K., Bahmid, N. A., & Castro-Muñoz, R. (2024). Recent advances in reinforced bioplastics for food packaging—A critical review. *International Journal of Biological Macromolecules, 263*, 130399. https://doi.org/10.1016/j.ijbiomac.2024.130399

Silverio, M. S., Piccoli, R. A. M., dos Reis, J. L. M. S., Gomez, J. G. C., & Baptista, A. S. (2024). Techno-economic feasibility of P(3-hydroxybutyrate) bioprocess with concentrated sugarcane vinasse as carbon and minerals source: An experimental and in silico approach. *Biomass Conversion and Biorefinery, 14*(2), 2071. https://doi.org/10.1007/s13399-022-02710-y

Singh, P., & Verma, R. (2020). Bioplastics: A green approach toward sustainable environment. In *Environmental microbiology and biotechnology: Volume 1, biovalorization of solid wastes and wastewater treatment*. Springer. https://doi.org/10.1007/978-981-15-6021-7_3

Singh, N., Singhania, R. R., Nigam, P. S., di Dong, C., Patel, A. K., & Puri, M. (2022). Global status of lignocellulosic biorefinery: Challenges and perspectives. *Bioresource Technology, 344*, 126415. https://doi.org/10.1016/j.biortech.2021.126415

Siracusa, V., & Blanco, I. (2020). Bio-polyethylene (bio-PE), bio-polypropylene (bio-PP) and bio-poly(ethylene terephthalate) (bio-PET): Recent developments in bio-based polymers analogous to petroleum-derived ones for packaging and engineering applications. *Polymers, 12*(8), 1641. https://doi.org/10.3390/polym12081641

Song, J., Lin, X., Wu, H., Huang, Z., Gan, T., Hu, H., Qin, Y., & Zhang, Y. (2024). Fabrication of biodegradable and cold-water-soluble starch/polyvinyl alcohol films as inner packaging materials of pesticides: Enhanced emulsification, dispersibility, and efficacy. *Carbohydrate Polymers, 328*, 121713. https://doi.org/10.1016/j.carbpol.2023.121713

Soo, X. Y. D., Muiruri, J. K., Wu, W. Y., Yeo, J. C. C., Wang, S., Tomczak, N., et al. (2024). Bio-polyethylene and polyethylene biocomposites: An alternative toward a sustainable future. *Macromolecular Rapid Communications, 45*, 2400064. https://doi.org/10.1002/marc.202400064

Sridevi, V., Surya, D. V., Reddy, B. R., Shah, M., Gautam, R., Kumar, T. H., et al. (2024). Challenges and opportunities in the production of sustainable hydrogen from lignocellulosic biomass using microwave-assisted pyrolysis: A review. *International Journal of Hydrogen Energy, 52*, 507–531. https://doi.org/10.1016/j.ijhydene.2023.06.186

Swetha, T. A., Ananthi, V., Bora, A., Sengottuvelan, N., Ponnuchamy, K., Muthusamy, G., & Arun, A. (2023). A review on biodegradable polylactic acid (PLA) production from fermentative food waste—Its applications and degradation. *International Journal of Biological Macromolecules, 234*, 123703. https://doi.org/10.1016/j.ijbiomac.2023.123703

Swetha, T. A., Bora, A., Ananthy, V., Ponnuchamy, K., Muthusamy, G., & Arun, A. (2024). A review of bioplastics as an alternative to petrochemical plastics: Its types, structure, characteristics, degradation, standards, and feedstocks. *Polymers for Advanced Technologies, 35*(6), e6482. https://doi.org/10.1002/pat.6482

Tan, E. C. D., & Lamers, P. (2021). Circular bioeconomy concepts—A perspective. *Frontiers in Sustainability, 2*, 701509. https://doi.org/10.3389/frsus.2021.701509

Taneepanichskul, N., Purkiss, D., & Miodownik, M. (2022). A review of sorting and separating technologies suitable for compostable and biodegradable plastic packaging. *Frontiers in Sustainability, 3*, 901885. https://doi.org/10.3389/frsus.2022.901885

Thapliyal, P. C. (2022). Utilization of chemical additives to enhance biodegradability of plastics. In *Biodegradability of conventional plastics: Opportunities, challenges, and misconceptions*. Elsevier. https://doi.org/10.1016/B978-0-323-89858-4.00006-3

Thiele, I., Santolin, L., Detels, S., Osele, R., Neubauer, P., & Riedel, S. L. (2024). High-cell-density fed-batch strategy to manufacture tailor-made P (HB-co-HHx) by engineered Ralstonia eutropha at laboratory scale and pilot scale. *Microbial Biotechnology, 17*(6), e14488. https://doi.org/10.1111/1751-7915.14488

Tsakalova, M., Onorato, C., Karathanasis, A. Z., & Deligkiozi, I. (2024). A circular economy framework for the assessment of bio-based value chains. *Sustainable Chemistry for the Environment, 6*, 100099. https://doi.org/10.1016/j.scenv.2024.100099

Udali, A., Chung, W., Talbot, B., & Grigolato, S. (2024). Managing harvesting residues: A systematic review of management treatments around the world. *Forestry: An International Journal of Forest Research, cpae041*. https://doi.org/10.1093/forestry/cpae041

Ungureanu, N., Vlăduț, V., & Biriș, S. S. (2022). Sustainable valorization of waste and by-products from sugarcane processing. *Sustainability (Switzerland), 14*(17), 11089. https://doi.org/10.3390/su141711089

Uthandi, S., Kaliyaperumal, A., Srinivasan, N., Thangavelu, K., Muniraj, I. K., Zhan, X., Gathergood, N., & Gupta, V. K. (2022). Microbial biodiesel production from lignocellulosic biomass: New insights and future challenges. *Critical Reviews in Environmental Science and Technology, 52*(12), 2197–2225. https://doi.org/10.1080/10643389.2021.1877045

Valle, C., Voss, M., Calcio Gaudino, E., Forte, C., Cravotto, G., & Tabasso, S. (2024). Harnessing agri-food waste as a source of biopolymers for agriculture. *Applied Sciences, 14*(10), 4089. https://doi.org/10.3390/app14104089

Vamsi, V. (2020). The valuable natural resources of the world, economic growth & the global environment. *International Journal of Engineering Development and Research, 8*(3), 389–421. Retrieved from https://rjwave.org/IJEDR/papers/IJEDR2003060.pdf

Van-Thuoc, D., Quillaguamán, J., Mamo, G., & Mattiasson, B. (2008). Utilization of agricultural residues for poly(3-hydroxybutyrate) production by Halomonas boliviensis LC1. *Journal of Applied Microbiology, 104*(2), 071003000434003. https://doi.org/10.1111/j.1365-2672.2007.03553.x

Venkatesh, G. (2022). Circular bio-economy—Paradigm for the future: Systematic review of scientific journal publications from 2015 to 2021. *Circular Economy and Sustainability, 2*(1), 231–279. https://doi.org/10.1007/s43615-021-00084-3

Vigneswari, S., Kee, S. H., Hazwan, M. H., Ganeson, K., Tamilselvan, K., Bhubalan, K., Amirul, A. A., & Ramakrishna, S. (2024). Turning agricultural waste streams into biodegradable plastic: A step forward into adopting sustainable carbon neutrality. *Journal of Environmental Chemical Engineering, 12*(2), 112135. https://doi.org/10.1016/j.jece.2024.112135

Virgin, I., Diaz-Chavez, R., Morris, E. J., Haileselassie, T., Tesfaye, K., De Cliff, S., Njau, K., Munganyinka, E., Muyambi, F., & Otim, M. O. (2022). *The state of the bioeconomy in eastern Africa: 2022*. Stockholm Environment Institute, East African Science and Technology Commission, BioInnovate Africa. Retrieved from https://www.researchgate.net/profile/Muyambi-Fortunate/publication/361525962_Status_of_the_East_African_Bioeconomy_report_2022/links/62b6b3516ec05339cca131d0/Status-of-the-East-African-Bioeconomy-report-2022.pdf

Visco, A., Scolaro, C., Facchin, M., Brahimi, S., Belhamdi, H., Gatto, V., & Beghetto, V. (2022). Agri-food wastes for bioplastics: European prospective on possible applications in their second life for a circular economy. *Polymers, 14*(13), 2752. https://doi.org/10.3390/polym14132752

Wang, K., Remón, J., Jiang, Z., & Ding, W. (2024). Recent advances in the preparation and application of biochar derived from lignocellulosic biomass: A mini review. *Polymers, 16*(6), 851. https://doi.org/10.3390/polym16060851

Widjaja, T., Nurkhamidah, S., Hendrianie, N., Rohmah, A. A. Z., & Setiawan, K. P. (2022). Optimization of acid-catalyst hydrolysis process in lactic acid production from rice husk by using Lactobacillus bacteria. *Journal of Hunan University Natural Sciences, 49*(3), 141–150. https://doi.org/10.55463/issn.1674-2974.49.3.15

Wu, J., Wei, X., Guo, P., He, A., Xu, J., Jin, M., Zhang, Y., & Wu, H. (2021). Efficient poly(3-hydroxybutyrate-co-lactate) production from corn stover hydrolysate by metabolically engineered *Escherichia coli. Bioresource Technology, 341*, 125873. https://doi.org/10.1016/j.biortech.2021.125873

Xu, Q., Lin, J., & Jiang, G. (2022). Synthesis, characterization and properties of soybean oil-based polyurethane. *Polymers, 14*(11), 2201. https://doi.org/10.3390/polym14112201

Yaashikaa, P. R., Kumar, P. S., Saravanan, A., Varjani, S., & Ramamurthy, R. (2020). Bioconversion of municipal solid waste into bio-based products: A review on valorisation and sustainable approach for circular bioeconomy. *Science of the Total Environment, 748*, 141312. https://doi.org/10.1016/j.scitotenv.2020.141312

Yan, J., Lv, Y., & Ma, S. (2022). Wheat bran enrichment for flour products: Challenges and solutions. *Journal of Food Processing and Preservation, 46*(11), e16977. https://doi.org/10.1111/jfpp.16977

Yong, T. K., Musa, M. N. S., Abdulla, R., Derman, E., Gansau, J. A., & Rajin, M. (2024). Synthesis and characterization of bioplastics, polyhydroxyalkanoates produced from sugarcane bagasse by using Bacillus cereus. *Biocatalysis and Agricultural Biotechnology, 58*, 103181. https://doi.org/10.1016/j.bcab.2024.103181

Yousuf, A., Pirozzi, D., & Sannino, F. (2019). Fundamentals of lignocellulosic biomass. In *Lignocellulosic biomass to liquid biofuels*. Elsevier. https://doi.org/10.1016/B978-0-12-815936-1.00001-0

Zhao, K., Qiao, Q., Chu, D., Gu, H., Dao, T. H., Zhang, J., & Bao, J. (2013). Simultaneous saccharification and high titer lactic acid fermentation of corn stover using a newly isolated lactic acid bacterium *Pediococcus acidilactici* DQ2. *Bioresource Technology, 135*, 481. https://doi.org/10.1016/j.biortech.2012

Zhao, Y., Lu, K., Xu, H., Zhu, L., & Wang, S. (2021). A critical review of recent advances in the production of furfural and 5-hydroxymethylfurfural from lignocellulosic biomass through homogeneous catalytic hydrothermal conversion. *Renewable and Sustainable Energy Reviews, 139*, 110706. https://doi.org/10.1016/j.rser.2021.110706

Zhou, G., Li, S., Meng, Q., Niu, C., Zhang, X., & Wang, Q. (2023). A new type of highly efficient fir sawdust-based super adsorbent: Remove cationic dyes from wastewater. *Surfaces and Interfaces, 36*, 102637. https://doi.org/10.1016/j.surfin.2023.102637

Bioethanol for "Bharat" (India): A Focused Journey Toward Sustainability

Sibabrata Mukherjee

Abstract

The Indian government has been strengthening its Ethanol-Blended Petrol (EBP) program in an effort to become more self-sufficient in the energy sector in a sustainable manner. Presently, Indian ethanol producers and oil marketing companies (OMCs) are striving to achieve a 20% ethanol blending target nationwide by 2025–2026, which calls for more than 1000 crore liters of ethanol. Apart from being renewable, ethanol when blended with petrol is also a better fuel for the environment than regular fossil-based petrol. Over the past few years, the country has witnessed a notable expansion in the ethanol sector, with over 100 new distilleries, and marked the capacity of 1623 crore liters by 2023–2024. The Indian government's diversification of feedstocks allows producers to ferment ethanol from a variety of sources, including sugarcane juice, syrup, starchy grains like surplus rice, maize, and damaged food grains, in addition to molasses, which was previously the primary raw material for bioethanol production. This allows flexibility for the producers to produce 1G ethanol. The work has also been done and is still being done in the area of 2G ethanol, which uses lignocellulosic feedstocks such as agro-residues to ferment fuel-grade ethanol.

Keywords

Bioethanol · Biofuels · Sustainable approach · Fermentation · Feedstock · Saccharification

S. Mukherjee (✉)
F2I Nutrition & Health Private Limited, a subsidiary of Leaf by Lesaffre, Mumbai, Maharashtra, India
e-mail: s.mukherjee@leaf.lesaffre.com

G. Mukherjee, S. Dhiman (eds.), *Value Addition and Utilization of Lignocellulosic Biomass*, https://doi.org/10.1007/978-981-96-2786-8_13

India with annual consumption of 249 million MT of fossil-based oil in 2023 was the fourth largest after the USA, China, and Europe (Powell et al., 2024; Davenport & Wayth, 2024). When only 12–13% of it is met by domestic sources, India must import oil from other nations to cover the remaining 87–88%, or the majority, of its demands. As per Indian Petroleum & Natural Gas Statistics 2022–2023, the import of crude oil during 2022–2023 was 232.70 million MT valued at ₹ 1,260,372 crore which is higher by 9.6% in quantity terms and 39.8% in value terms vis-a-vis 2021–2022. Apart from the significant erosion of foreign exchange linked to imports, burning such enormous quantities of fossil fuels pollutes the environment and raises worries about human health and carbon emissions. In this context, the Indian Govt. has prioritized renewable energy to become more self-sufficient in the energy field, with a lower carbon footprint and the expectation of a cleaner and greener environment that will enhance farmers' income. Since microbial fermentation is efficiently used to manufacture ethanol from plant-originated feedstocks, mainly sugarcane molasses and starchy grains, ethanol is renewable. The government's Ethanol-Blended Petrol (EBP) program is one such effort to address these difficult problems by taking an environmentally friendly strategy. Because ethanol burns entirely and emits no carbon dioxide when it burns, ethanol-blended petrol is a superior fuel to pure fossil fuels such as conventional petrol or gasoline or motor spirit. India wants to use ethanol produced domestically to achieve a 20% ethanol blend with petrol by the ethanol supply year (ESY) 2025–2026. Thus, 20% less carbon emissions will result from using E20, or 20% more ethanol in place of petrol. A 113-octane fuel ethanol is gaining popularity as an oxygenate additive to regular petrol and may eventually replace compounds that pollute the environment, including methyl *t*-butyl ether (MTBE) (Solution Buggy, 2023).

The chemical formula for ethanol, commonly referred to as ethyl alcohol (CH_3CH_2OH, C_2H_5OH, or C_2H_6O), can be written as well as EtOH, where Et stands for ethyl. It is an organic compound that comprises one hydroxyl group (OH), two carbon atoms, and five hydrogen atoms. It was defined by Nicolas-Theodore de Saussure in 1808, and Antoine Lavoisier provided a description of it. The second most fundamental alcohol is ethanol, which has carbon atoms that have undergone sp^3 hybridization. The ethanol structural formula was published by Archibald Scott Couper in 1858. Apart from being mixed with petrol and diesel, ethanol finds application in the production of alcoholic beverages and downstream chemicals. Extra-neutral alcohol (ENA) is the term for the 96% strength of ethanol complying with other standard limits used in alcoholic beverages. Although ethanol is widely utilized in personal care products like sanitizer, it additionally serves as an organic solvent in the chemical industry. Ethanol has bactericidal properties that make it advantageous in the pharmaceutical industry as an antiseptic and disinfectant. In 2003, the Indian EBP effort was introduced, with the aim of achieving a 5% blend with petrol, or E5, through research programs that proved the concept. The National Policy on Biofuels 2018 included an indicated aim of achieving 20% (E20) ethanol blending in petrol by the ethanol supply year (ESY) 2030 under the EBP program; therefore, this proportion is expected to rise. The National Policy on Biofuels was then modified by the Union Cabinet in 2022, pushing forward the fuel firms' deadline to raise the ethanol percentage in petrol to 20% from ESY 2030 to ESY 2025. About 1016 crore liters of ethanol is needed to accomplish the 20% blending target by 2025; the overall amount of ethanol needed, including for other purposes, is 1350 crore liters. If the plant runs at 80% efficiency, this means that approximately 1700 crore liters of ethanol-producing capacity must be in place by 2025 (PIB Delhi, 2023).

The industry has grown significantly in the last few years in the Indian ethanol market due to driving forces like increased blending mandates, bolstering the country's energy security, financial support for ethanol producers, feedstock diversification, development of the market for quick adaptation of EBP, and commitment to reducing carbon footprints through cleaner and greener energy. It results in a notable growth in the number of ethanol distilleries: in the last 9–10 years, the number of molasses-based distilleries has increased by more than twice as much as their capacity. After 2018, the nation saw the commissioning of about 150 new grain distilleries, a few of which are still in various phases of installation. The Department of Food & Public Distribution (DFPD) reports that as of August 31, 2024, India's capacity to produce ethanol increased from 518 crore liters in 2017–2018 to an astounding 1623 crore liters in 2023–2024 (ChiniMandi, 2024).

The most effective and sustainable approach to making ethanol is microbial fermentation. When oxygen is scarce, yeast, a unicellular fungi, known for ethanol fermentation, would ferment sugars to produce ethanol, carbon dioxide, and energy. India, an agriculturally rich nation, produces a substantial quantity of sugarcane and various grains, including rice, maize, and others. Producing ethanol from excess and/or damaged grains or sugarcane-based feedstocks, like molasses, juice, or syrup, may be a sustainable means of improving the financial situation of the farmers. Following consideration of the agricultural data, the Indian Govt. approved a range of feedstocks for the manufacturing of ethanol. For the 2024–2025 ethanol supply year, the central Govt. of India has given sugar mills and distilleries permission to make ethanol from sugarcane juice, syrup, and B-heavy molasses. Furthermore, according to the most recent circular in August 2024, the government authorized the disbursement of 23 lakh tons of surplus rice from the FCI stock for the manufacturing of fuel-grade ethanol. Additionally, maize and damaged food grains (DFG) can be used by ethanol producers to produce ethanol for the OMCs.

Distilleries that employ sugarcane as their primary source of sugar use juice, syrup, and various grades of molasses, such as B and/or C, which contain free fermentable sugars. When yeast is inoculated with feedstocks that have optimal sugar levels, the sugars are converted by the yeast into energy, carbon dioxide, and ethanol in the presence of favorable circumstances. Following fermentation, the fermented wash undergoes distillation, allowing the ethanol to be recovered at an appropriate concentration and purity. On the other side, enzymatic hydrolysis is required to transform starches—which are bigger, nonfermentable sugar polymers—into fermentable sugars when starchy grains like rice or maize are utilized as feedstocks in grain-based ethanol plants. This is typically a two-step process called liquefaction assisted by alpha-amylase and saccharification assisted by glucoamylase enzyme. Simultaneous saccharification and fermentation (SSF) is the term used to describe the simultaneous occurrence of saccharification and fermentation. The yeast uses fermentable sugars produced by enzymatic saccharification while SSF is occurring. In order to extract the ethanol from the fermented wash, which is a mixture of water and other fragments, etc., it is then distilled in a manner similar to that of the molasses fermented wash. Grain distilleries create a solid stream at the end of the process that is typically rich in protein and is absent from molasses distilleries. This stream is composed of unfermented parts of the grains, yeast biomass, etc. called distillers dried grains and solubles (DDGS). DDGS is a byproduct of grain distilleries, is made by drying, and is mostly supplied to producers of animal feed.

1G ethanol is bioethanol made from starchy substrates like rice, maize, and other starchy grains, or from sugarcane molasses, juice, or syrup. Nevertheless, cellulose and hemicelluloses from agricultural and agro-industrial wastes such as rice and wheat straw, sugarcane bagasse, maize cobs and stover, cotton stalk, empty fruit bunches (EFB) from oil palm are used as a source of sugar in 2G bioethanol technology.

The upstream and midstream processing stages of 2G ethanol production are essentially different from those of 1G ethanol production. To ensure that the substrate is adequately exposed to the enzymes, a preliminary treatment step is necessary for the biomass of interest, which is rich in lignocellulose. Different technology suppliers have experimented with a variety of pretreatment techniques, including steam treatment, alkali treatment, and acid pretreatment. Generally, the hydrolysis of cellulosic (C6) and hemicellulosic (C5) polymers is accomplished by enzymes like cellulase and xylanase, respectively. It is possible to ferment both C6 and C5 carbohydrates into ethanol using specially developed yeast for 2G ethanol production. Fuel ethanol is produced by distilling the fermented wash after fermentation (Fig. 1).

The Indian Govt. recognized and introduced additional feedstocks over conventional 1G ethanol feedstocks for the manufacturing of bioethanol in the modified National Policy on Biofuels. These feedstocks include sugarcane molasses (B-heavy and C-heavy), sugarcane juice, sugar, sugar syrup, and sugar-containing materials such as sugar beetroot and sweet sorghum; food grains classified as damaged (DFG), such as broken rice, unfit for human consumption, and food grains during the surplus phase as proclaimed by the National Biofuel Coordination Committee (NBCC); and starchy materials like corn, cassava, rotten potatoes, agro-food/pulp industry waste. Furthermore, agricultural leftovers (rice straw, cotton stalk, maize cobs, sawdust, bagasse, etc.) and lignocellulosic biomass in the form of grasses are included (PIB Delhi, 2024a, 2024b).

In the ethanol supply year (ESY) 2023–2024 (November–October), the cumulative ethanol blending from November 2023 to August 2024 indicated ~13.6% ethanol blending with over 545 crore liters of ethanol. On the other hand, India's monthly ethanol blending has hit 15% and above in months since May 2024 consecutively for the 4 months. The percentage of blending in August was 15.8%, depicting the clear intention to achieve the target as per the EBP program (PIB Delhi, 2024a, 2024b).

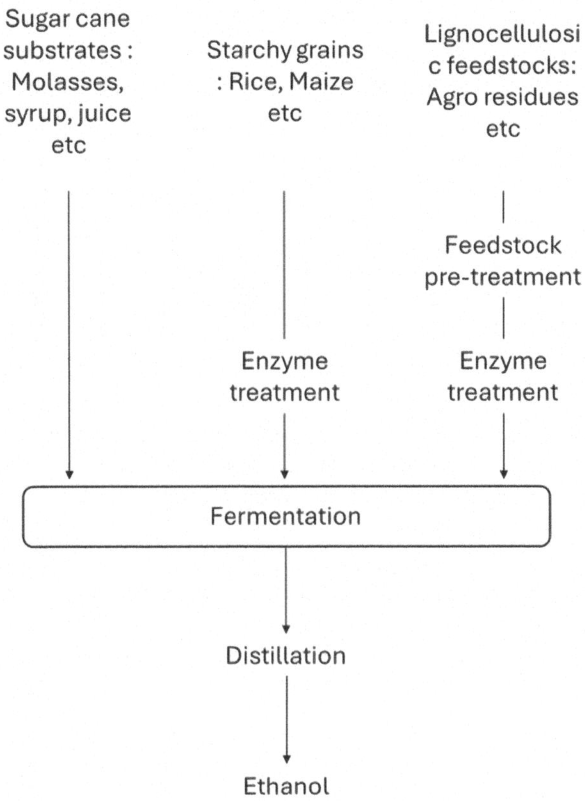

Fig. 1 Ethanol production from different types of feedstocks

Historically, sugarcane-based feedstocks, mostly molasses, have produced the majority of the ethanol given to OMCs under the EBP program. For instance, in ESY 2021–2022 and ESY 2022–2023, sugarcane-based distilleries supplied 83% and 73% of the ethanol needed by the OMCs, respectively, with the remaining 17% and 23% coming from grain-based distilleries. For the first time in India's history, during the ESY 2023–2024, grain-based distilleries supply more ethanol than sugarcane-based distilleries (53% vs. 47%, respectively). Up until ESY 2022–2023, rice (DFG/FCI/broken rice) has been the main grain feedstock utilized in the manufacturing of gasoline-ethanol. Following the government's 2023 suspension of the FCI rice supply toward fuel ethanol, maize has replaced rice as the primary grain feedstock as of ESY 2023–2024. Out of the entire fuel ethanol provided to OMCs in ESY 2023–2024 by grain-based distilleries, it yields around 135 crore liters of ethanol produced from maize. That means that maize accounts for about 64% of the grain-based ethanol supplied to OMCs. In ESY 2021–2022 2 years prior, maize made up none of the ethanol used in EBP. Overall, ethanol blending to gasoline has increased from 38 crore liters in ESY 2013–2014 to over 500 crore liters in ESY 2022–2023 under the EBP program (Fig. 2).

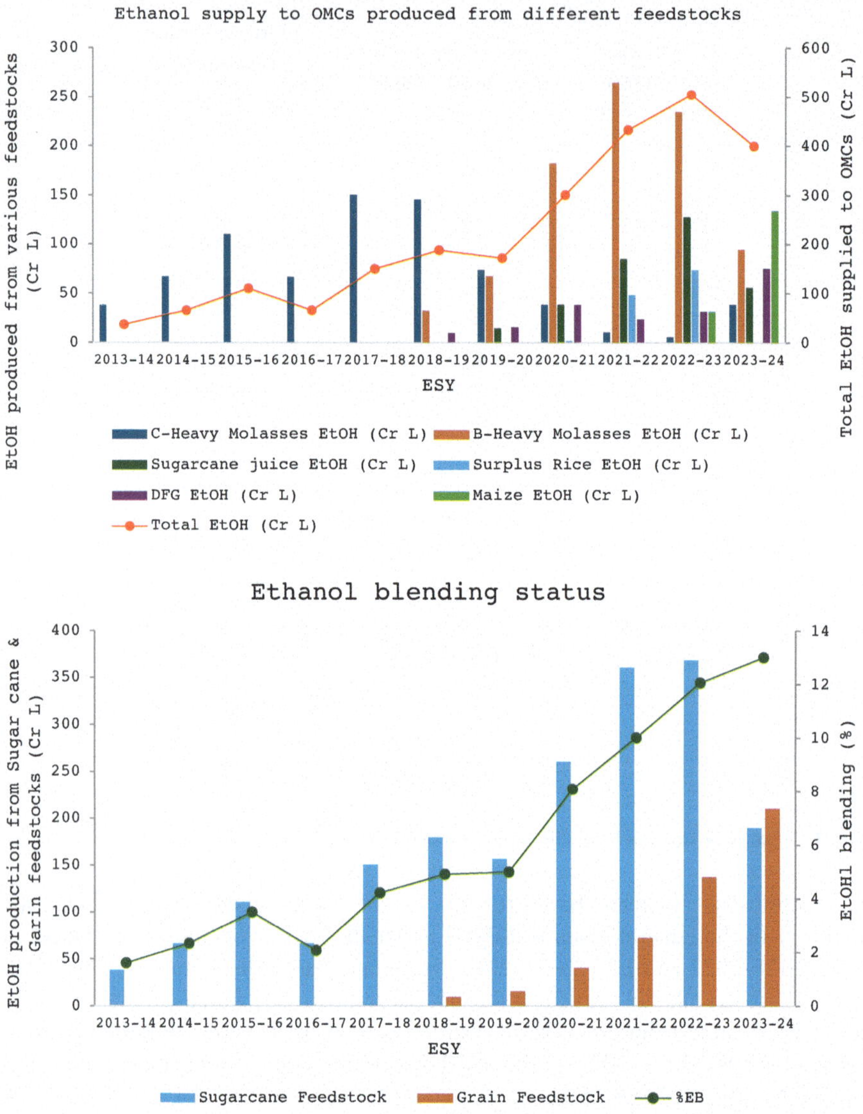

Fig. 2 Ethanol supply years are from December to November and December to October for ESY 2022–2023 and November to October for ESY 2023–2024. The data for ESY 2023–2024 is until June 2024

The Indian government started the Pradhan Mantri JI-VAN Yojana in August 2024 with the goal of stepping up advanced biofuel initiatives, particularly 2G ethanol. Biofuels made from lignocellulosic feedstocks, such as algae, synthesis (syn) gas, industrial waste, and residues from forestry and agriculture, will now fall under its expanded purview. The strategy will also help "brownfield projects" and "bolt-on" plants, enabling them to make better use of their current infrastructure and increase their viability to meet the E20 needs timely. The program will give top priority to project proposals that emphasize innovation and make use of a variety of feedstocks and new technologies. Its goals include giving farmers a profitable return on their agricultural wastes, reducing pollution in the environment, generating jobs locally, and enhancing India's energy independence and security. The Government of India has implemented a number of measures to meet the 20% ethanol blending target by ESY 2025–2026 and continues with firm determination. These include a comprehensive roadmap for ethanol production throughout the nation, a wider range of feedstock options for ethanol fermentation, ethanol pricing, a reduction in the GST rate to 5% on ethanol for the EBP program, an amendment to the Industries (Development & Regulation) Act to allow for free movement of ethanol across states for EBP, an interest subvention scheme to enhance and intensify the nation's capacity for ethanol production, and regular floating of expressions of interest (EoI) by OMCs for ethanol procurement (PIB Delhi, 2024a, 2024b).

1 Conclusion

The last few years have seen a massive expansion of the Indian ethanol business due to the growing need for domestic fuel, specifically ethanol, through the EBP program. The producers of ethanol worked together to maximize their output from both types of main raw materials, such as starchy grain-based feedstocks and sugarcane, in order to reach the nation's blending target. In the past 1.5 years, the manufacturers of grain-based ethanol have adapted to the global feedstock shift from rice to maize with remarkable agility. Grain-based distilleries provide OMCs with more ethanol during the ESY 2023–2024 than do sugarcane-based distilleries (53% vs. 47%, respectively). More than 500 crore liters of ethanol have been blended nationwide, representing a total 13.6% share. India's monthly ethanol blending rate reached 15% for last few months of ESY 2023–2024, according to this data. For instance, the August' 24 blending percentage of 15.8% showed a clear commitment to meet the objective as per the EBP program, and the Indian ethanol market is expected to grow continuously over the next coming years.

References

ChiniMandi. (2024). *Ethanol production capacity in India increases to 1,623 crore litres*. Retrieved from https://www.chinimandi.com/ethanol-production-capacity-in-india-increases-to-1623-crore-litres/

Davenport, J., & Wayth, N. (2024). *Statistical review of world energy* (73rd ed.). Energy Institute. Retrieved from https://www.energyinst.org/statistical-review

PIB Delhi. (2023). *Centre extends timeline for disbursement of loan/completion of ethanol projects up to 30 Sep 2023*. Retrieved from https://www.pib.gov.in/PressReleasePage.aspx?PRID=1935491#:~:text=Hence%2C%20the%20national%20ethanol%20production,has%20increased%20by%208%20times

PIB Delhi. (2024a). *Increase in blending of ethanol*. Retrieved from https://pib.gov.in/PressReleaseIframePage.aspx?PRID=2043042#:~:text=In%20order%20to%20achieve%2020,GST%20rate%20to%205%25%20on

PIB Delhi. (2024b). *Department of Food and Public Distribution, Government of India working on four pillars for 100 days action plan*. Retrieved from https://pib.gov.in/PressReleasePage.aspx?PRID=2055957

Powell, L., Sati, A., & Tomar, V. (2024). *India's energy profile: View from the South*. Retrieved from https://www.orfonline.org/expert-speak/india-s-energy-profile-view-from-the-south

Solution Buggy. (2023). *How to establish ethanol factory in India?* Retrieved from https://www.chinimandi.com/how-to-establish-ethanol-factory-in-india/

The manufacturer's authorised representative in the EU is Springer Nature Customer Service Centre GmbH, Europaplatz 3, 69115 Heidelberg, Germany. If you have any concerns regarding our products, please contact ProductSafety@springernature.com

Printed and bound by CPI Group (UK) Ltd, Croydon, CR0 4YY
01/10/2026
02225942-0003